£60.

40

3

26

OXFORD MONOGRAPHS ON CLASSICAL ARCHAEOLOGY

Edited by

MARTIN ROBERTSON
JOHN BOARDMAN
JIM COULTON
DONNA KURTZ

(E 113) Statue of Seneb, 51.280. Cairo Museum, JdE 51280

DWARFS IN ANCIENT EGYPT AND GREECE

Véronique Dasen

CLARENDON PRESS · OXFORD

1993

Oxford University Press, Walton Street, Oxford OX2 6DP

Oxford New York Toronto
Delhi Bombay Calcutta Madras Karachi
Kuala Lumpur Singapore Hong Kong Tokyo
Nairobi Dar es Salaam Cape Town
Melbourne Auckland Madrid
and associated companies in
Berlin Ibadan

Oxford is a trade mark of Oxford University Press

Published in the United States
by Oxford University Press Inc., New York

British Library Cataloguing in Publication Data
Data available

Library of Congress Cataloging in Publication Data
Dwarfs in ancient Egypt and Greece / Véronique Dasen.
(Oxford monographs on classical archaeology)
Revision of author's thesis (Ph.D.)—University of Oxford, 1988.
Includes bibliographical references and index.
1. Dwarfs. 2. Palaeopathology. 3. Dwarfs—Egypt—History.
4. Dwarfs—Greece—History. 5. Dwarfs in art. 6. Egypt—
Antiquities. 7. Greece—Antiquities. I. Title. II. Series.
GN69.3.D37 1993 573'.8—dc20 92–35446
ISBN 0–19–814699–X

Typeset by Best-set Typesetter Ltd., Hong Kong
Printed in Great Britain
on acid-free paper by
St. Edmundsbury Press, Bury St. Edmunds

TO MY FATHER
TO ANDREAS, RAPHAËL, AND NICOLAS

The way I shall take, gentlemen, in my praise of Socrates, is by similitudes. Probably he will think I do this for derision; but I choose my similitude for the sake of truth, not of ridicule. For I say he is likest to the Silenus-figures that sit in the statuaries' shops; those, I mean, which our craftsmen make with pipes and flutes in their hands: when their two halves are pulled open, they are found to contain images of gods. And I further suggest that he resembles the satyr Marsyas. Now, as to your likeness, Socrates, to these in figure, I do not suppose even you yourself will dispute it; but I have next to tell you that you are like them in every other respect.

Plato, *Symposium*, 215A–B (transl. W. R. M. Lamb, Loeb 1925)

PREFACE

This book is a modified version of my doctoral thesis submitted to the University of Oxford in 1988. This research started in the form of a *Mémoire de Licence* in classical archaeology entitled *Les nains d'Athènes. Essai sur la représentation de l'altérité physique* (unpublished), which was written under the supervision of C. Bérard and presented at the University of Lausanne (Switzerland) in 1982.

During the course of my research many academic institutions, scholars, and friends have helped me in various ways. Specific debts are acknowledged in the text. For information and suggestions I am most grateful to O. el-Aguizy, J. Bourriau, Y. Harpur, G. Le Cuyot, Ph. Mudry, G. Pinch, G. Robins, J. F. Romano, B. B. Shefton, H. Sourouzian-Stadelmann, C. Sourvinou-Inwood, W. Vycichl, and H. Whitehouse. Dr S. Braga, Dr A. E. H. Emery, and Dr R. Smith helped me with medical diagnoses. Particular thanks are due to Dr Ruth Wynne-Davies for her constant willingness to discuss medical questions throughout the preparation of the typescript. For information on the status of short people today, I am very grateful to Pamela Rutt, chairman of the British Restricted Growth Association. Nicholas Horsfall read the text in draft, corrected my English, and made many valuable suggestions. Andreas Tuor accompanied me in my various travels, helped me with photographs and commented on the text.

I owe special thanks to my D.Phil. supervisors, Sir John Boardman and John Baines, for their continued interest in my work and practical support. I am especially grateful to John Baines, who welcomed my incursion into the field of Egyptology; my work is largely indebted to him for his unfailing assistance.

For granting me facilities to study and take photographs of objects in collections in their care, I should like to thank the following:

S. Miller (Athens, Agora Museum), M. Schmidt (Basle, Antikenmuseum), L. Giuliani, G. Platz (Berlin, Antikenmuseum), R. Krauss (Berlin, Ägyptisches Museum), H. Kischkewitz, I. Kriseleit (Berlin, Pergamonmuseum), M. Saleh (Cairo Museum), the late A. Kadry (Egyptian Antiquities Organisation), O. Keel (Fribourg, Institut Biblique), I. Saverkina (Hermitage, Greek Department), A. O. Bolschakov (Hermitage, Egyptian Department), B. Adams (University College London), B. Cook (British Museum, Department of Classical Antiquities), A. J. Spencer (British Museum, Department of Egyptian Antiquities), S. Besques, A. Pasquier (Louvre, Department of Greek and Roman Antiquities), J. L. de Cenival, E. Delange (Louvre, Department of Egyptian Antiquities), D. C. Kurtz (Oxford, Beazley Archive), M. Vickers (Ashmolean Museum, Department of Greek and Roman Antiquities), H. Whitehouse (Ashmolean Museum, Department of Egyptian Antiquities).

It is a pleasure to acknowledge the financial assistance of FTR (Fabriques de Tabacs Réunies, Neuchâtel), the ASFU (Association Suisse des Femmes Universitaires), the Fonds National Suisse, the Fondation van Walsem (Lausanne), the Société Académique Vaudoise (Lausanne), the British Council, the Zunftgesellschaft zu Schmieden (Bern), and, in Oxford, of the Meyerstein Fund, the Craven Committee, the Committee for Graduate Studies, and the Governing Body of Lincoln College.

CONTENTS

x CONTENTS

LIST OF PLATES

2   Faience (h. 7.5). Berlin, SM (West), 7759. Photo A. Tuor.

3   Faience (h. 4.7). Oxford, Ashmolean Museum, 1890.357. Photo Museum.

pl. 7

1   Bronze (h. 10.4). Athens, NM, 614. Photo DAI, Athens (See ch. 6, n. 133).

2   Bronze (h. 12). Samos, Vathy Museum, B 353. Photo DAI, Athens (See ch. 6, n. 135).

3a, b   Faience (h. 18). London, BM, 26267. Author's photographs.

pl. 8

1   Papyrus. Berlin, SM (East), P 3128. Photo Museum (See ch. 6, n. 69).

2a–c   Faience (h. 7.2). London, BM, 26316. Author's photographs.

pl. 9

1a, b   Philae, columns I and K of the temple of Hathor (Ptolemaic). PM VI. 248. Photos A. Tuor (See ch. 6, n. 175).

2   Dendara, mammisi of Nectanebo. PM VI. 104. Photo A. Tuor (See ch. 6, n. 145).

3   Saqqara, 'Bes chambers', room 14 (max. h. 1.50). After J. E. Quibell, *Excavations at Saqqara (1905–1906)* (Cairo, 1907), pl. XXVII, 2 (See ch. 6, n. 149).

pl. 10

1   Stela, limestone (h. 30; w. 90). London, BM, 1178. Photo Museum (See ch. 6, n. 187).

2   Stela, limestone (h. 26.5; w. 47.7; th. 9). Brooklyn Museum, 58.98. Photo Museum (See ch. 6, n. 102).

pl. 11

1a, b   Papyrus. Brooklyn Museum, 47.218.156. Photos Museum (See ch. 6, n. 89).

2   Terracotta (h. 49). London, BM, 61296. Photo Museum (See ch. 6, n. 194).

3   Terracotta (total h. 40.3; h. of Bes figures 19.7). Leiden, Rijksmuseum, F 1975/11.2. Photo Museum (See ch. 6, n. 191).

pl. 12

1a–c   Faience (h. 7.6). Virginia Museum, 84.47. Photos Museum.

2   Faience (h. 7.6). London, BM, 11211. Author's photograph.

3   Faience (h. 8.2). London, BM, 63475. Author's photograph.

4   Faience (h. 5.3). London, BM, 11210. Author's photograph.

pl. 13

1   Faience (h. 6.2). London, BM, 69524. Author's photograph.

2   Faience (h. 6.4). London, BM, 54000. Author's photograph.

3a, b   Faience (h. 7). Tokyo, private. Photos courtesy of K. Ishiguro.

pl. 14

1a, b   Faience (h. 6.1). London, BM, 60205. Author's photographs.

2   Faience (h. 2.4). Fribourg, Institut Biblique, 1609. Photo A. Tuor.

3   Faience (h. 2.7). Fribourg, Institut Biblique, 1608. Photo A. Tuor.

pl. 34

1   (E *d* 138) Limestone (h. 42.3). London, UC, 16523/6. Author's photograph.

2   (E 146) Wood (h. 18.4). Liverpool University, E 7081. Photo Museum.

3a, b   (E 197) Wood (h. 10.9; w. 4.3). Paris, Louvre, E 14696. Photos Museum.

pl. 35

1a, b   (E 195) Ivory (h. 6.9). London, BM, 58409. Photos Museum.

2   (E 198) Boxwood (h. 5.9). Boston, MFA, 48.296. Photo Museum.

3   (E *d* 200) Calcite (h. 19.5). New York, MMA, 17.190.1963. Photo Museum.

pl. 36

a   (E 202) Boat, calcite (h. 67.5; w. 70), from the tomb of Tutankhamun. Cairo Museum, JdE 535. Photo Oxford, Griffith Institute.

b   Id. Photo Cairo Museum.

c   Id. Photo Oxford, Griffith Institute.

pl. 37

1   (E 203) Figure vase, calcite (h. 18). Oxford, Ashmolean Museum, 1911.407. Author's photograph.

2   (E 205) Figure vase, calcite (h. 20). London, BM, 30459. Photo Museum.

3   (E 204) Figure vase, terracotta (h. 9). London, BM, 29935. Author's photograph.

4   (E 206) Figure vase, calcite (h. 10). London, UC, 15758. Author's photograph.

pl. 38

1   Askos. Paris, Louvre, G 610. Photo Museum (See ch. 12, n. 24).

2   Plate fr. Athens, NM, ACR 1073. Photo Museum (See ch. 12, n. 26).

3   Pyxis fr. Boston, MFA, 10.216; *ARV* 81. Photo Museum (See ch. 12, n. 27).

4   Cup. Rome, Vatican, 16552; *ARV* 916.183, Para 430. Photo Museum (See ch. 12, n. 29).

pl. 39

1   Skyphos. Boston, MFA, 20.18. Photo Museum (See ch. 12, n. 32).

2   Chous. Taranto, Ragusa coll., 126. After I. MacPhee, *AK*, 22/1, 1979, pl. 15.2 (See ch. 13, n. 125).

3   Krater. Milan, Museo Archeologico, C 408. Photo Museum (See ch. 15, n. 156).

4   Lekane. Berlin, SM (West), 3366. Photo Museum (See ch. 15, n. 164).

pl. 40

1   Chous. Athens, NM, 12139. Photo Museum (See ch. 15, n. 168).

2   Lekythos. Boston, MFA, 00.351; *ARV* 723.1. Photo Museum (See ch. 15, n. 169).

3a, b   Terracotta, figure vase (h. 10.3). Paris market. Photos D. Widmer, Basle (See ch. 5, n. 55).

pl. 41

a–c   (G *d* 1) Pyxis (h. 3.9). Paris, Louvre, CA 1707. Photos Museum.

pl. 42

a, b (G 3) Skyphos (h. 7.5). Paris, Louvre, G 617. Photos Museum.

pl. 43

a (G 4) Rhyton, donkey head (h. 20). Ferrara, Museo Nazionale, 2561. Photo Museum.

b, c Id. Photos A. Lezzi-Hafter.

pl. 44

a–d (G 5) Aryballos (h. 9). Paris, Louvre, CA 2183. Photos Museum.

pl. 45

(G 6) Pelike (h. 15). Agrigento, Museo Civico, ex Giudice 638. Photo Museum.

pl. 46

1 (G 7) Cup fr. (dm. 19.2). Ferrara, Museo Nazionale, 20363. Photo Museum.

2a (G 8) Pelike (h. 24.1). Boston, MFA, 76.45. After Caskey/Beazley, i. 52, no. 59, pl. 27.

b Id. Photo Museum.

pl. 47

1 (G 9) Stamnos fr. (h. 8.5; w. 12.5). Erlangen, University, I 707. Photo Museum.

2 (G 10) Pelike (h. 14). St Petersburg, Hermitage, b 1621. Photo Museum.

pl. 48

1 (G 11) Cup. Formerly Munich, Preyss coll. Photo Beazley archive.

2 (G 12) Pelike (h. 12). Laon Museum, 37.1031. Photo Museum.

pl. 49

1 (G 13) Cup fr. (h. 7.5). Athens, Agora Museum, P 2574. Photo Museum.

2 (G 14) Chous (h. 9.5). Dresden, Albertinum, ZV 1827. Photo Museum.

pl. 50

1 (G 15) Cup fr. (h. 4.5). Hamburg, Museum für Kunst und Gewerbe, 1984.456. Photo Museum.

2 (G 17) Cup fr. (dm. 9.5). Oxford, Ashmolean Museum, 1938.312. Author's photograph.

pl. 51

a, b (G 16) Skyphos (h. 6). Munich, Antikensammlungen, 8934. Photos Museum.

pl. 52

a (G 18) Oinochoe (h. 18.3). Oxford, Ashmolean Museum, 1971.866. Photo Museum.

b Detail. Author's photograph.

pl. 53

1 (G 19) Cup fr. (h. 4.6; w. 14.6). Paris, Louvre, CA 5909. Photo Museum.

2 (G 20) Cup fr. (h. 22.5; w. 6.5). Todi, Museo Civico, 471. Photo Museum.

3 (G 21) Oinochoe fr. (h. 3.7). Tübingen, University, 1616. Photo Museum.

pl. 54

a, b (G 22) Skyphos (h. 7.6). Yale University, R. Darlington Stoddart coll., 160. Photos Museum.

pl. 55

1 (G 23) Bell krater (h. 29.1). Zurich market, Arete Gallery. Photo Arete Gallery.

2 (G 24) Loutrophoros (h. 89). Athens, NM, 2563. Photo Museum.

pl. 56

1 (G 26) Askos (h. 12.5). Basle, Antikenmuseum, Z 303. Photo Museum.

2 (G 31) Oinochoe (h. 20). Taranto, Ragusa coll., 7. Photo Museum.

3a, b (G 27) Oinochoe (h. 17.3). Melbourne, National Gallery, 90/5. Photos Museum.

pl. 57

1 (G 33) Pelike. Taranto, Ragusa coll., 127. Photo Museum.

2 (G 34) Oinochoe (h. 21). Toledo, Museum of Art, 67.136. Photo Museum.

pl. 58

a–d (G 40) Volute krater (h. 66). Florence, Museo Etrusco, 4209. Photos Museum.

pl. 59

1a–c (G 41) Aryballos (h. 7.8). New York, MMA, 26.49. Photos Museum.

2 (G 42) Cup, detail (h. 16.8). Taranto, Museo Nazionale, I.G. 4435. Photo Museum.

pl. 60

1 (G 45) Cup fr. (dm. c. 18). Berlin, SM (West), F 1785. Photo Museum.

2 (G 46) Cup fr. (h. 2). Berlin, SM (West), Brommer coll. 249. Photo F. Brommer.

3 (G 47) Hydria (h. 34). Paris, Louvre, F 44. Photo Museum.

pl. 61

1a, b (G 48) Hydria (h. 31.1). Rome, Villa Giulia, 50425. Photos Museum.

2 (G 53) Clay altar (h. 13.2). Corinth Museum, M.F. 8953. Photo Museum.

pl. 62

1 (G d 56) Amphora (h. 32.4). Formerly Castle Ashby. Photo Museum.

2a, b (G 60) Rhyton (h. 17). St Petersburg, Hermitage, b 1818. Photo Museum.

pl. 63

1 (G 61) Hydria (h. 22.5). Bologna, Museo Civico, 169. Photo Museum.

2 (G 62) Hydria, fr. (h. 8.8). Athens, Agora Museum, P 8892. Photo Museum.

3a, b (G 66) Rhyton, figure vase (h. 29.7). Bonn, University, 545. Photos Museum.

pl. 64

a, b (G 67) Rhyton (h. 24). Compiègne Museum, 898. Photos Museum.

pl. 65

1 (G 70) Amphora (h. 30.2). Brussels, Musées Royaux, R 302. Photo Museum.

2 (G 74) Pyxis, lid (h. 5.5). Athens, NM, 17714. Photo Museum.

pl. 76
    (G 114) Amphora (h. 88). Foggia Museum, 132723. Photo Museum.

pl. 77
1a, b (G 137) Terracotta (h. 7.5). London, BM, 89. Author's photographs.
2   (G 138) Terracotta (h. 9). London, BM, 93. Author's photograph.
3a, b (G 139) Terracotta (h. 8). London, BM, 94. Author's photographs.

pl. 78
1   (G 145) Terracotta, figure vase (h. 15). London, BM, 86. Photo Museum.
2   (G 146) Terracotta (h. 17). London, BM, 88. Photo Museum.
3a, b (G 154) Terracotta (h. 19). Kassel, Staatl. Kunstsammlungen, S 55. Photos Museum.

pl. 79
1a–c (G 155) Wood (h. 21.9). Samos, Vathy Museum, H 43. Photos DAI, Athens.
2   (G 156) Terracotta (h. 8.1). Samos, Vathy Museum, RB 77-643,1. Photo DAI, Athens.
3   (G 162) Terracotta (h. 14.7). Taranto, Museo Nazionale, 4960. Photo Museum.

pl. 80
1   (G 183) Terracotta (h. 10). Carthage, Museum, 895.12. Photo Museum.
2   (G 202) Terracotta (h. 9.2). Frankfurt, Liebieghaus, 454. Photo Museum.
3   (G 206) Terracotta (h. 7.5). Munich, Antikensammlungen, 7563. Photo Museum.
4   (G 210) Wood (h. 24.5). Paris, Louvre, N 846. Photo Museum.

LIST OF FIGURES

Drawings by the author: Figs. 1.1, 5.1, 5.2, 7.2, 7.4, 9.1, 9.5, 9.9, 9.16, 9.20, 9.23.
Drawings by A. Tuor: Figs. 3.1, 7.3, 9.22, 15.1. Fig. 7.1, courtesy of the Institut d'égyptologie V. Loret, Lyons.

ABBREVIATIONS

For periodicals and series, abbreviations follow those of the *Index Medicus* for medical journals, of the *Lexikon der Ägyptologie* for Egyptology, and of the *Année Philologique* for the classical world.

ABV	J. D. Beazley, *Attic Black-Figure Vase-Painters* (Oxford, 1956).
Add	L. Burn and R. Glynn, *Beazley Addenda* (Oxford, 1982).
Add²	T. H. Carpenter *et al.*, *Beazley Addenda²* (Oxford, 1989).
African Pygmies	L. L. Cavalli-Sforza (ed.), *African Pygmies* (London etc., 1986).
El-Aguizy, 'Dwarfs'	O. el-Aguizy, 'Dwarfs and Pygmies in Ancient Egypt', *ASAE* 71 (1987), 53–60.
Altenmüller, *Apotropaia*	H. Altenmüller, *Die Apotropaia und die Götter Mittelägyptens*, Diss. (Munich, 1965).
Ancient Egypt	B. G. Trigger *et al.*, *Ancient Egypt: A Social History* (Cambridge, 1983).
ARV	J. D. Beazley, *Attic Red-Figure Vase-Painters²* (Oxford, 1963).
Atlas	W. Wreszinski, *Atlas zur altaegyptischen Kulturgeschichte* (Leipzig, 1923–38).
Atlas of Skeletal Dysplasias	R. Wynne-Davies *et al.*, *Atlas of Skeletal Dysplasias* (Edinburgh etc., 1985).
Ballabriga, 'Nains'	A. Ballabriga, 'Le Malheur des nains. Quelques aspects du combat des grues contre les pygmées dans la littérature grecque', *REA* 83 (1981), 57–74.
Ballod, *Prolegomena*	F. Ballod, *Prolegomena zur Geschichte der zwerghaften Götter in Ägypten* (Moscow, 1913).
Boardman, *ABFH*	J. Boardman, *Athenian Black Figure Vases* (London, 1974).
Boardman, *ARFH* i	J. Boardman, *Athenian Red Figure Vases: The Archaic Period* (London, 1975).
Boardman, *ARFH* ii	J. Boardman, *Athenian Red Figure Vases: The Classical Period* (London, 1989).

Bonnet, *Reallexikon*	H. Bonnet, *Reallexikon der ägyptischen Religionsgeschichte* (Berlin, 1952).
Book of the Dead	T. G. Allen, *The Book of the Dead or Going Forth by Day* (Chicago, 1974).
Brommer, *VL*	F. Brommer, *Vasenlisten zur griechischen Heldensage*[3] (Marburg, 1973).
Brunner-Traut, *Tanz*	E. Brunner-Traut, *Der Tanz im alten Ägypten* (Glückstadt etc., 1938) (ÄF 6).
Bruyère, *Deir el Médineh*	B. Bruyère, *Rapport sur les fouilles de Deir el Médineh (1934–1935)*, iii (Cairo, 1939) (FIFAO 16).
Burkert, *GrRel*	W. Burkert, *Greek Religion*, trans. J. Raffan (Oxford, 1985; Stuttgart, 1977).
CAF	T. Kock, *Comicorum Atticorum Fragmenta* (Leipzig, 1880–8).
Caskey/Beazley	L. D. Caskey and J. D. Beazley, *Attic Vase Paintings in the Museum of Fine Arts* (Boston and London, 1931–63).
CT	A. de Buck, *The Egyptian Coffin Texts* (Chicago, 1935–61). Translation: R. O. Faulkner, *The Ancient Egyptian Coffin Texts* (Warminster, 1973–8).
CVA	*Corpus Vasorum Antiquorum*.
DA	C. Daremberg and E. Saglio (eds.), *Dictionnaire des antiquités grecques et romaines* (Paris, 1877–1919).
Daressy, *Statues*	G. Daressy, *Statues de divinités* (CG 38001-39348) (Cairo, 1905–6).
Dasen, 'Dwarfism'	V. Dasen, 'Dwarfism in Egypt and Classical Antiquity: Iconography and Medical History', *Med. Hist.* 32 (1988), 253–76.
Dawson, 'Pygmies'	W. R. Dawson, 'Pygmies and Dwarfs in Ancient Egypt', *JEA* 24 (1938), 185–9.
Deubner, *Att.Feste*	L. Deubner, *Attische Feste* (Berlin, 1932).
Diseases in Antiquity	D. R. Brothwell and A. T. Sandison (eds.), *Diseases in Antiquity* (Springfield, Ill., 1967).
EAA	*Enciclopedia dell'arte antica, classica e orientale* (Rome, 1958–85).
FGrH	F. Jacoby (ed.), *Die Fragmente der griechischen Historiker* (Berlin and Leiden, 1923–58).
Gardiner, *Grammar*	A. Gardiner, *Egyptian Grammar*[3] (Oxford, 1957).

Gruppe, *GrMyth*	O. Gruppe, *Griechische Mythologie und Religionsgeschichte* (Munich, 1906) (HdA V, 2).
Haspels, *ABL*	C. H. E. Haspels, *Attic Black-Figured Lekythoi* (Paris, 1936).
Hayes, *Scepter*	W. C. Hayes, *The Scepter of Egypt: A Background for the Study of Egyptian Antiquities in the Metropolitan Museum of Art* (New York, 1953–9).
Hemberg, *Kabiren*	B. Hemberg, *Die Kabiren* (Uppsala, 1950).
Heraion	P. Zancani Montuoro and U. Zanotti-Bianco, *Heraion: Alla Foce del Sele* (Rome, 1951–4).
Hornemann, *Types*	B. Hornemann, *Types of Ancient Egyptian Statuary* (Copenhagen, 1951–69).
Hornung, *Conceptions*	E. Hornung, *Conceptions of God in Ancient Egypt*, trans. and rev. by J. Baines (London, 1983; Darmstadt, 1971).
Hornung/Staehelin	E. Hornung and E. Staehelin, *Skarabäen und andere Siegelamulette aus Basler Sammlungen* (Mainz am Rhein, 1967).
Kaplony, *Inschriften*	P. Kaplony, *Die Inschriften der ägyptischen Frühzeit* (Wiesbaden, 1963–4).
Kassel/Austin	C. Austin and R. Kassel, *Poetae Comici Graeci* (Berlin, 1983–6).
Krall, 'Bes'	J. Krall, 'Ueber den ägyptischen Gott Bes', in O. Benndorf and G. Niemann, *Das Heroon von Gjölbaschi-Trysa* (Vienna, 1889), 72–96.
Kunze/Nippert, *Genetics*	J. Kunze and I. Nippert, *Genetics and Malformations in Art* (Berlin, 1986).
LÄ	W. Helck *et al.* (eds.), *Lexikon der Ägyptologie* (Wiesbaden, 1975–86).
Lanzone, *Dizionario*	R. V. Lanzone, *Dizionario di mitologia egizia* (Turin, 1882–1975).
Lepsius, *Denkmaeler*	C. R. Lepsius, *Denkmaeler aus Aegypten und Aethiopien* (Berlin [1849–59], cited by Abt.).
Lepsius, *Denkmaeler: Ergänzungsband*	E. Naville (ed.), *Ergänzungsband* (Leipzig, 1913).
Lichtheim, *Literature*	M. Lichtheim, *Ancient Egyptian Literature* (Berkeley, Calif., 1973–80).
LIMC	*Lexicon Iconographicum Mythologiae Classicae*, i– (Zurich and Munich, 1981–).
LSJ	H. G. Liddell and R. Scott, *A Greek–English Lexicon*², rev. H. S. Jones (Oxford, 1940).

Michailidis, 'Bès'	G. Michailidis, 'Bès aux divers aspects', *BIE* 45 (1963–4), 53–93.
Musée égyptien du Caire	M. Saleh and H. Sourouzian, *Musée égyptien du Caire: Catalogue officiel* (Mainz am Rhein, 1987).
Nilsson, *GGR*³	M. P. Nilsson, *Geschichte der griechischen Religion*³, i (Munich, 1967) (HdA V, 2, 1).
Nofret	D. Wildung and S. Schoske, *Nofret, die Schöne*, i (Mainz am Rhein, 1984); B. Schmitz, *Nofret, die Schöne*, ii (Mainz am Rhein, 1985).
Ortner/Putschar, *Pathological Conditions*	D. J. Ortner and W. G. J. Putschar, *Identification of Pathological Conditions in Human Skeletal Remains* (Washington, DC, 1981).
Para	J. D. Beazley, *Paralipomena*² (Oxford, 1971).
Piankoff/Rambova, *MythPap*	A. Piankoff and N. Rambova, *Mythological Papyri* (New York, 1957) (ERT 3).
PM	B. Porter and R. L. B. Moss (later with E. W. Burney, later with J. Málek), *Topographical Bibliography of Ancient Egyptian Hieroglyphic Texts, Reliefs and Paintings*, i– (Oxford, 1927–51,² 1960–).
Preller/Robert, *GrMyth*	L. Preller, *Griechische Mythologie*⁴, rev. with add. by C. Robert (Berlin, 1887–1926).
Ranke, *PN*	H. Ranke, *Die ägyptischen Personennamen* (Glückstadt etc., 1935–77).
Raven, *Pataekos*	M. J. Raven, 'A Puzzling Pataekos', *OMRO* 67 (1987), 7–17.
RE	A. Pauly, G. Wissowa, *et al.*, *Pauly's Real-Encyclopädie der classischen Altertumswissenschaften, neue Bearbeitung*, i– (Stuttgart, 1894– ; Munich, 1972–).
Richter/Hall	G. M. A. Richter and L. F. Hall, *Red-Figured Athenian Vases in the Metropolitan Museum of Art* (New Haven and London, 1936).
Roeder, *Bronzefiguren*	G. Roeder, *Ägyptische Bronzefiguren* (Berlin, 1956) (Staatl. Museen zu Berlin, Mitteilungen aus der ägyptischen Sammlung 6).
Romano, 'Bes'	J. F. Romano, *The Bes-Image in Pharaonic Egypt*, Ph.D. thesis (New York Univ., 1989).
Romano, 'Origin'	J. F. Romano, 'The Origin of the Bes-Image', *Bull. of the Egyptological Seminar*, 2 (1980), 39–56.

Roscher

W. H. Roscher (ed.), *Ausführliches Lexikon der griechischen und römischen Mythologie* (Leipzig, 1884–1937).

Ruffer, *Palaeopathology*

M. A. Ruffer, in R. L. Moodie (ed.), *Studies in the Palaeopathology of Egypt* (Chicago, 1921).

Rühfel, *Kinderleben*

H. Rühfel, *Kinderleben im klassischen Athen* (Mainz am Rhein, 1984).

Rupp, 'Zwerg'

A. Rupp, 'Der Zwerg in der ägyptischen Gemeinschaft. Studien zur ägyptischen Anthropologie', *CdE* 40/80 (1965), 260–309.

Sandman Holmberg, *Ptah*

M. Sandman Holmberg, *The God Ptah* (Lund, 1946).

Sinn, *Antidoron*

U. Sinn, 'Zur Wirkung des ägyptischen "Bes" auf die griechische Volksreligion', in D. Metzler *et al.* (eds.), *Antidoron. Festschrift für J. Thimme zum 65. Geburtstag* (Karlsruhe, 1983), 87–94.

Smith, *Art and Architecture*

W. S. Smith, *The Art and Architecture of Ancient Egypt*² rev. with add. by W. K. Simpson (Harmondsworth, 1984).

Sourdive, *La Main*

C. Sourdive, *La Main dans l'Égypte pharaonique: Recherches de morphologie structurale sur les objets égyptiens comportant une main* (Berne, 1984).

Thompson, *Motif Index*

S. Thompson, *Motif Index of Folk-Literature* (Copenhagen, 1955–8).

Tran Tam Tinh, *Bes*

Tran Tam Tinh, *LIMC* iii (1986), 98–108, s.v. Bes.

Trendall, *LCS*

A. D. Trendall, *The Red-Figured Vases of Lucania, Campania, and Sicily* (Oxford, 1967).

Trendall, *PhV*²

A. D. Trendall, *Phlyax Vases*² (London, 1967) (BICS suppl. 19).

Trendall/Cambitoglou, *RVAp*

A. D. Trendall and A. Cambitoglou, *The Red-Figured Vases of Apulia* (Oxford, 1982).

Vandier, *Manuel*

J. Vandier, *Manuel d'archéologie égyptienne* (Paris, 1952–78).

Vandier d'Abbadie, 'Singes'

J. Vandier d'Abbadie, 'Les Singes familiers dans l'ancienne Égypte, i, l'Ancien Empire', *RdE* 16 (1964), 147–77.

Wb

A. Erman and H. Grapow, *Wörterbuch der ägyptischen Sprache* (Leipzig, 1926–63).

Weeks, *Anatomical Knowledge*

K. R. Weeks, *The Anatomical Knowledge of the Ancient Egyptians and the Representation of the*

	Human Figure in Egyptian Art, Ph.D. thesis (Yale University, 1970).
Wilson, 'Bes'	V. Wilson, 'The Iconography of Bes with Particular Reference to the Cypriot Evidence', *Levant*, 7 (1975), 77–103.
Winter, *TK*	F. Winter, *Die Typen der figürlichen Terrakotten* (Berlin and Stuttgart, 1903).

MUSEUMS

Athens, NM	Athens, National Museum
Berlin, SM (East)	Berlin, Pergamonmuseum
Berlin, SM (West)	Berlin, Antikenmuseum and Ägyptisches Museum
Boston, MFA	Boston, Museum of Fine Arts
London, BM	London, British Museum
London, UC	London, University College
New York, MMA	New York, Metropolitan Museum of Arts

A dash after a museum reference indicates that no inventory number is available.

General Introduction

Every society produces minorities, which may be ethnic, linguistic, social, religious, or physical. While most types of minorities may change, physical anomalies, especially congenital, have occurred with similar frequency throughout human history.

In ancient Egypt and Greece physical beauty, defined in terms of proportion, was highly admired, even to excess. What happened to those who conformed neither to these 'ideal proportions' nor to norms of human appearance? Were they rejected as beings outside the world of creation, who disturbed the established order and should be killed or hidden? Were they objects of curiosity and derision? Or were they associated with higher powers and integrated within the socio-religious system of the community? How did the ancients explain the causes of congenital physical defects?

This book examines the case of dwarfism. This growth disturbance is mainly due to genetic mutations, and ancient cultures all had to confront the same incidence of this phenomenon. The study is based on iconography, but not exclusively; it also includes a discussion of the limited amount of extant epigraphic, literary, and skeletal evidence. The spectacular forms of dwarfism were always a focus of interest, and it is the physical disorder which was depicted most in antiquity, from predynastic times in Egypt to the end of the Roman Empire. The abundance of pictorial sources, which comprise tomb paintings and reliefs, vase paintings, statuary and minor arts, contrasts, however, with the paucity of written evidence.

By analysing different models of behaviour towards the physically malformed, from integration to rejection, this study is meant to contribute to the more general discussion on acceptance of physical deviance, and hence of tolerance in human societies.[1]

I compare the attitudes to dwarfs in two ancient civilizations, in Egypt and in Greece. This choice is due partly to the absence of material from other cultures, especially Near Eastern, which seem generally to have avoided representing physical malformation. Both ancient Egypt and Greece produced a relatively large quantity of pictures which permit us to approach the position of short-statured people in myth and daily life. A comparative approach is profitable on two levels. First, the characteristics of each society appear more sharply, thus prompting many new questions that can be asked of familiar material. Second, the transmission of a number of motifs and beliefs from Egypt to Greece can be detected and analysed in depth. The pictures produced in each culture

---

[1] For studies on other types of physical marginality, see Kunze/Nippert, *Genetics*, esp. pp. v–vii, 3–13 (with bibliog.).

conform to specific iconographic conventions, but their forms and symbolism are not totally unrelated. Both Egyptian and Greek cultures belong to the Mediterranean cultural area and were in contact for a long time. Greek authors, such as Herodotus, often refer to their Egyptian or Phoenician neighbours, and acknowledge their cultural relationship. Since I am by training a classicist, this study of a specific aspect of Egyptian culture has posed a stimulating challenge, to which I hope I have responded successfully. I originally planned to add a third part on dwarfs in the Hellenistic and Roman world, but the abundance and the complexity of Egyptian and Greek material led me to set this aside for the time being.

The book consists of the following parts. A medical introduction presents the different types of dwarfism known today, so as to determine the features which may be identified in art. An examination of the physical problems associated with restricted growth also helps the assessment of the position of short-statured people in antiquity. This introduction is completed by a review of skeletal remains from Egypt and Greece. The presentation is then similar in the Egyptian and Greek sections. The analysis of the terminology of dwarfism gives an initial insight into the notion of restricted growth. This is followed by a presentation of the artistic conventions governing the representations of dwarfs in each iconographic system, and by a review of the different types of short stature shown in art. A discussion of the general attitude to the physically disabled in written and iconographic sources introduces the study of the position of dwarfs in myth and in daily life.

The scheme of presentation for Egyptian and Greek evidence differs on a few points. The Egyptian material covers about 3,000 years, the Greek only about three centuries. The Egyptian material is therefore divided according to the principal historical periods (Predynastic and Early Dynastic periods, Old, Middle, and New Kingdoms—with a separate section on the Amarna period—Late and Graeco-Roman periods), while the Greek sources, which consist mainly of Athenian archaic and classical vase-painting, are presented thematically. This chronological disproportion is, however, compensated by the very different rate of change in the two civilizations. Except for the Amarna period, Egyptian art and culture changed less during 3,000 years than Greece and South Italy did between 600 and 300 BC. The relationship of literary sources to iconography is also sharply distinct in each culture. In particular, the transmission of myths varies: Egyptian myths are seldom fully narrated, and cannot be analysed by themselves,[2] while Greek literary and iconographic accounts of a single myth, which conform to different traditions, are discussed in separate sections.[3]

[2] J. Assmann, 'Die Verborgenheit des Mythos in Ägypten', GM 25 (1977), 7–43; J. Baines, 'Egyptian Myth and Discourse: Myth, Gods, and the Early Written and Iconographic Records', JNES 50 (1991), 81–105 (with earlier bibliog.).

[3] Burkert, GrRel 8. See also F. Graf, Griechische Mythologie (Munich and Zurich, 1985); J. Bremmer (ed.), Interpretations of Greek Mythology (London and Sydney, 1987) (with earlier bibliog.).

Each part of the book is accompanied by a select bibliography and a catalogue of pictorial evidence divided into two sections: (i) reliefs and drawings, (ii) statuary. As the material from Egypt and Greece differs in type and number, the conventions of presentation differ. The Egyptian catalogue consists of representations of human dwarfs only. It was not possible in a work of this length to include lists of Bes and Ptah-Pataikoi, which are far too numerous. The chronology and the spelling of Arabic place names are based on J. Baines and J. Málek, *Atlas of Ancient Egypt* (Oxford, 1980). Chronological adjustments are made after R. Krauss, *Sothis- und Monddaten* (Hildesheim, 1985). Since this book is intended for both egyptologists and classicists, all Egyptian material quoted is transcribed and translated. Personal names are vocalized, and the vocalization of the most frequent terms is given in the notes. These transcriptions are not the result of original research but conform to current practice. The spelling of the names of tomb-owners usually follows the Topographical Bibliography (abridged PM), or was revised by John Baines. When not otherwise mentioned all Egyptian translations are by John Baines. The Greek catalogue lists depictions of human and divine dwarfs. Vase-painting is divided thematically (human dwarfs, pygmies, Cercopes, dwarf Hephaistos).

The position of dwarfs and malformed people in ancient communities has never been studied comprehensively. The realism of the representations, especially Egyptian, has appealed to a number of scholars, but mainly for its contribution to medical history (Ber, Leca, Regnault, Ruffer, Schrumpf-Pierron, Silverman, Tolstoï);[4] different types of growth disorders have been identified in art, but these diagnoses need to be revised in the light of present medical terminology and of an increasing body of evidence. Classicists usually discuss whether these figures represent pathological cases or 'portraits', caricatures or actors mimicking a deformity (Bartsocas, Becatti, Binsfeld, Grmek, Metzler, Rizzo, Zinserling). Few scholars go beyond these problems of identification to question, for example, the relations between artistic conventions and history of mentalities, between pathology and the apotropaic repertory (Badawi, Fischer).

Previous studies on Egyptian dwarfs are based principally on Old Kingdom material (el-Aguizy, Dawson, Junker, Montet, Naster, Rupp, Weeks), while evidence on Greek dwarfs has never been collected systematically (Binsfeld, Lippold, Metzler, Pottier, Robertson, Shapiro). Works on other minorities have produced useful observations, in particular studies on the elimination of abnormal births and eugenics (Germain, Glotz, Schmidt), works on the status of the blind (Buxton, Esser), of blacks (Raeck, Snowden), of twins (Baines), and of slaves (Himmelmann). Apart from the general works of Bolkestein and Hands on social assistance in antiquity, the history of the status of the disabled remains to be written, especially for ancient Egypt. I should also mention works on modern attitudes to disability, such as the pamphlets on practical questions

---

[4] The works cited here are listed in the bibliography.

produced by the British Restricted Growth Association,[5] which contribute helpful insights.

The role of dwarfs in Egyptian myths has occasionally been examined by scholars. The best known is the dwarf god Bes. His various forms and functions have been analysed in several articles (e.g. Altenmüller, Bruyère, Krall, Michailidis, Wilson) and in a few comprehensive studies (Ballod, Romano). Amuletic Ptah-Pataikoi, however, have never been discussed at length, except for their medical identification, or their association with Ptah (Hückel, Parrot, Raven, Sandman Holmberg, Spiegelberg, Vassal). Since each of these gods deserves a thesis by himself, I have restricted myself to what they may contribute to the general understanding of the phenomenon 'dwarf' in Egyptian culture. I also suggest a few hypotheses on the nature of these gods in the light of recent discoveries and of general progress of research in Egyptian religion. I should also mention here the articles by Malaise,[6] Matzker,[7] and Hawass,[8] which appeared too late to be included in this book, and bring useful complementary information.

The position of dwarfs in Greek myths has not attracted much attention. Discussion of literary sources about the tale of the pygmies and cranes belongs to a long tradition of scholarship (Ballabriga, Gusinde, Hennig, Janni, Monceaux, Waser, Wüst), but works on the iconographic evidence are relatively few (Becatti, Cèbe, Freyer-Schauenburg, Minto, Richter, Štal), and do not question their contribution to Greek ethnography. The tale of Heracles and the Cercopes has been studied briefly by Zancani Montuoro and Zanotti-Bianco (1954), who concentrate on its iconography. Only archaic kourotrophic demons have been examined in excavation reports, and are the subject of specific articles (Caubet, Furtwängler, Hadzisteliou Price, Sinn). The possible kinship of the Kabeiroi, Daktyloi, and Telchines to Hephaistos and to Indo-European dwarf smiths has been mentioned by several authors (Blinkenberg, Friedländer, Hemberg, Malten, Wilamowitz), but most scholars leave it aside as an insignificant element (cf. Detienne, Nilsson).

---

[5] *Coping with Restricted Growth*, (n.p., 1980); *The Layman's Guide to Restricted Growth*[2] (Rickmansworth, 1983). See also e.g. E. Tietze-Conrat, *Dwarfs and Jesters in Art* (London, 1957); F. Adanos and H. Scheugl, *Showfreaks and Monsters* (Cologne, 1974); J. O. Weiss, 'Social Development of Dwarfs', in E. T. Hall and C. L. Young (eds.), *Proceedings of a Conference on Genetic Disorders: Social Service Interventions* (Pittsburgh, 1977), 56–61; M. Monestier, *Les Nains* (Paris, 1977); id. *Les Monstres* (Paris, 1978); R. Lebeck, *Riesen, Zwerge, Schauobjekte* (Dortmund, 1979); P. Darmon, 'Autrefois, les nains', *L'Histoire*, 19 (Jan. 1980), 48–57; N. Kreibich and J. M. Parkin, 'The Problems of People with Severely Restricted

Growth', *The Practitioner*, 229 (1985), 278–9; C. Lecouteux, *Les Nains et les elfes au Moyen Age* (Paris, 1988).

[6] M. Malaise, 'Bès et les croyances solaires', in S. Israelit-Groll (ed.), *Studies in Egyptology Presented to Miriam Lichtheim* (Jerusalem, 1990), 680–729.

[7] I. Matzker, 'Gruppierung von Patäken anhand von Merkmalsvergleichen', in *Festschrift Jürgen von Beckerath zum 70. Geburtstag* (Hildesheim, 1990), 199–207.

[8] Z. Hawass, 'The Statue of the Dwarf *Pr-n (j)-ˁnh(w)* recently discovered at Giza', *MDAIK* 47 (1991), 157–62.

I

MEDICAL CONTEXT

I

Typology of Growth Disorders

Dwarfism is commonly defined as an abnormally short stature over three standard deviations below the mean height of a population of the same age and sex.[1] In western countries this includes persons of 1.50 m. and under; most affected people are between 1 m. and 1.40 m. in height, but some can be as short as 70 cm.[2] Disorders of the growth process may have many causes, ranging from genetic defects to hormone deficiencies, metabolic and nutritional diseases, which induce a great variety of skeletal abnormalities.[3] Since the mid-1970s about a hundred forms of dwarfism with distinct etiology and physical features have been identified.[4]

Growth disturbances can be divided into two main categories: a disproportionate type, where restricted growth affects the limbs, or the trunk, or both in varying degrees, and a proportionate type, where the whole body is involved and remains small. Body proportions are conventionally defined by the measurement of the upper to lower segment ratios (U/L), taken from the top of head to pubis, and from pubis to heel. Normal adults have a U/L of approximately 0.95 and an arm span equal to their total

[1] See the standards established by J. M. Tanner et al., 'Standards from Birth to Maturity for Height, Weight, Height Velocity and Weight Velocity: British Children', Arch. Dis. Childhood, 41 (1966), 454–71, 613–35; A.-M. E. Nehme et al., 'Skeletal Growth and Development of the Achondroplastic Dwarf', Clin. Orthop. 116 (1976), 8–23; W. A. Horton et al., 'Standard Growth Curves for Achondroplasia', J. Paediatr. 93 (1978), 435–8.

[2] Tom Thumb, for example, is reported to have been 63.5 cm. tall (25 in.). For the sizes of other famous court and fair dwarfs, see the table in M. Monestier, Les Monstres (Paris, 1978), 202–3.

[3] For tables showing the various causes of dwarfism, see e.g. R. W. B. Ellis and R. G. Mitchell, Disease in Infancy and Childhood (Edinburgh and London, 1973), 205–6; R. Smith, Biochemical Disorders of the Skeleton (Boston and London, 1979), 66–7, tables 2.IV–V; id. 'Disorders of the Skeleton', in Oxford Textbook of Medicine, ii (Oxford, 1983), 17.33, table 16; S. D. Frasier, Pediatric Endocrinology (New York, 1980), 67, table 3.8.

[4] The main authorities are: V. A. McKusick, Heritable Disorders of Connective Tissue[4] (St Louis, 1972); J. W. Spranger et al., Bone Dysplasias. An Atlas of Constitutional Disorders of Skeletal Development (Philadelphia, 1974); H. Moll, Atlas of Pediatric Diseases (Philadelphia etc., 1976); D. W. Smith, Recognizable Patterns of Human Malformations[2] (Philadelphia, 1976); D. Bergsma (ed.), Birth Defects Compendium[2] (Basingstoke and London, 1979); P. Maroteaux, Bone Diseases of Children (Philadelphia, 1979); D. L. Rimoin and R. S. Lachman, 'The Chondrodysplasias', in A. E. H. Emery and D. L. Rimoin (eds.), Principles and Practice of Medical Genetics, i (Edinburgh, 1983), 703–35; Atlas of Skeletal Dysplasias.

height (fig. 1.1a).[5] The deviation of U/L from the average defines the type of dwarfism present: disproportionate (short limbs or short trunk), or proportionate.

If we include all types of dwarfism, the rate of occurrence is of one abnormal birth in 10,000 live births. In most conditions growth disturbances are due to fresh genetic mutations, both parents being normal. These disorders occur sporadically and the offspring have a 50 per cent chance of inheriting them when one parent is affected (e.g. achondroplasia, hypochondroplasia). Endocrine disorders are usually acquired (e.g. hypopituitarism, hypothyroidism), while nearly all the metabolic dwarfisms are acquired (e.g. nutritional dwarfism) and only a few inherited (e.g. mucopolysaccharidoses). Constitutional dwarfism, such as pygmyism, is genetically transmitted. In the most frequent types (e.g. achondroplasia, hypochondroplasia), men and women seem to be equally affected.

In specific or severe cases, the condition can be detected at birth (e.g. achondroplasia, diastrophic and metatropic dwarfism, spondylo-epiphyseal dysplasia congenita); in other or milder forms, it becomes progressively evident during infancy or even as late as adolescence (e.g. hypochondroplasia, hypothyroidism, mucopolysaccharidoses, spondylo-epiphyseal dysplasia tarda).[6] Growth disorders may be linked with infantilism (e.g. hypopituitarism, hypothyroidism). In adulthood, the individual can show severe dwarfing, and handicaps can be increased by derived or associated defects (e.g. club-foot, coxa vara, genu valgum or varum, deafness, root compression leading to paraplegia). The life-span of short-statured persons is generally reduced in comparison to that of normal-sized individuals because of their greater vulnerability to infections, diseases, and accidents.

For each group I present the most common conditions, and hence those most likely to have been depicted in the past. I concentrate on the visual features which may allow their identification in art, keeping in mind that ancient writers and artists could not differentiate between very similar disorders and only noted the most striking features of the prototypes; some conditions (e.g. hypopituitarism) could not easily be depicted within iconographic conventions (see table on 14–15).

DISPROPORTIONATE SHORT STATURE

Short Limbs, Normal Trunk

The most common type of dwarfism is achondroplasia, which is due to a dominant genetic mutation (fig. 1.1b). Its incidence is of about one in 34,000 live births, and most

---

[5] See Rimoin/Lachman (n. 4 above) 705–6, fig. 51.3 (cf. Leonardo da Vinci's model of ideal human proportions). In Blacks, who have longer limbs, the U/L ratio is about 0.85 for adults.

[6] See P. Maroteaux, 'The Chondrodystrophies Detectable at Birth', in F. C. Frasier and V. A. McKusick (eds.), *Congenital Malformations. Proceedings*

of the Third International Conference, The Hague, The Netherlands, 7–13 Sept. 1969 (Amsterdam and New York, 1970), 222–6, esp. 224, table 1; Rimoin/ Lachman (n. 4 above), 714–25, table 51.1A (conditions identifiable at birth) and B (identifiable in later life). On prenatal diagnosis, see *Atlas of Skeletal Dysplasias*, pp. xiii–xiv.

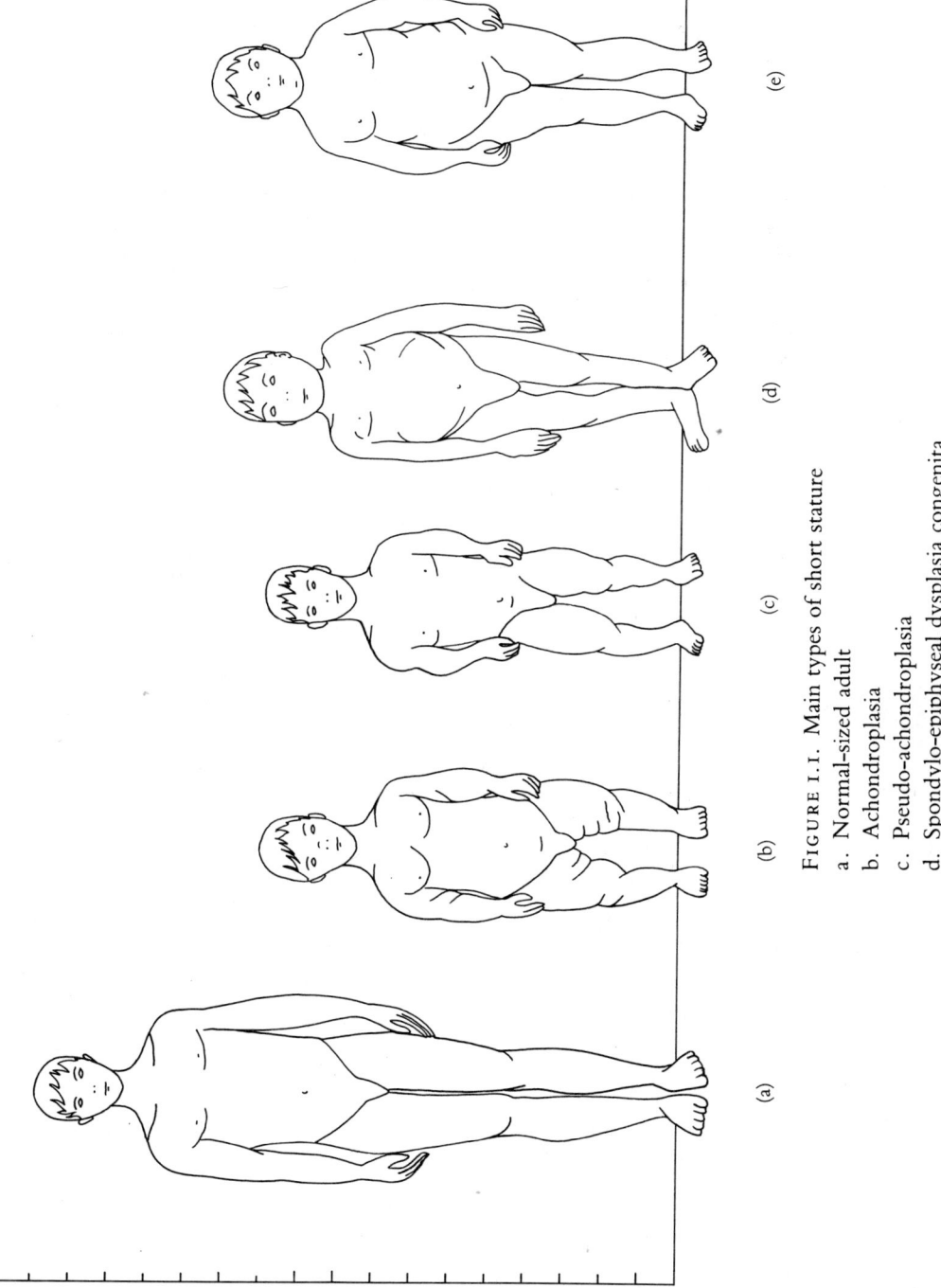

FIGURE 1.1. Main types of short stature

a. Normal-sized adult
b. Achondroplasia
c. Pseudo-achondroplasia
d. Spondylo-epiphyseal dysplasia congenita
e. Hypothyroidism

(a) (b) (c) (d) (e)

cases (over 80 per cent) are new mutants.[7] The disorder involves the ossification process of the cartilaginous bones, initiated in the uterus. The limb shortening is severe, while the trunk is almost normal in length. The head shows a large cranial vault and small facial bones (pl. 1.1–2). The prominent forehead is due to the normal development of the cranium from membranous bone, while the base of the skull ossifies in cartilage and remains shorter; the small foramen magnum may produce hydrocephalus. The nasal bridge is depressed, and protruding jawbones are common. A pronounced pelvic tilt induces lumbar lordosis with very prominent abdomen and buttocks (pl. 1.1).[8] There may be associated thoracolumbar kyphosis, and occasionally scoliosis. The fingertips reach only to the top of the thigh (greater trochanter) or even the hips (iliac crest). Joint mobility at the shoulders and the elbows is limited; the arms do not extend normally and fall in a stiffened way to the sides. The legs are bowed (genu varum) with folds of skin due to the abnormal shortening of the bones.

At birth, the disturbance is usually easily recognizable, or, in milder cases, becomes evident during the first year of life. The specific features are the shortness of the extremities in relation to the normal-length trunk, the shallow thoracic cage, and the enlarged head. Excess soft tissue is folded.

This growth disorder is not associated with metabolic defects; mental development is usually unimpaired. Genital organs, despite persistent legends, are normal in size, and sexual maturity occurs at the normal time, although the small pelvis creates obstetric problems.[9] Agility is often well developed, but sacral tilt, combined with progressive curvature of the legs, induces a waddling gait which increases with age. Derived complications, such as osteoarthritis, deafness, and limping are common. The main problems are due to spinal complications, such as spinal cord or root compression which can lead to paraplegia. The mean adult height in males is 131 ± 5.6 cm., and in females 124 ± 5.9 cm.[10]

In hypochondroplasia, a milder type of dwarfism as common as achondroplasia, the proportions of the body are similar, but the skull is not involved, and the head and the face are normal (pl. 1.3).[11] The trunk is only mildly affected by lumbar lordosis; genu varum may occur. The child may appear normal until 2 or 3 years of age.

Other similar disorders involve principally the limbs, but can be differentiated by detailed clinical, biochemical, and radiographic investigations; their occurrence is much less common than for the two types just described (2 to 4 per million population).

[7] R. J. M. Gardner, 'A New Estimate of the Achondroplasia Mutation Rate', *Clin. Genet.* 11 (1977), 31–8.

[8] See *Atlas of Skeletal Dysplasias*, 181–99, esp. figs. 18.1–4.

[9] Cf. the skeleton of an achondroplastic woman who died in childbirth in Ortner/Putschar, *Pathological Conditions*, 330, fig. 515.

[10] Smith (n. 4 above), 188; *Atlas of Skeletal Dysplasias*, 184, figs. 18.5–8.

[11] See R. Wynne-Davies *et al.*, 'Achondroplasia and Hypochondroplasia, Clinical Variation and Spinal Stenosis', *J. Bone and Joint Surg.* 63B (1981), 508–15; *Atlas of Skeletal Dysplasias*, 200–12, esp. figs. 19.1–3.

I mention only mesomelic dwarfism, a disorder characterized by very short lower extremities, often associated with club-foot,[12] and metatropic dwarfism,[13] a very rare disorder which produces a markedly short stature, with very short limbs, associated with scoliosis or kyphosis. In both types the skull is not affected and facial appearance is normal. The adult height is between 1.10 m. and 1.20 m.[14] There are other types of dwarfism with excessively short limbs, such as thanatophoric dwarfism and achondrogenesis, but they are lethal at birth.[15]

Short Limbs, Short Trunk

In other types of short-limbed dwarfism, like pseudo-achondroplasia (fig. 1.1c)[16] or diastrophic dwarfism,[17] the whole skeleton is small and may be affected by joint laxity, or severe contractions of the joints, genu varum, valgum or recurvatum, talipes equinovarus, and club-hand. For pseudo-achondroplasia, the mean adult height is between 82 and 130 cm., for diastrophic dwarfism, between 80 and 140 cm.

Osteogenesis imperfecta, a disorder characterized by extremely brittle bones, also affects the whole body.[18] Multiple fractures in the limbs induce a short stature; it may be associated with varied deformities, such as club-feet, bowing of long bones, contracture of the joints, and scoliosis. The head is large with a triangular face. The disorder is often not identifiable at birth, except when the child is born with fractures. In severe cases growth ceases at 3 or 4 years of age, but mental development is usually normal.

Short-Trunk Conditions

A disproportionate short stature may also result from an extreme shortening of the trunk due to a flattening of vertebrae, often accentuated by hidden flexion of the hips, coxa vara and marked lumbar lordosis. The neck is short, the chest barrelled and the limbs appear long in relation to the trunk. Hands and feet are normal in size. In spondylo-epiphyseal dysplasia congenita (fig. 1.1d),[19] the skull is mildly affected, the face is flat, and cleft palate is occasionally present at birth. Dislocated hips, associated with club-foot, lead to a waddling gait. Premature osteoarthritis is frequent and paraplegia may occur. The disorder becomes apparent around 2 and 3 years of age, and

---

[12] *Atlas of Skeletal Dysplasias*, 223–8, esp. fig. 22.1.

[13] Under 1 per million population; *Atlas of Skeletal Dysplasias*, 274–88, esp. fig. 28.1.

[14] Cf. also the group of metaphyseal dysplasia (normal facial appearance, bowed legs, coxa vara) (incidence: 3 to 6 per million population); *Atlas of Skeletal Dysplasias*, 131–77, esp. figs. 10.1–2, 11.1, 12.1, 14.1–2, 15.1, 16.1–2.

[15] See e.g. Maroteaux (n. 6 above), 225, fig. 6 (thanatophoric dwarfism); *Atlas of Skeletal Dysplasias*, 311–21.

[16] Incidence: *c.*4 per million population; *Atlas of*

Skeletal Dysplasias, 239–57, esp. figs. 26.1–8.

[17] Incidence: *c.*1 per million population; *Atlas of Skeletal Dysplasias*, 258–73, esp. figs. 27.1–4.

[18] Incidence: 1 in 20,000 live births. See R. Smith et al., *The Brittle Bone Syndrome* (London, 1983); *Atlas of Skeletal Dysplasias*, 411–30, esp. figs. 54.1–3.

[19] Incidence: 2–4 per million population. See R. Wynne-Davies, 'Two Clinical Variants of Spondylo-Epiphyseal Dysplasia Congenita', *J. Bone and Joint Surg.* 64B (1982), 435–41; *Atlas of Skeletal Dysplasias*, 61–98, esp. figs. 5.1–3, 6.1.

in milder types (SED tarda),[20] between 5 and 10 years of age. The mean adult height varies between 84 and 128 cm.

In mucopolysaccharidoses, the disorder is due to an abnormal breakdown of complex carbohydrates which accumulate in the tissues and affect the whole metabolism, involving bone, cartilage, skin-meninges, blood vessels, heart valves, and other tissues. Six different types are distinguished with similar skeletal changes in various degrees.[21] The most typical are Morquio's and Hurler's syndrome. In Morquio's syndrome,[22] the stature is severely reduced, with a marked lumbar lordosis associated with kyphosis. Facial features are malformed: the nasal bridge is flattened, the wide mouth has thick lips, and jawbones are prominent. The neck is very short and the sternum protruding. Walking can be difficult because of hip and leg deformities (genu valgum), and the mobility of the joints is reduced. Mental retardation is minimal or absent. In Hurler's syndrome,[23] these symptoms are associated with a protuberant abdomen due to an enlarged spleen and liver. Intelligence declines early and can be severely deficient.

In all these types, the appearance of the newborn infant is usually normal. Growth and mental retardation become evident between 1 and 5 years of age. The life-span is generally reduced because of cardiac, respiratory and/or neurological complications due to spinal cord compression. In Morquio's type, death can occur before the age of 40, and it is often earlier in the other types (in Hurler's syndrome survival is rare beyond fifteen to twenty years). Stiffness of the joints, deafness and blindness are the most common derived disorders. The mean adult height is around 100 cm.

Pott's disease, a progressive disorder due to tuberculosis of the spine which may cause a shortening of the trunk, and hence restricted growth, should also be mentioned here.[24]

PROPORTIONATE SHORT STATURE

Dwarfism with proportionate short stature is related to congenital or acquired deficiencies of the pituitary and thyroid glands.[25]

Hypopituitarism is due to a deficiency of human growth hormone related to a reduced secretion of the pituitary gland. It can be also secondary to tumours or lesions in the pituitary complex. In most cases, the etiology remains unclear. The incidence of the disturbance is of about one in 25,000 live births. The adult has a well-proportioned

[20] Incidence: 7–8 per million population. See R. Wynne-Davies and C. Hall, 'Spondylo-Epiphyseal Dysplasia Tarda with Progressive Arthropathy', *J. Bone and Joint Surg.* 64B (1982), 442–5; *Atlas of Skeletal Dysplasias*, 99–128, esp. figs. 7.1, 8.1–2, 9.1–4.

[21] Incidence: 3 per million population. See R. Smith, *Biochemical Disorders of the Skeleton* (Boston and London, 1979); R. E. Stevenson, in D. Bergsma (ed.) (n. 4 above), 727–34, no. 674–80; *Atlas of Skeletal Dysplasias*, 343–84.

[22] Ibid. 372–84, esp. figs. 47.1–3.

[23] Ibid. 343–9, esp. figs. 41.1–4.

[24] See D. Morse *et al.*, 'Tuberculosis in Ancient Egypt', *Amer. Review of Resp. Diseases*, 90 (1964), 524–41; Ortner/Putschar, *Pathological Conditions*, 145–9, 166–76.

[25] See Frasier (n. 3 above); J. C. Job and M. Pierson, *Pediatric Endocrinology* (New York, 1981).

diminutive stature which gives a childish appearance (pl. 1.4). The doll-like face shows a bulging forehead, a short nose, round cheeks and a small chin. The voice tends to be high-pitched. Hands and feet are pudgy; the muscles are usually poorly developed and the trunk tends to a slight obesity. A secondary hormone deficiency (gonadotrophin) may impair sexual development and lead to infantilism. The mean adult height is about 1.30 m. The newborn child is often normal in size. Growth retardation becomes evident progressively between 6 months and 4 years of age, but mental development is usually normal.

On the other hand, a thyroid dysfunction may induce pronounced physical and mental retardation, with a general slowing-down of the metabolism. In endemic cretinism, the disorder is associated with goitre and is due to a very low concentration of iodine in drinking water. In severe cases (congenital hypothyroidism) it can lead to complete cretinism. The adult retains infantile proportions (in some cases with short limbs) with a distended abdomen accentuated by kyphosis (fig. 1.1e). The head is large and shows myxoedema: the face has coarse features, thick lips, a large protruding tongue, a wrinkled forehead and an apathetic expression. Infantilism is complete: intellectual and emotional development is retarded, and sexual maturity is not reached.

This disorder is usually recognizable at birth or becomes evident during early infancy. The typical coarse facial features are present, especially the protruding tongue in a constantly open mouth and the bulging, half-closed eyes, which give a frog-like appearance. The hair is dry, sparse and shaggy, and the scaly skin pale yellow. An umbilical hernia is always present. The infant is lethargic, lacks appetite and shows psycho-motor retardation. These symptoms may appear progressively in late infancy. The general slowing-down of the metabolism increases with advancing age. The life-span is usually reduced; death occurs on average at 30 years of age, in the past generally due to tuberculosis. The mean adult height is about 1.20 m.

Pygmyism (or constitutional dwarfism)

Pygmies may be placed in this category since their physical features resemble those of persons affected by a growth hormone deficiency. The best-known examples are the pygmy tribes of Central Africa, who live as hunter-gatherers in the rain forest. They are traditionally divided in four major groups, eastern (Zaire, Ituri forests), central (Zaire, north of lake Leopold), western (Zaire, Central African Republic, Cameroon, Gabon), and southern (Rwanda, Burundi, Zambia).[26] Today these groups together number approximately 200,000 people, but they must have been more numerous earlier, and scientists usually regard them as 'relics of a much larger population' which 'covered a much wider area'.[27] They have various local names, such as Aka or Babinga

[26] L. L. Cavalli-Sforza, 'Demographic Data', in *African Pygmies*, 23–44, fig. 2.1; id., 'Evaluation of the State of Research', ibid. 362–4, fig. 26.1. See also S. Bahuchet (ed.), *Pygmées de Centrafrique* (Paris, 1979) (SELAF Études Pygmées iii). They are distinct from the Bushmen farther south and from savanna hunters.

[27] Cavalli-Sforza (n. 26 above), 20.

TABLE. *Clinical features of the main types of dwarfism*

Dysplasia	Head and neck	Stature/body proportions	Diagnosis/age
Achondroplasia	large head, bulging forehead, low nasal bridge, prominent jawbones, ±hydrocephalus	short limbs/normal trunk, lumbar lordosis, ±scoliosis, ±kyphosis, ±genu varum, limitation of elbow extension, skin folds due to excess soft tissue	identifiable at birth
Diastrophic dwarfism	normal skull and face, cystic swelling of the ear ('cauliflower ears'), ±cleft palate	short limbs/short trunk, progressive contractures of joints, talipes equinus or equinovarus, scoliosis	identifiable at birth
Hypochondroplasia	normal face or slight prominence of forehead	short limbs/normal trunk, lumbar lordosis, ±genu varum, ±limitation of elbow extension	identifiable in later life (2−3 years)
Hypopituitarism	±normal head, 'doll-like' facial features	proportionate short stature	identifiable in later life (between 6 months and 4 years)
Hypothyroidism	myxoedema, short nose, thick lips, large protruding tongue	±proportionate short stature, ±kyphosis, distended abdomen	identifiable at birth and in later life
Mesomelic dwarfism	normal skull and face	short limbs/normal trunk, severe shortening of lower extremities, ±talipes equinovarus, ±dislocated knees and ankles	identifiable at birth
Metatropic dwarfism	normal skull and face	very short limbs/normal trunk, severe scoliosis or kyphosis	identifiable at birth
Mucopolysaccharidoses	low nasal bridge, thick lips, prominent jawbones, ±hydrocephalus (Hurler's syndrome)	±normal limbs/short trunk, thoracolumbar lordosis, ±kyphosis, ±scoliosis, stiffness of joints, ±with contractures, joint laxity, coxa valga, genu valgum	identifiable in later life
Osteogenesis imperfecta	small triangular face	short limbs/short trunk, multiple fractures in the limbs, joint laxity, contractures, ±scoliosis	identifiable at birth and in later life (3−4 years)

TABLE. *Continued*

Dysplasia	Head and neck	Stature/body proportions	Diagnosis/age
Pott's disease	normal skull and head	normal limbs/short trunk, laxity and subluxation of joints, angular kyphosis	identifiable in later life
Pseudo-achondroplasia	normal skull and face	very short limbs/short trunk, genu varum, valgum or recurvatum, joint laxity, ±scoliosis	identifiable in later life (2–3 years)
Spondylo-epiphyseal dysplasia, congenita	flat face, short neck ±cleft palate	±normal limbs/short trunk, short barrel chest, scoliosis or kyphoscoliosis, coxa vara, ±genu valgum, varum or recurvatum, ±club-foot	identifiable at birth
Spondylo-epiphyseal dysplasia, tarda	normal head	normal limbs/short trunk, ±scoliosis	identifiable in later life (between 5 and 10 years)

Source: D. L. Rimoin and R. S. Lachman, 'The Chondrodysplasias', in A. E. H. Emery and D. L. Rimoin (eds.), *Principles and Practice of Medical Genetics*, i (Edinburgh, 1983), 714–25, table 51.1, and *Atlas of Skeletal Dysplasias*.

(western group), Twa (central and southern group), and Mbuti or Bambuti (eastern group).[28] They have the lowest average stature on earth, usually well below 1.60 m. Eastern pygmies from the Ituri forest are the smallest group (pl. 2.1): the mean adult height is under 1.45 m. in males, and under 1.36 m. in females. They are thin and muscular, with a lighter skin-colour than their negroid neighbours, but similar facial characteristics. The breadth of their hips and the volume of their skulls are only mildly reduced, and thus the head may appear relatively large.[29] The persistence of these specific physical features suggests that pygmyism may be due to a genetic adaptation to the equatorial environment.[30] Recent research has shown that pygmies have a deficiency in an insulin-like growth factor which induces their typical proportionate short stature.[31]

[28] Cavalli-Sforza, 'Anthropometric Data', in *African Pygmies*, 81–93, tables 5.1–3; id., 'Evaluation of the State of Research', ibid. 389–94.

[29] Id., ibid. 90, notes that the word for pygmies in Central Africa is Babinga, which means 'big head'.

[30] See L. L. Cavalli-Sforza, 'Biological Research on African Pygmies', in G. A. Harrison (ed.), *Population Structure and Human Variation* (Cambridge, 1977), 273–84; J. Hiernaux, 'Long-Term Biological Effects of Human Migration from the Savanna to the Equatorial Forest: A Case Study of Human Adaptation to a Hot and Wet Climate', ibid. 187–217.

[31] T. J. Merimee et al., 'Dwarfism in the Pygmy', *N. Engl. J. Med.* 305/17 (1981), 965–8; T. J. Merimee and D. L. Rimoin, 'Growth Hormone and Insulin-Like Growth Factors in the Western Pygmy', in *African Pygmies*, 167–77.

2

Palaeopathology

The diagnosis of a disturbance of development can be established principally by inspection of the spine, the long bones, and the skull, including an examination of the pituitary cavity, which may show an enlargement due to a tumour. This diagnosis is often prevented by incomplete human remains which can easily mislead; for example, fragmentary remains of children might be mistaken for those of proportionate dwarfs. Similarities between different conditions, such as those included in the short-limbed category, may not allow an accurate diagnosis, but only a statement of a range of possible disorders.[1] Remains of abnormal still-births and infants are extremely rare, and most of the individuals listed here are adults who exhibit a mature form of their growth disturbance.

The majority of dwarf remains have been found in Egypt. This is due to the good preservation of Egyptian dead bodies thanks to artificial and natural mummification.[2] There has been also more work on Egyptian palaeopathology. Most skeletons, however, have not been thoroughly studied and published; as Weeks notes, one usually has to rely on brief descriptions in excavation reports, often without drawings or photographs (S *d* 4, 5, *d* 8–10, *d* 13–16), and it is necessary to take the author's diagnosis on trust.[3] Some specimens, like the skeletons found by Amélineau in Abydos, were not kept; in other cases, the present location is unknown (S 1, *d* 4–6, *d* 8–10, 12–17). In Greece, where burial customs were different, most skeletal remains were deposited in unfavourable climatic and geological conditions.[4] The paucity of evidence is also due to the relative lack of scholarly interest in classical osteo-archaeology.[5]

---

[1] On the identification of skeletal dysplasias, see esp. Ortner/Putschar, *Pathological Conditions*, 329–45. See also D. R. Brothwell, *Digging up Bones*[3] (London, 1981), 161–3, 167–74; S. Živanović, *Ancient Diseases* (London, 1982), 125 ff.; K. Manchester, *The Archaeology of Disease* (Bradford, 1983), 27–9.

[2] On the preservation of dead bodies in Egypt, see A. and E. Cockburn (eds.), *Mummies, Disease and Ancient Cultures* (Cambridge, 1983) (with earlier bibliog.).

[3] Weeks, *Anatomical Knowledge*, 196–7.

[4] See J. L. Angel, 'Skeletal Changes in Ancient Greece', *Amer. J. Phys. Anthrop.* NS 4 (1946), 69–70, and, more recently, J. H. Musgrave, 'Appendix C: The Human Remains from the Cemeteries', in M. R. Popham *et al.* (eds.), *Lefkandi*, i (London, 1980) (BSA), 429–46.

[5] On Greek palaeopathology, see M. D. Grmek, *Les Maladies à l'aube de la civilisation occidentale* (Paris, 1983), 86 ff.

DISPROPORTIONATE SHORT STATURE

Most extant skeletal remains of dwarfs that have been diagnosed are identified as achondroplastic, or as belonging to the short-limbed category (S 1–3, 5–7, 14, 16–9). This may reflect the higher incidence of that type of disorder, which affects bone formation and resorption, and induces easily recognizable skeletal changes.

Short Limbs, Normal Trunk

The skeleton of the Badarian period (c.4500 BC) is the earliest example of dwarfism found in Egypt (S 1; fig. 2.1). As Jones observed, the skull and clavicles do not present any specific skeletal change, but the long bones are reduced, also in diameter, and the left humerus is particularly short.[6] Jones attributed these abnormalities to achondroplasia, but recognized that 'this is not an example of achondroplasia as we usually know it'.[7] The articular surfaces of the shoulder and elbow joints have marks of attrition, which may denote arthritism. Wynne-Davies suggests that the man might have been affected by pseudo-achondroplasia or another short-limbed condition, such as multiple epiphyseal dysplasia because the humeral head shows the typical 'hatchet' shape of an epiphyseal disorder.[8]

The remains of two dwarfs from the first dynasty tomb of King Semerkhet at Abydos (S 7) present the characteristic features of achondroplasia. The skull vault is normal, but with a shortened base and a marked midfacial depression stressed by prognathism. The long bones of the dwarfs are significantly short. The tibiae are slightly bowed, the humerus is thick with broadened ends, as is the remaining humerus of a dwarf from the tomb of King Djer at Abydos (S 6).[9] Similar features may be seen in the predynastic femur and tibiae now in Cambridge (S 2).

The skeleton of a male adult dwarf buried in a subsidiary grave of the first dynasty at Saqqara (S 3; pl. 2.2) has very short tibiae and bowed fibulae. For Emery these malformations indicate rickets, but Weeks suggests that the deformity may be due to achondroplasia.[10]

The skull of a woman 20 to 25 years old (S 17) was discovered with over eighty incomplete skeletal remains in a tomb chamber below the temple of Tuthmosis IV in Thebes. Seligmann first attributed the poorly developed nasal bones to cretinism, but Keith showed that the typical foreshortening of the cranial basis, prominent forehead and arrested growth of the nasal bones are characteristic of achondroplasia.[11]

[6] E. W. A. H. Jones, 'Studies in Achondroplasia', *J. of Anat.* 66 (1932), 569–73.

[7] Jones, ibid. 573.

[8] R. Wynne-Davies, personal communication. On multiple epiphyseal dysplasia, see *Atlas of Skeletal Dysplasias*, 19–35, esp. figs. 2.45–8 (radiographs of malformed humerus).

[9] See Ortner/Putschar, *Pathological Conditions*, 331–2, figs. 518–21.

[10] W. B. Emery, *Great Tombs of the First Dynasty*, ii (Cairo and London, 1954), 36, no. 58 (diagnosis established after examination by A. el-Batrawi, Cairo University); Weeks, *Anatomical Knowledge*, 198–9, 2.

[11] C. G. Seligmann, 'A Cretinous Skull of the Eighteenth Dynasty', *Man*, 12 (1912), 17–18; A. Keith, 'Abnormal Crania-Achondroplastic and Acrocephalic', *J. of Anat. Physiol.* 47 (1913), 189–206.

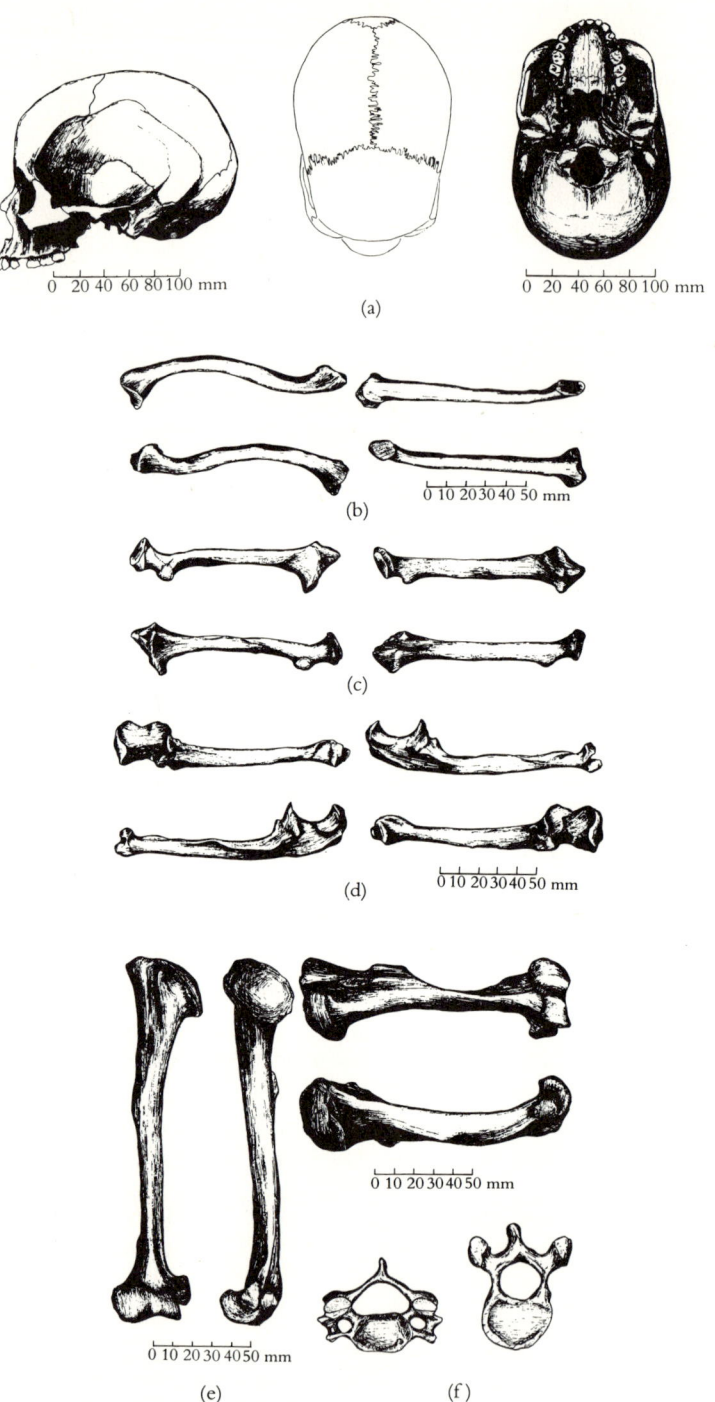

FIGURE 2.1 (S 1) El-Mustagidda. Cemetery 2200/3500, tomb 3510. (a) Skull. (b) Clavicles. (c) Radii. (d) Ulnae. (e) Humeri. (f) Vertebrae.

There are also three unpublished skeletal remains which are known from excavators' reports only, and which may belong to the short-limbed category. Papillault examined the skeleton of a dwarf found by Amélineau in the tomb complex of King Djer (S 5).[12] He described typical changes in the skull, the enlarged vault, with a depressed midfacial area, and a very short basioccipital and foramen magnum. He also noted that the long bones have irregular flaring heads and are extremely reduced; the tibia measures 21.5 cm. and the femur 26.3 cm. in maximum length,[13] and he concluded that the dwarf was probably achondroplastic (length of coffin: 1.14 m.).

In the Old Kingdom tomb of Ipi at Beni Hasan, Garstang recorded the burial of a dwarf in a shaft (S 14). The dwarf may be identified as the small man depicted on the south wall of the tomb, standing beneath the chair of Ipi with a dog (E 60).[14] Garstang stated that 'an anatomical examination of his bones did not show any signs of disease, but the back was rounded, the vertebrate bones were ossified to the curved position, and the limb bones were smaller than those of ordinary humanity';[15] this is suggestive of short-limbed dysplasia.

At Asyut, Chassinat and Palanque found the skeleton of a dwarf buried in a Middle Kingdom reused coffin (S 16). They briefly described the malformations of the skeleton (about 92 cm.). These may indicate achondroplasia: the skull has a bulging forehead, the trunk is abnormally long, the legs are thin, and hands and feet are atrophied.[16] They could not examine the backbone, which had disintegrated almost completely.

Short Limbs, Short Trunk

Garstang found near Beni Hasan (Speos Artemidos) the skeleton of an infant buried in a cartonnage case (S 18), which he first misinterpreted as the remains of a mummified monkey. As Gray demonstrated, this is most likely a case of osteogenesis imperfecta.[17] The bones are friable and very light; the long bones are deformed, especially in the lower extremities (antero-lateral bowing). The vault of the skull is enlarged, with multiple ossification centres (Wormian bones), and vertically elongated eye orbits. The teeth are also peculiar, with abnormally developed roots.

Bartsocas describes the only known specimen of a Greek dwarf, the skeleton of a

---

[12] G. Papillault, 'Crânes d'Abydos, étude anthropologique', in E. Amélineau, *Les Nouvelles Fouilles d'Abydos, 1897–1898*, ii (Paris, 1905), 730.

[13] Cf. the similar length of the femur of the dwarf from the tomb complex of King Semerkhet (S 6): 23.5 cm.; Ortner/Putschar, *Pathological Conditions*, 332.

[14] J. Garstang, *The Burial Customs of Ancient Egypt* (London, 1907), 38–9.

[15] Ibid. 41.

[16] E. Chassinat and C. Palanque, *Une Campagne de fouilles dans la nécropole d'Assiout* (Cairo, 1911) (MIFAO 24), 14.

[17] P. H. K. Gray, *Mummies and Human Remains: Catalogue of Egyptian Antiquities in the British Museum*, i (London, 1968), 13–14, no. 24; id., 'A Case of Osteogenesis Imperfecta, Associated with Dentinogenesis Imperfecta, Dating from Antiquity', *Clin. Radiol.* 20 (1969), 106–8; Ortner/Putschar, *Pathological Conditions*, 338.

woman about 35–40 years old, which was found in Gouvalari (S 19).[18] He attributes the missing clavicles and the abnormal position of the humeri in front of the thorax to cleidocranial dysplasia, a very rare congenital disorder which is characterized by marked skeletal changes in the skull and the clavicles.[19]

Short-Trunk Conditions

Two early dynastic humeri (S 11) show abnormal heads with severe malformation of the articular surface (porosity), and failure in the development of the epiphysis, which suggests a diagnosis of pseudo-achondroplasia or mucopolysaccharidoses (Hunter's or Morquio's syndrome).[20]

PROPORTIONATE SHORT STATURE

Pituitary Disorders

Proportionate dwarfism due to endocrine disorders presents very similar features to those of children or of constitutional dwarfs, such as pygmies, and so may be difficult to diagnose.[21]

Garstang describes briefly the abnormal characteristics of the skeleton of Seneb, daughter of Aty, who was buried at Beni Hasan (S d 15). The lady was of small stature (about 1.40 m.), 'but well-formed', which is suggestive of pituitary disorders.[22] Garstang notes that the child's size of the coffin may have misled robbers; they ignored the tomb's contents, which contained gold and silver jewels, precious cosmetic objects, and models of servants.[23] The skeleton found by Amélineau in the tomb complex of King Djer (S d 4) may similarly belong to a pituitary dwarf or to a child (length of coffin: 82 cm.).[24]

---

[18] C. S. Bartsocas, 'Stature of Greeks of the Pylos Area during the Second Millenium BC', *Hippocrates Magazine*, 2 (1977), no. 2, 157–60, fig. 1; id., 'An Introduction to Ancient Greek Genetics and Skeletal Dysplasias', in *Skeletal Dysplasias: Third International Clinical Genetics Seminar, Athens 1982* (New York, 1982), 11–12, fig. 10.

[19] Incidence: less than 1 per million population. See Ortner/Putschar, *Pathological Conditions*, 338–40; *Atlas of Skeletal Dysplasias*, 584–95, esp. fig. 77.1 (absence of clavicles). Cf. J. Rogers, 'Mesomelic Dwarfism in a Romano-British Skeleton', *Palaeopathology Newsletters*, 55 (1986), 6–10, for a possible case of mesomelic dwarfism in Roman Britain (3rd cent. AD).

[20] Ortner/Putschar, *Pathological Conditions*, 336–7 (mucopolysaccharidoses); R. Wynne-Davies, personal communication (pseudo-achondroplasia).

[21] Cf. the description of a female skeleton of the Roman period (4th cent. AD) found at Gloucester; C. Roberts, 'A Rare Case of Dwarfism from the Roman Period', *J. of Paleopathology*, 2 (1989), 9–21. Two possible New World pituitary dwarfs are also described by Ortner/Putschar, *Pathological Conditions*, 302–4.

[22] Garstang (n. 14 above), 41.

[23] See the tomb's contents in Garstang (n. 14 above) 113–14, pl. v, figs. 102–7.

[24] E. Amélineau, *Les Nouvelles Fouilles d'Abydos, 1897–1898*, i (Paris, 1904), 104.

Pygmyism

No skeleton of a pygmy has been found in Egypt. Smith describes nine abnormally small skulls dating to the third dynasty (S 12).[25] They belong to adults, all but one of them female. Their infantile features might lead to their identification with pygmies, but they do not exhibit any negroid characteristics, such as prognathism, which suggests that their disturbance is related to pituitary disorders. The author does not give the provenance of these skulls, nor does she explain their attribution to the third dynasty, and no conclusion can be drawn concerning the incidence of pituitary disorders in ancient Egypt.

[25] H. D. Smith, A Study of Pygmy Crania, Based on Skulls found in Egypt', *Biometrika*, 8 (1921), 262–6.

II

EGYPT

3

Terminology

Three Egyptian words for abnormally short people, real and mythical, are known: *dng*, *nmw*, and *ḥwʿ*. These terms are usually accompanied by a determinative which depicts a disproportionate dwarf with a long trunk and short limbs .

*Dng/*dlg (d3ng, d3g)*  is the earliest attested term.[1] It is found in two Old Kingdom texts, the letter of King Pepy II to Harkhuf, and a spell in the Pyramid Texts, both dating to the sixth dynasty, where it may refer to ethnically short people.

The most informative text is the letter of congratulation sent by King Pepy II (*c.*2246–2152 BC) to Harkhuf, a high official who had successfully led a fourth trading expedition to the southern kingdom of Yam. Harkhuf had the king's letter carved beside the accounts of his journeys on the façade of his tomb in the necropolis of Aswan.[2] We read that he brought back many precious exotic products, such as incense, ebony, and ivory.[3] He also had an unusual present from the land of *Akhtiu*, 'of the horizon-dwellers(?)', a small man called a '*dng* of the god's dances', whom the king longed to see:

You have said in this dispatch of yours that you have brought all kinds of great and beautiful gifts, which Hathor mistress of Imaau has given to the *ka* of King Neferkare, may he live forever. You have said in this dispatch of yours that you have brought a *dng* of the god's dances from the land of the horizon-dwellers, like the *dng* whom the god's seal-bearer Werdjededba brought from Punt in the time of King Izezi. You have said to my Person that his like has never been brought by anyone who did Yam previously.

. . . Come north to the residence at once! Hurry and bring with you this *dng* whom you brought from the land of the horizon-dwellers live, hale, and healthy, for the dances of the god, to gladden the heart, to delight the heart of King Neferkare, may he live forever! When he goes down with you into the ship, get worthy men to be around him on deck, lest he fall into the water! When he

[1] *Wb* v. 470. 5–7 (vocalized for convenience *deneg*). On these various spellings, see K.-J. Seyfried, *LÄ* vi (1986), s.v. Zwerg, 1432 n. 4.

[2] Qubbet el-Hawa; PM v. 237, façade (1)–(4). K. Sethe, *Urkunden des aegyptischen Altertums*, i, *Urkunden des alten Reiches*[2] (Leipzig, 1933), 120–31; Lichtheim, *Literature*, i. 26–7; A. Roccati, *La Littérature historique sous l'Ancien Empire égyptien* (Paris, 1982), 206–7.

[3] See the list of these gifts in the account of the third journey; Lichtheim, *Literature*, i. 26, 4–5. On imports from southern countries, see A. Lucas, *Ancient Egyptian Materials and Industries*,[4] rev. with add. by J. R. Harris (London, 1962), esp. 32–3 (ivory), 90–7 (incense), and 434–6 (ebony).

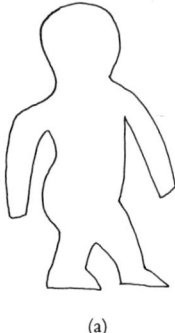      

(a)   (b)   (c)

FIGURE 3.1. Aswan, tomb of Harkhuf

lies down at night, get worthy men to lie around him in his tent. Inspect ten times at night! My Person desires to see this *dng* more than the gifts of the mine-land and of Punt!

When you arrive at the residence and this *dng* is with you live, hale, and healthy, my Person will do great things for you, more than was done for the god's seal-bearer Werdjededba in the time of King Izezi, in accordance with my Person's wish to see this *dng*.[4]

A *dng* thus appears as a being of great interest, highly praised because of his rarity; such a person had not been seen in Egypt since the time of King Izezi, more than a century earlier, when the seal-bearer Werdjededba had brought one back from Punt. Pepy II warmly congratulates Harkhuf and promises to reward him munificently. The physical characteristics of these beings, such as skin colour, size, or bodily proportions, are not described, but determinatives show small men with disproportionate features (fig. 3.1a–c).[5] The king's injunctions to watch Harkhuf's *dng* day and night on the ship suggest that he might try to escape and was regarded as wild. The king, who was still a child, may also have expressed an exaggerated solicitude.

From the routes of the expeditions of Werdjededba and Harkhuf, scholars usually agree that these *dng* may have been pygmies.[6] The earlier *dng* was brought from Punt, a country probably located along the Red Sea in a zone comprising east Sudan, north Eritrea and north-east Ethiopia,[7] while the second came, not from Yam, located in the present Dongola province, or further south,[8] but from the land of *Akhtiu*, 'of the horizon-dwellers', at the south-eastern limits of the known world, or in a realm of

---

[4] Trans. Lichtheim, *Literature*, i. 26–7 (rev. J. Baines).

[5] Dasen, 'Dwarfism', 258–9, fig. 2b.

[6] See e.g. Dawson, 'Pygmies', 185–6; A. Lansing, 'The Egyptian Expedition 1933–34', *BMMA* 29 (1934), part ii, 34–5; Hayes, *Scepter*, i. 222; Seyfried (n. 1 above); el-Aguizy, 'Dwarfs', 54.

[7] K. A. Kitchen, *LÄ* iv (1982), s.v. Punt, 1199–200; B. J. Kemp, 'The African Hinterland', in *Ancient Egypt*, 136–7.

[8] See D. O'Connor, 'The Locations of Yam and Kush and their Historical Implications, *JARCE* 23 (1986), 27–50 (with earlier bibliog.).

myth.[9] This half-legendary place could have designated the far-away homeland of pygmies. At present, most pygmy tribes move within the forest regions of Central Africa, in the Congo basin. Their location in antiquity is difficult to establish, but they probably lived further north, much closer to the Nile, and retreated with the contraction of tropical forests and swamps.[10] Before 1933 pygmies were still seen in the marshes of the White Nile on the Bahr el-'Arab and Bahr el-Ghazal, as Crazzolara reported.[11]

Yet Harkhuf and Werdjededba need not have travelled so far south to obtain the small men. Most likely, they did not meet any pygmy tribes, but acquired the specimens by trading in Punt and Yam. Pygmies have always had very ambivalent relations with their black neighbours. They exchange game against items which they cannot find in the forest, such as iron and salt. Neighbouring people tend to keep them in semi-bondage to cultivate their fields,[12] but they also respect them as 'spirits of the forest', famous for their dancing and their magical powers.[13] Pygmies may have been esteemed for these special qualities already in antiquity. Some were perhaps kept at the courts of the rulers in Yam and Punt, as has been customary at African courts,[14] and they may have been offered as prestige gifts to Egyptians.[15] Their capture would imply the existence of very early contacts between lower Nubian countries and sub-Saharan Africa. This hypothesis is supported by the fact that the word *dng* is paralleled by the Amharic *denk* 'dwarf',

[9] See C. Kuentz, 'Autour d'une conception égyptienne méconnue: l'*Akhit* ou soi-disant horizon', *BIFAO* 17 (1920), 121–90, esp. 132 ff.

[10] L. L. Cavalli-Sforza, 'African Pygmies: An Evaluation of the State of Research', in *African Pygmies*, 361–4, fig. 26.1. On the probable progressive desiccation in Egypt during the Old Kingdom, see K. W. Butzer, *Studien zum vor- und frühgeschichtlichen Landschaftswandel der Sahara* (Wiesbaden, 1959), 93–6; id., *Early Hydraulic Civilization in Egypt* (Chicago, 1976), 26–38. See also S. Adam, 'La Nubie, trait d'union entre l'Afrique centrale et la Méditerranée', in *Histoire générale de l'Afrique*, ii (Paris, 1980), 251–2.

[11] P. Crazzolara, 'Pygmies on the Bahr el Ghazal', *Sudan Notes and Records*, 16 (1933), 85–8.

[12] On these complex relationships, see e.g. Crazzolara (n. 11 above), who reports that pygmies' lives were also threatened by the belief that their bodies contained a powerful 'medicine'. See also L. Demesse, *Changements économiques et sociaux chez les pygmées Babinga* (Paris, 1978), 1 ff.; C. M. Turnbull, 'Survival Factors among M'buti and other Hunters of the Equatorial African Rain Forest', in *African Pygmies*, 113 ff.

[13] On pygmies' dancing, see in particular P. Schebesta, *Die Bambuti-Pygmäen vom Ituri*, ii. 2 (Brussels, 1941), 252–9. On their magical powers, see e.g. E. de Rosny, *Les Yeux de ma chèvre* (Paris, 1981), 127–8, 327–8, 348–9, 351 (initiation of healers, Cameroon).

[14] On pygmies and dwarfs as court jesters in African kingdoms, see e.g. G. Schweinfurth, *The Heart of Africa*, ii (London, 1873), 126–7: 'those who had been attached to the Niam-niam expeditions, whenever they described the variety of wonders about the splendour of the courts of the cannibal kings, never omitted to mention the dwarfs who filled the office of court buffoons.' Cf. also Kimenya, the dwarf messenger of King Kamrasi in J. H. Speke, *Journal of the Discovery of the Source of the Nile* (Edinburgh and London, 1863), 550–1. For representations of these court dwarfs in Benin (14th–15th cent. AD), see A. Duchâteau, *Benin, Kunst einer Königskultur* (Paris, 1989), 45, and the figs. on 12–13, 16–17, 114–15.

[15] The same happened with European explorers. Schweinfurth, ibid. 67–8, thus exchanged one of his dogs for 'a specimen of the little Akka people' living at the court of King Munza.

which suggests that it could have been an African term that came to Egypt with the small men.[16]

Some pygmy tribes may have lived in other parts of Africa. Nonnosus, in the time of Justinian, and the Arab writer al-Nuwairi (fourteenth century AD) mentioned small people on the island of Nu'man in the northern Red Sea.[17] No trace of them, however, has yet been found on the present island of Nu'man on the coast of Arabia. The view presented by Thomson and Randall-MacIver at the beginning of this century that a prehistoric race of pygmies lived in Egypt has no scientific basis and must be rejected.[18]

For Weeks, the fact that the determinative of the word *dng* depicts a pathological dwarf is a major argument against its meaning 'pygmy'. Since the hieroglyph does not characterize these *dng* by some ethnic feature, he considers them as Egyptian dwarfs of a special rank.[19] Yet Egyptians may not have differentiated between indigenous and ethnically short people, and they may have used the same class indicator for both. The sign of a malformed dwarf was also more distinctive than that of a proportionate one in the script. An inscription in the temple of Montu at Karnak thus uses the standard dwarf logogram to describe southern short people: in a row of Nubian offering-bearers, the picture of a disproportionate dwarf designates the 'dwarfs of the southern lands' (pygmies?) who bring tribute to Ptolemy VI.[20]

The second occurrence of a *dng* is in the Pyramid Texts, religious spells which secured the welfare of the king in the hereafter (*c.*2300–2150 BC).[21] The meaning of the term is more difficult to define here. In one utterance, the king asks the boatman of the gods to ferry him to the sky:

(§ 1188) O you who ferry over the righteous boatless as the ferryman of the Field of Rushes, I am deemed righteous in the sky and on earth, I am deemed righteous in this Island of earth to which I have swum and arrived, which is between the thighs of Nut! (§ 1189) I am that pygmy (*dng*) of the dances of god (*jb3w ntr*), who diverts the god in front of his great throne![22]

[16] W. Vycichl, 'Amharique *denk* "nain", Egyptien *d-n-g*', *Annales d'Éthiopie*, 2 (1957), 248–9. On the resemblances of other African names to the Egyptian *dng* (*dank, dinki, donku, dinka*), see D. Olderogge, 'Migrations et différenciations ethniques et linguistiques', in *Histoire générale de l'Afrique*, i (Paris, 1980), 311–15; el-Aguizy, 'Dwarfs', 54 n. 2. I am grateful to D. Dixon for this suggestion.

[17] Nonnosus, *FGrH* iv. 180 = Phot. *Bibl.* 3.3a. 21–38. For a discussion of this location and its possible relation to Punt, see A. Nibbi, 'Punt and Pygmies in the Northern Red Sea', *DE* 2 (1985), 27–36, and n. 19.

[18] A. Thomson and D. Randall-MacIver, *The Ancient Races of the Thebaid* (Oxford, 1905), 87. *Contra*: e.g. Weeks, *Anatomical Knowledge*, 209.

[19] K. Weeks, 'Art, Word, and the Egyptian World View', in K. Weeks (ed.), *Egyptology and Social Sciences* (Cairo, 1979), 72–3.

[20] PM ii. 3, base (8); K. Sethe, *Urkunden des aegyptischen Altertums*, viii. 1, *Thebanische Tempelinschriften aus der griechisch-römischen Zeit* (Berlin, 1957), 36 (c), l. 3; J. Dümichen, *Geographische Inschriften altägyptischer Denkmäler*, i (Leipzig, 1865), pl. XXXI.

[21] Pyramid Texts of Pepy I, Merenre, Pepy II, spell 517, § 1189. K. Sethe, *Die altägyptischen Pyramidentexte*, ii (Leipzig, 1910), 163; R. O. Faulkner, *The Ancient Egyptian Pyramid Texts* (Oxford, 1969), 191; Lichtheim, *Literature*, i. 48.

[22] Trans. Faulkner, ibid.

The king is thus identified with a *dng* who performs an *jb3* dance like the 'dwarf' brought back by Harkhuf. Nothing suggests that a pygmy rather than a pathological dwarf was meant in the spell. The term has no determinative, apart from an abbreviation in the tomb of Merenre: 𓀐. As the scene takes place in a supernatural world, this *dng* could be a purely mythical dwarf figure, for example an early form of the god Bes, who is closely connected with dancing and with solar theology in later periods.

The root *dng* is also used for an edible plant and an ear deformity.[23] The latter usage seems to denote an external abnormality as well as an internal deficiency; it is perhaps related to the fact that dwarfs often suffer from ear defects such as deafness, or it may refer metaphorically to dwarf ears or hearing. The word *dngngs*, attested in the Graeco-Roman period, includes the root *dng* as a component.[24] *Dngngs* designates the uraeus snake, symbol of the burning eye of Re, worn on the forehead of the king and of various goddesses, such as Buto, Nekhbet, and Isis, and it could have some connection with the solar affinities of dwarfs.

In the Middle and New Kingdoms, *dng* occurs as a personal name in the forms of *d3g*, *dlg* for men, and *d3gt*, *dngt* for women.[25] These names do not seem, however, to have described physical appearance. The deceased are not represented with dwarfish features, nor are they referred to as foreigners.[26] This point is significant because short stature was not regarded as an ordinary physical deficiency which would be suppressed, but as a divine mark.[27] If the deceased was a dwarf, he would probably have liked his malformation to be shown on his stela, as happened with Djeho (E 84; pl. 26.2). For the same reason, it is unlikely that this name was prophylactic.[28] On the contrary, when applied to normal-sized people, it may have alluded to the magical or moral qualities attributed to dwarf gods like other theophoric names.[29]

An interesting exception is the Late Period depiction of a malformed woman, called *dg-nt*, on a small obelisk in Brooklyn (E *d* 83; pl. 26.1). Her name might refer to her peculiar physical form. The woman seems to suffer from a rare type of short stature, short-trunk dwarfism; her trunk and neck are very short, her arms appear unusually long, and her nose and jaw are curiously protruding.[30] De Meulenaere first suggested

[23] *Wb* v. 470.2–4.

[24] *Wb* v. 470.8–11.

[25] Ranke, *PN* i. 396.3 (Middle and New Kingdoms); 396.5 (Middle Kingdom); ii. 180.

[26] For common nicknames denoting physical deficiencies and foreign origins, see the list of Ranke, *PN* ii. 177–80.

[27] A few cases of concealed physical malformation are revealed by skeletal remains, e.g. a woman suffering from elephantiasis was depicted with a perfect slim body in her tomb at Asyut; E. Chassinat and C. Palanque, *Une Campagne de fouilles dans la nécropole d'Assiout* (Cairo, 1911) (MIFAO 24), 21–2, no. 5, pl. II, figs. 1 (statuette) and 4 (mummy).

[28] Hypothesis of D. Silverman, 'Pygmies and Dwarves in the Old Kingdom', *Serapis*, 1 (1969), 57.

[29] Cf. P. Vernus, *LÄ* iv (1982), s.v. Namengebung, 326–33, esp. 328–9 (theophoric names) and 330 (nicknames referring to physical appearance). See also H. Brunner, *LÄ* i (1975), s.v. Blindheit, 829–30, for names like 'the blind one' which may similarly refer to moral qualities rather than to physical reality.

[30] See diagnosis below 42.

that her name might mean 'dwarf of Neith',[31] a reading which is supported by the fact that dwarf gods are found associated with the goddess Neith of Sais from the New Kingdom on.[32] For Koenig, other theophoric names with the compound *dg*, such as *dg-B3stt*, *dg-ḫnsw* and *dg-Dḥwtj*, may similarly allude to the relationship of dwarfs with other gods.[33] De Meulenaere and Yoyotte, however, have recently refuted this interpretation by demonstrating that *dg3* is more likely to mean 'offshoot', 'offspring', and is not related to dwarfism.[34]

The term *nmw (nmj, nmt)* 〰〰 〰〰 is attested from the Middle Kingdom on, mostly in magical and religious texts.[35] It refers to human and supernatural dwarfs, both with short-limbed bodies.

The most complete description of a *nmw* occurs in a spell from the Late New Kingdom Harris Magical Papyrus.[36] The divine *nmw* is clearly characterized as an achondroplastic dwarf. He is a god 'whose face is big (ʿ3), whose back is long (q3) and whose legs are short (ḥwʿ)' (U 8.9).[37] The dwarf gods Bes and Ptah, who are characterized by similar physical proportions, are also called *nmw*. In a New Kingdom spell from Deir el-Medina, the picture of a divine *nmj* shows a small man, drawn in profile, with a large head and short curved limbs, like a Ptah-Pataikos: ⌇.[38] In another New Kingdom papyrus, formerly in the Borchardt collection, the determinative of the divine *nmj* is a frontal, squat figure with hands on hips and bandy legs like Bes, but without his feather crown: ⌇.[39] *Nmw* also occurs as the caption to depictions of human dwarfs with curved legs and small arms, as in the Middle Kingdom tombs of Khety and Baqt I (E 67, 68; fig. 9.21); it describes the achondroplastic dwarf Djeho in the biographical text carved on his sarcophagus (E 84; pl. 26.2). No text, however, associates a *nmw* with southern countries, or with any foreign land.

This word has no other meanings, except in the variant *nmw.t* 'bird trap',[40] which has no connection with dwarfism and may be etymologically unrelated.

Nmw or *nmj* is found in personal names of the eighteenth dynasty and Late Period.[41]

[31] H. De Meulenaere, 'Un Sacerdoce spécifique de Basse-Egypte', *CdE* 40/80 (1965), 254–5.

[32] R. el-Sayed, *La Déesse Neith de Saïs, Importance et rayonnement de son culte*, i (Cairo, 1982) (BdE 86), 10–11. On this association, see also below 50–1, 77, 87.

[33] Y. Koenig, *Le Papyrus Boulaq 6* (Cairo, 1981) (BdE 87), 70.

[34] H. De Meulenaere and J. Yoyotte, 'Deux Composants "natalistes" de l'anthroponymie tardive, 2, DG3, "rhizome, racine, rejeton"?', *BIFAO* 83 (1983), 112–22, esp. 113, no. 7.

[35] *Wb* ii. 267.4–6 (vocalized *nemu* or *nemi*).

[36] Spell U 8-V 9.13; H. O. Lange, *Der magische Papyrus Harris* (Copenhagen, 1927), 72 and 74, ll.

1–4; G. Roeder, *Der Ausklang der ägyptischen Religion* (Zurich, 1961), 175.

[37] Trans. Dawson, 'Pygmies', 188.

[38] J. Černý and G. Posener, *Papyrus hiératiques de Deir el-Médineh*, i (Cairo, 1978) (DFIFAO 8), 9–10, verso 5, 5–6.

[39] Černý/Posener, ibid. 9–10. See a similar determinative in Boulaq Papyrus 6; Koenig (n. 33 above), recto vii. 2.

[40] *Wb* ii. 267.7.

[41] Ranke, *PN* i. 204.10; ii. 179. For the 18th-dyn. stela, see K. Dyroff and B. Pörtner, *Ägyptische Grabsteine und Denksteine aus süddeutschen Sammlungen*, ii (Strasburg, 1904) 71, no. 132, pl. 9 (name of a scribe), and S. R. K. Glanville, 'Records of a Royal

As with *dng*, this name does not seem to have been used for people of short stature or of foreign origin; most likely it was also a theophoric epithet.

In some personal names, the word *nmw* is combined with the name of the dwarf god Ptah, as in the name *Ptḥ-sḏm-p3-nm* 'Ptah who listens, the dwarf', which labels the depiction of a short-limbed dwarf under a Ptolemaic miniature sarcophagus (fig. 7.1).[42]

Ḥwˁ 𓀀𓃾𓏤 is attested as far back as the Old Kingdom.[43] This term usually designates someone or something short, as in the Harris Magical Papyrus, where it describes the shortness of the legs of the dwarf god.[44] In the New Kingdom Borchardt papyrus, it clearly designates the god himself: the dwarf is described alternatively as *nmw*, 'the dwarf', and as *ḥwˁ*, 'the Short One'; both words are accompanied by the same determinative showing a Bes-like figure.[45]

The reconstruction by Fischer of [𓀀𓃾] 𓏤 as part of a title with a dwarf determinative on an Old Kingdom stela is now abandoned.[46] Fischer has shown that this is a title of unknown reading beginning with *s*, which also occurs on a limestone offering basin from Saqqara.[47]

In a medical context *ḥwˁ* denotes physical deficiencies like short-sightedness or a weakened heart, but is not found describing a short man.[48] *Ḥwˁ* is unknown as a personal name.

It has also been suggested that the words *jwḥw*, *ḏnb* and *jw* may describe specific types of dwarfism. *Jwḥw* 𓂋𓃾𓀀𓃾 [49] occurs on several Old Kingdom monuments, twice associated with short-statured persons: in a title of the dwarf Seneb in Giza, 'leader of the *jwḥw*' (E 113c 2), and as a caption above the small but well-proportioned Pepy leading pet animals in the tomb of Ty in Saqqara (E 29a; pl. 20.1). Weeks suggests that this term could define Pepy's specific affliction and denoted a hypopituitary dwarf.[50] As Seneb was overseer of both 'the dwarfs in charge of linen', *ḥrp* 𓀀 *sšrw* (E 113b 2;

Dockyard of the Time of Tuthmosis III: Papyrus British Museum 10056', *ZÄS* 66 (1931), 107; id., ibid. 68 (1932), 34 and 13 n. 22 (name of a dockyard worker).

[42] Wood (l. 11.5 cm.; w. 5.5 cm.; h. 3.5 cm.), Institut d'égyptologie V. Loret, Lyons, I.E 677 (1046); W. Spiegelberg, 'Ägyptologische Mitteilungen, III, Zu dem Typus und der Bedeutung der als Patäken bezeichneten ägyptischen Figuren', *SBAW* 2 (1925) 9 (with fig.); J. Quaegebeur *et al.*, 'The Memphite Triad in Greek Papyri', *GM* 88 (1985), 28, fig. 1 (with further examples of similar names).

[43] *Wb* iii. 51–2 (vocalized *hua*). B. Gunn, 'The Egyptian for "short"', *RecTrav* 39 (1921), 101–4; Weeks, *Anatomical Knowledge*, 214–15.

[44] See n. 36 above.

[45] See n. 39 above. See also the description of a

dwarf god in the Papyrus New York 35.9.21, col. 26, 12–13; J.-C. Goyon, 'Les Dernières Pages des "Urkunden mythologischen Inhalts"', *BIFAO* 75 (1975), 362–3. I am very grateful to J.-C. Goyon for this reference.

[46] Cairo Museum, CG 1652; H. G. Fischer, 'Chroniques. Monuments of the Old Kingdom in the Cairo Museum', *CdE* 43/86 (1968), 310–12; id., 'Five Inscriptions of the Old Kingdom', *ZÄS* 105 (1978), 48.

[47] See the discussion of these titles below 131–2.

[48] H. von Deines and W. Westendorf, *Grundriss der Medizin der alten Ägypter*, vii. 2, *Wörterbuch der medizinischen Texte* (Berlin, 1962), 591.

[49] *Wb* i. 57.15 (vocalized *iuhiu*).

[50] Weeks, *Anatomical Knowledge*, 215.

c 1), and of *jwḥw*, this would imply that Egyptians made a careful distinction between disproportionate and proportionate dwarfs. However, as Junker first noted, *jwḥw* is more likely to be a title occasionally borne by dwarfs.[51] It is also held by full-sized servants like Ipet in the mastaba of Djedmankh, and by the official Ankhu in an inscription in the Wadi Hammamat.[52] As the dwarf Pepy is shown leading pet animals, the title may have described custodians of dogs, a function often performed by dwarfs in Old Kingdom households.[53] It is tempting to relate *jwḥw* to the onomatopoeic terms *jwjw* and *jw* ('howler') which denote dogs from the Middle Kingdom onwards.[54] The title of Seneb, 'leader of the *jwḥw*', may have been the equivalent of *3ṯw n mnjww ṯzmw*, 'supervisor of hounds' keepers', which was held by Middle Kingdom officials.[55] As a title, *jwḥw* may also have referred not to animal-tending, but to a more important office in which the dwarf Pepy is not shown engaged.[56]

The words *dnb* and *jw* refer to physical anomalies which are associated with short stature in two Middle Kingdom monuments. In the Beni Hasan tombs of Khety and Baqt I, a dwarf, captioned *nmw*, is followed by a man of similar size with club-feet who is termed *dnb* 'the crooked one' (E 67, 68; fig. 9.21).[57] The limb ratio of the lame man is not abnormal, but he may suffer from a type of restricted growth not associated with severe bodily malformations. In the tomb of Baqt I (E 68; fig. 9.21), the row is completed by a third man, a hunchback called *jw*, a rare word which may designate his hump.[58] His physical proportions are unusual; his neck and his trunk are shortened, while his back has a characteristic angular kyphosis. He may suffer from a tuberculous deformity of the spine (Pott's disease).[59] It is, however, not possible to assert that *dnb* and *jw* characterized specific types of dwarfism.[60] As Weeks notes, the similar size of these malformed attendants does not imply that they were short-statured; they may be shown at the same scale as the *nmw* dwarf to indicate that they had the same status in the household, or for compositional reasons only.[61]

To sum up, three words designate a dwarf: *dng*, *nmw*, and *ḥwꜤ*. The former two do not occur in sources of the same period, and it does not seem possible to establish their specific meanings. One may only point out the following distinct elements:

[51] H. Junker, *Giza*, v (Vienna, 1941), 10–11. See also Sourdive, *La Main*, 94–5; Seyfried (n. 1 above), 1432 n. 7.

[52] L. Borchardt, *Das Grabdenkmal des Königs Ne-user-reꜤ* (Leipzig, 1907), 122; G. Goyon, *Nouvelles Inscriptions rupestres du Wadi Hammamat* (Paris, 1957), 14–15, 57–8, no. 23. See also Kaplony, *Inschriften*, i. 375.

[53] So for Seyfried (n. 1 above), 1432 n. 7.

[54] On both terms, see H. G. Fischer, *LÄ* iii (1980), s.v. Hunde, 77.

[55] Cf. Fischer, ibid. 78 n. 26.

[56] So Junker (n. 51 above); Sourdive, *La Main*, 95.

[57] *Wb* v. 576.2–6; Weeks, *Anatomical Knowledge*, 215–16. For *dnb* denoting a crooked nose, see J. H. Breasted, *The Edwin Smith Surgical Papyrus*, i (Chicago, 1930), 249–50, case 12, gloss B, 1.2; von Deines/Westendorf (n. 48 above), 1004. Cf. also the Middle Kingdom personal name, *ḥtj dnb* 'the bow-legged one'; Ranke, *PN* i. 278.4.

[58] *Wb* i. 43.11.

[59] Ruffer, *Palaeopathology*, 42–3; D. Morse *et al.*, 'Tuberculosis in Ancient Egypt', *Amer. Review of Resp. Diseases*, 90 (1964), esp. 526.

[60] *Contra*: el-Aguizy, *Dwarfs*, 53–4.

[61] Weeks, *Anatomical Knowledge*, 216.

Dng is found in contexts alluding to southern countries (but not always), and it may have denoted a pygmy, as in the letter of Pepy II. It might also have applied to indigenous dwarfs, like the short woman on the Brooklyn obelisk (E *d* 83; pl. 26.1). *Nmw* and *ḥwꜥ* appear interchangeable, as though they were synonyms. *Nmw* may more specifically have denoted physical disproportions, especially those of achondroplasia, while *ḥwꜥ* may have referred to an abnormal shortness only. Both terms do not relate to a specific geographic location.

These words could also have implied non-physical distinctions. Weeks has suggested that they may have defined social position: *dng* may have meant a dwarf of a higher status, a god's dancer, as in the Pyramid Texts, while *nmw* described a lower ranking dwarf, a servant, like those standing beside their lord at Beni Hasan (E 67, 68; fig. 9.21).[62] This interpretation does not, however, fit the evidence. *Nmw* frequently designates a dwarf god in religious and magical texts, and, in the Late Period, the term also describes a sacred dancer, the dwarf Djeho (E 84; pl. 26.2).

[62] Ibid. 213–14; id. (n. 19 above).

4

Iconographic Conventions

In Egyptian art the rendering of the human body conformed to a rigid system of conventions which remained largely unchanged throughout the dynastic period. The normal human figure has ideal proportions, fixed in a canon applied through guidelines or squared grids. The body has two relatively equal parts,[1] and is divided by six horizontal guidelines in the Old Kingdom, eighteen squares to the hair line from the Middle Kingdom on, and twenty-one to the eyeline from the twenty-fifth dynasty onwards.[2] Robins has demonstrated that this canon closely follows natural proportions, which were characterized by long distal segments in the limbs, as in Blacks. These proportions may be emphasized; she notes that in some figures 'the combined height of the femur and the tibia occupies more than half the body height, which is a greater proportion than in real life'.[3] A significant non-realistic feature is that the lower leg (tibia) is generally shown longer than the femur, although in life the reverse is true. This elongation was associated with a reduction of thighs and buttocks, probably for aesthetic reasons.[4]

In this system, forms for dwarfs became very early standardized. A review of the stock figures showing similar physical forms, such as a small or malformed body, helps us to define the specific combination of features which characterized pathological dwarfs in Egyptian conventions. Some conditions are so accurately observed that a medical diagnosis can be attempted.

---

[1] Average upper to lower segment ratio (U/L), taken from the top of head to pubis and from pubis to heel: 1/0.95; ratio head/body: 1/7.

[2] See H. Schäfer, *Principles of Egyptian Art* (Oxford, 1986), 326–34; G. Robins, *Egyptian Painting and Relief* (Aylesbury, 1986), 27–37, 43–51.

[3] G. Robins, 'Natural and Canonical Proportions in Ancient Egyptians', *GM* 61 (1983), 18–20 and 23. U/L of Black adults: 0.85. See D. L. Rimoin and R. S. Lachman, 'The Chondrodysplasias', in A. E. H. Emery and D. L. Rimoin (eds.), *Principles and Practice of Medical Genetics* (Edinburgh, 1983), 705–6.

[4] Robins (n. 3 above), 20–3.

IDENTIFICATION

Minor Figures

In tomb reliefs, variations of scale reflect the importance of the personages. Minor figures, such as members of the household, craftsmen, or herdsmen are always shown smaller than the tomb-owner and his wife.[5]

In most instances, pathological dwarfs are clearly differentiated from other lower-ranking figures. They are not only small, but show typical physical disproportions: a large head, a long trunk and short limbs, slightly bent.[6] Often they have a lordosis and a paunchy belly, enhanced by a kilt that slopes at the waist, as in depictions of wealthy, fat men (e.g. E 32, 52d; figs. 9.9, 9.18). The trunk may be as long as the legs or longer, especially in tombs of the sixth dynasty.[7]

Their relation to miniature, but full-sized servants is revealing. Dwarf attendants are usually significantly smaller than the figures standing on the same base line. In Old Kingdom scenes, their heads reach the level of the third guideline of a normal figure, above the waist,[8] or the second, below the waist.[9] Some dwarfs are thus as little as half the size of the other servants, or much smaller.[10] To preserve the balance of the composition, the space left free above the head of the dwarf may be filled with inscriptions (e.g. E 24; pl. 19.2, fig. 9.14), or the figure may carry a tall object (e.g. E 11a, 39; figs. 9.2, 9.6); the dwarf may also be placed in a sub-register (e.g. E 29b, 59a, 75b; pls. 20.2, 25, fig. 9.13a).

A dwarf standing alone in a register may be shown at a larger scale than the minor figures depicted in other parts of the relief, like the dwarf Redji in the tomb of Mereri (E 50), probably to indicate his higher position in the household; the identification of a pathological condition relies only on the rendering of abnormal physical proportions.[11] The decoration of the tomb of the dwarf Seneb offers a similar compromise between two conventions. Seneb's rank had to be expressed in terms of scale by a significantly large size, but his physical malformation, which was a specific constituent of his person, also had to be shown. The result is striking: Seneb's abnormal morphology is accurately rendered, but he is as tall as his servants (E 41e; fig. 9.19b), even sometimes slightly taller (E 41b; fig. 9.19a).[12]

The size of dwarfs may also be adapted to other figures, such as pet animals, or contexts, such as chairs and palanquins. The composition stresses their ambivalent relation to animals. Dwarfs may be smaller than the monkey or dog they lead (e.g. E 61c; fig. 9.15b), or only slightly taller (e.g. E 29a–b, 52b; pl. 20.1–2, fig. 9.11). Often

[5] Schäfer (n. 2 above), 230–4; Robins (n. 2 above), 19.

[6] Head/body: 1/5 (instead of 1/7 in normal-sized figures).

[7] See e.g. U/L of the dwarf in the tomb of Mereruka (E 52d; fig. 9.9): 2.2/1.

[8] See e.g. E 15, 20a, 43; pl. 18.2, fig. 9.7. Ratio

dwarf/normal-sized adult: 1/1.5–1.6.

[9] See e.g. E 11a, 17, 39, 59a; figs. 9.2, 3, 6, 13a. Ratio dwarf/normal-sized adult: 1/1.6–1.8.

[10] See. e.g. the very small dwarfs E 45, 61a, 62b, 72; figs. 9.15a, 17.

[11] Head/body: 1/4.9; U/L: 2.9/1.

[12] U/L: 1.9/1 and 2/1; head/body 1/4.5.

they are placed in confined spaces reserved for pet animals; they stand under the chair where the deceased is seated, under his palanquin, or in small friezes between two normal registers.[13]

The identification of dwarfs with slight or no bodily malformations, as in pituitary disorders, is more problematic. When human miniatures stand alone in a register, it is impossible to differentiate between conventional and pathological smallness. In the tomb of Inpuhotpe (E *d* 42), oxen are led by two minute men who are described as dwarfs in Porter/Moss (iii. 107). This identification is uncertain because the figures are proportionately small; it is not possible to distinguish between dwarfs and full-grown attendants represented on a small scale, leading over-sized oxen.[14] Thus, dwarfism can be confidently identified only when an unusual smallness is associated with some physical disproportion, even crudely rendered, and contrasts with the size of minor figures standing in the same line. In statuary, I consider the identification uncertain when it is based on the rendering of a single unusual physical feature, such as an overlarge head (E *d* 112) or a short trunk (E *d* 135).

Children

In Egyptian art, children are usually characterized by three distinct features which occur separately or are combined: they are naked, with a well-proportioned body, and a lock of hair hangs on the right side of the head; often they also hold a hand to their mouth.

Dwarfs may have one or several of these features, but they are distinguished by bodily disproportions and/or indications of adulthood, such as kilts or beards. For example, in the tomb of Wehemka (E 17; fig. 9.3) a child and a dwarf stand in the same line of servants. Both are a little over half the size of the other figures, but the child is slightly taller than the dwarf; he is naked, with limb ratios resembling those of normal figures, while the dwarf is clad in a kilt, like an adult, and is slightly malformed.[15] Similarly, the minute man tending a dog and a monkey under the palanquin of Ty is probably a pathological dwarf (E 29a; pl. 20.1). Unlike children, he wears a pointed kilt, a garment of high-ranking servants; he has a name, Pepy, and an obscure title, *jwḥw*, which indicates that he held a specific office in the household. A pathological deformity is suggested by his unusual slimness and by the slight shortening of his legs in relation to his trunk. He is, moreover, in charge of pet animals, a function often reserved for dwarfs. In the temple of Amenophis III at Soleb, a beard similarly identifies a small slender dancer as a short-statured adult (E 73).

---

[13] Under the chair: E 44, 60, 61a, 62b, 70; pl. 24.3, fig. 9.15a. Under his palanquin: E 20a, 29a, 37, *d* 46, 51, 52b; pls. 20.1, 22.2, fig. 9.11. Between two normal registers: E 34, 49a, 52c; pls. 21, 23a. Compare the row of dogs below the palanquin of Waᶜtetkhethor (E 52c, pl. 23a).

[14] See also the proportionate(?) dwarfs from the tomb of Hetepniptah (E *d* 40) and Niᶜankhkhnum (E *d* 47). Cf. the opposite convention (over-sized servants lead minute cattle): Schäfer (n. 2 above), 230.

[15] Child: U/L: 0.9/1; head/body 1/7.1. Dwarf: U/L: 1.37/1; head/body 1/6.3.

A few statuettes blend the physical attributes of childhood and dwarfism. They may depict young dwarfs, or express the ambiguity of dwarfs' status. Thus, the short-statured woman on Tutankhamun's calcite boat is characterized as a short-limbed dwarf with turned-in feet (talipes equinovarus), but she is also naked, with a sidelock, like a youthful figure (E 202; pl. 36a–c). Similarly, a Middle Kingdom statuette shows a naked dumpy man, with small limbs, who puts a hand to his mouth, like a child (E 128; pl. 32.1).[16] The frequent nakedness of dwarfs may also reflect the interest of artists in depicting unusual morphologies.

When no abnormal or adult features are shown, the identification is very uncertain. For example, in the tomb of Nefermaᶜet a naked male figure tends three pet animals (E d 14). His function is typical for a dwarf, but, as he is slender and well-proportioned, it is not possible to diagnose a pathological condition. Though he has no sidelock, he could be a young boy.[17] Similarly, Middle Kingdom figurines showing squatting naked men with flat elongated skulls, and no conspicuous malformation, could represent children or proportionate dwarfs.[18]

Other Physical Malformations

Malformed, but normal-sized minor figures are usually shown taller than dwarfs, like the hunchbacked attendants in the tombs of Seshemnufer I (E 18; pl. 19.1a) and of Ty (E 29b; pl. 20.2). In a few examples, however, they are depicted with dwarfs at the same scale, perhaps to indicate their similar position in the household, or for compositional reasons, like the hunchback and the lame man in the tombs of Baqt I and Khety at Beni Hasan (E 67–8; fig. 9.21). Since dwarfism may be associated with another physical disorder, such as a hump or club-feet, it is not possible to make a certain identification here.

A well-known problematic case is that of the queen of Punt. In the temple of Hatshepsut at Deir el-Bahari, she is shown with her husband receiving Egyptian officials.[19] While the Puntites are slim and well proportioned, she is very fat with dwarfish proportions, a large head, short legs and a pronounced lordosis.[20] These features resemble those on an ostracon in Berlin, probably a sketch made after the Deir

---

[16] See also E 147, 153, and the figurines touching their ears or head with one or both arms (E 168, 170–2, d 175, 193; pl. 33.4).

[17] Seen as a dwarf by e.g. L. Klebs, *Die Reliefs des Alten Reiches* (Heidelberg, 1915), 32, no. 3; O. Koefoed-Petersen, *Catalogue des bas-reliefs et peintures égyptiens* (Copenhagen, 1956), 14–15, no. 2; Rupp, 'Zwerg', 300, 1 H. *Contra* (boy): see e.g. W. M. F. Petrie, *Medum* (London, 1892), 26; Dawson, 'Pygmies', 187 n. 5; Smith, *Art and Architecture*, 81.

[18] See e.g. E d 141, d 148, d 191, d 196, d 199–200; pl. 35.3, fig. 9.23.

[19] Temple of Hatshepsut, Deir el-Bahari, PM ii. 344, middle colonnade, south half (v); A. Mariette, *Deir el-Bahari* (Leipzig, 1875), pl. 5 (line drawing). For a photograph of the relief, see E. L. B. Terrace and H. G. Fischer, *Treasures of the Cairo Museum* (London, 1970), no. 21, 101–2; *Musée égyptien du Caire*, no. 130a.

[20] U/L: 1.95/1; head/body: 1/5.8. For the features of Puntites in general (brown skins, no negroid features, bearded, long hairstyle), see e.g. D. O'Connor, 'New Kingdom and Third Intermediate Period', in *Ancient Egypt*, 270–1.

el-Bahari relief.[21] These bodily malformations have led a few scholars to suggest that she was an achondroplastic dwarf, or a pygmy of 'Khoisanid' type, with racial steatopygia, who was shown on the same scale as the other figures because of her rank, like the dwarf Seneb (E 41; figs. 9.19a–d).[22] However, the queen's figure presents a number of features which do not fit ethnic or pathological dwarfism; she has no frizzy hair or flat nose, her arms are of normal length, with well-formed hands. It seems more likely that the queen of Punt had some disease which Egyptian artists tried to render as realistically as possible. For some scholars, she suffered from muscular dystrophy, a disorder involving a progressive atrophy of the muscles, often associated with obesity; this condition is inheritable, which would explain the similar but less pronounced appearance of the queen's daughter, who might be at an earlier stage of the disease.[23] Other scholars, like Ghalioungui, Terrace, and Fischer explain her appearance by extreme obesity (lipodystrophy, or Dercum disease),[24] a feature which was highly appreciated in African society, and is still a mark of wealth and power in many modern cultures.[25] In any case, whatever was the cause of the unusual appearance of the queen of Punt, she probably was not a dwarf.

Caricatures

The borderline between faithful representation and caricature may be tenuous, especially in the spirited sketches which are found on New Kingdom ostraca. Draughtsmen freely depicted human figures in unusual poses or with abnormal bodies which cannot be diagnosed, but may have been inspired by real malformations, like the plump flautist

[21] Berlin, SM (West) 21442 (h. 14 cm.; w. 8 cm.); E. Brunner-Traut, *Die altägyptischen Scherbenbilder (Bildostraka) der deutschen Museen und Sammlungen* (Wiesbaden, 1956), 75–6, no. 76, pl. xxviii; W. Kayser, *Ägyptisches Museum* (Berlin, 1967), 64–5, no. 729.

[22] See e.g. P. Richer, *Le Nu dans l'art*, i (Paris, 1925), 190–2 (dwarf); B. Schrumpf-Pierron, 'Les Nains achondroplasiques dans l'ancienne Égypte, *Aesculape*, 24/9 (1934), 230–1 (dwarf); R. Watermann, *Bilder aus dem Lande Ptah und Imhotep* (Cologne, 1958), 91–103 (Bushwoman); R. Herzog, *Punt* (Glückstadt, 1968) (ADAIK 6), 58–61 (Khoisanid type).

[23] H. Pöch and P. E. Becker, 'Eine Muskeldystrophie auf einem altägyptischen Relief', *Der Nervenarzt*, 26/12 (1955), 528–30; E. Brunner-Traut, 'Die Krankheit der Fürstin von Punt', *WdO* 2 (1957), 307–11; ead., 'Noch einmal die Fürstin von Punt', in *Festschrift zum 150jährigen Bestehen des Berliner Ägyptischen Museums* (Berlin, 1974), 71–85.

[24] S. Tolstoï, *Étude des représentations pathologiques dans l'art égyptien* (Paris, 1938), 66–9; P. Ghalioungui, 'Sur deux formes d'obésité représentées dans l'Égypte ancienne', *ASAE* 49 (1949), 303–16; A.-P. Leca, *La Médecine égyptienne au temps des pharaons* (Paris, 1971), 265–9; Terrace/Fischer (n. 19 above), 102; H. Helwin, 'Gehörte die Königin von Punt zu den chondrodystrophen Zwergen?', *Gegenbaurs Morph. Jahrb.* 120/2 (1974), 280–9; *Musée égyptien du Caire*, no. 130a. For alternative diagnoses (rickets, double-sided hip dislocation), see Kunze/Nippert, *Genetics*, 41, fig. 42.

[25] J. H. Speke, *Journal of the Discovery of the Source of the Nile* (Edinburgh and London, 1863), 209–10, describes one of the princesses of Karague thus: 'She could not rise; and so large were her arms that, between the joints, the flesh hung down like large, loose-stuffed puddings.' Her husband explains 'that this is all the product of those (milk) pots: from early youth upwards we keep those pots to their mouths, as it is the fashion at court to have very fat wives.' See ibid. 231, descriptions and measurements of similar 'wonders of obesity'.

with a hump on a New Kingdom ostracon in Cairo, who could represent a dwarf hunchback, or be a caricature of a normal-sized musician.[26] The crude rendering and the absence of context do not allow a certain identification.[27]

MEDICAL DIAGNOSIS

Representations of human and divine dwarfs are often painstaking, and in many cases a medical diagnosis can be attempted.

Disproportionate dwarfism, the most common type of short stature, is the most frequently depicted. It is also the most easy to diagnose certainly. The earliest examples are small ivory figurines from the predynastic sites of Ballas and Naqada. They represent naked male and female dwarfs with strikingly short arms and severe leg deformities (E 88–9, 91–5; pl. 27.1–2).[28] Two ivory figures of female dwarfs from the great deposit at Hierakonpolis (Early Dynastic Period) have similar physical characteristics, but they are dressed and wear elaborate wigs (E 106–7; pl. 27.4). A third has short lower limbs, with bowing involving the entire lower leg, but relatively long arms (E 108; pl. 28.1a, b). These features could be suggestive of Morbus Paget, a disorder of old age involving the spine, the limbs, and a dilation of the cranium,[29] but this diagnosis is unlikely since an old person would probably not be shown naked in Egyptian iconography.

Early two-dimensional depictions date to the first dynasty; they are found on the funerary stelae made for the dwarfs buried in subsidiary graves around the royal tombs at Abydos. The surface of the stelae is often damaged and the carving is crude, but dwarfs are identifiable by typical bodily disproportions. They are always shown standing, exhibiting the massiveness of their trunks in relation to their limbs. Back deformities may be marked (e.g. E 3, 8; pl 17.3), and are sometimes associated with a plump abdomen (E 6; pl. 17.2). No figure, however, has a conspicuously large head and no precise diagnosis can be made. Skeletal remains reveal that various types of short-limbed dwarfism were probably rendered by the same representational type. Thus, a stela in Berlin (E 1) was found with the skeleton of a short-limbed dwarf (S 5) in the tomb complex of King Djer. Similary, two stelae in London (E 8; pl. 17.3) and Philadelphia (E 9) were found with two incomplete skeletons (S 7) of achondroplasts in the tomb complex of King Semerkhet. Two humeri from the same period may

[26] Limestone (h. 19 cm.; w. 10 cm.), Cairo Museum, CG 25040; G. Daressy, *Ostraca* (Cairo, 1901) (CGC), 9, pl. VIII.

[27] See also the dwarfish males on two New Kingdom ostraca; J. Vandier d'Abbadie, *Catalogue des ostraca figurés de Deir el Médineh*, ii (Cairo, 1937), 98, no. 2473 (IFAO 3214), 190, no. 2870 (IFAO 3944).

For other possible examples of caricatures, see A. M. Badawi, 'Le Grotesque: Invention égyptienne', *Gazette des Beaux-Arts*, 66 (1965), 189–98; E. Brunner-Traut, *LÄ* iii (1980), s.v. Karikatur, 337–9.

[28] U/L: 2.07/1.

[29] Suggestion of Dr S. Braga, Inselspital, Bern (personal communication).

show a case of mucopolysaccharidoses (S 11); the precise provenance of these bones is unknown, but they were probably related to one of the stelae listed here.

On Old Kingdom reliefs, most depictions of dwarfs conform to a similar short-limbed model. Naturalistic details are rendered, such as the thick thighs, the muscular trunk suggested by broad shoulders, or the curved and stubby arms showing the bowing of the bones in the limbs (e.g. E 13, 18, 37; pls. 18.1, 19.1b, 22.2). Seated dwarfs have peculiar irregularities of the spine, such as humps which may indicate kyphosis, a relatively frequent spine deformity (E 36a, 49a). Some features, however, are not realistic. For example, the lower leg (tibia and fibula) is often particularly short (e.g. E 13; pl. 18.1), although in reality the femur is more affected.[30] Like other minor figures, some dwarfs are partially bald (e.g. E 36a, 51, 52a, b; pl. 22.1); this feature could indicate the bulging foreheads of achondroplastic dwarfs, but this is contradicted by the fact that they have normal faces, like normal-sized men, as if they were all hypochondroplastic.

In statuary, particular anatomical details, such as the bowing in the limbs, are also very accurately rendered. This feature is well shown in the statue of Khnumhotpe who exhibits a marked genu varum associated with proximal overgrowth of the fibula (E 117; pl. 29.2); his very short curved arms, with squat hands, fall stiffly to the sides, because of the limited range of motion at the elbow.[31] His facial features are idealized, like those of Seneb and other Old Kingdom dwarfs (E 113c, 115–17; pls. 28.2, 29.1–3). It is interesting to note that Seneb's three children are not shown as inheriting the disorder of their father, although the chances of passing the condition on to the children are 50 per cent in couples with one affected person; they may also have been dwarfs, but they are not shown as such.[32]

In the Middle Kingdom, this short-limbed model is slightly modified. The most typical features of achondroplasia, snub-noses and bulging foreheads, appear in statuary (e.g. E 144; pl. 33.3). In small faience figurines, the lips may be emphasized, which gives the dwarfs a negroid appearance (e.g. E 133, 192; pl. 32.4). The rendering of their skulls is also new; instead of having round heads, they have curiously flat skull tops, as in the dwarf Ita (E 197; pl. 34.3).[33] This feature cannot be well explained in medical terms; rather, it recalls the elongated skull, often adorned with a scarab-beetle, of the god Ptah in the form of a dwarf, who may have emerged in that period.

From the New Kingdom onwards, amuletic Ptah-Pataikoi reproduce exactly the features of achondroplasia. In most depictions, the proportions are realistic: the curved arms reach only to hip level, the legs are bowed with skin folds in the thighs; the figures

---

[30] A.-M. E. Nehme *et al.*, 'Skeletal Growth and Development of the Achondroplastic Dwarf', *Clin. Orthop.* 116 (1976), 17.

[31] For a detailed description, see Ruffer, *Palaeopathology*, 38–9.

[32] On other possible diagnoses of Seneb

(achondroplasia, hypochondroplasia, metaphyseal chondrodysplasia), see Kunze/Nippert, *Genetics*, 16–17, fig. 9.

[33] See Kunze/Nippert, *Genetics*, 18–19, fig. 12 (achondroplasia).

also show genu varum with lateral curvature (e.g. pl. 12.1).[34] For some scholars, the baldness and small facial bones of the god suggest that he represented a foetus.[35] It is, however, more likely that, as a dwarf, this deity embodied a being young and old at the same time. Many figurines show the god with a sidelock of youth, standing on crocodiles and holding venomous animals, like the child Horus (e.g. pl. 12.2–3). In other depictions he looks like an adult, sometimes with a beard (pl. 12.4), or a wrinkled forehead.

In the Late Period, achondroplastic features are best shown in the relief of the dwarf Djeho (E 84; pl. 26.2). He has a large cranial vault with an elongated skull and a small button nose, a slightly plump trunk, and thick thighs; his arms are very short and the hands reach only to the hips (iliac crest).[36]

Other types of short-limbed dwarfism may be identified, such as diastrophic dwarfism, which is characterized by very short lower limbs, often severely contracted. An Old Kingdom statuette thus depicts a dwarf harpist with very small long bones, and abnormal ear lobes, large and ossified, which may be associated with this disorder (E 116; pl. 29.3). This detail, however, does not necessarily indicate a pathological condition. Many Egyptian figures have such large ears without any medical cause; these large ears could also have a symbolic meaning in a musician. At el-ᶜAmarna, the pair of dwarf attendants of Princess Mutnedjmet have very bowed legs with club-feet, short arms with a strikingly stiff pose, and club-hands(?) which may be a rendering of the contraction of the joints (E 75–9; pl. 25, fig. 9.27). They could also be affected by pseudo-achondroplasia, which induces similar crippling malformations (fig. 1.1c).

Hypothyroidism (fig. 1.1e) has been often recognized in representations of the dwarf god Bes. As a short-limbed dwarf, Bes is characterized by a large head, a long stout trunk and short bandy legs. His human appearance conforms to the short-limbed model used for Ptah-Pataikoi.[37] On a Late Period bicephalic amulet, a Bes and a Ptah-Pataikos stand back to back (pl. 8.2); they have distinct head-dresses, the *Atef* crown for Ptah, and the feather crown for Bes, but they share the same achondroplastic body. Some particular features, like his large squat nose, his protruding tongue, and his obesity, often associated with infantile genitals, led a few scholars to think that he represented

[34] See J. Parrot, 'Sur l'Origine d'une des formes du dieu Phtah', *RecTrav* 2 (1880), 129–33; F. Regnault, 'Les Nains dans l'art égyptien', *Bull. Soc. Franç. d'Hist. de la Méd.* 20 (1926), 135–50; R. Hückel, 'Über Wesen und Eigenart der Pataiken', *ZÄS* 70 (1934), 103–7; Schrumpf-Pierron (n. 22 above), 223–38; Tolstoï (n. 24 above), 37–40; T. Dzierżykray-Rogalski and E. Pomińska, 'La Statuette de Ptah-Patèque des collections du Musée égyptien du Caire', *Africana Bulletin*, 13 (1970), 109–11. For pseudo-achondroplasia, see F. N. Silverman, 'De l'Art du diagnostic des nanismes dans l'art', *J. Radiol.* 63 (1982), 133–40. For other alternative diagnoses (spondylo-epiphyseal dysplasia, metaphyseal dysostosis), see Kunze/Nippert, *Genetics*, 18–19, figs. 10–11.

[35] P. A. Vassal, 'La Physio-Pathologie dans le panthéon égyptien: Les Dieux Bès et Phtah, le nain et l'embryon', *Bull. Mém. Soc. Anthr. Paris*, 7 (1956), 168–81.

[36] U/L: 1.6/1; head/body: 1/5.2.

[37] See Kunze/Nippert, *Genetics*, 20–5, figs. 13–15 (achondroplasia).

a dwarf affected by myxoedema.[38] This diagnosis, however, seems unlikely, since hypothyroidism is usually associated with severe mental and metabolic disorders which do not fit the lively nature of Bes. This interpretation also neglects New Kingdom and later depictions of the god with an erect or large phallus (pl. 4.1).[39] For other scholars, the high plumed head-dress of the god and his facial tattoo patterns indicate that his model was an African pygmy.[40] In fact, Bes is a hybrid demonic creature who cannot be diagnosed medically. As Romano has demonstrated, the god is composed of human and animal features; he has the proportions of a short-limbed dwarf, mixed with the round ears of a lion, its tail, and even its ruff.[41]

Short-trunk dwarfism (fig. 1.1d) is almost never shown. It might be identified in the unusual picture of a woman on a small Late Period obelisk (E *d* 83; pl. 26.1).[42] Her trunk is very short, with a slight lordosis producing a prominent belly. Her arms are of unrealistic length and reach mid-tibia. She also has a depressed nasal bridge and protruding jawbones, giving her an almost simian appearance. These unusual features may represent a racial characteristic, but no positive conclusion can be drawn from her name *dg-nt*.[43]

Tuberculous osteomyelitis of the spine, or Pott's disease, a disorder which was probably present in Egypt as early as predynastic times, may be identified in two representations only.[44] In the tomb of Nikauisesi, a dwarf attendant exhibits a short trunk, a protruding chest and a hump which may be due to that disease (E 54c). The hunchback in the tomb of Baqt I shows similar characteristics (E 68; fig. 9.21); he is captioned *jw*, a term which may describe his malformation.[45] In the tomb of Ty, a man leading two dogs in a sub-register also has a shoulder protruding at a curious angle (E 29b; pl. 20.2); however, no conspicuous shortening in the trunk is present, and the deformity may be due to scoliosis, as suggested by Morse.[46]

Very few dwarfs with proportionate short stature have been identified. As noted, this is partly due to the fact that these disorders are difficult to diagnose within Egyptian iconographic conventions. It was necessary to differentiate such dwarfs from minor figures shown at the same scale, either by adding some bodily disproportion, or by

[38] F. Regnault, 'Le Dieu égyptien Bès était myxoedémateux', *Bull. Soc. Anthr. Paris*, 4th ser. 8 (1897), 434–9; id. (n. 34 above), 143–6; A. Ber, 'Déité de l'Égypte ancienne, Bès eut-il pour modèle un nain hypothyroïdien?', *Organorama*, 10/4 (1973–4), 25–30.

[39] See also the New Kingdom faience figurine in Brooklyn Museum, 16.580.13; Romano, 'Origin', 45, fig. 6. For the Graeco-Roman period, see e.g. Tran Tam Tinh, *Bes*, 99, no. 9, pl. 75; 102, no. 45, pl. 80.

[40] See in particular L. Keimer, 'Un Bès tatoué?', *ASAE* 42 (1943), 159–61.

[41] Romano, 'Origin', 39–50. See also Tolstoï (n.

24 above), 44–7.

[42] U/L: 1/0.87.

[43] See above 29–30.

[44] See D. Morse *et al.*, 'Tuberculosis in Ancient Egypt', *Amer. Review Resp. Diseases*, 90 (1964), esp. 524–8 on pictorial evidence; D. Morse, 'Tuberculosis', in *Diseases in Antiquity*, 249–71. See also M. D. Grmek, *Les Maladies à l'aube de la civilisation occidentale* (Paris, 1983), 261–6, and K. Manchester, 'Tuberculosis in Antiquity: An Interpretation', *Med. Hist.* 28 (1984), 162–73.

[45] See the discussion above 32.

[46] Morse *et al.* (n. 44 above), 526; Morse (n. 44 above), 261, fig. 4C.

indicating their adult role. Thus, a case of hypopituitarism may be identified in the short man leading an ox in the tomb of Ireru because of his relative smallness and of his slightly abnormal proportions (E 23; fig. 9.12).[47] Similarly, the small animal-tender Pepy in the tomb of Ty at Saqqara is probably a dwarf because of his adult name, dress and function (E 29a; pl. 20.1). In the tomb chamber of Wa꜀tetkhethor, twelve female dwarfs with different pathologies walk in a row (E 52c; pl. 23a, b); all have normal facial features. Ten have a disproportionate short stature, with broad shoulders and prominent buttocks, while two are conspicuously slim with a trunk of normal length, and may depict, like Pepy, persons affected by pituitary disorders.[48]

The absence of securely identified skeleton or depiction of pygmies tends to confirm that ethnic dwarfs were a great rarity in dynastic times. A possible example is an ivory toy from el-Lisht cemetery, which shows dancing dwarfs (E 122; pls. 30.2, 31a, b).[49] The physical characteristics of the figures are very unusual: they are proportionately small in size, muscular, with a slight spinal deformity and projecting buttocks; their genitals are made conspicuous, which is very rare in Egyptian statuary. The treatment of their faces, with grimacing and wrinkled foreheads, stresses their alien appearance. Yet they do not exhibit clear ethnic characteristics from which we might conclude that they are pygmies and not short Egyptians. No diagnosis can be drawn from the bowing of their legs, which is due to their dancing; their flat noses could indicate a negroid origin as well as a pathological abnormality.

Another possible case of a pygmy is the small black woman who seems to perform a Nubian dance on a tambourine from Akhmim (E 86; fig. 9.28a). She substitutes for the leonine god Bes, who is depicted on the other side of the tambourine. Borchardt identified her with a southern *dng* because of her black skin.[50] She may, however, also be a black dwarf, possibly suffering from short-trunk dwarfism, as her abnormally long arms suggest. The miniature man dancing at the *sed*-festival on a relief at Soleb (E 73) might also be a pygmy; he is labelled as from Punt, but no certain identification can be made from the sketchy drawing available.[51] Some scholars have referred to a Late Period bronze figurine in Cairo as depicting a pygmy (E *d* 207), but this diagnosis is uncertain since the figure exhibits no particular ethnic features, apart from a protruding belly and a possible lordosis.[52]

[47] U/L: 1.25/1; head/body 1/5.3 (instead of 1/6.7 in the full-sized figures).

[48] Cf. also the proportionate dwarfs in the tomb of Amenemhet (E 66; fig. 9.20) and from Tell-Basta (E 82; fig. 9.26).

[49] See e.g. A. Lansing, 'The Egyptian Expedition 1933–1934', *BMMA* 29 (1934), part ii, 34; M. Stracmans, 'Les Pygmées dans l'ancienne Égypte', in *Mélanges G. Smets* (Brussels, 1952), 626–7; Hayes, *Scepter*, i. 222; Rupp, *Zwerg*, 295; *Musée égyptien du Caire*, no. 90.

[50] L. Borchardt, 'Die Rahmentrommel im Museum

zu Kairo', *Mélanges Maspéro*, i (Cairo, 1935–8) (MIFAO 66), 4. See also H. Wild, 'Une Danse nubienne d'époque pharaonique', *Kush*, 7 (1959), 81–3.

[51] I am very grateful to J. Leclant for showing me a drawing of this unpublished relief.

[52] See e.g. G. Daressy, 'Statuette grotesque égyptienne', *ASAE* 4 (1903), 124–5; W. R. Dawson, 'Pygmies, Dwarfs and Hunchbacks in Ancient Egypt', *Ann. Med. Hist.* 9 (1927), 320, fig. 25; Schrumpf-Pierron (n. 22 above), 236, fig. 22.

To sum up, throughout the dynastic period Egyptian artists used a similar model derived from the physical features of the most common condition, short-limbed dwarfism. Even the notion of a pygmy was encoded in the script by a disproportionate dwarf.

The characteristic features of this condition are in general acutely observed and selected (the big head, the long trunk in relation to the short limbs), and realistic details often are added, such as the bowing in the limbs, heavy thighs and buttocks, and back deformities. From the Middle Kingdom on, bulging foreheads and snub-noses are also indicated.

These models imply that most representations of dwarfs are not 'portraits', but use a generic form. The features of specific individuals may be present in only a few cases, when unusual morphologies were shown for which no archetype existed, as in the cases of the diminutive Pepy in the tomb of Ty (E 29a; pl. 20.1) and of the two crippled dwarfs from el-ᶜAmarna (E 75–9; pl. 25, fig. 9.27).

Three features of the standard form for dwarfs are not realistic, but have an iconographic meaning:

First, in most depictions, especially in the Old Kingdom, short people have normal faces, as if they were all hypochondroplastic. Since achondroplasia has a similar incidence to hypochondroplasia, this presentation is probably purely conventional. It may be related to the fact that dwarfs are hardly caricatured in Egyptian art. Short persons are usually dressed in kilts which hide their leg deformities; Seneb even had a special palanquin with large side-panels concealing his short legs (E 41g; fig. 9.19c). When naked, dwarfs do not have over-large genitals, nor show signs of emaciation. They may only have a large belly, a 'genre' feature often associated with dwarfism, perhaps to demonstrate that they were well fed and had a good standing in the household; they may also have lacked normal exercise.

Second, dwarfs may be shorter than figures of similar status, often impossibly so. In tomb reliefs, dwarfs may be shown 1/1.6 smaller than full-grown adults, or even twice as short. If taken literally, this would mean that if an Egyptian man measured about 170 cm.,[53] a dwarf would be as small as 106 cm. (1/1.6) or even 85 cm. (1/2), which is shorter than actual average (between 100 and 140 cm.). This unrealistic smallness was artistically valuable; it added life to the picture by highlighting the dwarf's abnormality. It may also indicate that the élite had a special liking for very small dwarfs, who were rarer. Dwarfs' tibiae and fibulae, in particular, are conventionally extremely short, while in reality the femur is more affected. This rendering was probably meant to emphasize the dwarf's oddity by contrast with the elongated tibia of the normal figures. The thick lips of several Middle Kingdom faience figurines are also not realistic; so far they are unexplained.

Thirdly, the flat skull of some figures does not seem to have a medical basis and may

[53] Robins (n. 2 above), 31.

be an invented convention. It was perhaps used by artists as a physical attribute stressing the solar affinity of human dwarfs, as an allusion to the scarab-beetle adorning the flat skull-top of the dwarf Ptah. The two short servants of the queen's sister at el-ᶜAmarna, a place where solar religion was prominent, have such skulls (E 75–9; pl. 25, fig. 9.27); their grandiose names ('the Sun' and 'For ever') emphasize this affinity. The dwarf Djeho, who also has such a flat skull, was a sacred dancer performing at Athribis and Heliopolis for the burial of the sacred bulls Apis and Mnevis, heralds of Ptah and Re (E 84; pl. 26.2).

No single archetype seems to have existed for proportionate dwarfs, whose identification is often uncertain. This is probably due to their greater rarity, and to the fact that they were not easily made distinctive. Egyptians recognized other conditions, such as diastrophic dwarfism, pseudo-achondroplasia, or Pott's disease, but, so far as is known, they did not formulate this knowledge in medical texts.

5

Anonymous Dwarf Gods

Egyptian religion was deeply concerned with maintaining the order of the created world, always threatened by the primeval chaos, outside and within creation, full of monstrous hybrid beings.[1] Dwarfs, though malformed, do not seem to have been assimilated to these disturbing and threatening creatures. Their liminality was made symbolically acceptable by association with positive religious concepts. From the New Kingdom on, and probably also earlier, they emerge as popular deities, best known in their forms of Bes and Ptah-Pataikoi, who are invoked in a host of magical practices to protect the living and the dead. Before analysing the material relating to these two gods, I present the magical and liturgical texts where dwarf gods are invoked as helpers, but are not named; these texts range in date from the New Kingdom to the Graeco-Roman period. The study of the epithets, attributes, and functions of these anonymous deities helps us to define the place reserved for dwarfs in Egyptian cosmogony.

MYTHICAL STATUS

In most magical texts, dwarf gods appear as manifestations of the sun-god Re. In several New Kingdom papyri, they are described as rising up to the sky and going down to the underworld, like Re. In a papyrus formerly in the Borchardt collection, the divine *nmw* is first greeted as: 'O this dwarf, the man in Heliopolis, the short one whose legs are between earth and sky', and, on the verso, as: 'O Re, [who is] half the dwarf of the sky, (half) the dwarf of the earth'.[2] In other spells, a divine *nmw* is adressed as 'the dwarf who is in the middle of the sky',[3] or the 'dwarf of the sky . . . , the great pillar who starts in the sky and ends in the underworld'.[4]

In Papyrus Salt 825 (Graeco-Roman Period), the magician invokes the eyes of a dwarf in an enumeration of magical powers related to Re: 'Blood of the eye. Heart of the

---

[1] See E. Hornung, 'Chaotische Bereiche in der geordneten Welt', *ZÄS* 81 (1956), 28–32; id., *Conceptions*, 172–85.

[2] J. Černý and G. Posener, *Papyrus hiératiques de Deir el-Médineh*, i (Cairo, 1978) (DFIFAO 8), 9–10.

[3] Deir el-Medina papyrus 4; Černý/Posener, ibid.

9, verso 5. 5–6.

[4] Harris Magical Papyrus, spell U 8.9–10; H. O. Lange, *Der magische Papyrus Harris* (Copenhagen, 1927), 74, ll. 1–4; G. Roeder, *Der Ausklang der ägyptischen Religion* (Zurich, 1961), 175.

monkey. Head of the uraeus. Eyes of the dwarf (*nmj*) of Upper and Lower Egypt' (9. 4–5).[5] These eyes are probably the sun and the moon, the cosmic eyes of the creator-god Re.[6] The cosmic eyes of this anonymous dwarf may be shown on a New Kingdom pilgrim flask depicting Bes with two eyes in his hands (pl. 6.1).[7]

In the next formula, the deity is addressed as 'the hidden one himself. Held in great awe, hidden although he is tall' (9. 5). This confirms that the *nmj* was a form of the sun-god; Re is commonly described as a hidden god, 'with hidden forms', whose true name is hidden.[8]

In the Demotic Magical Papyrus of London and Leiden, a dwarf god is again identified with a hidden god, probably Re: 'I am the noble child who is in the house of Re; I am the noble dwarf (*nmw*) who is in the [blocked] cavern (*tph.t d3.t*)' (11. 6–7).[9] The sun-god is often invoked as 'the Lord of the caverns' of the underworld, 'who belongs to his mysterious and hidden caverns'.[10] The 'blocked cavern' could also refer to the tomb of Osiris, perhaps in a Memphite sanctuary, as Borghouts suggests.[11] This hypothesis is supported by the fact that, like Re, dwarf gods may be fused with Osiris in the underworld.[12] In the papyrus formerly in the Borchardt collection, the *nmw* is thus 'the king of the underworld, the ruler of the Two Banks, the lord of the corpse in Heliopolis (Osiris)'.[13]

In spell 164 of the *Book of the Dead*, a *nmw* is invoked with another deity to protect the body of the deceased: 'The mysterious lion is the name of the one; Son, the Dwarf (or: Son of the Dwarf), (is the name of the) second'.[14] The first god may be Re, often described as a lion in his night form, while the dwarf may represent the morning form of Re, or Horus, the archetypal son. Both deities are depicted as dwarfs in the

[5] E. Drioton, 'La Cryptographie du Papyrus Salt 825', *ASAE* 41 (1942), 124–6; P. Derchain, *Le Papyrus Salt 825 (BM 10051), Rituel pour la conservation de la vie*, i (Brussels, 1965), 141, 175–6, nn. 117–23; ibid. ii. 12.

[6] On the cosmic eyes of Re (and of Horus), see Bonnet, *Reallexikon*, s.v. Mondauge, 472–4; ibid. s.v. Sonnenauge, 733–5; W. Helck, *LÄ* iv (1982), s.v. Mond, 193–4. See e.g. the New Kingdom hymn in J. Assmann, *Ägypten, Theologie und Frömmigkeit einer frühen Hochkultur* (Stuttgart, 1984), 273, no. 109: 'Deine beiden Augen sind Sonne und Mond, dein Kopf ist der Himmel, deine Füsse sind die Unterwelt.'

[7] Faience flask (h. 15.3 cm.), from Gurob; Oxford, Ashmolean Museum, 1890.897; W. M. F. Petrie, *Illahun, Gurob and Hawara* (London, 1891), 17, pl. XVII, 9; Ballod, *Prolegomena*, 48, fig. 37; J. Baines and J. Málek, *Atlas of Ancient Egypt* (Oxford, 1980), fig. on 217; Romano, 'Bes', ii, no. 91.

[8] See e.g. E. Ledrain, 'Le Papyrus de Luynes', *RecTrav* I (1870), 91: 'O Re, seigneur de vérité . . .

celui qui est caché, dont on ne sait pas le lieu où il est.' On this epithet in general, see Hornung, *Conceptions*, 117.

[9] F. Ll. Griffith and H. Thompson, *The Demotic Magical Papyrus of London and Leiden*, i (London, 1904), 82–3.

[10] A. Piankoff, *The Litany of Re* (New York, 1964) (ERT 4), 22, no. 1, 24, no. 23.

[11] J. F. Borghouts, *The Magical Texts of Papyrus Leiden I 348* (Leiden, 1971), Excursus I, 194–7, esp. 196–7. Cf. *Wb* v. 364a, iii: 'Ort der Toten, der Osiris, Räume der Unterwelt (Amduat)', and 366: 'Kapelle im Tempel als Höhle des Gottes.'

[12] On the New Kingdom fusion of Re with Osiris, see J. G. Griffiths, *LÄ* iv (1982), s.v. Osiris, 629, vii; Hornung, *Conceptions*, 93–6.

[13] Černý/Posener (n. 2 above), 9.

[14] Trans. T. G. Allen, *The Book of the Dead or Going Forth by Day* (Chicago, 1974), 160, S 2. For the reading 'Son of the Dwarf', see P. Barguet, *Le Livre des morts des anciens Égyptiens* (Paris, 1967), 237.

accompanying vignette, which is described in the directions for recitation (fig. 7.4).[15] Their attributes, the crown with twin feathers and the falcon's head, stress their solar qualities, as el-Aguizy notes.[16]

Dwarf gods are also identified with Horus as a child and as a sky-god, especially on the stelae known as Horus-cippi which were very popular in the Late Period (pl. 3.2).[17] These monuments are covered with magical texts alluding to the healing of the young god, stung by a scorpion in the Delta. His powerless mother Isis appealed to Thoth, the master of magic, who descended from the solar barque and brought the child back to life with the help of other gods.[18] By sympathetic magic, poisoned or sick humans are assimilated to Horus and invoke the help of the divine child, who is occasionally described as a dwarf. Thus, a formula on the Béhague statue base and the Metternich stela states that 'the protection of Horus is that great dwarf (*nmw*) who goes through the underworld (or: the Two Lands) in the twilight' (iv. f 9–10).[19] The dwarf may also represent Re himself, or the moon, the left eye of the sun-god.[20]

The Late Period statue of Djeho is carved with similar spells.[21] The dwarf god is identified with the man invoking the healing power of the water poured over the magical inscriptions. He is addressed as a '*nmw* who drinks water', or 'who is protected by the water', probably that which flowed over the stela.[22] He is also closely associated with the moon: he is next greeted as 'the dwarf held (i.e. protected) by a baboon', probably Thoth in his manifestation as the moon, as the following statement suggests: 'the monkey (*gf*) is the name of the moon.'[23]

This assimilation of dwarfs with Horus is repeated in the Ptolemaic temple of Horus at Edfu. A text describes the birth of the divine child thus: 'A lotus emerged in which there was a beautiful child who illuminated the earth with rays of light, a bud in which was a dwarf (*nmw*) whom Shu liked to see.'[24]

[15] *Book of the Dead*, spell 164; papyrus, London, BM 10257/21 (Ptolemaic Period); Barguet, ibid. fig. on 236; R. O. Faulkner, *The Ancient Egyptian Book of the Dead* (London, 1985), fig. on 163.

[16] El-Aguizy, 'Dwarfs', 57 n. 6.

[17] Stela, black steatite (h. 18.6 cm.); London, BM 36250; E. A. W. Budge, *The Mummy, A Handbook of Egyptian Funerary Archaeology*[2] (Cambridge, 1925), 472 (8), pl. 38.

[18] For a complete set of texts, see C. E. Sander-Hansen, *Die Texte der Metternichstele* (Copenhagen, 1956). On Horus stelae in general, see K. C. Seele, 'Horus on the Crocodiles', *JNES* 6 (1947), 43–52; Bonnet, *Reallexikon*, s.v. Horusstele, 317–18; L. Kákosy, *LÄ* iii (1980), s.v. Horusstele, 60–2; I. E. S. Edwards, ibid. s.v. Krankheitsabwehr, 760 nn. 8–10.

[19] Béhague statue base, Leiden, Rijksmuseum, F 1950/8.2 (dyn. 30); A. Klasens, *A Magical Statue Base (Socle Béhague) in the Museum of Antiquities at Leiden* (Leiden, 1952) (OMRO 33), 56, 94. Metternich stela,

New York, MMA, Fletcher Fund 1950 (dyn. 30); N. E. Scott, 'The Metternich Stela', *BMMA* 9 (1951), 201–17; Sander-Hansen, ibid. 72, l. 223.

[20] See Klasens, ibid. 94; Borghouts (n. 11 above), 146 n. 347.

[21] Cairo Museum, CG 46341 (dyn. 30); E. Jelínková-Reymond, *Les Inscriptions de la statue guérisseuse de Djed-Her-le-sauveur* (Cairo, 1956) (BdE 23), 44 (90)–(91).

[22] See the discussion of these two possible readings in Jelínková-Reymond, ibid. 44 n. 9. On the healing properties of the water poured on magical formulae, see Seele (n. 18 above), esp. 48.

[23] See Klasens (n. 19 above), 94.

[24] Le Marquis de Rochemonteix, *Le Temple d'Edfou*, i[2] (1st edn. Paris, 1897, MMAF 10), rev. with add. by S. Cauville and D. Devauchelle (Cairo, 1984) (IFAO X, 1) 289, pl. xxixb, 'chapelle du trône de Ra, paroi nord (4)'; Y. Koenig, *Le Papyrus Boulaq 6* (Cairo, 1981) (BdE 87), 69.

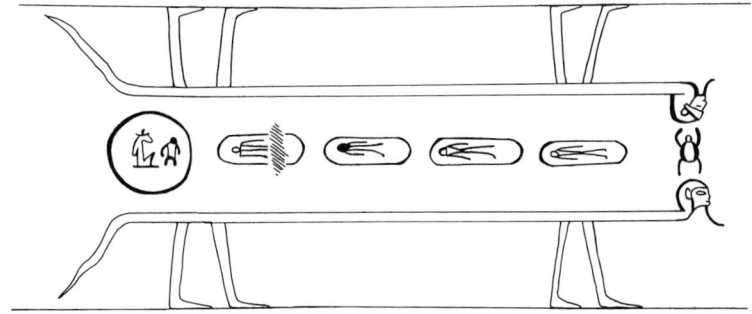

FIGURE 5.1. Papyrus, The Hague, Meermanno-Westreenianum Museum, 37

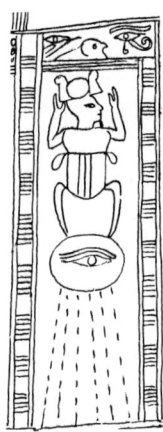

FIGURE 5.2. *Book of the Dead*, spell 165. Papyrus, London, BM 10257/22

FIGURE 5.3. Sarcophagus, Turin Museum

The identification of dwarfs with Re and Horus is not explained by a specific myth. It relies upon a symbolic analogy. Egyptians believed in cosmic correspondences which expressed aspects of the manifold powers of the gods.[25] Thus, dwarfs were identified with the sun-god in his youthful form as Horus because of their ambiguous physical appearance, infantile and mature at the same time, like a god who is newly born but already wise and experienced.[26] From the New Kingdom on, it is also expressed by a

[25] See e.g. B. Mathieu, 'Les Hommes de larmes. A propos d'un jeu de mots mythique dans les textes de l'ancienne Égypte', in *Hommages à François Daumas* ii ((Montpellier, 1986), 499–509 (on the play upon *rmṯ*, 'mankind' and *rmjt* 'tears').

[26] Cf. also the dwarfish youthful god dancing on a lotus flower on a relief from Edfu in Warsaw, National Museum; K. Michalowski, *L'Art de l'ancienne Égypte* (Paris, 1968), 326, fig. 131 (Nefertum?).

symbolic play upon an image. Dwarfs are assimilated to the scarab-beetle Khepri, which incarnates the morning sun-god.[27] Both types of beings have a similar shape, with a large trunk and short, curved limbs, and their silhouettes are interchangeable. A twenty-first-dynasty papyrus clearly shows this equation: in the solar disk, beside a ram-headed figure symbolizing the ageing Re, a dwarf stands in the place of the sacred beetle (fig. 5.1).[28] Anonymous dwarf-figures with beetle-shaped chests, and even the legs of the insect occur in other contexts, such as vignettes of the *Book of the Dead* (figs. 5.2–3),[29] and in representations of the god Bes.[30] It is best shown in amuletic figurines of Ptah-Pataikos, conventionally crowned with a scarab-beetle, who is, like Khepri, strongly associated with regeneration concepts (e.g. pls. 12.2–4, 13.3, 14.1).

This analogy between dwarfs and scarabs occurs again in a maxim in the Ptolemaic instruction text of Papyrus Insinger; both beings are invoked as hypostases of a major god, Re or Horus: 'The small scarab [is great] through its secret image, the small dwarf is great because of his name' (24.8–9).[31]

Dwarf gods may be related to deities other than Re and Horus. Neith, in particular, seems to have been associated with a dwarf god in the Late and Graeco-Roman Periods. A few magical spells mention that she is protected by an amulet in the shape of a dwarf: 'A dwarf (*nmw*) of faience, fallen into the water, the one who is [at] the neck of Neith—beware you of her!'[32] A spell in the Ptolemaic temple of Horus at Edfu similarly invokes 'the protection of that dwarf of faience who guards the neck of Neith'.[33]

[27] On the scarab as the morning form of Re, see Hornung/Staehelin, *Skarabäen*, 13–15; R. Giveon, *LÄ* v (1984), s.v. Skarabäus, 968–9. Cf. Assmann (n. 6 above), 129, no. 29: 'Chepre am Morgen, Re am Mittag, Atum am Abend.' For the biological context, see S. I. Bishara, 'Biology and Identification of Scarab Beetle', in W. A. Ward, *Studies on Scarab Seals*, i (Warminster, 1978), 87–101.

[28] The Hague, Meermanno-Westreenianum Museum, 37; M. Heerma van Voss, 'Een mythologisch papyrus in den Haag', *Phoenix*, 20 (1974), 332–3, fig. 93; id., 'Zwei ungewöhnliche Darstellungen des ägyptischen Sonnengottes', *Visible Religion*, 4–5 (1985–6), 73–5; Dasen, 'Dwarfism', 264, fig. 4. I am very grateful to A. Nibbi for showing me this document.

[29] *Book of the Dead*, spell 165; papyrus, London, BM 10257/22 (Ptolemaic Period); Barguet (n. 14 above), fig. on 238; Faulkner (n. 15 above), fig. on 164. Sarcophagus, Turin Museum (dyn. 21); Lanzone, *Dizionario*, pl. 250. See also the Late Period dwarfish 'pantheistic scarab' carved on the statue of Djeho in Jelínková-Reymond (n. 21 above), 22 and 37, and that illustrated in Lanzone, *Dizionario*, pl. 166, 3.

[30] See e.g. the Late Period bronze figurines in Cairo Museum, CG 38696, 38698, and 38699;

Daressy, *Statues*, pl. XXXVII (the wings of the beetle are clasped in the back). These figurines are listed by Roeder, *Bronzefiguren*, 101, § 147a; 102–3, § 148a.

[31] Trans. Lichtheim, *Literature*, iii. 204. See also F. Lexa, *Papyrus Insinger, Les enseignements moraux d'un scribe égyptien du 1er siècle* (Paris, 1926), i. 77, and ii. 63. On the general prescription 'not to slight smallness', and its relation to Hellenistic gnomologia, see also M. Lichtheim, 'Observations on Papyrus Insinger', in E. Hornung and O. Keel (eds.), *Studien zu altägyptischen Lebenslehren* (Fribourg and Göttingen, 1979), 183–306 (OBO 28); id., *Late Egyptian Wisdom in the International Context* (Fribourg and Göttingen, 1983) (OBO 52), 163.

[32] Horus-cippus, Cairo Museum, CG 9431 *bis* (Late Period); G. Daressy, *Textes et dessins magiques* (Cairo, 1903) (CGC), 40–1, left side of seat, 48–51; R. el-Sayed, *La Déesse Neith de Saïs* (Cairo, 1982) (BdE 86), i. 131; ii. 468 no. 640. Trans. Borghouts (n. 11 above), 154 n. 370 (with refs. to two stelae in Cairo Museum, CG 9403 and JdE 47280 with similar but incomplete formulae).

[33] PM vi. 162 (312), girdle wall, inner face, north wall, west part, top register; E. Chassinat, *Le Temple d'Edfou*, vi (Cairo, 1931) (MMAF 23), 149, l. 43; id. ibid. x. 2 (Cairo, 1960) (MMAF 27), pl. 149; el-Sayed (n. 32 above), i. 131; ii. 592, no. 950.

Two inscriptions suggest that this dwarf god may have had a specific temple and a priest in her main cult-centre at Sais. A twenty-second-dynasty stela from Sais states that a field of ten arurae is given by 'the supervisor (?) of the House (the temple?) of the dwarf of Neith . . . , the great one, the mother of the god'.[34] The relief depicts a dwarf standing behind the goddess, facing the king (pl. 3.1); his figure should be that of a god, perhaps of the 'dwarf of Neith' mentioned in the inscription, since on the divine side of a relief no human is normally shown facing the same way as a deity. On a Late Period statue from Sais, a title seems to confirm that a particular priest served the cult of this dwarf; Pairkep, the 'prophet (ḥm nṯr) of Amun-Re', 'servant of Neith, the great mother of gods', is also called 'the prophet of the dwarf'.[35] This hypothesis is supported by an allusion to Neith in the Harris Magical Papyrus; the dwarf god is said to come from the 'womb of Neith', an expression which may denote her temple in Sais, as suggested by Roeder.[36] As el-Aguizy suggests, these anonymous dwarf gods could be the hypostases of a major god revered at Sais, such as Amun, Horus, or Osiris.[37] The name dg-nt borne by the malformed woman on the Late period Brooklyn obelisk (E d 83; pl. 26.1) might also associate a real short person with the cult of that goddess.

The association of Neith with dwarfs may be due to their similar 'warrior' imagery, which aimed to repel malignant forces. Two crossed arrows, a bow, and a shield are the standard attributes of Neith as a 'terrifying' goddess,[38] whereas dwarf gods, in the form of Bes especially, are similarly armed with a sword and a shield in the Graeco-Roman period (e.g. pls. 10.1–2, 11.2).

The earth-god Geb is also found wearing an amulet in the shape of a dwarf. The only source is the New Kingdom Magical Turin Papyrus 1993 which says: 'For he (the patient) is the faience dwarf (nmt); it was at the neck of Geb, while Neith was afraid of it.'[39] This association is probably related to the role of Geb as healer of the bite of snakes and sting of scorpions.[40]

These dwarf gods, like other divine beings, do not have merely human proportions. In many spells, they are described as a 'great dwarf',[41] short, but rising from earth to sky,[42] as if they could be small and gigantic at the same time. The dwarf in the Harris

---

[34] Cairo Museum, CG 28731 (h. c.29 cm.); W. Spiegelberg, 'Neue Schenkungsstelen über Landstiftungen an Tempeln', ZÄS 56 (1920), 59–60, pl. vi; D. Meeks, in E. Lipiński (ed.), State and Temple Economy in the Ancient Near East, ii (Louvain, 1979), 674, no. 26.0.6.

[35] Cairo Museum, provisional no. 427; R. el-Sayed, 'Deux aspects nouveaux du culte à Saïs', BIFAO 76 (1976), 91–100, esp. 93–5, pl. xvib; id. (n. 32 above), i. 131; ii. 443, no. 564b.

[36] Roeder (n. 4 above), spell V 9.5.

[37] El-Aguizy, 'Dwarfs', 57 n. 12.

[38] G. Daressy, 'Neith, protectrice du sommeil', ASAE 10 (1910), 177–9; Bonnet, Reallexikon, s.v.

Neith, 512–13, figs. 128–9; S. Sauneron, J. Yoyotte, 'La Naissance du monde selon l'Égypte ancienne', in La Naissance du monde (Paris, 1959) (SourcesOr 1), 31, 72; R. Schlichting, LÄ iv (1982), s.v. Neith, 393–4.

[39] W. Pleyte and F. Rossi, Papyrus de Turin (Leiden, 1869–76), 124, l. 14; el-Sayed (n. 32 above), i. 131; ii. 372, no. 381 (trans. Borghouts, n. 11 above, 154 n. 370).

[40] Bonnet, Reallexikon, s.v. Geb, 202–3.

[41] See e.g. the Metternich stela and the Béhague statue base (n. 19 above).

[42] See e.g. the papyrus formerly in the Borchardt collection: Černý/Posener (n. 2 above), 9.

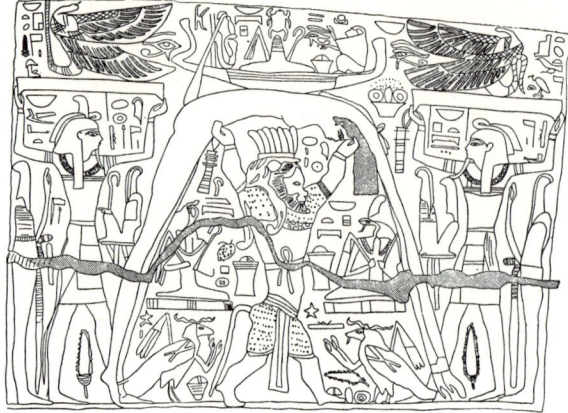

FIGURE 5.4. Sarcophagus, Vienna, Kunsthistorisches Museum 270

Magical Papyrus transforms into 'a giant of seven cubits' (*c*.3.50 m.), who can also enter a naos of half a cubit (*c*.0.26 m.).[43] In the papyrus formerly in the Borchardt collection, the dwarf Re is a giant 'of a million cubits', the conventional distance between earth and sky.[44] This characteristic is also expressed in the Late Period Brooklyn Magical Papyrus, where the dwarf god Bes appears as a 'giant of a million cubits' who 'carries the sky with his powerful arms'.[45] A sarcophagus in Vienna illustrates this supernatural ability: Bes stands in the normal place of Shu, and holds the sky with his short, curved arms (fig. 5.4).[46]

FUNCTION

As hypostases of major deities, the function of these anonymous dwarf gods is essentially protective. Women in childbirth invoke their help, as in Papyrus Leiden I 348. In the spell 'of the vulva', the woman in pain shouts 'to the man for a dwarf-statue of clay'.[47] In the spell 'of the dwarf' the magician prays to a *nmw* from heaven, who appears as the messenger of Re:

---

[43] Roeder (n. 4 above), spell U 9.2–3; spell V 9.7–10.

[44] Černý/Posener (n. 2 above), 10.

[45] S. Sauneron, *Le Papyrus magique illustré de Brooklyn, Brooklyn Museum 47.218.156* (Brooklyn, 1970), 23, 4.4.

[46] Sarcophagus, Vienna, Kunsthistorisches Museum, 6270, from the cachette of Deir el-Bahari; outside of the case, left shoulder part (middle/late dyn. 21); P. Virey, *La Religion de l'ancienne Égypte* (Paris, 1910), 187, fig. 12 (the scene is now badly damaged). I am very grateful to E. Hornung for this reference. See also the sarcophagus of Ankhrui from Hawara; W. M. F. Petrie, *Hawara, Biahmu and Arsinoe* (London, 1889), 21, pl. II, outside of lid, fourth register.

[47] Spell 31, 12.8; trans. Borghouts (n. 11 above), 29.

O good dwarf, come, because of the one who sent you—for that is Pre (Re) . . . Come down placenta, come down placenta, come down! . . . Look, Hathor will lay her hand on her with an amulet of health! I am Horus who saves her! To be recited four times over a dwarf of clay placed on the brow of the woman who is giving birth while suffering (spell 30, 12.2–6).[48]

These small gods also protect humans, especially young children, against diseases and attacks of venomous animals. In the New Kingdom Papyrus Boulaq 6, the magician calls for the help of a *nmj* to fight the poison of a snake which is represented by the goddess Bastet. The body of the man is assimilated to Re and to a dwarf:

It has gone up into your heart as Bastet, it is great, the one which has entered into the womb of the dwarf and the man. Come out, O poison! Come down Bastet who is in the womb of the dwarf and the man . . . Come down, O poison which is in the heart of Re, in the heart of the dwarf, of the man.[49]

On Horus-cippi, dwarf gods are invoked to provide the healing of humans identified with the divine child.[50] A vignette on a magical stela in Cairo illustrates this protection; an anonymous dwarf god is crowned with the *udjat*-eye of Horus (or Re), a divine attribute which symbolizes his power over evil.[51]

Like Re, dwarf gods also protect Osiris. In the Harris Magical Papyrus, the *nmw* guards 'the powerful corpse (Osiris) which rests in On (Heliopolis); the powerful Lord of life (Osiris) who rests in Dedu (Busiris)'.[52] Similarly, they 'keep whole' the body of the deceased, as in spell 164 of the *Book of the Dead*. They also guard humans identified with Osiris. In the Harris Magical Papyrus, the magician commands: 'Watch over him by day, watch over him by night! Protect him as you protected Osiris against the one whose name is hidden on the day of the burial in Heliopolis'.[53] In Papyrus Deir el-Medina 4, a *nmw* is similarly asked to 'come and save N. daughter of M., as you saved the drowned one (Osiris) on the day of the burial'.[54]

To sum up, dwarfs, identified with Re, Horus, and the moon, were integrated within the Egyptian pantheon. Their silhouette was assimilated to that of the sacred scarab-beetle Khepri; their physical malformation was not regarded as a disquieting attribute, but as a divine one. This symbolic correspondence seems to have passed beyond dynastic Egypt into the Greek world. The most striking piece of evidence is an archaic Corinthian statuette which depicts a man blending dwarfish proportions and the appearance of a beetle (pl. 40.3a, b).[55] He has a large trunk, with small, incomplete, or crudely

---

[48] Trans. Borghouts (n. 11 above), 29.

[49] Koenig (n. 24 above), 68, recto vi. 4–5; 74, recto vii. 2.

[50] See e.g. the Metternich stela and the Béhague statue base (n. 19 above).

[51] Cairo Museum, CG 9430; Daressy (n. 32 above), 38, reverse, second row, no. 2, pl. XI. On the power of the eye of Horus, see e.g. J.-C. Goyon,

Rituels funéraires (Paris, 1972), 148–50.

[52] Roeder (n. 4 above), spell U 8.10–11.

[53] Ibid. spell U 8.11–12.

[54] Černý/Posener (n. 2 above), 9.

[55] Terracotta, figure vase (h. 10.3 cm.); Paris market (formerly J. Lauffenburger coll.); J. Chamay, J.-L. Maier, *Céramiques corinthiennes, Collection du docteur Jean Lauffenburger* (Geneva, 1984), 140–1.

modelled legs which represent the limbs of the insect; his back is divided by two lines which evoke its carapace. The provenance of the statuette is unknown (Sicily?). It may have come from a tomb or from a sanctuary like the group of archaic dwarfs which are also inspired by Egyptian dwarf gods.[56]

This connection between dwarfs and scarabs is still expressed in an astrological calendar from Egypt of the late second century AD. The creation of all kinds of human deformities is attributed to a redoubtable deity: 'this deity causes long old age, until a man be bent by old age; he produces hunchbacks or makes men bent by sickness, he causes *dwarfs* (νάνους) to be born and *monstrosities* (τέρατα) *shaped like a beetle*, and persons with no eyes, and like a beast and dumb, and deaf and toothless (viii. 225−8).'[57] Here, however, dwarfs and beetles are no longer associated with divine powers; they are rejected as monsters.

---

[56] See below ch. 13, *sub* Kourotrophic Demons.
[57] *P.Oxy*, iii, no. 465, 133 (text) and 137 (italics added).

6

Bes

The dwarf god Bes ⌡𝑙𝐾 (coptic BHC, greek *Βησα*, *Βησᾶς*, or *Βήσας*) is one of the most familiar tutelary deities of the Egyptian pantheon. He emerges in the Middle Kingdom and becomes very popular from the New Kingdom on, mainly as a protector of the household. He is best known through his abundant iconography which occurs on a wide range of objects of various social levels.

Bes is characterized by his monstrous appearance, mingling animal and human features; in his most typical form, he has dwarfish proportions, with short bent limbs, a head-dress of feathers, and a lion's mane and tail. These elements are displayed in his determinatives, 𝔐 and 𝐾, an animal skin which may refer to his hybrid nature.[1]

Written evidence is scarce. It consists mostly of captions giving the names and epithets of the god. Bes seems to have featured in legends of Re, Horus, and Hathor, but these are never fully narrated; we must be content with scattered allusions to his mythical role in magical, votive and ritual texts of different periods.

NAME

The identity of the god is complex. Bes seems to have been employed as a generic name for different dwarf deities with the same physical appearance. Several other names are found denoting these figures. The most ancient is Aha (*ʿḥ3* or *ʿḥ3j*), which goes back to the Middle Kingdom.[2] Bes (*bs*),[3] Segeb (*sgb*),[4] and Soped (*spd*)[5] appear in the New Kingdom; Hit (*ḥjtj* with variants) and Tettenu (*tttnw*) in the Late and Graeco-Roman

---

[1] Gardiner, *Grammar*, sign-list, F 27 and 28.

[2] H. Altenmüller, *LÄ* i (1975), s.v. Aha, 96–8; id. *Apotropaia*, i. 152. For texts, see *CT*, spell 388, v. 58b; spell 1100, vii. 416c.

[3] H. Altenmüller, *LÄ* i (1975), s.v. Bes, 720.

[4] An underworld gatekeeper: Valley of the Kings, tomb 15 (Tausert); PM i². 2, 529, hall, E (14)–(15);

Krall, 'Bes', 87, no. 81, fig. 32; Romano, 'Bes', ii, no. 146.

[5] Soped is a special case: it is probably not a proper name but the name of another marginal god assimilated to Bes. See Bonnet, *Reallexikon*, s.v. Sopdu, 743; R. Giveon, *LÄ* v (1984), s.v. Sopdu, 1109 n. 34.

Periods.[6] Some occur as elements in personal names, Aha from the Middle Kingdom on, Bes especially in the Graeco-Roman Period.[7]

The meanings of the most common names, Aha and Bes, shed some light on the nature of the god. Thus, Aha, 'the fighter', describes his function as a repeller of evil.[8] Although the meaning of the name Bes is not elucidated satisfactorily, two hypotheses suggested by Bruyère may be mentioned. First, the name could be related to *bs*, 𓃀𓊃𓊮 'flame',[9] which would refer to the fiery aspect of Bes as a hypostasis of Re; a circle of flames symbolizes triumph over enemies, as depicted in Late Period magical scenes (pl. 11.1a, b). Bes could also have been associated with *bz*, 𓃀𓂋 'to introduce, to be initiated',[10] which would allude to the use of masking in the cult of the god. A third etymology, which assimilated Bes to a panther, was much favoured at the end of the nineteenth century. It relied on the misreading *b3sw* 𓃀𓃭𓏲 'panther' (instead of *b3 šmꜥ*), which identifies a fantastic feline creature in the Beni-Hasan tomb of Baqt III, and is now rejected.[11] The standard words for panther, *b3* and *3bj*, have no possible connection with Bes.[12]

Could these names denote the various powers of a single dwarf deity, or do they really describe different gods with the same visual form? The first hypothesis is supported by the fact that great gods, especially Re, Amun, and Osiris, are commonly addressed by a multiplicity of names and epithets expressing their varied capacities; Bes might similarly have had many names corresponding to his different functions.[13]

The second hypothesis has more in its favour. Egyptian religion includes other 'families' of divine beings, especially among deities associated with childbirth, like Bes. There are seven (or twelve) Hathors, four Meskhenets and twelve (or fourteen) Tawerets.[14] Usually these gods appear in a group; they have the same physical form and

---

[6] Add also *spt, mfdt, mnw, ꜥmꜥm*. For a full list of these names and their variants, see Krall, 'Bes', 77; Ballod, *Prolegomena*, 11–14, 24–36; Romano, 'Bes', i. 13 n. 49.

[7] Ranke, *PN* i. 44.6, 11–12, and 359.5; Ballod, *Prolegomena*, 61–2; Altenmüller, *Apotropaia*, i. 152. Add M. A. Leahy, '"Harwa" and "Harbes"', *CdE* 55/109–10 (1980), 47–62 (on *Ḥrbs*, 'face of Bes'); M. Thirion, 'Notes d'onomastique, *RdE* 34 (1982–3), 110 (on a Ptolemaic or Roman *wn-bs*).

[8] Ranke, *PN* i. 103.9.

[9] *Wb* i. 476.1–6; Bruyère, *Deir el Médineh*, 95 n. 7.

[10] *Wb* i. 473–4.1–4; Bruyère, *Deir el Médineh*, 96 'maître des secrets'.

[11] PM iv. 151, hall (2)–(6), top register; P. E. Newberry, *Beni Hasan*, ii (London, 1893) (ASE 2), pl. iv.

[12] *Wb* i. 7.11, 415.7–11. See the discussion in F. W. v. Bissing, 'Miscellen', *ZÄS* 40 (1902–3), 97–8; Ballod, *Prolegomena*, 18–19 and 21; G. Jéquier, 'Nature et origine du dieu Bès', *RecTrav* 37 (1915), 114–15.

[13] So for Krall, 'Bes', 77; Bruyère, *Deir el Médineh*, 93. On the names of major gods, see Hornung, *Conceptions*, 86–91. See e.g. the 75 names of Re in E. Hornung, *Das Buch der Anbetung des Re im Westen (Sonnenlitanei)* (Basle and Geneva, 1976), 56–9, 61–96.

[14] See Bonnet, *Reallexikon*, s.v. Hathor, 282, s.v. Meschenet, 458, s.v. Nilpferdgöttin, 532. In particular, see F. W. v. Bissing and H. P. Blok, 'Eine Weihung an die sieben Hathoren', *ZÄS* 61 (1926), 83–93; C. Desroches Noblecourt, 'Un "lac de turquoise"', *MonPiot* 47 (1953), 23–30; W. Helck, *LÄ* ii (1977), s.v. Hathoren, sieben, 1033.

attributes, and they seem to share a single identity. Inscriptions, however, reveal that they were differentiated. For example, on Late and Graeco-Roman monuments each of the twelve Tawerets has her own name and supervises one month of the year; the goddess guarding the month Phamenoth was called Ta-Weret 'The Great One', an epithet which later was used by the Greeks to designate all types of hippopotamus-goddess.[15] It is tempting to suggest that in the Graeco-Roman Period Bes was similarly used as a generic name for all types of hybrid dwarf gods. No text indicates that this group consisted of a specific number of deities, but it is a possibility.

Like each of the twelve Tawerets, each type of Bes may have had a role of its own. The scarcity of inscribed monuments limits attempts to define their particular forms and functions. Ballod tried to identify the specific attributes (head-dress, knives, musical instrument) and activities (fighting, dancing) of different groups, but he found many exceptions and ambiguous examples.[16] However, this group of deities presents enough common physical features and functions to be analysed together, and, like most Egyptologists, I use here for convenience the generic term 'Bes', 'Bes-gods', or 'Bes-figures' to describe them.

ICONOGRAPHIC TYPOLOGY

The iconography of Bes varied little during the dynastic period. I briefly present here the main steps of its chronological development, which are based upon the thorough studies of Ballod and Romano.[17]

There is no certain Old Kingdom depiction of Bes, perhaps in part because representations of deities from that period are very rare. Only two reliefs and a statuette may relate to an early form of the god; all three wear feline masks and shoulder-length wigs or manes.[18] The face of the personage from the pyramid temple of Sahure is carved with lines recalling the facial folds of snarling lions, as in New Kingdom figures of Bes.

---

[15] G. Daressy, 'Thouéris et Meskhent', *RecTrav* 34 (1912), 189–93. See also R. Gundlach, *LÄ* vi (1986), s.v. Thoueris, 495, on the possible New Kingdom dating of the name.

[16] Ballod, *Prolegomena*, 71–85. See also Bonnet, *Reallexikon*, s.v. Bes, 103–4; Altenmüller, *Apotropaia*, i. 152–5; id., *LÄ* ii (1977), s.v. Hit, 1226–7; Raven, *Pataekos*, 15–16; Romano, 'Bes', i. 18–19. For example, in the birth-house of Trajan at Dendara Bes-figures are differentiated by distinct names (*Bs, Ḥ3tjtj*), but have similar poses and attributes; PM vi. 103, colonnade, architrave; F. Daumas, *Les Mammisis de Dendara*, (Cairo, 1959), 271–85, pl. xcv a.

[17] Ballod, *Prolegomena*, 36–70; Romano, 'Origin', 39–56; id., 'Bes'. See also Wilson, 'Bes', and Tran Tam Tinh, *Bes*.

[18] 1: Relief, Leipzig Museum, 2095; L. Borchardt, *Das Grabdenkmal des Königs Sa3ḥu-reʿ*, ii (Leipzig, 1913), pl. 22; Sourdive, *La Main*, 112–13, pl. xxix, fig. 2; Romano, 'Bes', ii, no. 1. 2: Relief, London, BM 994; PM iii². 1,309; J. Capart, 'Note sur un fragment de bas-relief au British Museum', *BIFAO* 30 (1930–1), 73–5 (with pl.); Baines, *Fecundity Figures*, 129–30, fig. 85; Sourdive, *La Main*, 48–52, 112–16, pls. xvii, xxix, fig. 1; Romano, 'Bes', ii, no. 3. 3: Limestone statuette, Berlin, SM (East) 18175; L. Borchardt, *Das Grabdenkmal des Königs Nefer-ir-k3-reʿ* (Leipzig, 1909), 70, fig. 78; Sourdive, *La Main*, 112–13, pl. xxix, fig. 3; Romano, 'Bes', ii, no. 2. Add a First Intermediate Period design amulet with a sketchy Bes; Romano, 'Bes', ii, no. 4.

The statuette has dwarfish proportions, while the two-dimensional figures are characterized by attributes, such as a pendulous breast, a distended belly, and a belt with strips,
which are reminiscent of fecundity figures and Bes-gods.[19]

The earliest representations of Bes date to the Middle Kingdom, a period from which
the range of evidence for religious practices is broader. The god occurs principally on
the so-called magic 'knives' or wands, but also on various other objects. He appears as a
lion-man, with a mane, round ears and a tail hanging between his legs; he has no
conspicuous physical malformation. He is usually standing, full-faced, holding a snake
in either hand. Occasionally the figure is female.[20] Inscriptions on two magic wands
give the name Aha (fig. 6.1a).[21] Several statuettes, made of wood, faience, or ivory, also
depict leonine demons, male or female, in the same pose as on magic wands; some have
dwarfish proportions (pl. 3.3).[22]

During the New Kingdom, Bes progressively acquired his most characteristic form.
His slender leonine silhouette changed into that of a bandy-legged dwarf with a protruding tongue, a paunchy belly, and a feather head-dress.[23] Early eighteenth-dynasty
depictions are still in the tradition of Middle Kingdom iconography. As on magic
wands, Bes stands frontally, hands on thighs, with splayed feet; his limbs are slender,
and horizontal lines on the torso seem to indicate animal ribs or muscles (pl. 4.2).[24] His
protruding tongue and ostrich-feather crown first appear during the reign of Hatshepsut
and Tuthmosis III, whereas kilts (pls. 5, 6.1, fig. 6.5a, b), jewellery, especially broad
collars, and occasionally wings (pl. 6.1), come in later, in the time of Amenophis
II–III.[25] Representations in partial or full profile are common; often the god is shown
grasping the root of his tail (pl. 4.1).[26] His leonine face appears as an ornamental device
on various objects, such as headrests, bowls, and mirrors.[27] The range of attributes of
Bes becomes gradually larger: in addition to snakes, he may hold weapons, especially

---

[19] Baines, *Fecundity Figures*, 129–30; Romano,
'Bes', i. 22–32.

[20] See Ballod, *Prolegomena*, 27–9, 38–41;
Altenmüller, *Apotropaia*, i, esp. 36–9; Romano,
'Bes', i. 33–57, ii, nos. 5–61.

[21] Ivory (l. 41), Berlin, SM (West) 14207; F.
Legge, 'The Magic Ivories of the Middle Empire',
PSBA 27 (1905), 136–8, pl. IV, fig. 4; Ballod,
Prolegomena, 27–9, fig. 2; Altenmüller, *Apotropaia*, ii.
11–12, no. 10; Romano, 'Bes', ii, no. 23. See also the
magic knife in Brussels, Musées Royaux, E 2673;
Altenmüller, *Apotropaia*, ii. 20–1, no. 20.

[22] Faience (h. 17.3 cm.), Baltimore, WAG 48.420;
Ballod, *Prolegomena*, 40, fig. 17; G. Steindorff,
*Catalogue of the Egyptian Sculpture in the Walters Art
Gallery* (Baltimore, 1946), 143, no. 624, pl. XCIV;
Romano, 'Bes', ii, no. 49. See the similar faience
figurine from Esna (h. 8.6 cm.), Liverpool Museum,
1977.110.2; J. Bourriau, *Pharaohs and Mortals*

(Cambridge, 1988), 112–13, no. 99; Romano, 'Bes',
ii, no. 48. See also the ivory statuette (h. 4 cm.),
Copenhagen, Ny Carlsberg Glyptothek, AEIN 1380;
M. P. Mogensen, *La Collection égyptienne*
(Copenhagen, 1930), 110, pl. 34; Romano, 'Bes', ii,
no. 46.

[23] See Ballod, *Prolegomena*, 41–53, 86–8; Romano,
'Origin', 43 ff.; id., 'Bes', i. 58–122.

[24] Faience (h. 3.1 cm.), Brooklyn Museum,
37.912 E; Romano, 'Origin', 43, fig. 3.

[25] Romano, 'Bes', i. 78–99. For a list of Bes in
kilts, see also Wilson, 'Bes', 78–80 nn. 19 and 20.

[26] Faience (h. 4.4 cm.), Brooklyn Museum,
16.426; Romano, 'Origin', 46, fig. 9.

[27] On Bes-mask motifs, see Ballod, *Prolegomena*,
91–3. See e.g. the Bes-head on a bowl in London,
BM 6533; T. G. H. James, 'Two Egyptian Plates of
the New Kingdom', *BMQ* 19/1 (1954), 16–17.

knives, various sceptres, amuletic signs, lotus or papyrus stems (pl. 4.3). His activities too are diversified: often he is dancing and playing a musical instrument, such as the tambourine, lute, or double flute (pl. 4.1; figs. 6.3, 6.5b).[28]

From the Third Intermediate Period on, his facial features acquire more of a grimace. His neckless head is framed by a moustache and cheek whiskers, a large tongue protrudes beneath a row of menacing teeth (pl. 6.2); his beard may end in curls. His attributes are slightly modified: his feather head-dress becomes very tall, while a leopard's skin, whose head may be seen on his chest, covers his shoulders. New poses appear in figurines: the god may be nursing a smaller figure of himself (pl. 7.3a), holding various quadrupeds (pl. 6.2) or sitting on the shoulders of a male or female figure (pl. 7.1–2).[29]

Late Period depictions of Bes do not differ significantly from earlier models. Iconographic changes consist mainly in a greater degree of stylization, in particular of the curly beard of the god (pl. 6.3); negroid facial features are often emphasized.[30] The main innovation consists in the creation of a new type of Bes-figure, the so-called 'pantheos'. In the growing tendency to syncretism, Bes fuses with major gods: he is transformed into a composite creature with a misshapen or full-sized body, two pairs of arms and wings, and various human or animal heads and parts (pl. 11.1a, b).[31]

In the Ptolemaic Period, older motifs are given new forms. Bes, the leonine fighter, is transformed into a warrior, naked or in a kilt, holding a shield and brandishing a sword near his head, sometimes killing a serpent grasped in his hand (pls. 10.1–2, 11.2).[32] His caricatured appearance inspired a derogatory Alexandrian saying: 'To stand like Bes: to stand gaping with one's mouth open, in a stupid way.'[33]

Beset, a female form of Bes, appears in the late Middle Kingdom, but remains rare until the Graeco-Roman Period.[34] In the Middle Kingdom, she has the same leonine form as her male counterpart, but with slightly different poses and attributes. As Romano points out, Beset usually has no tail and holds her legs tightly together (fig. 6.2); she grasps not only snakes, like the males, but also hares and lizards; she wears a larger range of ornaments, such as collars, wristlets, and anklets.[35] In the Graeco-Roman Period, she has plump dwarfish proportions and a tall head-dress of feathers, like Bes; but her head is fully human, with round cheeks and long hair, often held in a turban.

[28] Wilson, 'Bes', 80; Romano, 'Origin', 46–7; id., 'Bes', i. 64–77.

[29] See Ballod, *Prolegomena*, 53–5, 88–9; Romano, 'Bes', i. 123–69.

[30] Ballod, *Prolegomena*, 56–61; Romano, 'Bes', i. 170–211, esp. 174–91.

[31] Romano, 'Bes', i. 148–51 (attributed to the Third Intermediate Period).

[32] See Ballod, *Prolegomena*, 61–9, 84–5, 89–90;

Tran Tam Tinh, *Bes*, 101–2, nos. 31–43, pls. 78–80.

[33] *Suda*, s.v. Βησᾶς· ἔστηκεν, οἷον ἀχανῆς· οὗτος ἔστηκεν ἀχανὴς καὶ παταγώδης καὶ ὑπόμωρος.

[34] For possible New Kingdom depictions of Beset, see K. Bosse-Griffiths, 'A Beset Amulet from the Amarna Period', *JEA* 63 (1977), 98–106. *Contra*: Romano, 'Bes', i. 64 n. 129.

[35] Romano, 'Bes', i. 47–8, 52–3. See also Altenmüller, *Apotropaia*, i. 38.

She is clearly a female figure and not an androgynous form of Bes.[36] She may be standing, in the guise of a warrior, or dancing, naked or in a dress, alone or with Bes in a pair. Occasionally she nurses a miniature figure of Bes.[37]

ORIGIN

Bes does not belong to the conventional Egyptian pantheon. As Baines notes, his liminality is indicated by his peculiar physical form which does not conform to the normal image of gods: he has a grotesque, malformed, and mostly naked body, while major gods are slim, healthy, and usually clothed.[38] His dancing and musical performances contrast with the dignified attitudes of major gods. His large, frontal, mask-like face is another uncommon feature. Apart from a few hieroglyphs, like ḥr ☒, a frontal face is found mainly in ornamental figures of the cow-eared Hathor, which often occur in the same context as Bes; Hathor heads decorate objects related to attire and music, such as mirrors and sistra, or buildings associated with birth, such as birth houses of Graeco-Roman temples.[39]

Do these unusual characteristics imply that Bes is a foreign import from the East, or from Africa, or is he an indigenous creation?[40] The eastern hypothesis is now generally abandoned. It relies essentially on a few representations of dwarf-like gods from Middle Bronze Age sites (early second millenium) in Anatolia, Mesopotamia, Syria, Phoenicia, and Palestine. Like the Egyptian Bes, these small gods are usually shown full-faced, hands on hips, with bandy legs. Often they have leonine features (mane, ears, rib-cage), like a limestone figurine from Byblos[41] and a bone plaque from Alaça Hüyük.[42]

These objects are contemporary with the Middle Kingdom in Egypt, but it does not follow that they represent ancestors of Bes. As Wilson has shown, Mesopotamian dwarf gods probably originated independently; they are related to the figure of Humbaba,

---

[36] H. Altenmüller, *LÄ* i (1975), s.v. Beset, 731. On a possible Bes-figure fusing a female body and a male head, see W. A. Ward, 'A unique Beset figurine', *Orientalia*, 41 (1972), 149–59. *Contra:* Romano, 'Bes', i. 16–17.

[37] For material, see Tran Tam Tinh, *LIMC* iii (1986), s.v. Besit, 112–14. See e.g. P. Perdrizet, *Les Terres cuites grecques d'Égypte de la collection Fouquet* (Paris, 1921), pls. XL (Bes and Beset), XLIII (Beset).

[38] Baines, *Fecundity Figures*, 128–9.

[39] On frontality, see H. Schäfer, *Principles of Egyptian Art*,[5] trans. and rev. by J. Baines (Oxford, 1986), 205–10. See e.g. the Hathor capitals in the birth-houses of Dendara in S. Sauneron and H. Stierlin, *Les Plus Beaux Temples égyptiens* (Geneva, 1980), 21, 69, and of Philae, ibid. 156 and 164.

[40] For a general survey of these theories, see Ballod, *Prolegomena*, 14–20, and Romano, 'Origin',

40–1.

[41] M. Dunand, *Fouilles de Byblos, II, 1933–1938* (Paris, 1958), 767, no. 15377, *Atlas*, pl. XCV (h. 5.6 cm.); Romano, 'Bes', ii, no. 53b. On the find context of this figurine, see E 150–d 190, and below 137, 141.

[42] H. Z. Kosay, *Ausgrabungen von Alaça Höyük* (Ankara, 1944), 31, no. Al-a 88, pl. XLIV; Romano, 'Bes', ii, no. 47. See also the leonine dwarf god on a cylinder from Ras Shamra (early 2nd millennium); C. F.-A. Schaeffer, *Corpus des cylindres-sceaux de Ras Shamra-Ugarit et d'Enkomi-Alasia*, i (Paris, 1983), 25–6, no. R.S.7.181. For Near-Eastern material, see A. Grenfell, 'The Iconography of Bes, and of Phoenician Bes-Hand Scarabs', *PSBA* 24 (1902), 21–40; C. T. de Vartavan, *Bes, The Bow-Legged Dwarf or the Ladies' Companion*, unpub. diss. (University of London, 1986), 1–20.

while the figurines found in Alaça Hüyük and Byblos may be Egyptian imports, or local copies of Egyptian gods.[43]

In later periods, however, eastern influences may be seen in the iconography of the god. Thus, in the New Kingdom, eastern motifs probably inspired the attire of Bes, in particular his kilt, with a long sash or apron, and his wings (pl. 6.1, fig. 6.5a, b).[44] The Late Period motif of Bes as a master of animals may also have been influenced by the East, especially Phoenicia. It is best illustrated by Graeco-Phoenician scarabs and gems which depict Bes shouldering a lion or an antelope, or holding pairs of inverted animals, such as lions, goats, and uraeus-snakes, sometimes with two gazelles sprouting from his head.[45] Later, in the archaic Greek world, this motif fused with the iconography of Heracles mastering the Nemean lion.[46]

The second hypothesis, which considers Bes as a native Sudanese deity, has been accepted much more widely. It is supported by temple texts of the Late and Graeco-Roman Periods which describe Bes as the Lord of southern countries, in particular *t3-stj* (Nubia), *Bwgm*, a southern country associated with Punt, and Punt itself.[47]

Apart from leonine elements, several physical characteristics of the god stress these southern associations. His negroid facial features, combined with his short bent limbs, recall the appearance of a pygmy. On a few statuettes Keimer distinguished facial tattoos, or scars similar to those of present-day Sudanese groups.[48] The mask-like treatment of the god's face is reminiscent of the ritual use of masking in African tribes, and some authors, like Wolff and Delpech-Laborie, have suggested that Bes could depict a pygmy deity or sorcerer.[49]

The traditional attributes and companions of the god are also suggestive of a southern provenance. From the New Kingdom on, Bes wears a head-dress made of ostrich

[43] Wilson, 'Bes', 83–4.

[44] Wilson, 'Bes', 84–6; Smith, *Art and Architecture*, 289 n. 15. Cf. the wings of the Assyrian god Pazuzu; H. Frankfort, *The Art and Architecture of the Ancient Orient*[4] (Harmondsworth, 1970), 195.

[45] For material, see principally J. Boardman, 'Near Eastern and Archaic Greek Gems in Budapest', *Bull. du Musée Hongrois des Beaux-Arts,* 32/3 (1969), 8–12, no. 3, figs. 8–13; A. Hermary, *LIMC* iii (1986), s.v. Bes (Cypri et in Phoenicia), 108–12, esp. 110, nos. 14–22, pl. 87. See also Wilson, 'Bes', 83, 88–9; Michailidis, 'Bès', 61–3.

[46] See A. M. Bisi, 'Da Bes a Herakles. A proposito di tre scarabei del Metropolitan Museum', *RStudFen* 8 (1980), 19–42, esp. pls. iii–v; Tran Tam Tinh, *Bes*, 107. As noted by Michailidis, 'Bès', 70, Bes and Heracles have in common the ability to repel evil.

[47] See H. Junker, *Der Auszug der Hathor-Tefnut aus Nubien* (Berlin, 1911), 86; F. Daumas, *Les Mammisis des temples égyptiens* (Paris, 1958), 139–43.

[48] L. Keimer, 'Un Bès tatoué?', *ASAE* 42 (1943), 159–61 and 534; id., *Remarques sur le tatouage dans l'Égypte ancienne* (Cairo, 1948) (MIE 53), 104.

[49] See H. F. Wolff, 'Die kultische Rolle des Zwerges im alten Ägypten', *Anthropos*, 33 (1938), 445–514, esp. 469 ff.; J. Delpech-Laborie, 'Le Dieu Bès, nain, pygmée ou danseur?', *CdE* 16/32 (1941), 251–4; Michailidis, 'Bès', 54. On the notion of pygmies as benevolent demons, see E. de Rosny, *Les Yeux de ma chèvre* (Paris, 1981), 340, 381; he reports that in Cameroon healers invoke genies of water who represent pygmies; these beings are small, black, with bushy hair, large eyes, and feet turned backward; they procure wealth for men and fertility for women. See also P. Schebesta, *Die Bambuti-Pygmäen vom Ituri*, ii. 3 (Brussels, 1950), 60–6; M. Sidibé, 'Légendes autour des génies nains en Afrique noire', *Notes Africaines*, 47 (1950), 100.

feathers, an exotic material imported from the South (pls. 5, 6.1, 7.3).[50] A very similar crown is worn by the goddess Anukis who is also closely associated with Elephantine and Nubia.[51] Occasionally, the dwarf god wears a single ostrich feather on the top of his head (pl. 9.1a), as do Nubians.[52]

The god is frequently associated with monkeys, many of which were brought from Nubia.[53] On scarabs, the animals stand around Bes in an attitude of worship,[54] while in statuary they familiarly sit on his shoulders, between his feet or against his legs, as they normally do with Nubian servants (pls. 6.2, 7.3a).[55] The two types of being may have had a similar association with the South. Often objects related to attire, such as kohl-pots or calcite unguent vessels, were made in the form of Bes (pl. 4.4) or of monkeys.[56] Sometimes the appearance of Bes blends with that of monkeys; thick protruding lips give the god ape-like facial features, while his mane evokes that of a baboon, especially in Middle Kingdom depictions.[57] This fusion appears on a New Kingdom ostracon in Hildesheim, which depicts the god with a large mane, a hairy body and a long flexible tail; he is ithyphallic, as monkeys often are.[58] This assimilation of dwarfs with monkeys is found in magic spells; in the Harris Magical Papyrus (Late New Kingdom), an anonymous dwarf god (*nmw*) is compared with an old monkey, and finally changes into an ape 'with the mane of a baboon'.[59]

The popularity of Bes in the southern kingdom of Napata-Meroë has been related to the fact that the god might have originated from Nubia; the most spectacular pieces of evidence are the monumental figures of Bes which adorned the columns of a temple of the reign of Taharqa (690–664 BC) at Napata (Gebel Barkal).[60] Yet, as this temple was dedicated to a queen, the presence of Bes was more likely due to his association with birth or Hathor.

---

[50] A. Lucas, *Ancient Egyptian Materials and Industries*,[4] rev. with add. by J. R. Harris (London, 1962), 28–9.

[51] D. Valbelle, *Satis et Anoukis* (Mainz am Rhein, 1981), 94, § 15; 96–7, § 17; 109, § 31 with figs. on 96 and 115.

[52] See e.g. the Bes-figures in the Graeco-Roman birth-room at Armant (Cleopatra the Great and Ptolemy XVI); PM iv. 157, exterior, (35)–(36); Krall, 'Bes', 80, fig. 61. Cf. the Nubians in the tomb of Huy in Gurnet Murᶜai; J. Leclant *et al.*, *L'Empire des conquérants* (Paris, 1979), fig. 101.

[53] E. Brunner-Traut, *LÄ* i (1975), s.v. Affe, 83–5.

[54] See the material collected by Grenfell (n. 42 above), esp. 29–30, figs. 25, 32, and by Hornung/ Staehelin, *Skarabäen*, 94.

[55] For Nubian servants with monkeys, see e.g. the figures in C. L. Woolley and D. Randall-MacIver, *Karanog* (Philadelphia, 1910), pl. XXI, fig. 108, and O. W. Muscarella (ed.), *Ancient Art, The Norbert Schimmel Collection* (Mainz am Rhein, 1974), no. 229.

[56] Cf. e.g. the monkey holding a kohl-pot (h. 10 cm.), Cairo Museum, CG 18576; *Nofret*, i. 126–7, no. 59. Diorite or calcite vessels in the shape of monkeys were offered in temples at Byblos from the Old Kingdom onwards; see P. Montet, *Byblos et l'Égypte, Quatre campagnes de fouilles à Gebeil* (Paris, 1928), 72–4, nos. 56–62, pls. XL–XLI.

[57] H. Altenmüller, *LÄ* i (1975), s.v. Aha, 96. See e.g. the ape-like features of the faience figurine in Cairo Museum, CG 38749 (h. 5 cm.); Daressy, *Statues*, 191–2, pl. XLI.

[58] Limestone (h. 11.1 cm.; w. 9 cm.), Hildesheim, Pelizaeus Museum, 5268; *Nofret*, ii. 122, no. 157 (a lion-man for B. Schmitz).

[59] Spell U 8.12–9.1; 9.4; V 9.10; G. Roeder, *Der Ausklang der ägyptischen Religion* (Zurich, 1961), 176–7.

[60] Krall, 'Bes', 91, figs. 92 and 95, nos. 16a–b; Ballod, *Prolegomena*, 53; Wolff (n. 49 above), 453.

There is also strong evidence in support of the third hypothesis, that of an indigenous origin of Bes. Southern epithets are not peculiar to Bes; they are applied to several other gods, such as Amum, Anukis, Min, and Hathor.[61] Bes could be a purely Egyptian creation expressing an *Egyptian* concept of the South, as does the goddess Anukis.

His southern attributes are symbolically significant. They reflect his function as a defender and a hypostasis of the sun-god. The south-eastern deserts were the mythical land where Re emerged at dawn; Bes, who contributed to this daily rebirth by deterring underworld enemies, was perhaps appropriately identified with that region. His simian features relate him to the mythical baboons, described as 'the eastern souls', who greet the rising sun with screeching and dancing.[62] His leonine appearance may assimilate him to the mythical lions, seated back to back, who guard the horizon over which the sun rises.[63]

Bes may have emerged from the world of native Egyptian demons, living on the margins of the created world, in the underworld and in the southern deserts. Like Bes, these liminal beings often have misshapen or composite bodies made of animal and human parts—like the serpo-feline creature, the human-headed snake, or the griffin who process on Middle Kingdom magic wands. Bes first appears as a member of this group of monstrous but benevolent beings (fig. 6.1a, b).[64]

The demonic origin of Bes is also suggested by his kinship with the formidable creatures who guard the doors of the underworld and attend the judgement of souls.[65] Some are drawn frontally to show that they belong outside the ordered cosmos, like the two mummiform beings in a row of infernal creatures who face the viewer on the shrine of Tutankhamun.[66] This monstrous appearance was often meant to be frightening; some demons have names referring to their fearful faces, such as 'combative of face', 'savage one of face', and 'black of face'.[67]

---

[61] For their association with Punt or Nubia, see in general Ballod, *Prolegomena*, 17, and in particular Junker (n. 47 above), 27 (Min), 28 (Horus, Hathor), 29 (Shu).

[62] L. Störk, *LÄ* iv (1982), s.v. Pavian, 915–20, esp. 917. See also J. Assmann, *Liturgische Lieder an den Sonnengott* (Berlin, 1969), 208 (Medinet-Habu, temple of Ramesses III), and S. Sauneron, 'L'Hymne au soleil levant des Papyrus de Berlin 3050, 3056 et 3048', *BIFAO* 53 (1953), 69 and 76 n. 60.

[63] Typically in vignettes to spell 17 of the *Book of the Dead*. See also e.g. the papyrus of Heruben (dyn. 21) in Piankoff/Rambova, *MythPap*, no. 1, 73, fig. 3, scene 3. On the symbolism of leonine demons, see C. de Wit, *Le Rôle et le sens du lion dans l'Égypte ancienne* (Leiden, 1951), 138–47.

[64] See H. G. Fischer, 'The Ancient Egyptian Attitude towards the Monstrous', in A. E. Farkas et al. (eds.), *Monsters and Demons in the Ancient and Medieval Worlds, Papers Presented in Honor of Edith Porada* (Mainz am Rhein, 1987), 17–19.

[65] For a general survey of these underworld creatures, see Fischer ibid. 20–1.

[66] A. Piankoff, *The Shrines of Tutankhamon* (New York, 1955) (ERT 2), fig. 41 (first register, fourth and fifth figures from the right). See also the row of bizarre frontal hybrid beings in the papyrus of Djed-Khonsuiufankh II, in Piankoff/Rambova, *MythPap*, no. 22, scene 11. Primordial beings are also malformed; see e.g. the creatures with four hands in E. Chassinat and F. Daumas, *Le Temple de Dendara*, vi (Cairo, 1965), pls. DLXXXII–DLXXXIII.

[67] J. Zandee, *Death as an Enemy* (Leiden, 1960), 206–8. See also the picturesque names of these fearful demons in the *Book of the Dead*, spell 125, 60–102.

Several mythological papyri show Bes as one of these beings. In a New Kingdom vignette to spell 28 of the *Book of the Dead*, he waves a knife before the deceased who invokes a lion (perhaps Aha?) in order to stop his heart from being taken away.[68] In the papyrus of Tahemetnet-Mut, he is an underworld doorkeeper; his two heads are crowned with cobras and biting snakes, and he holds two knives on his knees; he is paired with a hippopotamus-headed demon and a dwarf-like child (pl. 8.1).[69]

Some physical characteristics and attributes of Bes reveal his kinship with another indigenous form: the fecundity figure. Bes often appears as an ageing adult, with a pendulous breast showing his reassuring, prosperous nature, and a protruding belly, stressed by a navel set very low (e.g. pl. 6.1).[70] Like fecundity personifications, he occasionally wears a belt with three strips, a garment of workmen.[71] He may grasp lotus and papyrus flowers (probably regeneration emblems),[72] or hold *ankh*- and *was*-signs, symbols of life and power which are closely associated with fecundity figures (pl. 4.3).[73]

According to Altenmüller, Bes may have originated in the Hermopolitan nome.[74] The name Aha occurs among the deities listed for that nome in the chapel of Senwosret I at Karnak,[75] and names compounded with Aha are very frequent from that area in the Middle Kingdom.[76] Hermopolis was also the main cult-centre of Thoth revered as a baboon, which might partly account for the link of Bes with monkeys, but this hypothesis is not supported by further evidence.

MYTHICAL STATUS

The demonic nature of Bes may be defined more precisely. As suggested by his presence in the *Litany of Re*, Bes seems to have been regarded principally as a hypostasis of the sun-god.[77] His identification with Re is revealed by his attributes.

[68] Papyrus of Neferuebenef; Krall, 'Bes', 87, fig. 80, no. 31; Ballod, *Prolegomena*, 29, fig. 4; T. G. Allen, *The Book of the Dead or Going Forth by Day* (Chicago, 1974), 38 n. 65; E. Hornung, *Das Totenbuch der Ägypter* (Zurich and Munich, 1979), 92, fig. 14.

[69] Berlin, SM (East) P 3128 (dyn. 21); *Nofret*, ii. 130–1, no. 163. For an almost identical scene, see e.g. the papyrus of Nesipautiutaui, Piankoff/Rambova, *MythPap*, no. 3, scene 4. A similar leonine figure faces the onlooker in the papyrus of Khonsu-Mes B; Piankoff/Rambova, *MythPap*, no. 17, scene 5. For further depictions of Bes as a fearful guardian of a pylon, see Krall, 'Bes', 87, figs. 81–2, nos. 32–3; Ballod, *Prolegomena*, 34–5.

[70] See in general Baines, *Fecundity Figures*, 30, 93–8, 118–22, esp. 127–31. Cf. the Bes-like figure on the Old Kingdom in the pyramid temple of Sahure (n. 18 above).

[71] See e.g. the tomb of Ipy, Thebes; PM i². 1, 316, hall (6) iii; Baines, *Fecundity Figures*, 128–9, fig. 84. Cf. the figure on the Old Kingdom relief in London (n. 18 above).

[72] See e.g. the dancing Bes in Bruyère, *Deir el Médineh*, 254–5, fig. 131; Smith, *Art and Architecture*, 166, fig. 57. See also the abacus in Dendara, Roman mammisi, PM vi. 104; Lepsius, *Denkmaeler*, iv. 83c; Krall, 'Bes', 81, fig. 62, no. 17, and the discussion by F. Daumas, *Les Mammisis des temples égyptiens* (Paris, 1958), 138–9.

[73] See e.g. the fecundity figures in the mortuary temple of Sahure at Abusir; PM iii². 1, 330 (16); Baines, *Fecundity Figures*, 84, figs. 43–4.

[74] Altenmüller, *Apotropaia*, i. 152, 155–6. See also Romano, 'Origin', 49.

[75] P. Lacau and H. Chevrier, *Une Chapelle de Sésostris Iᵉʳ à Karnak*, i (Cairo, 1956), 228, no. 647.

[76] Ranke, *PN* i. 44.6, 11–12; R. Anthes, *Die Felseninschriften von Hatnub* (Leipzig, 1928), nos. 10, 12, 15, 19, 21, 25, 39, 41–2 (ꜥḥ3-nḫt and ꜥḥ3-ḥtp).

[77] A. Piankoff, *The Litany of Re* (New York, 1964) (ERT 4), 91, 151, no. 25 (papyrus of Taudjare).

From the New Kingdom on, Bes is frequently associated with solar emblems, such as the *udjat*-eye, the solar disk or scarab, and the uraeus-snake. For example, a New Kingdom pilgrim bottle depicts a winged Bes surrounded with solar disks and *ankh*-signs (pl. 6.1); he holds two *neb*-baskets with *udjat*-eyes, symbols of the sun and the moon, the cosmic 'eyes of the dwarf' invoked in Papyrus Salt 825.[78] On an ivory tray dating to the reign of Tuthmosis III, Bes holds up *ankh*-signs below the solar winged scarab.[79] In Third Intermediate Period amuletic disks, Bes is also crowned with the *udjat*-eye, or the eye is carved behind his head.[80]

Several Late Period bronze statuettes depict Bes standing on top of a papyrus shaft, raising a sword in his right hand, and holding in his left the uraeus or a falcon, symbols of Re and Horus.[81] The god may also be surrounded with uraei, as on a Roman Period amulet (pl. 6.3).[82] Depictions of Bes holding or standing on animals symbolizing enemies, such as the gazelle or the pig (pl. 6.2),[83] derive from his power to avert malignant forces; often a gazelle with tied legs appears behind his head-dress of feathers with the same meaning (pl. 7.3b).[84] Figurines of Bes standing on two lions, two rams or two sphinxes may evoke the sun-god rising above the lions of the horizon.[85] This fusion of Bes with major gods is shown in a few late bronzes depicting a god with a full-sized body, but surmounted with the mask of Bes, associated with the white crown or the double feathers, and sometimes with the sidelock of youth of Horus.[86] Some figurines are labelled as Amum, Min, and Horus.[87]

This role of Bes as hypostasis of Re becomes greatly extended in the Late and Ptolemaic Periods. In statuary, on magical stelae and papyri, Bes becomes 'pantheos', a compound figure built up with attributes of several gods, especially those of the creator-

[78] P. Derchain, *Le Papyrus Salt 825 (BM 10051), Rituel pour la conservation de la vie*, i (Brussels, 1965), 141, ix, 4–5.

[79] Oxford, Ashmolean Museum, 1872.287 (l. 5.8 cm.; w. 4.8 cm.); Krall, 'Bes', 79, fig. 60b, no. 7.

[80] See C. Müller-Winkler, *Die ägyptischen Objekt-Amulette* (Fribourg and Göttingen, 1987) (OBO, SA 5), 47 and 102, pl. XII, nos. 213–16. See also a Bes-head crowned with an *udjat*-eye in W. M. F. Petrie, *Amulets* (London, 1914), 41, no. 190p, pl. XXXIV.

[81] See the figurines in Cairo Museum, CG 38718-19 and 38721; Daressy, *Statues*, 184–5, pl. XL. On the type in general, see Roeder, *Bronzefiguren*, 445, § 610a–b.

[82] See a similar figurine in Petrie (n. 80 above), 41, pl. XXXIV, no. 189g.

[83] See also e.g. the bronze staff (h. 18 cm.), Leiden, Rijksmuseum, E XVIII.116; Krall, 'Bes', 89, fig. 84, no. 56; Roeder, *Bronzefiguren*, 446, fig. 664; Romano, 'Bes', i. 143–4.

[84] Similar gazelles are depicted behind two head-dresses in the Michailidis coll.; Michailidis, 'Bès', 61, pls. X–XI. See also Krall, 'Bes', 89, fig. 83b, no. 60.

[85] Roeder, *Bronzefiguren*, 95–6, § 136f (on lions); Michailidis, 'Bès', 71–2, fig. 25 (on rams, crowned with *udjat*-eye); M. Werbrouck, 'Les Multiples Formes du dieu Bès', *BMRAH* 3rd ser. 11 (1939), 78, fig. 4 (holding a lion).

[86] See e.g. two Bes statuettes with twin feathers and sun disk in the Michailidis coll.; Michailidis, 'Bès', 75, fig. 31 and pl. XIX. See also the bronze statuette with a sidelock of youth in Berlin, SM 2489; M. Werbrouck, 'A Propos du dieu Bès', *Egyptian Religion*, 1 (1933), 30–1, fig. 4.

[87] See e.g. the bronze statuette (h. 14.1 cm.), Baltimore, WAG 48.1537; Steindorff (n. 22 above) 143–4, pl. XCV, no. 625 (with white crown, labelled as Harpocrates). A similar figurine is labelled as Min in Cairo Museum, CG 38836 (h. 14.2 cm.); Daressy, *Statues*, 208, pl. XLIII; Daressy ibid. mentions a third figurine in Liverpool Museum which is labelled as Amun. See also Romano, 'Bes', i. 190 n. 443.

god Amun-Re.[88] The god appears usually as a full-sized man with a Bes-head, crowned with ram's horns, a tall *Atef* head-dress, and uraei surrounding the solar disk. Small animal heads (usually seven or eight) sprout on either side of his head (pl. 11.1a, b).[89] The god may retain a dwarfish body, but with one or two animal heads, usually of ram, dog, or falcon, which may be paired back to back.[90]

His body, often ithyphallic, has one or two pairs of arms and wings, a falcon back which may end in a crocodile tail, and jackal-headed feet; emblems of the sun-god, such as uraei and lions' heads project from his body (pl. 11.1a, b).[91] Occasionally his body is covered with plain or *udjat*-eyes, or a pair of *udjat*-eyes surround his head.[92] These eyes may represent the cosmic eyes of the dwarf, or the 'thousand eyes' of a creator-god.[93]

His power over evil is displayed by his weapons (spears, knives, arrows) and divine attributes, such as the *was*-sceptre, the *nekhakha*-flail and the *ankh*-sign; he may also be surrounded by a circle of flames representing the triumphing fire of uraei. Sometimes he stands on an oval, made of a snake biting its tail (*ouroboros*), which encloses pictures of dangerous enemies, such as swamp and desert animals (crocodiles, hippopotami, lions, scorpions, and snakes) mastered by the god (pl. 11.1a).[94] On Graeco-Egyptian amuletic gems, the oval may contain an apotropaic spell.[95]

Some scholars identify this composite god with a syncretic deity, 'the god comprehending all gods'.[96] Others regard him as a substitute for various apotropaic gods, like Soped or Hormerty,[97] or for a specific supreme god, like Amun-Re.[98] The spell accompanying the depictions in the Brooklyn Papyrus supports the latter hypothesis. Bes is invoked as a powerful multiple manifestation (*b3w*) of Amun-Re; in his quality as a primordial god, he is the 'King of the gods, Lord of the sky, of earth, underworld,

[88] See in general Roeder, *Bronzefiguren*, 46–50, 91–4, 100–4; A. Delatte and P. Derchain, *Les Intailles magiques gréco-égyptiennes* (Paris, 1964), 126–31.

[89] Papyrus, Brooklyn Museum, 47.218.156, p. 3A, first vignette, p. 6A, second vignette; S. Sauneron, *Le Papyrus magique illustré de Brooklyn, Brooklyn Museum 47.218.156* (Brooklyn, 1970), figs. 2–3. On the different types of animal head, see the table in Roeder, *Bronzefiguren*, 92, § 134e (bull, cat, crocodile, falcon, hippopotamus, ibis, jackal, lion, monkey, ram, vulture).

[90] See e.g. Roeder, *Bronzefiguren*, 94, § 134g and 100–4.

[91] On the different elements composing the god, see the tables in Roeder, *Bronzefiguren*, 48, § 68f; 94, § 134g (arms, wings, uraeus, etc.).

[92] See e.g. the bronze statuette in Paris, Louvre, N 5140; Tran Tam Tinh, *Bes*, 103, no. 58b, pl. 82.

[93] Sauneron (n. 89 above) 11; E. Otto, *LÄ* i (1975), s.v. Auge, 559–60; W. Helck, *LÄ* i (1975), s.v. Augenschminke, 567.

[94] Hornung, *Conceptions*, 178–9. See e.g. the stelae in Cairo Museum, CG 9428–9; G. Daressy, *Textes et dessins magiques*, 36–7, pl. x. See also the stela in Hanover, Kestner Museum, 1935.200.688; Altenmüller (n. 3 above), fig. on 722. For a list of these animals, see Roeder, *Bronzefiguren*, 92–3, § 134f.

[95] See e.g. Delatte/Derchain (n. 88 above), 131–6, nos. 166–7, 171–4, and esp. 137, no. 177: 'Préserve-moi de tout mal'. See also the inscription: 'May Hor-Merty give life and health' on the statuette in Copenhagen, National Museum, 6623 (formerly Hilton Price 4438); F. W. v. Bissing, 'Miszellen, Zur Deutung der "pantheistischen Besfiguren"', *ZÄS* 75 (1939), 130–2, figs. 1, 2.

[96] E. A. W. Budge, *The Gods of the Egyptians*, i (London, 1904), fig. *contra* 492.

[97] Delatte/Derchain (n. 88 above), 130–1.

[98] Sauneron (n. 89 above), 16 ff.; J. Assmann, *Ägypten, Theologie und Frömmigkeit einer frühen Hochkultur* (Stuttgart, 1984), 281–2.

water and mountains' (4. 3), who breathes life into the created world.[99] As the god Shu, Bes becomes a 'giant of a million cubits' who 'carries the sky with his powerful arms' (4. 4), as illustrated on a sarcophagus in Vienna (fig. 5.4).[100] His monstrous appearance is terrifying and destroys every malevolent power, 'every male or female enemy, sow and guardian of the underworld, eater of the West, any danger coming from a dog or a bitch, from a dead man, from a dead woman' (4. 5, 8; 5. 2, 3).

The small heads, often very roughly depicted, which sprout on either side of the god's head or head-dress have been variously interpreted. Sauneron demonstrated that they do not belong to Bes, but represent the cohort of demons (conventionally seven) which were sent by major gods, usually to cast disease and death.[101] Like the 'seven striding spirits' depicted on a stela in Brooklyn (pl. 10.2),[102] these demons may have human bodies and various animal heads, but their presence may also be suggested by their heads only. These small heads often crown the sphinx Tithoes, who is described as 'the supervisor of the messengers of Sakhmet' (or of Neith and Bastet) on several Graeco-Roman monuments. The heads projecting out of the head of Bes probably had the same meaning; they represented the god as the chief of a similar troop of dangerous demons, ready to drive off enemies.

FUNCTION

Bes is primarily a helpful deity concerned with the protection of the household, but not exclusively. His tutelary functions involve at least five spheres of human life: the protection of women and childbirth, of sleep, of warfare, of the dead, and the celebration of music, dancing, and wine. In most roles, he is closely associated with two goddesses of the family, Hathor and Taweret.

Women and Childbirth

From the Middle Kingdom on, Bes commonly holds the office of a lady's attendant, watching over health and beauty. His hybrid image, comic and fierce at the same time, gives an exotic touch to all kinds of toilet objects, such as boxes and hairpins.[103] Very

[99] Sauneron (n. 89 above), 23.

[100] On Bes-Shu, see D. Jankuhn, *Das Buch 'Schutz des Hauses'* (Bonn, 1972), 88; D. Kurth, *Den Himmel stützen* (Brussels, 1975), 86–8. For iconographic and literary parallels, see above ch. 5, n. 46. Cf. the similar ability of Greek heroes to be dwarfs and giants at the same time.

[101] See S. Sauneron, 'Le Nouveau Sphinx composite du Brooklyn Museum et le rôle du dieu Toutou-Tithoès', *JNES* 20 (1960), 269–87, esp. 277–85. On these messengers in general, see D. Meeks, 'Génies, anges, démons en Égypte', in *Génies, anges et*

démons (Paris, 1971) (SourcesOr 8), 44–9.

[102] Limestone stela (h. 26.5 cm.; w. 47.7 cm.; depth 9 cm.), Brooklyn Museum, 58.98; Sauneron, ibid. esp. 276 ff., pls. XIV–XV; Tran Tam Tinh, *Bes*, 101, no. 33d, pl. 79.

[103] See e.g. J. Garstang, *El Arabah* (London, 1901) (BSAE 6), 5, pls. IV and XI (Middle Kingdom box); W. M. F. Petrie, *Gizeh and Rifeh* (London, 1907) (BSAE 13), 20–1, pl. XXIV (Middle Kingdom box); Romano, 'Origin', 44, fig. 5 (New Kingdom box); J. Vandier d'Abbadie, *Catalogue des objets de toilette égyptiens, Musée du Louvre* (Paris, 1972), 52–3, no.

often he occurs on items related to cosmetics, such as spoons and kohl-pots; he may hold the vessel (pl. 4.4), or the vessel has his form.[104] His formidable countenance was probably meant to deter malevolent spirits from harming the eyes of the lady of the house, a role of particular importance in a country where eye diseases, due to infections, were very frequent.[105] On the handle of a mirror, his grinning face had a similar role; it repelled the evil charms attracted by a too graceful appearance, or cast by a resentful rival.[106]

But the most prominent function of Bes was to assist women in the most vulnerable phase of their life, during pregnancy and delivery. These processes were especially critical in pre-modern societies with high infant and maternal mortality; there was also the archetypal fear of an abnormal birth.[107] Magic spells express these apprehensions and invoke a multitude of divine helpers to calm the anxiety and pains of the mother.[108]

Very early Bes appears in contexts related to childbirth. The first examples are Middle Kingdom magic wands carved in hippopotamus ivory, which were meant to drive away evil spirits (fig. 6.1a, b); their curved shape, which may derive from real knives or from the throwing sticks used to hunt birds in the marshes, suggests that they were regarded as magical weapons.[109]

These knife-amulets are decorated on one or both sides with a procession of apotropaic demons, often with fantastic animal forms. Weird beings appear near the leonine Bes, such as serpo-feline creatures, griffins, and the monstrous hippopotamus-goddess, mingling crocodile and leonine parts. These demons display their protective power by grasping or biting snakes, brandishing knives, or attacking enemies. They defend the nightly journey of the sun-god through the underworld, and ensure his daily rebirth at dawn.[110] Occasionally an inscription invokes protection for the owner of the 'knife',

154 (Saite Period box). See also A. Eggebrecht (ed.), *Ägyptens Aufstieg zur Weltmacht* (Mainz am Rhein, 1987), 289, no. 238 (New Kingdom hairpin).

[104] See e.g. Krall, 'Bes', 83, fig. 72, no. 19 (New Kingdom spoon); Vandier d'Abbadie, ibid. 55–9, nos. 160–73 (New Kingdom to Ptolemaic pots).

[105] G. Lefebvre, *Essai sur la médecine égyptienne de l'époque pharaonique* (Paris, 1956), 66–88; M. Helbing, *Der altägyptische Augenkranke, sein Arzt und seine Götter* (Zurich, 1980), 30–1; M. D. Grmek, *Les Maladies à l'aube de la civilisation occidentale* (Paris, 1983), 49 n. 62. On the healing properties of eye cosmetics, see e.g. C. Müller, *LÄ* v (1984), s.v. Schminke(n), 665–6 n. 4.

[106] See e.g. G. Bénédite, *Miroirs* (Cairo, 1907) (CGC), pl. IV, 44.017 (Old Kingdom?); Vandier d'Abbadie (n. 103 above), 170–1, no. 760 (Late Period?). For venomous charms, see e.g. Lefebvre (n. 105 above), 50: 'autre [drogue] pour faire que tombent les cheveux: un (ver-)ânâret cuit et bouilli avec de l'huile de ben. A mettre sur la tête de la

femme haïe' (P. Ebers 474).

[107] On gynaecology see Lefebvre (n. 105 above), 89–115; A. T. Sandison, *LÄ* ii (1977), s.v. Frauenheilkunde und -sterblichkeit, 295–7. A.-P. Leca, *La Médecine égyptienne au temps des pharaons* (Paris, 1971), 334–5, describes two mummies of women who died in childbed. See also the high rate of maternal mortality in Tell el-Dab'a cemetery (Palestine, c.1700 BC); J. Leclant, 'Fouilles et travaux en Egypte et au Soudan, 1976–77', *Orientalia*, 47 (1978), 271.

[108] See e.g. A. Erman, *Zaubersprüche für Mutter und Kind aus dem Papyrus 3027 des Berliner Museums* (Berlin, 1901); H. v. Deines, *et al.*, *Grundriss der Medizin der alten Ägypter*, iv. 1, *Übersetzung der medizinischen Texte* (Berlin, 1958), 291–5; Meeks (n. 101 above), 36–44.

[109] For real throwing sticks, see Hayes, *Scepter*, i. 248–9 and 279, fig. 181 (top left, second row).

[110] On the solar myth, see Altenmüller, *Apotropaia*, i. 82 ff., 136–77.

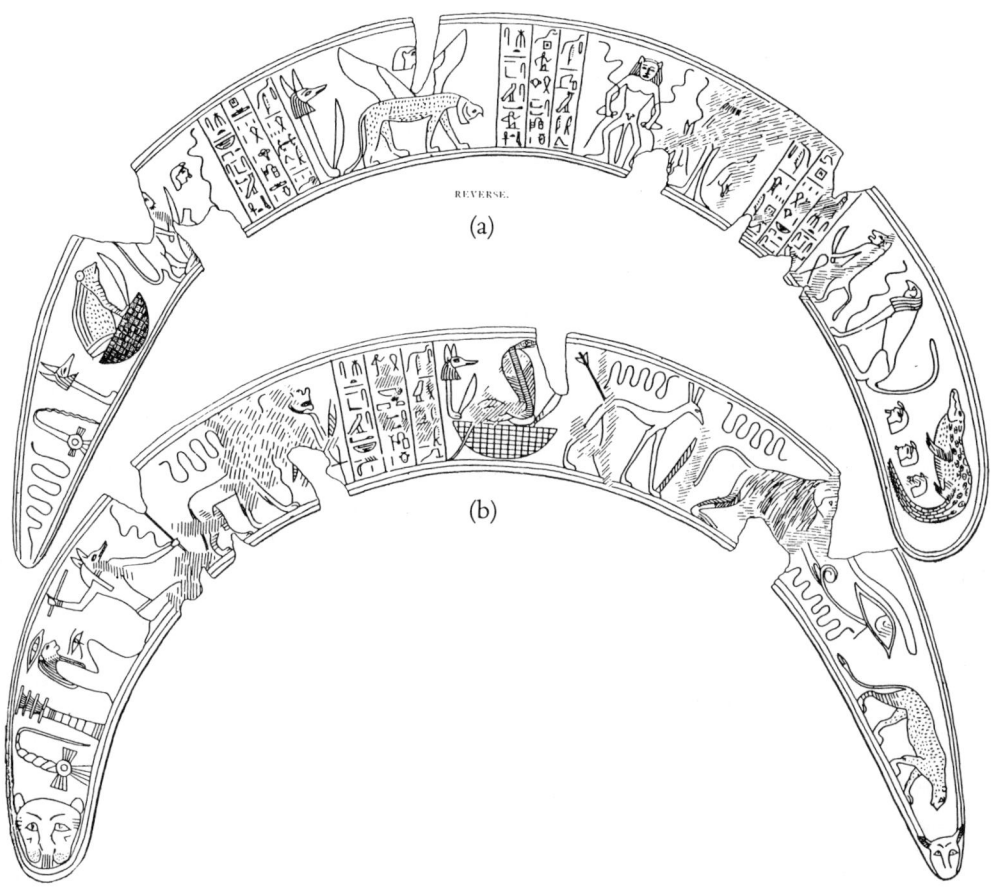

REVERSE.

(a)

(b)

FIGURE 6.1. 'Magic knife' (l. 41 cm.), from el-Lahun (Kahun). Berlin, SM (West) 14207

usually a child, assimilated by sympathetic magic to Re. A 'spell of Aha', inscribed besides a figure of the god on an ivory wand in Berlin, states: 'I have come to bring protection to Seneb, daughter of Senebseme, repeating life' (fig. 6.1a, b); on the same side of the wand other helpers twice repeat his spell in the plural: 'The many protectors say: we come to bring protection to Seneb, daughter of Senebseme, repeating life'.[111]

The amulets were probably used when spells were recited. Hayes noted that their ends are often worn away, as if they had been employed to mark magically safe places, such

[111] Trans. F. Legge, 'The Magic Ivories of the Middle Empire', *PSBA* 27 (1905), 136–8. For further examples, see Legge, ibid. 297–303; id., *PSBA* 28 (1906), 159–70. On these spells in general, see Altenmüller, *Apotropaia*, i. 64–78; id., 'Ein Zauberspruch zum "Schutz des Leibes"', *GM* 33 (1979), 7–12.

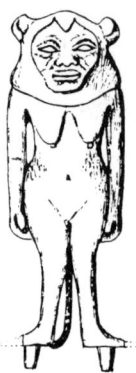

FIGURE 6.2. Wood (h. *c.*13 cm.). Present location unknown

as the birth-seat. The object may also have been laid under the bed of the patient to repel snakes and other venomous animals, or placed over the mother or the child's body.[112]

Other objects were associated with the performance of these rites. In a tomb at the Ramesseum Quibell found four fragmentary knife-amulets with several items of similar magical meaning, such as a bronze uraeus, part of a magical rod, female fertility figurines, statuettes of a lion and of baboons, and a female Bes-figurine of normal stature, holding two bronze snake wands;[113] these artefacts were probably used during ritual incantations. Papyri found in a box in the tomb included medico-magical texts about parturition and newborn children. As Gardiner observed, this set of objects may have composed the 'professional outfit of a magician and medical practitioner'.[114] Petrie made a similar discovery at el-Lahun (Kahun). A hole in the floor of a house contained a dwarfish female Bes-figurine (fig. 6.2)[115] with a pair of ivory clappers; a full-sized Bes-mask, made of painted canvas, was discovered in another room of the house.[116] These objects could have belonged to the town physician, midwife, or magician, who might have assumed the role of Bes to perform the rites.

Papyrus Leiden I 348 prescribes the use of figurines of dwarfs during the recitation of spells: utterance 30 states that the magician must speak 'four times over a dwarf of clay placed on the brow of a woman who is giving birth while suffering' (12.6).[117] These

[112] Hayes, *Scepter*, i. 248. On manual rites, see also Altenmüller, *Apotropaia*, i. 178–87; id., 'Ein Zaubermesser des Mittleren Reiches', *SAK* 13 (1986), 26–7; Bourriau (n. 22 above), 110 and 114.

[113] J. E. Quibell, *The Ramesseum* (London, 1898) (BSA 2), 3, pl. III. See also Bourriau (n. 22 above), 110–11, and 113, no. 100 (bronze cobra wand).

[114] A. H. Gardiner, *The Ramesseum Papyri* (Oxford, 1955), 1. For a description of the papyri, see ibid. 8 ff.

[115] Wood (h. *c.*13 cm.), present location unknown; W. M. F. Petrie, *Kahun, Gurob and Hawara* (London, 1890), 30, no. 14, pl. VIII; Romano, 'Bes', ii, no. 50.

[116] For the clappers and the Bes-mask, see Petrie, ibid. 30, nos 13, 13a, 27, pl. VIII; A. R. David, *The Pyramid Builders of Ancient Egypt* (London, etc., 1986), 136–7, figs. 8–9.

[117] J. F. Borghouts, *The Magical Texts of Papyrus Leiden I 348* (Leiden, 1971), 29; see also spell 31, 12.9.

clay figurines were probably very crude, and none has been identified, but Bes-amulets may have served in that context (e.g. pl. 4.1–2); they were perhaps worn as necklace pendants by pregnant women.

Middle Kingdom demons also watched over the feeding of the child. An infant feeding-cup from el-Lisht is decorated with the same procession of apotropaic figures (Bes-gods, water-turtle, lions, snake, a serpo-feline creature) as contemporary magic wands.[118]

In the New Kingdom, Bes frequently occurs in reliefs and paintings relating to childbirth. At Deir el-Medina, animated Bes-figures decorate the walls of enclosed platform beds ('salles du lit clos'); they dance, play the double flute or tambourine, blades sprout from their feet.[119] These enclosed beds may have served as domestic shrines for the cult of the household deities who ensured a safe delivery. In other houses, these rooms were painted with scenes referring to childbirth; Bruyère found figures of women dancing or nursing a child, and marsh scenes alluding to the birth of Horus.[120] In the workmen's village at el-ᶜAmarna, similar depictions were discovered in the front rooms of the houses. One painting represents four Bes-gods dancing in a row before Taweret whose sa-sign symbolizes protection; in another house, the scene shows a procession of rejoicing girls and women.[121] In the palace of Amenophis III at Malqata, a frieze of Bes-figures enlivens the walls of the bedroom.[122]

Troops of Bes-gods occur in various other forms in bedrooms, probably to bestow protection upon fertility. Thus Bes, paired with Taweret or lions, adorns the footboards of beds found in the joint burial of Yuya and Thuya, and in the tomb of Tutankhamun (pl. 5).[123] Their poses and attributes are similar to those in wall paintings and on furniture; they beat tambourines or grasp snakes and large knives; occasionally blades emerge from their feet. Some lean on sa-signs; some, usually winged, carry neb-baskets with ankh-, was-, and sa-signs. The glazed tiles with Bes-figures from the palace of Ramesses II at Qantir may have belonged to a similar domestic context, like the few extant depictions of women, pools, and gardens.[124]

Bes-gods also assisted women after childbirth, during the mother's period of seclusion and purification. This time was spent in a lightly built hut or 'birth arbour' depicted on

[118] Faience (diam. 8 cm.), New York, MMA 44.4.4; Fischer (n. 64 above), 17–18, pl. III. fig. 11; Romano, 'Bes', ii, no. 58.

[119] Bruyère, Deir el Médineh, 57–60, figs. 131, 133, 136, 202; Romano, 'Bes', ii, nos. 152–6.

[120] See B. Bruyère, 'Un Fragment de fresque de Deir el Médineh', BIFAO 22 (1923), 121–33; J. Vandier d'Abadie , 'Une Fresque civile de Deir el Médineh', RdE 3 (1938), 27–35; Bruyère, Deir el Médineh, 59–60, figs. 145, 157, 182, pls. IX–X.

[121] See B. J. Kemp, 'Wall Paintings from the Workmen's Village at el-ᶜAmarna', JEA 65 (1979), 47–53.

[122] Smith, Art and Architecture, 289, figs. 286–7.

[123] Footboard of bed, ebony (l. 184 cm.); Cairo Museum, JdE 62016; PM i². 2. 576; H. Carter and A. Mace, The Tomb of Tut-ankh-Amen, i (London, 1923), 113, pl. XLIX; Romano, 'Bes', ii, no. 120. See also JdE 62015; PM i². 2. 576; Romano, 'Bes', ii, no. 119. Tomb of Yuya and Thuya, Cairo Museum, CG 51109–10; J. E. Quibell, The Tomb of Yuaa and Thuiu (Cairo, 1908) (CGC), 50–1, pls. XXVIII–XXXI.

[124] W. C. Hayes, Glazed Tiles from a Palace of Ramesses II at Kantir (New York, 1937), 38–41, fig. 11, pls. XII–XIII; Romano, 'Bes', ii, nos. 135–41.

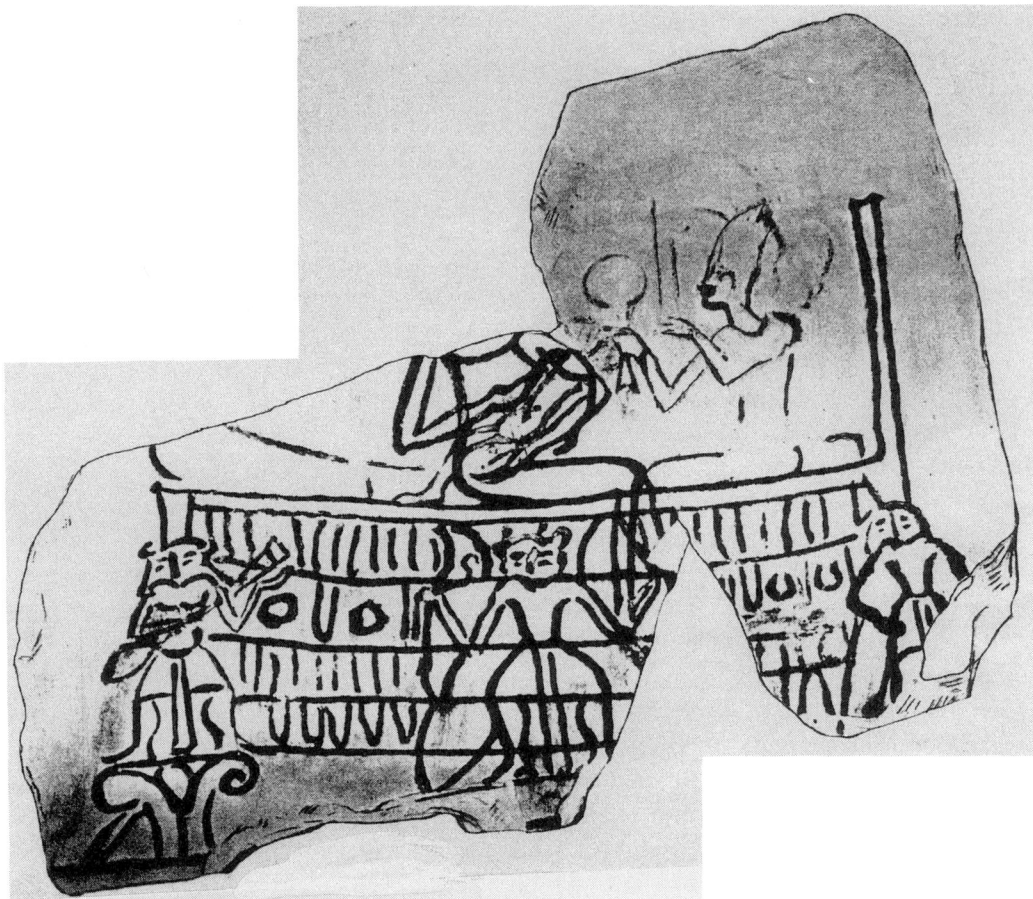

FIGURE 6.3. Ostracon (h. 12 cm.; w. 15 cm.), from Deir el-Medina. Berlin, SM (West) 21451

Deir el-Medina ostraca.[125] Bes-gods adorn the feet of the bed on which the mother nurses her newborn child; often they play a musical instrument, the flute or the lute, or hold a knife and a snake to avert evil (fig. 6.3).[126] Pairs of wooden bed feet, carved in the form of Bes, were excavated at Deir el-Medina.[127] A wall painting in the tomb of

[125] E. Brunner-Traut, 'Die Wochenlaube', *MIO* 3 (1955), 11–30; id., *LÄ* vi (1986), s.v. Wochenlaube, 1282–4.

[126] Ostracon (h. 12 cm.; w. 15 cm.), from Deir el-Medina; Berlin, SM (West) 21451; E. Brunner-Traut, *Die altägyptischen Scherbenbilder (Bildostraka) der deutschen Museen und Sammlungen* (Wiesbaden, 1956), 69, no. 65, pl. xxv. For parallels, see e.g. J. Vandier d'Abbadie, *Catalogue des ostraca figurés de Deir el Médineh (Nos. 2256 à 2722)* (Cairo, 1937), nos. 2340, 2344, 2346–7, 2353; G. Pinch, 'Childbirth and Female Figurines at Deir el-Medina and el-ᶜAmarna, *Orientalia*, 52 (1983), esp. 406–7, and n. 6.

[127] Brussels, Musées Royaux, E 7415–7416 (h. *c.*60 cm.); Bruyère, *Deir el Médineh*, 98, fig. 32; Werbrouck (n. 85 above), 79–82, figs. 9–10. See also two pairs in the Michailidis coll. (h. 65/67 cm.); Michailidis, 'Bès', 67, 69, 71, pls. xiv–xv.

Sennefer depicts a large Bes-figure standing next to a couch, between a candle and various boxes, which may imply that statuettes of Bes, made of wood or faience, were also set beside the bed.[128]

This protective function of Bes passed into official iconography. The god occurs in the cycle of temple reliefs showing the divine birth of the king; he stands with Taweret and other helpers below the royal bed in the Theban temples of Hatshepsut and Amenophis III.[129]

In partnership with Taweret, Bes-gods protected young children against hidden dangers. They appear on the furniture made for royal children, like princess Sitamun; the sides of her small chair are decorated with a group of Bes-gods waving knives to drive away malevolent spirits, and with hippopotamus-gods leaning on *sa*-signs; on the back, a winged Bes, with knives sprouting from his feet, holds in each outstretched hand a *sa*-sign and a basket filled with two pairs of *ankh*- and *sa*-signs.[130]

From the Third Intermediate Period on, Bes's link with conception is shown mainly in bronze and faience statuettes, placed on top of papyrus shafts, which belonged to staves or to items of furniture. They depict Bes seated on the shoulders of a woman who may have a small child in her arms or clinging to her legs.[131] In a few examples the woman seems to be pregnant; she may also stand on a gazelle,[132] probably to show her triumph over evil forces, or on a frog, symbol of the fertility goddess Heqet, as on a bronze in Athens (pl. 7.1).[133] Often Bes plays a musical instrument, usually the lute; he may also be armed with a sword. As Spiegelberg suggested, these statuettes may be votive offerings consecrated by women who wanted to conceive children, or who wished to express gratitude for a successful delivery.[134] Bes may also be seated on the shoulders of musicians, male or female, who play the double-flute or beat the tambourine, which again alludes to the celebration of childbirth. On a bronze figurine from Samos, the god himself is striking a tambourine (pl. 7.2);[135] the piece was found in

[128] Deir el-Medina, tomb 99; PM i². 1. 205, inner room (10); Bruyère, *Deir el Médineh*, 108, fig. 39; Romano, 'Bes', ii, no. 67.

[129] Deir el-Bahari, temple of Hatshepsut; E. Naville, *The Temple of Deir el Bahari*, ii (London, 1894), pl. 51; Ballod, *Prolegomena*, 42, fig. 20. Luxor, temple of Amenophis III; PM ii. 326, birth-room (152) ii (3); Krall, 'Bes', 79, fig. 58; H. Brunner, *Die Geburt des Gottkönigs* (Wiesbaden, 1964), 102, pl. 9, scene IX.

[130] Wood, Cairo Museum, CG 51112; Quibell (n. 123 above), 52–3, pls. XXXV–XXXVII. Cf. the similar decoration of the other chairs, Cairo Museum, CG 51111 and 51113; Quibell, ibid. 52–4, pls. XXXII, XXXIV, and XLI–XLII.

[131] For material, see Roeder, *Bronzefiguren*, 97–8, § 138; W. Spiegelberg, 'Die Weihestatuette einer Wöchnerin', *ASAE* 29 (1929), 162–5; K. Parlasca,

'Zwei ägyptische Bronzen aus dem Heraion von Samos', *MDAI(A)* 68 (1953), 127–36; See also Romano, 'Bes', i. 144–6, 163.

[132] See the bronze in Paris, Louvre, E 22874; Tran Tam Tinh, *Bes*, 105, no. 93c, pl. 86. A similar figurine is in Brussels, Musées Royaux, E 6755 (h. 11 cm.); Werbrouck (n. 85 above), 78 and 80, fig. 5; Parlasca, ibid. 132, no. 4.

[133] Bronze (h. 10.4 cm.), Athens, NM 614; Parlasca (n. 131 above), 133, no. 16, Beil. 46, 3. See also the bronze, Paris, Louvre, E 5891; Tran Tam Tinh, *Bes*, 105, no. 93b, pl. 85.

[134] Spiegelberg (n. 131 above), 164; Roeder, *Bronzefiguren*, 442–3, § 607.

[135] Bronze (h. 12 cm.), Samos, Vathy Museum, B 353; Parlasca (n. 131 above), pl. XII; U. Jantzen, *Samos*, viii (Bonn, 1972), 14, pl. 18. For parallels, see Parlasca, ibid.

the early archaic votive deposit of the Heraion, a sanctuary especially concerned with the protection of mother and child.[136] Other figurines depict Bes nursing a miniature figure of himself (pl. 7.3a).[137]

From the Late Period on, Bes is more closely associated with the infant Horus. On Horus-cippi, his protective mask surmounts the head of the young god, strengthening his magical power (pl. 3.2).[138] Bronze and wooden statuettes also depict Bes carrying a child, probably Horus, seated on his left shoulder.[139] In one statuette, Bes is clearly a substitute for Isis: he is seated on a throne and suckles a child in the same pose as the goddess.[140]

In Graeco-Roman reliefs and statuary, Bes is very often accompanied by women, either dwarfish or full-sized, playing music and dancing.[141] His protective role is displayed by his warrior attire, composed of a kilt, a sword, and a round shield.[142] The shield might have become one of his principal attributes because its shape evoked that of the tambourines associated with rejoicing at childbirth. In several terracottas Bes seems to play the role of a child. As a newborn baby, he is fed by Beset, or he is involved in juvenile activities, such as going to school.[143] His appearance is often similar to that of Horus-Harpocrates; he holds a jar and puts a hand to his mouth.[144]

In birth-houses, Bes-figures become a very common decorative element; they appear on abaci and architraves, as at Dendara (pl. 9.2),[145] alone or in a frieze, sometimes alternating with Hathor heads.[146] They attend the birth of the god,[147] or they surround the newborn child seated on the primordial lotus flower.[148]

[136] See below ch. 13, *sub* Kourotrophic Demons.

[137] See the similar figurines in Cairo Museum, CG 38728–30; Daressy, *Statues*, 187–8, pl. XL.

[138] See e.g. the stelae in Cairo Museum, CG 9401–32; G. Daressy, *Textes et dessins magiques* (Cairo, 1903) (CGC), 1–43, pls. I–XI. See also W. D. van Wijngaarden and B. H. Stricker, 'Magische Stèles', *OMRO* 22 (1941), 6–38, esp. pls. I–IV.

[139] See e.g. the statuette in London, BM 61206; H. R. Hall, 'An Egyptian St Christopher', *JEA* 15 (1929), 1, pl. I. There is a similar faience statuette of a dwarf (-god?) carrying a child (h. 4.8 cm.), London, UC 33528, unpublished.

[140] Michailidis, 'Bès', 56, fig. 8 (no indication of provenance, size or material); Third Intermediate Period according to J. F. Romano (personal communication).

[141] See e.g. Perdrizet (n. 37 above), pls. XLII–XLIV, and Tran Tam Tinh, *Bes*, 101–2, nos. 31a–b, 36b, 38t, pl. 78. See also the lamps with Bes and Isis(?); Ballod, *Prolegomena*, 99, fig. 113; Tran Tam Tinh, *Bes*, 102, no. 48, pl. 81.

[142] See e.g. W. Weber, *Die ägyptisch-griechischen Terrakotten* (Berlin, 1914), 162, nos. 256–9, pl. 25; Perdrizet (n. 37 above), pl. XLI.

[143] Tran Tam Tinh, *Bes*, 104, nos. 70, 72–6, pls. 83–4; id., *LIMC* iii (1986), s.v. Besit, 113, nos. 14–15, pl. 91.

[144] Tran Tam Tinh, *Bes*, 104, no. 71, pl. 84.

[145] Dendara, mammisi of Nectanebo; J. Leclant et al., *L'Égypte du crépuscule* (Paris, 1980), 89, fig. 70; Tran Tam Tinh, *Bes*, 99, no. 2; Dasen, 'Dwarfism', pl. 3b.

[146] For material, see Tran Tam Tinh, *Bes*, 99, nos. 1–3 and 5, pl. 74. See also e.g. Edfu: PM vi. 171, birth-house, ii, court (27)–(30); 175, corridor (106), 176 (118), 177, west side, columns; Kom-Ombo: PM vi. 199, birth-house, part of frieze, position unknown.

[147] See Dendara, mammisi of Nectanebo I; PM vi. 105; Daumas (n. 16 above), 9, no. 31, pl. II. Dendara, mammisi of Trajan; PM vi. 104–5, (10)–(16), third register; Lepsius, *Denkmaeler*, iv. 82b; Krall, 'Bes', 79, fig. 59; Daumas, ibid., 108, pl. LIX.

[148] See e.g. Armant, birth-room (Cleopatra the Great and Ptolemy XVI); PM iv. 157, exterior, (35)–(36); Krall, 'Bes', 80, fig. 61. Dendara, birth-house of Trajan; PM vi. 103, colonnade, architrave; Daumas (n. 16 above), 271–85, pl. XCV.

The most striking archaeological evidence for the connection of Bes with female sexuality is the sanctuary found by Quibell at Saqqara. The walls were decorated with tall clay figures of the god in high relief, standing, holding snakes and knives, surrounded by nude women (pl. 9.3).[149] Many phallic and 'erotic' ex-votos were found in the rooms and nearby.[150] Pinch suggests that these chambers might have been slept in by people who wanted to conceive children.[151]

This belief in Bes as the guardian of women was deeply rooted. It is still expressed in a Greek spell on a third- or fourth-century AD stela from Memphis. The stela is decorated with a crude depiction of a dwarf god fusing the attributes of Bes and Ptah-Pataikoi: he stands on crocodiles, brandishing a sword to his head in his right hand and holding a snake in his left.[152] His functions identify him as Bes. The god is addressed to as 'the great Lord of women's wombs' (μήτρας γυναικῶν θεὸς μέγιστος), protector (προστάτης), guardian (φύλαξ), healer (θεραπευτής), sower (σπορεύς), feeder (τροφεύς), and awakener (ἀφυπνηστής).

Sleep

Another function of Bes was to guard people during sleep, as he guarded Re against underworld enemies. Sleep was seen as a dangerous time: in a state close to death, man was abandoned to malignant forces which could cause terrifying visions and cast disease.[153]

Many major and minor deities, such as Neith, Anubis, Hathor, Isis, Nephthys, and hippopotamus-gods, could be invoked as helpers, but Bes seems to have been the most popular.[154] From the Middle Kingdom on, head-rests were placed under the magical protection of Bes. His face adorns the sides of the object, or his figure appears on the base or central pillar; he may be calmly leaning on a *sa*-sign, or leaping, brandishing knives, biting snakes, with blades sprouting from his feet.[155] Often he is paired with

[149] Room 14 (h. of the larger figure 1.50 m.); J. E. Quibell, *Excavations at Saqqara (1905–1906)* (Cairo, 1907), 28, pl. XXVII, 2; Ballod, *Prolegomena*, 80–1, fig. 93; Keimer (1948) (n. 48 above), pl. XXV. For the other figures, see Quibell, ibid. 12–14, 28, pls. III, XXVI–XXIX.

[150] Quibell, ibid. 13; P. Derchain, 'Observations sur les erotica', in G. T. Martin, *The Sacred Animal Necropolis at North Saqqâra* (London, 1981), 166 ff., pls. 24–9. See also G. T. Martin, '"Erotic" figurines: the Cairo Museum material', *GM* 96 (1987), 71–84.

[151] G. Pinch, *Votive Offerings to Hathor in New Kingdom Temples*, D.Phil. thesis (Oxford University, 1984 (in press)), 440.

[152] Limestone stela (h. 14.7 cm.; w. 10 cm.) Michailidis coll.; G. Michailidis, 'Le Dieu Bès sur une stèle magique', *BIE* 42–3 (1960–2), 65–85, pls. Ia and II. I am very grateful to P. M. Fraser for suggesting this date.

[153] S. Sauneron, 'Les Songes et leur interprétation dans l'Égypte ancienne', in *Les Songes et leurs interprétations* (Paris, 1959) (SourcesOr 2), 19–21; J. Zandee, *Death as an Enemy, According to Ancient Egyptian Conceptions* (Leiden, 1960), 82–4; Hornung, *Conceptions*, 180. See e.g. J. Černy and A. H. Gardiner, *Hieratic Ostraca* (Oxford, 1957), pl. XXXVII, recto 5–6, and A. H. Gardiner, *Hieratic Papyri in the British Museum, Third Series, Chester Beatty Gift*, i. 19 (recto 10. 10) and 77 (verso 14. 5).

[154] G. Daressy, 'Neith, protectrice du sommeil', *ASAE* 10 (1910), 177–9.

[155] See in general S. Schott, 'Eine Kopfstütze des Neuen Reiches', *ZÄS* 83 (1958), 141–4; H. G. Fischer, *LÄ* iii (1980), s.v. Kopfstütze, 688–9 and n. 19. For material, see the limestone, terracotta, and wooden head-rests excavated in Deir el-Medina; Bruyère, *Deir el Médineh*, 227–35, figs. 118–19, pl. XXIV. See also the ivory head-rest from the tomb

Taweret, Hathor, or other demons with leonine or dwarfish bodies. Occasionally spells ask the god to protect life and health, as on a Middle Kingdom headrest in London.[156] His figure also appears on miniature amuletic head-rests.[157]

In the Graeco-Roman period, Bes became an oracular god, a role which might derive from his special power over the spirits of the night, or may simply be common to many contemporary gods. At Abydos, visitors came to the abandoned temple of Seti I (Memnonion) and wrote questions on a scrap of papyrus; then they spent the night in a room of the temple so as to have an oracular dream.[158] The procedure was probably similar to that prescribed in the Greek magical Papyrus London 122:

On your left hand draw Besa in the way shown to you below. Put around your hand a black cloth of Isis and go to sleep without giving answer to anyone. The remainder of the cloth wrap around your neck (65–9).

Then follows the text of invocation for a dream or a direct vision. The depiction on the hand evoked Bes as a warrior:

What you draw is of this sort: a naked man, standing, having a diadem on his head, and in his right hand a sword that by means of a bent [arm] rests on his neck, and in the left hand a wand (105–9).

To dissolve the vision, the text instructs: 'If he reveals to you, wipe off your hand with rose perfume' (109–10).[159]

Warfare

Bes could also ensure triumph over foes made of flesh and blood, and he appears on a few items related to warfare. A full-faced Bes-figure is incised on a Second Intermediate Period leather archer's brace,[160] and six Bes-heads appear on various parts of a chariot

of Tutankhamun (h. 19.2 cm.), Cairo Museum, JdE 62023; PM i². 2, 576; *Musée égyptien du Caire*, no. 184; and the wooden head-rest from a Theban tomb, Cairo Museum, JdE 6269; C. Desroches Noblecourt, *Le Grand Pharaon Ramsès et son temps* (Montreal, 1985), no. 40.

[156] London, BM 35807; Ballod, *Prolegomena*, 24–7, fig. 1; Altenmüller, *Apotropaia*, i. 153. See also the inscribed head-rests from Deir el-Medina; Bruyère, *Deir el Médineh*, 228–9, 233, fig. 118, pl. xxiv.

[157] See the terracotta amulet (h. 3.3 cm.), from Alexandria, Michailidis coll.; Michailidis, 'Bès', 71, fig. 24. A similar one is in Berlin, SM 18248 (h. 3.3 cm.), unpub.

[158] See Amm. Marc. 19. 12. 3–4, and the discussion in P. Perdrizet, G. Lefebvre, *Les Graffites grecs du Memnonion d'Abydos* (Nancy, 1919), pp. xix–xxiii.

[159] Trans. H. Dieter Betz (ed.), *The Greek Magical Papyri* (Chicago and London, 1986), 147–8. See also K. Preisendanz (ed.), *Papyri Graecae Magicae*, ii (Leipzig, 1931), 10, 48, viii. 64–110. For a detailed commentary on this depiction, see C. Bonner, *Studies in Magical Amulets, Chiefly Graeco-Egyptian* (Ann Arbor and London, 1950), 108–9.

[160] G. Brunton, *Mostagedda and the Tasian Culture* (London, 1937), 128, pls. lxxiv, no. 1c and lxxv, no. 49; Romano, 'Bes', ii, no. 59.

from the tomb of Tutankhamun.[161] The widespread Graeco-Roman motif of Bes as a warrior, and more particularly as a horse-rider, may be related to that belief.[162]

The Dead

As a helper of birth, Bes also contributed to the rebirth of the deceased in the blissful Fields of Reeds. Many dangers had to be overcome in the underworld before achieving a second birth; the deceased had to pass through gateways guarded by terrifying demons, and was judged before forty-two assessor gods in the Judgement Hall of Osiris.

The role of Bes was mainly to protect the body of the deceased, just as he guarded the body of the living. In spell 28 of the *Book of the Dead*, he is asked to defend the heart of the deceased against attacks of enemies; in spell 163 he is invoked with Neith to prevent the putrefaction of the body.[163] Amuletic figurines of Bes may have been placed in the wrappings of the mummy to ensure a successful regeneration. Other objects decorated with Bes-figures, such as head-rests or magic wands, were perhaps left in the tomb to serve a similar purpose, as Altenmüller suggests.[164]

Miscarriages and infant deaths seem to have been placed under the special care of Bes. A child's coffin of the Third Intermediate Period was decorated with the god's face,[165] and four large wooden statuettes of Bes (h. *c.* 50 cm.), hollowed out behind, contained the remains of human foetuses.[166]

This funerary role of Bes is best illustrated by a relief in the west Osiris chapel at Dendara (fig. 6.4).[167] Osiris lies on a funerary bed; he is flanked by a kneeling Hathor at his head, and by the frog goddess Heqet at his feet. Two birds fly over his body, one of them a *ba*, the other a kite as a manifestation of Isis, who is conceiving Horus. Bes stands under the couch in the company of three deities, two solar cobras, and the ibis-headed Thoth holding the *udjat*-eye.[168]

The Celebration of Music, Dancing, and Wine

Music and dancing are the main occupations of Bes from the New Kingdom on. The god is shown playing the double flute (fig. 6.5b), the lyre (pl. 9a), or the lute (fig. 6.3),

[161] Cairo Museum, JdE 61989; M. A. Littauer and J. H. Crouwel, *Chariots and Related Equipment from the Tomb of Tut'ankhamun* (Oxford, 1985), 19, 28, 33, pls. XVI, XVIII, XIX, XXXIV–XXXV, XL–LI; Romano, 'Bes', ii, no. 123A–C.

[162] Tran Tam Tinh, *Bes*, 104, no. 77a–d, pl. 84.

[163] On the similar funerary role of Ptah-Pataikoi, see below 97 (spell 164).

[164] H. Altenmüller, 'Ein Zaubermesser aus Tübingen', *WdO* 14 (1983), 37–8.

[165] E. R. Ayrton *et al.*, *Abydos*, iii (London, 1904), 52, pl. XXVIII, 5–6 (MEEF 25).

[166] C. Gaillard and L. Lortet, *La Faune momifiée de l'ancienne Égypte*, ii (Lyons, 1905), 201–5, figs. 83–4;

F. Drilhon, 'Un Foetus humain dans un obélisque égyptien en bois', in *Archéologie et médecine, VIIèmes rencontres internationales d'archéologie et d'histoire, Antibes, Octobre 1986* (Juan-les-Pins, 1987), 515–16.

[167] Dendara, Roman mammisi; PM vi. 96, inner room; A. Mariette, *Dendérah*, iv (Paris, 1873) (37)–(41), west Osiris chapel, pl. 88; Lanzone, *Dizionario*, pl. 285.

[168] Compare with the demons holding knives and snakes under the bed of Osiris in the vignette to spell 182 of the *Book of the Dead*; E. Hornung, *Das Totenbuch der Ägypter* (Stuttgart and Zurich, 1979), 389 and 520, fig. 90.

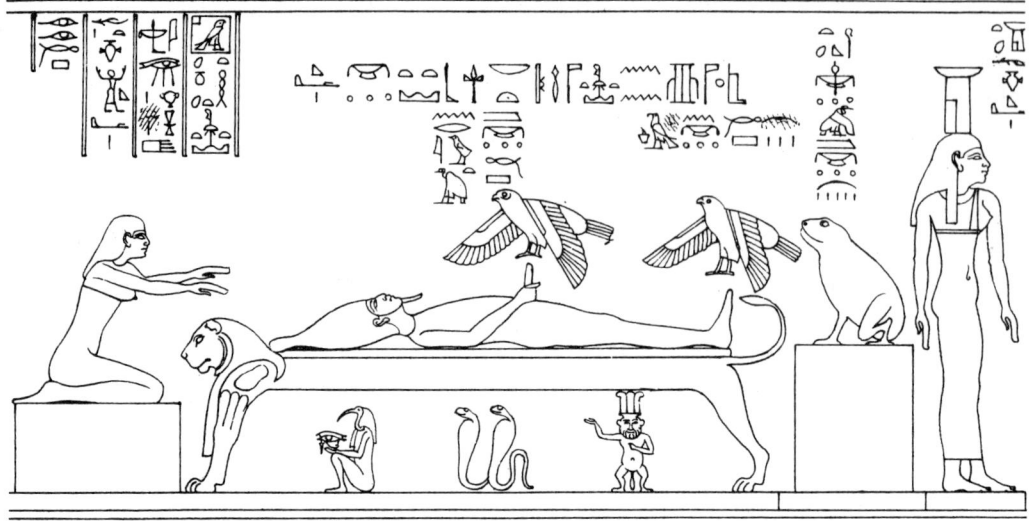

FIGURE 6.4. Dendara, Roman mammisi. PM vi. 96, inner room

and beating circular percussion instruments, drums or tambourines (pls. 4.1, 7.2, 9.1a, b; fig. 6.5b). In Graeco-Roman statuettes, he is occasionally singing, with a hand to his mouth.[169] Bes also occurs as an amulet or a body decoration (tattoo?) on female dancers and musicians.[170]

This strong relationship with music relates to the protective role of Bes during and after delivery. The tambourine, in particular, is an instrument commonly associated with the celebration of birth from the New Kingdom on.[171] On a mythical level, the performance of Bes may have evoked that of baboons greeting the daily birth of the sun-god.

In the Graeco-Roman period, Bes is also associated with the sistrum, an instrument sacred to Hathor which was believed to avert sorrow and fury, and to aid regeneration.[172] The sistrum handle may be in the form of Bes, who is sometimes paired back to back with Beset or satyrs.[173] The presence of Bes perhaps referred to his

[169] See e.g. Michailidis, 'Bès', 69, figs. 21–2.
[170] For material, see Keimer (1948) (n. 48 above), 40–4, pls. XX–XXIII, XXV.
[171] See C. Ziegler, 'Tambours conservés au Musée du Louvre', *RdE* 29 (1977), 203–14; id., *Catalogue des instruments de musique égyptiens, Musée du Louvre* (Paris, 1979), 31–40, esp. 39–40. Cf. the seven Hathors with round tambourines in Philae; H. Junker and E. Winter, *Das Geburtshaus des Tempels der Isis in*

Philä (Vienna, 1965), 220–3, pls. 417–18. Cf. the two Late Period tambourines with human and divine dwarfs (E *d* 85, 86, fig. 9.28).
[172] See H. Brunner, 'Das Besänftigungslied in Sinuhe', *ZÄS* 80 (1955), 7; F. Daumas, 'Les Objets sacrés de la déesse Hathor à Dendara', *RdE* 22 (1970), 69–73.
[173] See e.g. Ziegler (n. 171 above), 58–61, nos. 76–80.

appeasing role in the myth of the 'Distant goddess', or 'Solar eye'.[174] The story recounts the return to Egypt of Hathor-Tefenet, the sun-god's daughter or 'eye', who had retreated to the Nubian deserts in the form of a fearful lioness. Re sent two groups of wily messengers to bring her back: the gods Shu and Thoth, disguised as monkeys, who lured the goddess with attractive promises, and Bes-gods who calmed and rejoiced her with their musical performances. This merry retinue is depicted on the columns of the temple of Hathor at Philae (pl. 9.1a, b).[175] Once back in Egypt, Hathor transformed herself into a gracious and gentle deity, either Bastet, patron of the home, or Hathor, the goddess of love and entertainment.

In the Late and Graeco-Roman Periods, several terracottas depict Bes holding wine jars, or dancing among bunches of grapes.[176] A Late Period or Ptolemaic bronze figurine of the god is carved with the name of Shesmu, 'the god of the wine-press'.[177] At first, this connection of Bes with wine and drunkenness seems to derive from Greek iconography, in particular that of satyrs. It may, however, express in a new iconography an old idea relating to the myth of the 'Solar eye'. Some texts say that Thoth dissipated the fury of Hathor-Tefenet by offering her wine spiced with a touch of magic spells; the *mnw* wine-jug thus became a significant ritual object in the cult of the goddess as the symbol of the restoration of harmony between mankind and gods.[178] A version preserved in New Kingdom sources, called 'The Destruction of Mankind', also gives a prominent role to the euphoric power of alcohol. Hathor had been sent down by Re to destroy mankind, and she had transformed herself into a bloodthirsty deity intent on killing every man on earth. To appease her murderous temper, Re brought 7,000 jugs of beer, dyed with red ochre from Elephantine, which the goddess drank, thinking it was blood.[179] These legends may be evoked in the epithets of Hathor, 'the Lady of drunkenness' or 'the Lady of the wine-jug'.[180] As a member of Hathor's retinue, Bes was probably closely involved in celebrating this pacifying quality of wine. A lost

[174] For a reconstruction of the myth, see principally Junker (n. 47 above); id., *Die Onurislegende* (Vienna, 1917); W. Spiegelberg, *Der ägyptische Mythus vom Sonnenauge* (Strasburg, 1917); Bonnet, *Reallexikon*, s.v. Tefnut, 770–4. For later versions (which do not mention Bes-gods), see S. West, 'The Greek Version of the Legend of Tefnut', *JEA* 55 (1969), 161–83; W. J. Tait, 'A Duplicate Version of the Demotic Kufi Text', *AcOr* 36 (1974), 23–37.

[175] Philae, temple of Hathor (Ptolemaic), cols. I and K; PM vi. 248; F. Daumas, 'Les Propylées du temple d'Hathor à Philae et le culte de la déesse', *ZÄS* 95 (1968), pl. v. For the decoration of the other columns, see ibid. 1–17, pls. I–VI.

[176] For material, see e.g. Cairo Museum, CG 38709 bis, 38710; Daressy, *Statues*, 182, pl. XL; Weber (n. 142 above), 163, no. 261, pl. 25; Perdrizet (n. 37

above), pl. XL; Michailidis, *Bès*, 64–6, figs. 18–19; Tran Tam Tinh, *Bes*, 102, no. 46d, pl. 81.

[177] Bronze (h. 21.8 cm.), from Disuq, Michailidis coll.; Michailidis, 'Bès', 64, pl. XIII.

[178] See Junker (n. 47 above), 4 ff., 44; Daumas (n. 172 above), 75–6; id., *LÄ*, ii (1977), s.v. Hathor, 1026–7; ibid. s.v. Hathorfeste, 1035; C. Meyer, *LÄ* vi (1986), s.v. Wein, 1176. See also E. Hornung, 'Pharao ludens', *Eranos*, 51 (1982), 479–516, esp. 481–6.

[179] E. Hornung, *Der ägyptische Mythos von der Himmelskuh* (Fribourg and Göttingen, 1982) (OBO 46), 39–40, ll. 61–100.

[180] e.g. Junker/Winter (n. 171 above), 221, l. 7 ('Lady of drunkenness'). See the discussion by Daumas (n. 178 above), 1027.

variant of the legend might have attributed to him the presentation of the *mnw* wine-jug. Remote traces could be seen in the word *Βησσίον* which was also used to denote a vessel, 'broader below and narrower above'.[181]

CULT

Like other minor gods, Bes does not seem to have received an official cult during the dynastic period, but he was probably worshipped in domestic shrines. At Deir el-Medina, the workmen's houses contained several items which were perhaps part of the equipment for the cult of family deities. Besides offering-tables and lampstands, Bruyère found various types of vessels, some dedicated to Taweret, some decorated with Bes or Hathoric images, which may have been filled with water for ritual ablutions.[182] The chapel was perhaps located in the enclosed beds ('salles du lit clos'), which were decorated with paintings related to childbirth; in one of these rooms, Bruyère found two Bes-masks, made of clay, which may have been placed at the door of the shrine.[183]

Private ceremonies could have included mummers acting the part of the god. The attributes of Bes, such as his outsize shaggy head, tail, and kilt, could be worn as garments by normal-sized or short-statured people. In particular, the town physicians or midwives could have used this costume to accomplish medico-magical rituals; the scene depicted on a jar from Deir el-Medina could represent the performance of such snake-charmers or magicians (fig. 6.5b).[184] Later documents, such as the Ashmolean and Cairo tambourines (E *d* 85, 86; fig. 9.28a, b), where girls and female dwarfs seem to interchange with Bes-gods in contexts related to childbirth, may support this hypothesis. However, apart from the Bes-mask found at el-Lahun, archaeological evidence on masking is very scarce, and these hypotheses remain provisional.[185]

In his quality as a protector of women and childbirth, Bes was also involved in major cults, mainly that of Hathor. Amulets and vessels in the form of the dwarf god have been found in relatively large numbers in Hathor shrines, such as Deir el-Bahari and Serabit el-Khadim; these objects were perhaps offered by people who wished to have children, as Pinch suggests.[186] Bes-gods are also closely associated with Hathor in Graeco-Roman birth-houses.

[181] LSJ, s.v. *Βησσίον*, 314; see e.g. Hesychius s.v. *Βησίον: ποτήριον*. Varro, *Ling.* 5, 119, reports that *nanus* denotes a grotesque vessel which could also be related to Bes.

[182] Bruyère, *Deir el Médineh*, 55, 60, 101–4, esp. figs. 35–7 (water-jars).

[183] Ibid. 58 and 276, fig. 148; Raven, *Pataekos*, 14. In the workmen's village at el-ʿAmarna, the front rooms of the houses may also have served as private shrines; Kemp (n. 121 above), 49.

[184] Jar (h. 85 cm.), from tomb 1348, present location unknown; B. Bruyère, *Rapport sur les fouilles de Deir el Médineh (1933–1934)* (Cairo, 1937) (FIFAO 14), 113–15, figs. 48–9; Romano, 'Bes', ii, no. 194B. See the discussions of Bruyère, ibid. 114–16; Bosse-Griffiths (n. 34 above), 102–4; Raven, *Pataekos*, 12.

[185] See n. 116 above. M. A. Murray, 'Ritual Masking', in *Mélanges Maspero*, i (Cairo, 1935) (MIFAO 66), 251–5; H. Wild, 'Les Danses sacrées de l'Égypte ancienne', in *Les Danses sacrées* (Paris, 1963) (SourcesOr 6), Note additionelle, 100–1; C. Seeber, *LÄ* iii (1980), s.v. Maske, 1196–9.

[186] See e.g. the calcite Bes-vessel from Serabit el-Khadim (h. 22.3 cm.), Oxford, Ashmolean Museum,

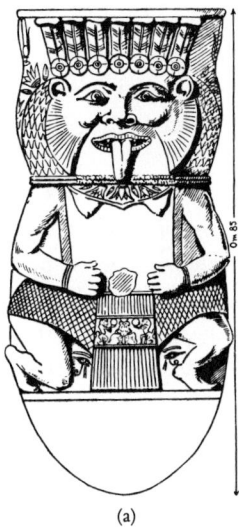
(a)

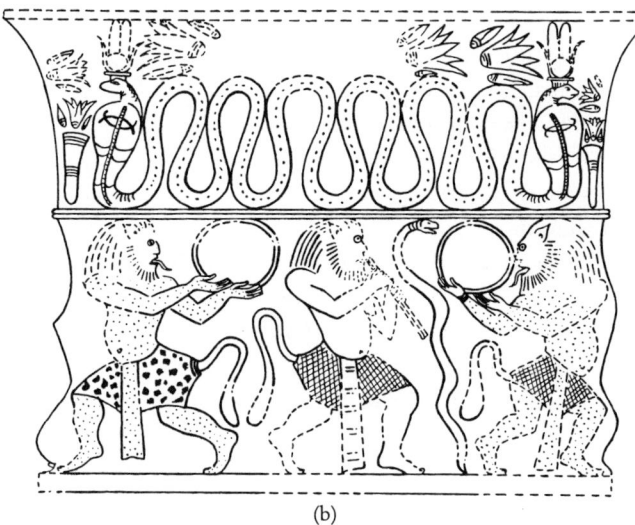
(b)

FIGURE 6.5. Jar (h. 85 cm.), from Deir el-Medina, tomb 1348. Present location unknown

In the Late Period, his protection seems to have received a more official acceptance: a free-standing stela depicts three large Bes-figures and one smaller, holding snakes and brandishing swords (pl. 10.1);[187] the gods may have bestowed protection upon a whole community.

In the Graeco-Roman Period, the cult of Bes as a fertility god seems to have been celebrated at Saqqara in the Bes sanctuary described above.[188] Other cult places of Bes probably existed in that period. Photius mentions a sanctuary of Bes in the Hermopolitan nome, at Antinoupolis, which was called Besantinoupolis in the third century AD;[189] a festival of Bes is also mentioned in the Heidelberg Papyrus.[190] These festivals may have involved phallic rites. A group of terracottas, some from the Saqqara sanctuary, depict two priests and two figures of Bes, or mummers in his guise, carrying a huge phallus (pl. 11.3);[191] on top of the phallus sits a dwarfish or childish figure of

1911.46; E. T. Leeds, 'Alabaster Vases of the New Kingdom from Sinai', *JEA* 8 (1922), 1–4, pl. II; Pinch (n. 151 above), esp. 515–17 and 535–8.

[187] Limestone stela (h. 30 cm.; w. 90 cm.), London, BM 1178; Tran Tam Tinh, *Bes*, 101, no. 32.

[188] See n. 149 above.

[189] Photius, *Bibl.*, 279.535b.41–2. Cf. Bonnet, *Reallexikon*, s.v. Bes, 108; Altenmüller, *Apotropaia*, i. 155–6.

[190] H. C. Youtie, *The Heidelberg Festival Papyrus* (Princeton, 1951), 189, and 201, no. 8. This is

perhaps a private practice which became an official festival as with the phallic procession performed in honour of Bastet; see J. Pirenne, *La Religion et la morale dans l'Égypte antique* (Neuchâtel, 1965), 143 n. 14.

[191] Terracotta (total h. 40.3 cm.; h. of Bes-figures 19.7 cm.), from Saqqara; Leiden, Rijksmuseum, F 1975/11.2; G. T. Martin, 'Excavations in the Sacred Animal Necropolis at North Saqqara', *JEA* 59 (1973), 11, pl. IX; id., *The Sacred Animal Necropolis at North Saqqâra* (London, 1981), 29, no 305, pl. 23;

Harpocrates with a small creature, a woman or a monkey, seated on his shoulder, beating a tambourine. Stricker has identified these processions with the 'Pamylia' or phallophoria described by Herodotus and Plutarch, where a phallic statue was carried in honour of Osiris to ensure fertility;[192] the dwarfish figure could represent Ptah, perhaps as a manifestation of Sokar-Osiris.

Bes was also involved in the cult of Apis, the sacred bull worshipped in Memphis as a manifestation of Ptah.[193] Several terracottas, such as a statuette in London (pl. 11.2),[194] depict Bes crowned with a naos containing a figure of Apis; elsewhere Apis appears behind the head-dress of the god.[195] A limestone Bes statue, now in the Louvre, was found in the court of the Memphite Serapeum.[196] The career of the fourth-century dwarf Djeho, who performed dances at the burial of the Apis and Mnevis bulls (E 84; pl. 26.2) also supports this association of dwarfs with Apis.

To sum up, from the Middle Kingdom onwards, Bes-gods appear as powerful demons, akin to underworld spirits and other liminal creatures, yet benevolent and close to men. They are associated with the south-eastern deserts symbolizing the physical margins of the world, where Re daily triumphs over evil. They also guard the margins of human lives: they bestow protection upon humans when life competes with chaotic forces, as during childbirth, sleep, and war; they attend transition periods, such as the seclusion stage following delivery, and the journey of the deceased through the underworld. Their multifarious powers are stressed by their ambivalent appearance, comic and frightening, young and old, human and animal at the same time. The popularity of Bes survived into Christian times. He faded gradually under Constantine, but folk legends allege that Bes-demons, *lakhia* or *djinn*, still haunt the ruins of Karnak and Abydos.[197]

Derchain, in Martin, ibid. 166 ff. See also an incomplete group in B. H. Stricker, 'Bes de danser', *OMRO* 37 (1956), 40, fig. 10. Add eight similar terracotta groups from Memphis (h. 12.5/17.4 cm.; l. 10.4/12.5 cm.), London, UC,-, unpub.

[192] Hdt. 2. 48; Plut., *De Is. et Os.*, 36, *Mor.*, 365b; Stricker, ibid. 39–43. See also Derchain, ibid. 169–70 (for the possible identification of Sem-priests).

[193] Apis is described as a manifestation of Re in a demotic text of the 3rd cent. BC; E. Kiessling, 'Die Götter von Memphis in griechisch-römischer Zeit', *AfP* 15 (1953), 26 n. 6. See also E. Otto, *Beiträge zur Geschichte der Stierkulte in Aegypten* (Leipzig, 1938), 27.

[194] Terracotta (h. 49 cm.), London, BM 61296; Lanzone, *Dizionario*, pl. 79, 1; Krall, 'Bes', 87, no. 35, fig. 73; Ballod, *Prolegomena*, 67, fig. 72.

[195] For material, see Tran Tam Tinh, *Bes*, 105, nos. 94a–b, 95, pl. 86. Add the terracotta from Mit

Rahina, Michailidis coll.; Michailidis, 'Bès', 61, pl. IX.

[196] Paris, Louvre, N 437; J.-P. Lauer and C. Picard, *Les Statues ptolémaiques du Sarapieion de Memphis* (Paris, 1955), 9, fig. 5; Tran Tam Tinh, *Bes*, 99, no. 16b, pl. 76.

[197] Cf. G. Maspero, *Ruines et paysages d'Égypte* (Paris, n.d. (1910?)), 151–2: 'La porte monumentale du Sud . . . sert de demeure à un *lakhia*, c'est-à-dire à un nain de grosse tête et de jambes cagneuses, agrémenté d'une barbe formidable. Il se promène le soir à la brume et il prend l'air aux alentours: si un étranger passant par là rit de sa figure grotesque, il lui saute à la gorge et il l'étrangle.' See also G. Legrain, *Louqsor sous les pharaons* (Brussels, 1914), 303. On Bes-demons at Abydos, see Perdrizet/Lefebvre (n. 158 above), p. xxiii; F. Lexa, *La Magie dans l'Égypte antique*, ii (Paris, 1925), 217–18; Meeks (n. 101 above), 52–5, adds on 54 n. 215 that Bes-amulets

The reasons for the transformation of the emaciated Middle Kingdom figure of Bes into a dwarf is not fully explained. An initial hypothesis is that the Old Kingdom notion of pygmies may have influenced this change. It is not impossible that the memory of the Old Kingdom *dng* was revived by New Kingdom contacts with the South, like the expedition of Hatshepsut to Punt. Bes might have acquired the form of an exotic dwarf to stress his relationship to the sun-god.[198] This metamorphosis may also have incorporated a belief in dwarfs as familiar protective beings which goes as far back as the Predynastic Period.

are still found in Coptic burials.

[198] For the link between Bes and dancing dwarfs or pygmies, see e.g. Ballod, *Prolegomena*, 38; Brunner-Traut, *Tanz*, 35–6; Michailidis, 'Bès', 70.

7

Ptah-Pataikoi

By the late New Kingdom, another dwarf god is commonly found in Egypt. He is clearly differentiated from Bes. He represents an achondroplastic dwarf, usually naked, with short bent limbs, a long trunk and a big flat-topped head; he is fully human, without foreign attributes, and does not carry weapons. Representations of this god are numerous, principally in amulets, and it is still unclear whether they depict various forms of one and the same god, or a group of dwarf gods, as with Bes.

NAME

The term 'Pataikos' was first used by Herodotus to describe representations of the god Ptah, equated with Hephaistos, in the form of a dwarf; these figures were seen by the Persian King Cambyses in Memphis:

Thus too he (Cambyses) entered the temple of Hephaistos and made much mockery of the image there. This image of Hephaistos is most like to the Phoenicians' *Pataikoi*, which the Phoenicians carry on the prows of their triremes. I will describe it for him who has not seen these figures: *it is in the likeness of a pygmy*. Also he entered the temple of the Kabeiroi, into which none may enter save the priest; the images here he even burnt, with bitter mockery. These also are like the images of Hephaistos, and are said to be his sons.[1]

Herodotus thus compares Egyptian figures with Phoenician gods, called Pataikoi, and with the Greek Kabeiroi. Very little is known about the Phoenician dwarfs, apart from the fact that they were placed on the prows of boats, probably as magical protection.[2] No representation of them has yet been identified, but they may well have been influenced by the Egyptian dwarf gods.

The word Pataikos is not known in Egyptian and its etymology is unclear. In the classical world, authors associated it with two verbs, πατεῖσται, to feed on, and ἀπατᾶν, to cheat, which may reflect the Greek image of dwarfs as disquieting and greedy

---

[1] Hdt. 3. 37 (trans. A. D. Godley, Loeb, 1921; italics added). On the identification of Hephaistos with Ptah, see J. Quaegebeur *et al.*, 'The Memphite Triad in Greek Papyri', *GM* 88 (1985), 26.

[2] See the *Suda* and Hesychius, s.v. πάταιχοι:

'Pataeci: gods whom the Phoenicians always put on the sterns (sic) of their ships.' See in general J. Ilberg, Roscher, iii (2) 1902–9, s.v. Pataikoi, 1675–7; H. Herter, *RE* xviii. 2 (1949), s.v. Pataikoi, 2550–5; G. Becatti, *EAA* v (1963), s.v. Pataikoi, 986.

beings.[3] Scholars now tend to derive the term Pataikos from the Egyptian Ptah, and some regard it as a possible diminutive form.[4] Ptah itself is a name of uncertain meaning, which may be related to the semitic verbs 'to engrave' or 'to open'.[5]

Egyptologists usually follow Herodotus' comparison and name the Egyptian dwarf figurines 'Pataikoi' or 'Ptah-Pataikoi', in order to distinguish them from the normal-statured form of Ptah. For convenience I follow this convention too, although neither name is really appropriate: Pataikos is a foreign term which did not describe Egyptian but Phoenician deities, while the compound form is rather cumbersome and over-simplifies the identity of the god.

While Bes was involved in several myths, such as that of the 'Distant goddess', no Egyptian text mentions Ptah-Pataikoi, who do not seem to have been part of a specific myth or of official texts and iconography. Anonymous dwarf gods invoked in magical spells may be identified with them, but none is explicitly assimilated to Ptah. The essential evidence for the nature and function of these deities is iconographic.

ICONOGRAPHIC TYPOLOGY

From the New Kingdom on, Ptah-Pataikoi are found principally as small amuletic figurines (1–9 cm. in height). Usually they have a loop on the neck or the back (e.g. pls. 13.1, 15.1b), or they stand on a rectangular base, often with a back pillar pierced at neck level (e.g. pl. 13.3); a few bronze examples seem to have been fitted onto a stand, or a staff, with a tenon.[6] Most figurines are made of faience, some of glazed steatite, ivory, or semi-precious stones, especially cornelian; a few are made of wood or bronze.[7] As a rule these dwarfs are male, but one female figurine crowned with a lotus flower may also depict a Ptah-Pataikos.[8]

---

[3] See below 215.

[4] See the discussions by Ilberg (n. 2 above); S. Morenz, 'Ptah-Hephaistos, der Zwerg', in *Festschrift für F. Zucker* (Berlin, 1954), 284; Bonnet, *Reallexikon*, 584; J. G. Griffiths, *LÄ* iv (1982), s.v. Patäke, 914–15 n. 5. J. Baines comments that this derivation is generally plausible, although Egyptian has no actual diminutive form (personal communication).

[5] See Sandman Holmberg, *Ptah*, 7–11; H. te Velde, *LÄ* iv (1982), s.v. Ptah, 1178 n. 11 ('the fashioner').

[6] *Bronze*: Cairo Museum, CG 38773 (h. 7.8 cm.); Daressy, *Statues*, 196, unpub. St Petersburg, Hermitage, 281 (h. 4.3 cm.), unpub. London, BM 11046 (h. 9.1 cm.), unpub. Paris, Louvre, E 19410 (h. 3.2 cm.), unpub. Cf. Roeder, *Bronzefiguren*, 442–5, § 607–9, on bronze figurines (esp. of Bes) inserted in staves.

[7] For *glazed steatite*, see e.g. Berlin, SM (West) 11383 (h. 6.8 cm.); R. Hückel, 'Über Wesen und Eigenart der Pataiken', *ZÄS* 70 (1934), 104, fig. 2. *Cornelian*: Paris, Louvre, AF 2542 (h. 1.9 cm.) and N 3701 (h. 3.7 cm.), unpub. *Jasper*: Paris, Louvre, E 34298 (h. 2.55 cm.), unpub. *Porphyry*: W. M. F. Petrie, *Amulets* (London, 1914), 38, no. 176a–b, pl. XXXI. *Ivory*: Cairo Museum, CG 38784 (h. 1.8 cm.); Daressy, *Statues*, 198. *Wood*: Cairo Museum, CG 38815 (h. 4.6 cm.); Daressy, *Statues*, 203, pl. XLII; Cairo Museum, CG 38821 (h. 6.5 cm.); Daressy, *Statues*, 205, pl. XLII. For a discussion of the various materials used for amulets, see C. Müller-Winkler, *Die ägyptischen Objekt-Amulette* (Fribourg and Göttingen, 1987) (OBO, SA 5), 485–7.

[8] *Ivory* (h. 6.7 cm.), Baltimore, WAG 71.507; Hornemann, *Types*, 852; R. H. Randall, *Masterpieces of Ivory from the Walters Art Gallery* (New York, 1985), 43, no. 17, pl. 7.

Petrie first divided the figurines into three groups showing the dwarf alone, with gods, and as a double or compound figure.[9]

In the simplest form, the dwarf is standing, naked, with a shaven or bald head, the arms hanging down to the sides, with closed fists (pl. 12.1). He has a youthful appearance, which may be stressed by a sidelock symbolic of youth on the right side of the head (pl. 12.2); he may also appear as an adult, with a wrinkled forehead, and even a beard (pl. 12.4). Most often he wears a broad collar (pls. 13.3a, 14.1a), sometimes also a wig and bracelets, and, very rarely, a pleated kilt.[10] Occasionally he is winged (pl. 13.3).[11] His set of attributes is very specific. The most common is the scarab-beetle, placed on his flat-topped head (pls. 12.2–4, 13.3), or as part of a head-dress in place of the sun-disk (pl. 14.1). These head-dresses vary: he may wear the *Atef* crown (with uraei) (pls. 8.2a, 14.1a), the cap with two plumes (with sun-disk),[12] more rarely the moon-disk.[13] Often he bites the tails of two snakes which frame his mouth like a moustache (pl. 12.3), and holds before his chest one or two knives or feathers (pls. 12.3, 13.3a, 14.1a), or more rarely snakes (pl. 13.2) and scorpions.[14] He may stand on two stylized crocodiles, with two falcons perched on his shoulders (pl. 13.2); occasionally two lions also stand on his sides.[15] All these attributes (lock of youth; scarab; head-dress; snakes, knives, or feathers; crocodiles) can be freely combined.

In more complex forms, deities appear on the sides and back of the amulets. Often the back is carved with a winged figure, depicting either the lioness-headed Sakhmet (pl. 14.1b), a goddess with sun-disk and cow's horns, holding two tall feathers, probably Hathor or Isis (pl. 14.2), or Maat, crowned with a feather (pl. 14.3). Nefertem may also stand with a lotus flower on top of his head (pl. 16.2b). Isis and Nephthys are usually found on either side of the dwarf, especially when the amulet has the shape of a miniature magical stela (pl. 13.2).

Some figurines are two- (pl. 13.1), three- or four-headed,[16] probably to increase the apotropaic power of the god. Some have two different faces: the dwarf's head is joined back to back with that of an animal, usually a falcon.[17] There is also a hybrid type of

[9] Petrie (n. 7 above), 38, no. 176.

[10] See e.g. faience (h. 3.6 cm.), Cairo Museum, CG 39270; Daressy, *Statues*, 317–18, pl. LIX (wig and bracelets). Faience (h. 5 cm.), Cairo Museum, CG 38813; Daressy, *Statues*, 203, pl. XLII (wig). Faience (h. 4.6 cm.), Cairo Museum, CG 38815; Daressy, *Statues*, 203, pl. XLII (wig). Faience (h. 2.8 cm.), Berlin, SM (West) 4065, unpub. (pleated dress).

[11] See also the statuette in R. Giveon, 'Fouilles et travaux de l'Université de Tel-Aviv, Découvertes égyptiennes récentes', *BSFE* 81 (1978), 12, fig. 12.

[12] See e.g. faience (h. 8 cm.), Cairo Museum, CG 38807; Daressy, *Statues*, 202, pl. XLII. Faience (h. 9.3 cm.), Cairo Museum, CG 38808; Daressy, *Statues*, 202, pl. XLII.

[13] See e.g. faience (h. 7.8 cm.), Cairo Museum, CG 38818; Daressy, *Statues*, 204, pl. XLII. Faience (h. 6.1 cm.), Fribourg, Institut Biblique, 1559, unpub.

[14] See e.g. the faience figurine (h. 7 cm.), Paris, Louvre, AF 2172, unpub. (snakes and scorpions).

[15] See e.g. the faience figurine (h. 5.4 cm.), Paris, Louvre, E 8783, unpub.

[16] See e.g. the bronze figurine (h. 4.8 cm.), New York, MMA 23.6.15; Hornemann, *Types*, 31 (two-headed). Faience (h. 2.4 cm.), London, BM 11261, unpub. (three-headed). Faience (h. 6.7 cm.), Cairo Museum, CG 38789; Daressy, *Statues*, 199, pl. XLII (four-headed).

[17] See e.g. the figurines in Cairo Museum, CG 38818–20; Daressy, *Statues*, 204–5, pl. XLII.

figurine depicting a dwarf identified as a Pataikos by his pose and attributes (clenched fists holding knives or feathers, crowned with a scarab-beetle or a head-dress), but with an animal head (falcon, ram, baboon) and a falcon's or crocodile's back (pl. 15.2).[18] Bronze pantheistic statuettes may have similar combinations of human and animal elements, but, apart from a dwarfish body, they do not have the head or the distinctive set of attributes of Ptah-Pataikoi.[19] It is thus unclear whether they should be related to Ptah-Pataikoi or to Bes.

Ptah-Pataikoi rarely occur in non-amuletic forms. They were not as favoured as Bes as a decorative motif. They do not appear in domestic architecture, nor in temple reliefs. Only a group of small flat faience figurines (h. 1.3–1.6 cm.) of the New Kingdom may have been used as inlay ornaments for jewellery or furniture. They depict naked dwarfs seen in profile, the arms hanging down along the body, the feet set apart; their lack of attributes makes an identification with Ptah-Pataikoi uncertain.[20]

The dwarf god may be identified in a few papyri and monuments. In a twenty-first-dynasty mythological papyrus, a dwarf figure with no specific attributes stands in the solar disk (fig. 5.1); he cannot be a Bes-god, who would have his readily stylized feather head-dress, and it is tempting to see him as a Ptah-Pataikos. The dwarf substitutes for the scarab-beetle as a morning form of Re, recalling strongly the function of Pataikoi as manifestations of youthful solar gods. A twenty-second-dynasty stela from Sais may also depict a Pataikos (pl. 3.1). A dwarf stands behind the goddess Neith, who faces the king; he has no attribute (scarab or snakes), but his identity is suggested by the fact that Ptah-Pataikoi, like anonymous dwarf gods in general, may be associated with Neith in magical spells and on Horus-cippi.[21]

A Graeco-Roman papyrus depicts another possible Ptah-Pataikos, shown in profile, grasping a snake in the right hand (pl. 16.1);[22] the dwarf is drawn on a grid of twelve squares and a half to the eye line, which suggests either that there was a canonical grid size for dwarfs, or that the figure was done by copying from a model. Anatomical details reveal Hellenic influence. Lines in the limbs render musculature: in the lower leg, they depict the kneecap (patella) and the bulge of the calf muscles (gastrocnemius),

[18] Bronze (h. 9.2 cm.), Oxford, Ashmolean Museum, 1965.172; *Ashmolean Museum, Exhibition of Antiquities and Coins Purchased from the Collection of the Late Captain E. G. Spencer-Churchill (11–23 October, 1965)*, [Oxford, 1965], 5, no. 10, pl. 1. See also the figurines in Cairo Museum, CG 38821–35; Daressy, *Statues*, 205–7, pl. XLII.

[19] See charts in Roeder, *Bronzefiguren*, 93–4, § 134g ('Zwergengreis mit Tierkopf'), and descriptions ibid. 100–4, § 143–8.

[20] See C. Herrmann, *Formen für ägyptische Fayencen* (Fribourg and Göttingen, 1985) (OBO 60), 40–1, nos. 130–4. See also W. M. F. Petrie, *Tell el Amarna* (London, 1894), pl. XVII, no. 275 'M'; id., *Hyksos and Israelite cities* (London, 1906) (BSA 12), pl. XXXVII, 57.

[21] For the inscription on the stela and the relationship of dwarfs with Neith, see above 50–1. For Neith and Ptah-Pataikoi on the same Horus-cippus, see e.g. the stela in Chicago, OI 16881; K. C. Seele, 'Horus on the Crocodiles', *JNES* 45, right side (6) and left side (1), pl. 1A.

[22] Papyrus, Berlin, SM (East) P 13558; A. Erman, 'Zeichnungen ägyptischer Künstler griechischer Zeit', *Amtliche Berichte aus den Königlichen Kunstsammlungen*, 30/8 (May 1909), 197–203, esp. 200, fig. 120. I am very grateful to H. Whitehouse for this reference.

FIGURE 7.1. Sarcophagus, wood (l. 11.5 cm.; w. 5.5 cm.; h. 3.5 cm.). Lyons, Institut
d'égyptologie V. Loret, I.E. 677

which are very prominent, while on the arm five lines indicate either the musculature
(deltoid, biceps, brachioradialis?) or folds of fat flesh.[23] On a miniature Ptolemaic
sarcophagus, a naked dwarf stands in the same pose with clenched fists; he is captioned
in demotic: *Ptḥ sḏm p3 nm(w)*, 'Ptah who listens, the dwarf' (fig. 7.1).

A precise dating of the amuletic figurines is difficult. Most objects in museum
collections have no provenance, and very few excavation reports provide accurately
dated contexts; types of faience and glaze (blue or green, dark or light) varied in the
course of time, but their attribution to specific periods is not yet clearly established.[24]
Stylistic features and inscriptions do, however, give some indications, and several major
stages in the development of the god's iconography can be discerned. New Kingdom
figurines are usually made of dark blue faience and depict dwarfs crowned with scarabs,
with broad collars, biting snakes, and holding knives or feathers (pl. 12.3).[25] During the
Third Intermediate Period, the head-dresses are more varied (pl. 14.1), and wings may
be added (pl. 13.3b).[26] The motif of the dwarf standing on crocodiles, with a falcon on
each shoulder (pl. 14.1), the whole often in the form of a small stela (pl. 13.2), is
borrowed from representations of 'Horus on crocodiles' (pl. 3.2), and is thus no earlier
than the twenty-fifth dynasty. Hybrid dwarfs, with animal elements (pl. 15.2), seem to

[23] On Hellenic anatomical renderings, see D. C.
Kurtz, *The Berlin Painter* (Oxford, 1983), esp. 18–36,
figs. 6 and 8.

[24] See in general A. Kaczmarczyk and R. E. M.
Hedges, *Ancient Egyptian Faience: An Analytical
Survey of Egyptian Faience from Predynastic to Roman
Times* (Warminster, 1983). On methodological
approaches, see also Müller-Winkler (n. 7 above),
32–4. The problem of dating scarabs is similar; cf.

Hornung/Staehelin, *Skarabäen*, 27.

[25] See also the figurine from a 20th-dyn. context in
D. Randall-MacIver and A. C. Mace, *El Amrah and
Abydos* (London, 1902) (MEES 23), pl. xlv, D 28.

[26] See e.g. the Ptah-Pataikoi from Illahun (22nd
and 23rd dyn.); Petrie (n. 7 above), 38, no. 176g
and p, pls. xxxi and xlvi (Atef crown). See also
the winged Pataikos from an Iron Age context in
Lachisch; Giveon (n. 11 above).

go back to the Late Period only. Crude or strongly stylized depictions are very difficult to date (pl. 8.2).[27]

ORIGIN

Egyptologists commonly relate the emergence of these amuletic figurines to two early beliefs. The first is the notion of dwarfs as protectors against snakes and harmful animals. This concept is clearly documented as far back as the Middle Kingdom, when Bes-gods are shown grasping or biting snakes (pl. 3.3, fig. 6.1a). Earlier traces can be found. An early dynastic ivory cylinder depicts four dwarfs and two full-sized men surrounded by crocodiles, lizards, a scorpion, and a bee (E 10; fig. 9.1); the scene may show the mastering of dangerous animals, or a combination of various apotropaic powers, as later on scarabs.[28] Sandman Holmberg thus claimed that Ptah-Pataikoi originated as amulets against fearsome creatures, and were associated later with various major gods, such as Ptah.[29] The second view assumes, on the contrary, that dwarfs had an early privileged relationship with Ptah, the patron of craftsmen. This hypothesis relies on the occurrence of dwarf jewellers in Old Kingdom tomb reliefs. The motif may have acquired a mythical dimension in later periods, and led to the creation of amuletic dwarf figurines representing the skilful 'sons' of Ptah.[30] Some scholars put forward a third hypothesis, based on Herodotus' testimony, which neglects New Kingdom evidence; they see Ptah-Pataikoi as a Late Period creation influenced by the introduction of foreign gods, such as the Phoenician Pataikoi or the Greek Kabeiroi.[31]

These hypotheses are not entirely satisfactory. They omit significant aspects of the symbolism of dwarfs, such as their solar affinities which are expressed repeatedly in magical texts and in iconography throughout the dynastic period. This concept of rejuvenation may have motivated the creation of dwarf amuletic figurines identified with Ptah, not as a craftsman, but as a creator-god, and likewise with Horus, Khons, Osiris, and other youthful and regenerative gods. Their role as guardians of small children may also be influenced by forerunners other than Bes, such as Middle Kingdom dwarf figurines (E 119–d 196; pls. 30–35.1, figs. 9.22–4). These small statuettes are not related to snakes, craft, or Ptah, but seem to be strongly connected with fertility, possibly as magical charms. They often strikingly resemble New Kingdom Pataikoi:

[27] See e.g. the Late Period figurine from Badari; G. Brunton, *Qau and Badari*, iii (London, 1930) (BSA 50), 22, pl. XLIII, tomb 4963, no. 71.

[28] On ambivalent aspects of snakes and lizards, see Raven, *Pataekos*, 12–14. On this object, see also below 105.

[29] Sandman Holmberg, *Ptah*, 184–5.

[30] See e.g. W. Spiegelberg, 'Ägyptologische Mitteilungen III. Zu dem Typus und der Bedeutung der als Patäken bezeichneten ägyptischen Figuren', *SBAW* 2 (1925), 8–11; P. Montet, 'Ptah Patèque et les orfèvres', *Revue Archéologique*, 40 (1952), 8–12; id., 'Ptah Patèque et les orfèvres nains', *BSFE* 11 (1952), 73–4; Bonnet, *Reallexikon*, s.v. Patäke, 584.

[31] Morenz (n. 4 above), esp. 281–5. See also F. W. v. Bissing, 'Bes-Kabeiros', *AfO* 13, 1/2 (1939), 63–4 (influence of Phoenician Pataikoi).

they have similar squat proportions, with flat-topped heads to support a scarab-beetle. Middle Kingdom stands in the shape of dwarfs have similar associations with fertility and protective concepts (E *d* 136–*d* 140, *d* 142–3; pl. 34.1, fig. 9.24). A limestone stand in Leiden, in particular, may show a possible early form of Pataikoi.[32] It depicts a naked dwarf, perhaps female, with short bent legs and a prominent belly, in the same pose as New Kingdom and Late Period Ptah-Pataikoi: in each fist she grasps a snake, symbol of negative forces. As Raven demonstrates, this combination of attributes—dwarfish proportions, pregnant appearance, and snake-charming—suggests that the figure served as an apotropaic figure for the protection of family life.

Several iconographic elements also indicate ancient links of Ptah-Pataikoi with the liminal world of demons. The abnormal appearance of the dwarfs and their attributes (knives, feathers, snakes) evoke the disquieting genii guarding the doors of the underworld. Some mingle a dwarfish body with an animal head, like the jackal-headed dwarf in the twenty-first-dynasty papyrus of Dirpu.[33] Often they bite snakes and hold small or large knives.[34] Underworld genii also hold feathers, usually of ostrich, probably related to Maat, which may symbolize their role as guardians of the ordered world.[35] One of these doorkeepers may be another forerunner of Ptah-Pataikoi, or a parallel form. He appears in several New Kingdom and Third Intermediate Period funerary vignettes. He has similar characteristics: he is usually naked, bald, with a strikingly flat-topped head, and he mingles a youthful appearance with slightly dwarfish proportions. He has the same attributes as Ptah-Pataikoi: he often grasps two knives and bites the tails of two snakes.[36] On the papyrus of Queen Henuttaui, the demon has turned-in feet, which enhances his physical oddity (fig. 7.2).[37] This demon may be associated with underworld creatures resembling Bes and Taweret. In the papyrus of Tahemetnet-Mut, he is seated behind a hippopotamus-headed god and a figure with Bes-like masks (pl. 8.1); in the papyrus of Khonsumes B, he sits behind a donkey seen from the back, and a full-face leonine figure, probably Bes.[38] The frequent association of this dwarfish figure with

[32] Leiden, Rijksmuseum, F 1984/11.3 (h. 29 cm.); Raven, *Pataekos*, fig. 1 and pl. 1, 7 ff. See also below 142.

[33] Piankoff/Rambova, *MythPap*, no. 6, scene 4.

[34] See e.g. the seated doorkeepers with big knives in G. Posener, *et al.*, *Dictionnaire de la civilisation égyptienne* (Paris, 1959), 74 (tomb of Sennedjem, Deir el-Medina).

[35] See e.g. the row of underworld demons in A. Piankoff, *The Litany of Re* (New York, 1964) (ERT 4), 69, no. 5 (holding an ostrich feather) and nos. 6–7 (biting snakes); on Maat, see Hornung, *Conceptions*, 213–16.

[36] See e.g. Piankoff/Rambova, *MythPap*, no. 11, scene 4 (papyrus of Khonsu-Renep). For similar, but seated figures: see Posener, *et al.* (n. 34 above), 74

(tomb of Sennedjem); G. Thausing and H. Goedicke, *Nofretari* (Graz, 1971), figs. 82 and 150 (tomb of Nofretari); Piankoff/Rambova, *MythPap*, no. 6, scene 7 (papyrus of Dirpu). Demon in a kilt: ibid. no. 3, scene 4 (papyrus of Nesipautiutaui); with a collar, in a kilt, holding a big knife and a whip(?): W. R. Dawson, 'Pygmies, Dwarfs and Hunchbacks in Ancient Egypt', *Ann. Med. Hist.* 9 (1927), 317, fig. 6 (papyrus of Ani).

[37] Papyrus of Queen Henuttaui, Cairo Museum, JdE 95887, special register number S.R. IV 992; A. Mariette, *Les Papyrus égyptiens du musée de Boulaq*, iii (Paris, 1876), pl. 20 (papyrus no. 23); Dawson (n. 36 above), 317, fig. 7; R. Watermann, *Bilder aus dem Lande Ptah und Imhotep* (Cologne, 1958), 46, fig. 22b.

[38] Piankoff/Rambova, *MythPap*, no. 17, scene 5.

FIGURE 7.2. Papyrus of Queen Henuttaui. Cairo Museum, JdE 95887, special register number S.R. IV 992

Bes-like demons, often in symmetrical compositions, suggests that he could be an early—or parallel—form of Ptah-Pataikos. The amuletic dwarfs do not seem to have influenced representations of this funerary demon, who appears in the same form in tombs of the Roman Period.[39] This demonic connection may explain the multiplicity of the Ptah-Pataikoi, who do not seem to have been regarded as independent deities, but as manifestations of greater gods, as I now show.

MYTHICAL STATUS

A closer study of the attributes of this dwarf god suggests that he was a manifestation of several major gods, in particular Ptah and Sokar, but also Horus, Amun-Re, Thoth, Min, and Osiris, in their aspects as youthful, creator, and regenerator gods.

Ptah and Sokar

Ptah, creator-god and patron of craftsmen, is usually depicted mummiform, with a tight cap, a straight beard, and a broad collar with or without a counterpoise (*menat*); he holds the *ankh* sign and *djed* and *was* staves.[40] He is also shown in the guise of a dwarf, as revealed by the attributes of Ptah-Pataikoi. The small gods generally have shaven skulls analogous with the close-fitting cap of Ptah, and they wear similar broad collars (pls. 13.3a, 14.1a); they are often associated with his companion, Sakhmet, and the youthful

[39] See e.g. the depiction of a two-headed naked demon, standing and holding snakes in the tomb of Petubastis at Qaret el-Muzawwaqa (1st cent. AD); A. Fakhry *et al.*, *Denkmäler der Oase Dachla* (Mainz am Rhein, 1982), 74, pls. 21c, 31a ('Doppelköpfiger

(Horus?-) Knabe mit einer Schlange in jeder Hand').
[40] On the iconography of Ptah, see Sandman Holmberg, *Ptah*, esp. 182–7; te Velde (n. 5 above), 1178–9. For material, see e.g. Daressy, *Statues*, 116–28, pls. XXIV–XXVII.

Memphite god Nefertem, who appear on the sides or back of the statuettes (pls. 14.1, 16.2). Inscriptions confirm this identification: some dwarf gods are called 'Ptah',[41] *Ptḥ-sḏm-p3 nm*, 'Ptah who listens, the dwarf' (fig. 7.1), or 'Ptah, the killer of snakes' (pl. 3.2). In the last case, the god is carved on a Horus-cippus; he stands like Ptah on a dais with steps. Similar dwarf figures appear on Horus-cippi, sometimes crowned with scarabs or *udjat*-eyes and holding snakes.[42]

The meaning of the relationship between Ptah and dwarfs is not easy to disentangle. Scholars have traditionally assumed that dwarfs were associated with Ptah because of their special artistic skill. There is, however, little supporting evidence for this idea, which could be influenced by our knowledge of dwarf or malformed smiths in Hellenic traditions.[43] Dwarfs appear as craftsmen in the Old Kingdom only, and only as jewellers. There is no evidence for dwarf metallurgists, or craftsmen in general, in later periods—but one could object that this silence is related to a more general change in decorum. Further, Ptah's association with handicrafts is uncertain before the Middle Kingdom; he first emerges as patron of sculpture, pottery and stoneworking, and not of metallurgy or of goldsmiths' work.[44] He is described as Lord of Mineral Wealth from the Ramesside period only, and this capacity may be due to his fusion with Tatenen.[45]

Another possible origin of the dwarfish form of Ptah could be his role as creator-god.[46] The 'half-formed' appearance of dwarfs may have embodied the continuing process of creation. Three aspects of Pataikoi support this hypothesis. In the first place, the small gods are usually crowned with a scarab, which represents the concept of 'coming into being', of emerging from the uncreated world. This symbolism is stressed by a second element, the foetus-like appearance of several figurines (pl. 12.1).[47] Thirdly, Pataikoi are often associated with the goddess Maat who personifies 'the pristine state of

---

[41] 'Ptah': Faience (h. 2.9 cm.), Berlin, SM (East) 15436, unpub. 'Ptah, giver of life': Faience (h. 2.9 cm.), Cairo Museum, CG 39237; Daressy, *Statues*, 310, pl. LIX. Faience, Fribourg, Institut Biblique, 1559 (n. 13 above). 'Ptah and Sakhmet, mistress of the [two] lands': Faience from Lachisch; Giveon (n. 11 above), fig. 12c.

[42] See e.g. the stelae in Cairo Museum: G. Daressy, *Textes et dessins magiques* (Cairo, 1903) (CGC), CG 9402, 5, obverse, second row, no. 1, pl. III (crowned with a scarab and holding snakes); ibid. CG 9409, 21, reverse, no. 8 (holding snakes); ibid. CG 9410, 24, left side, no. 3; ibid. CG 9430, 38, reverse, second row, no. 2, pl. XI (crowned with an *udjat*-eye). See also the stela in Chicago, OI 16881; Seele (n. 21 above), 45, right side (6) (holding snakes), and base (3), pl. 1A.

[43] A few scholars have tried to make these Mediterranean traditions totally coherent, as for example K. Aterman, 'Why did Hephaestus limp?',

Amer. J. Dis. Child 109 (1965), 381–92, who seeks to show that Hephaistos was depicted as a dwarf primarily because he derived from the Egyptian Ptah-Pataikoi.

[44] See Sandman Holmberg, *Ptah*, 50–6; R. Drenkhahn, *Die Handwerker und ihre Tätigkeiten im alten Ägypten* (Wiesbaden, 1976), 46 n. 24; E. J. Brovarski, *LÄ* v (1984), s.v. Sokar, 1059 (VI).

[45] H. A. Schlögl, *Der Gott Tatenen* (Fribourg and Göttingen, 1980) (OBO 29), 37–8; id., *LÄ* v (1984) s.v. Tatenen, 238–40.

[46] On this aspect of Ptah, see Sandman Holmberg, *Ptah*, 31 ff.; S. Sauneron and J. Yoyotte, 'La Naissance du monde selon l'Égypte ancienne', in *La Naissance du monde* (Paris, 1959) (SourcesOr 1), 62–7.

[47] On this aspect of Pataikoi, see P. A. Vassal, 'La Physio Pathologie dans le panthéon égyptien: les dieux Bès et Phtah, le nain et l'embryon', *Bull. Mém. Soc. Anthr. Paris*, 7 (1956), 168–81, esp. 171–3.

the world' at the time of creation, and its perfect harmony.[48] The goddess was closely related to Ptah as a creator-god; he is commonly invoked as the 'Lord of Maat',[49] the guardian of the ordered world, and he may be shown standing on the Maat-sign or a Maat-shaped dais.[50] The goddess appears on the back pillar of the Pataikoi, crowned with an ostrich feather, holding two tall feathers (pl. 14.3); the feathers held by the dwarf gods may be of similar shape, but of miniature size (pl. 13.3?).[51] Thus, Ptah-Pataikoi may embody both the concept of creation, symbolized by the beetle, and of protection of the ordered world, symbolized by Maat.

Pataikoi also merged with Sokar, a deity closely associated with Ptah. Patron of craftsmen and god of the dead, Sokar is traditionally depicted in human form with a falcon's head.[52] Several dwarf figurines have falcon-heads like him, and a naked Pataikos published by Lanzone is labelled 'Ptah-Sokar'.[53]

Solar, Lunar, and Juvenile Gods

The small gods were also equated with Horus, principally in his form as a young sun-god, emerging from a lotus flower. The similar physical proportions of dwarfs and children probably influenced this assimilation. In several amuletic figurines this resemblance is stressed by the addition of a sidelock of youth (pl. 12.2). Yet the dwarfs are never shown putting a finger to their mouth, a gesture which was perhaps reserved for 'real' children.

Two attributes in particular stress the identification of Ptah-Pataikoi with the morning sun-god: the scarab placed on their heads, which symbolizes the daily regeneration of Re, and the lotus flower, which has a similar symbolism of rejuvenation, and occasionally alternates with the sacred insect.[54] This substitution may be observed on an unusual figurine of a female dwarf crowned with a lotus flower, probably to identify her with the newly born sun-god; she may be a female doublet of the Pataikoi, a pairing found with many Egyptian gods, especially in the Late Period.[55]

In the Late Period, the fusion with Horus is complete: the dwarf substitutes for the child on miniature Horus-cippi; he stands on crocodiles, and is often surrounded by Isis and Nephthys (pl. 13.2).[56] The original scheme is slightly modified: while Horus grasps harmful animals, such as snakes, scorpions, lions, and oryx (cf. pl. 3.2), the dwarf holds

---

[48] Hornung, *Conceptions*, 213, and on Maat in general 213–16, 279.

[49] See e.g. these epithets in a New Kingdom hymn to Ptah in Lichtheim, *Literature*, ii. 109–10.

[50] Te Velde (n. 5 above), 1179. See e.g. the bronze statue of Ptah (h. 28 cm.), Cairo Museum, CG 38446; Daressy, *Statues*, 121, pl. XXVI.

[51] See e.g. the figurines in Cairo Museum, CG 38801 and 38813; Daressy, *Statues*, 201 and 203, pl. XLII.

[52] See Bonnet, *Reallexikon*, s.v. Sokaris, 723–7;

Brovarski (n. 44 above), 1062 (x); Sandman Holmberg, *Ptah*, 132 nos. 2–3, 133 nos. 1–2.

[53] Lanzone, *Dizionario*, pl. 99, 2. On falcon-headed gods, see n. 17 above.

[54] Hornung/Staehelin, *Skarabäen*, 15, and 164 n. 4, with further examples of substitution lotus/scarab.

[55] See n. 8 above. On female doublets, see Hornung, *Conceptions*, 83–5, 218.

[56] On this motif, see C. Bonner, *Studies in Magical Amulets, Chiefly Graeco-Egyptian* (Ann Arbor and London), 1950, 140 ff.

knives or feathers; he may also bite snakes, which Horus never does, possibly because that act was reserved for demonic beings. The falcons, otherwise normally perched on papyrus stalks, now stand on his shoulders.

This syncretistic relationship with Horus is more conspicuous in Hellenistic art. Many depictions, principally in terracotta, show Harpocrates in the guise of an ithyphallic, ageing dwarf with a lock of youth and a headband with buds.[57] These figurines allude to the regeneration symbolism of Pataikoi in a new iconography. Some associations, such as with a frog, symbol of the birth-helper Heqet, clearly refer to their older function as family guardians.[58]

Ptah-Pataikoi were also identified with Amun-Re. Several figurines wear Amun's head-dress, consisting of a flat-topped cap with two plumes and the solar disk.[59] Some statuettes are ram-headed, with the head occasionally turned backwards (pl. 15.1).[60] As Piankoff suggested, this unusual type may represent the day and night aspects of the sun-god: the dwarf's body could refer to its morning form, the ram's head to its evening form.[61] Specific solar symbols also occur on a few figurines, such as the cobra running along the back of a figurine in Cairo,[62] or the lions surrounding a statuette in the Louvre.[63]

Inscriptions on the plinth or the back pillar of the figurines stress these solar associations. Some depict the sun-god in his three forms: as the scarab Khepri of the morning, the shining disk of midday, and the bent old man of the evening (figs. 7.3a–b).[64] Some include symbols of creator-gods, such as the lotus bud, probably evoking the primordial lotus from which the young sun-god emerged.[65] The plinth of an amulet in New York is carved with the royal name *Mn-k3-r^c*, which is probably a cryptogram for the name of Amun (fig. 7.3c).[66] Another amulet in New York juxtaposes a lotus bud with two long-tailed monkeys (fig. 7.3d), which may evoke those greeting the rising sun, as well as

[57] For material see esp. W. Weber, *Die ägyptisch-griechischen Terrakotten* (Berlin, 1914), 71–9, 100–5, pls. 12–14; E. Breccia, *Monuments de l'Égypte gréco-romaine. Terrecotte figurate greche e greco-egizie del Museo di Alessandria*, i–ii (Bergamo, 1926–34); H. Philipp, *Terrakotten aus Ägypten* (Berlin, 1972).

[58] On frogs and dwarfs in Hellenistic Egypt, see e.g. H. Wrede, 'Ägyptische Lichtsbräuche bei Geburten', *JbAC* 11–12 (1968–69), 83–93; P. Derchain, 'Miettes, A propos d'une grenouille', *RdE* 30 (1978), 65–6.

[59] See n. 12 above.

[60] See the ram-headed figurines in Cairo Museum, CG 38829–35; Daressy, *Statues*, 206–7, pl. XLII. See also the figurine with a ram's head turned backward (h. 5.3 cm.), London, UC –; A. Piankoff, 'A Pantheistic Representation of Amon in the Petrie Collection', *AE* (1935), 49–51, fig. 1a–b.

[61] Piankoff, ibid. I am very grateful to E. Hornung for this reference.

[62] Cairo Museum, CG 38813 (n. 10 above). See also CG 39270 (n. 10 above) with an uraeus on the forehead.

[63] See n. 15 above.

[64] See also the inscription in A. Fabretti *et al.*, *Regio Museo di Torino* (Turin, 1882), 46, no. 584: 𓂋𓀭𓏥 .

[65] See Cairo Museum, CG 38803, 38805, 38817, 38819, 38820, 38830; Daressy, *Statues*, 201, 204–5, 207. On the symbolic meaning of the lotus, see H. A. Schlögl, *Der Sonnengott auf der Blüme. Eine ägyptische Kosmogonie des Neuen Reiches* (Basle and Geneva, 1977) (ÄH 5); Hornung/Staehelin, *Skarabäen*, 164–5.

[66] On cryptograms, see Hornung/Staehelin, *Skarabäen*, 174–8, esp. 175 on Menkheperre.

(a)

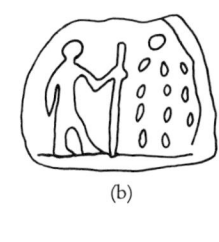

(b)

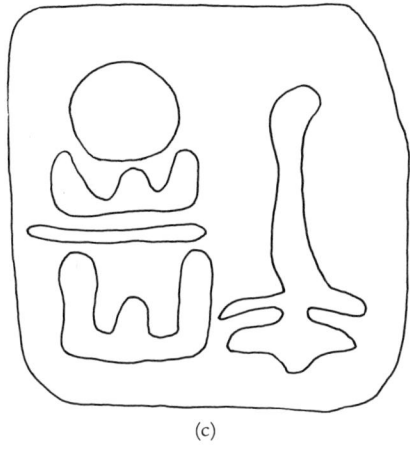

(c)

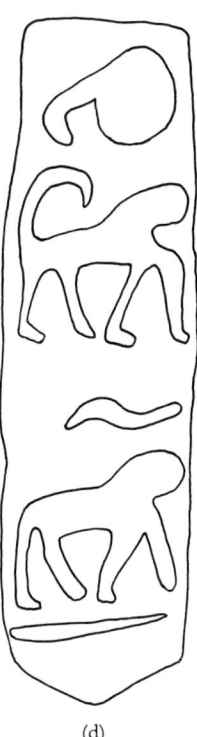

(d)

FIGURE 7.3. Inscriptions
a. Faience (h. 5.6 cm). Berlin, SM (West) -
b. Faience (h. 4.1 cm.). Berlin, SM (West) 17594
c. Faience. New York, MMA 17.194.2448
d. Faience. New York, MMA 44.4.34

refer to the general auspiciousness of monkeys and their fertility associations.[67] These apes may also allude to the moon, as suggested by the inscription on the Late Period statue of Djeho: 'the monkey (*gf*) is the name of the moon'.[68]

This connection with the moon is also shown by figurines of Ptah-Pataikoi merging

[67] Hornung/Staehelin, *Skarabäen*, 106, note that only baboons appear on scarabs, but, ibid. 108, ascribe a general regeneration symbolism to long-tailed monkeys.

[68] See above 48.

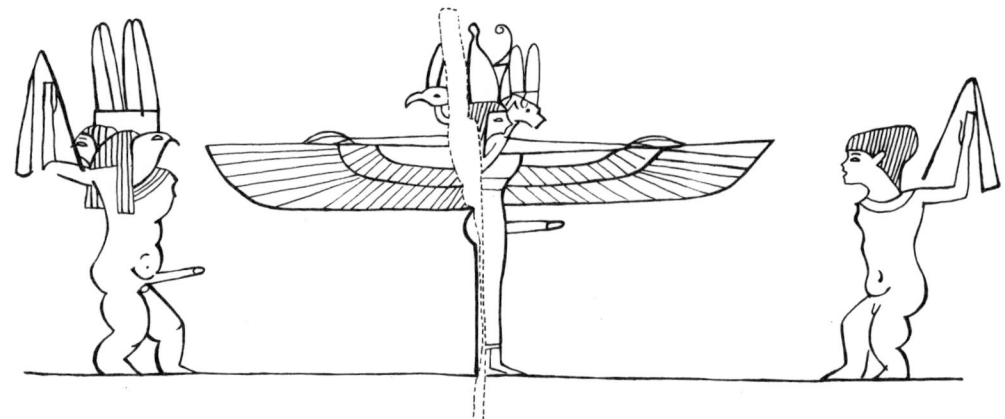

FIGURE 7.4. *Book of the Dead*, spell 164. Papyrus, London, BM 10257/21

with the youthful Khons.[69] Like Ptah, Khons is usually shown mummiform with a broad collar, but he has a sidelock of youth like Horus, with whom he is occasionally equated. As a moon-god, he is characterized by a moon-disk. This head-dress is worn by several Ptah-Pataikoi.[70] Comparable figures of baboon-headed dwarfs probably represent Thoth, another moon-god (pl. 15.2a).[71]

Funerary and Regeneration Gods: Osiris and Min

Several attributes associate Ptah-Pataikoi with Osiris, the lord of the underworld. The dwarfs may wear the *Atef* crown (pls. 8.2a, 14.1a); often they are flanked by his sisters Isis and Nephthys (pl. 13.2). This identification is probably related to the conception of rebirth in the underworld and is strongly expressive of regeneration. Like scarabs, dwarf gods may have been assimilated to Osiris equated with the nightly rejuvenated Re,[72] as they fused with Sokar, a god sometimes identified with the rising sun, 'who dwells in death but possesses potential life'.[73] Their foetus-like appearance may have reinforced their rejuvenating quality.[74]

[69] On Horus/Khons: Bonnet, *Reallexikon*, s.v. Chons, 140–4, esp. 143; H. Brunner, *LÄ* i (1975), s.v. Chons, 960–3, esp. 962 n. 3.

[70] For material, see n. 13 above.

[71] See also the figurines in Cairo Museum, CG 38823–26; Daressy, *Statues*, 205–6, pl. XLII. On Thoth and baboons, see Hornung/Staehelin, *Skarabäen*, 107.

[72] On the relation between scarabs and Osiris, see Hornung/Staehelin, *Skarabäen*, 14, esp. n. 31.

[73] See Brovarski (n. 44 above), 1059 (v) and 1061 (VIII).

[74] Mummies of two still-born babies were deposited in the tomb of Tutankhamun, perhaps because they were associated with similar beliefs. See H. Carter, *The Tomb of Tut-Ankh-Amen*, iii (London etc., 1933), 27–8, and D. E. Derry, ibid. 167–9; F. Filce Leek, *The Human Remains from the Tomb of Tut'ankhamun* (Oxford, 1972), 21–3; F. Drilhon, 'Un Foetus humain dans un obélisque égyptien en bois', in *Archéologie et médecine, VIIèmes rencontres internationales d'archéologie et d'histoire, Antibes, Octobre 1986* (Juan-les-Pins, 1987), 499–521, esp. 517–18, with further examples of foetus burials.

The fusion of Ptah-Pataikoi with Min, a fertility god often associated with Amun, and connected with funerary rituals, expresses these rejuvenating notions.[75] A few drawings, especially in vignettes of the *Book of the Dead*, depict dwarfs with the flagellum and erect phallus of Min, as in spell 164 (fig. 7.4).[76] The text of the spell describes their protective function; they are invoked with the goddesses Sakhmet-Bastet and Mut to protect the deceased whose body may perish: '(Then) he (the deceased) shall be divine among the gods in the god's domain and shall not be kept away forever and ever. His flesh and his bones shall stay sound like (those of) one who is alive; indeed he shall not die.'

The directions for recitation say: 'To be said over (an image of) Mut having three faces . . . with one dwarf standing before her and (one) behind her, facing her, wearing twin plumes and with upraised arm, having two faces—one like the face of a falcon, the other like a human face—(and a scourge and a phallus).'[77] These attributes, however, have not been found in amuletic figurines.

This funerary aspect of the small gods may explain the protective presence of a Ptah-Pataikos on a miniature wooden sarcophagus (fig. 7.1).

FUNCTION

The function of the amulets seems to have concerned both the living and the dead. Like Horus-cippi, Ptah-Pataikoi guarded the living, especially small children, against unpredictable negative forces. They could be worn around the neck, like the faience dwarfs prescribed in magical spells, perhaps only at critical times. Miniature models of Horus-cippi were perhaps set up in houses or in gardens, as Bonner suggested, possibly to chase away dangerous animals like snakes and scorpions.[78] Like the sacred beetle, Ptah-Pataikoi also carried a strong symbolism of regeneration, so crucial in Egyptian funerary beliefs. Some figurines were found in burials, or were associated with death (fig. 7.1). Their rejuvenating qualities are stressed by the signs carved under their plinths to ensure health and fertility, such as the *udjat*-eye of Horus.[79] These eyes occur with other symbolic figures, such as fish (tilapia?), lotus buds, lions, and less easily identified quadrupeds (fig. 7.3a).[80] Inscriptions describe Ptah as 'giver of life'.[81] Some figurines may have been votive offerings; the back pillar of an amulet in Cairo has a standard new year's wish.[82]

---

[75] Bonnet, *Reallexikon*, s.v. Amun, 31–2; R. Gundlach, *LÄ* iv (1982), s.v. Min, 138–9 (at Thebes). See e.g. the invocation of Min in an embalming ritual in J.-C. Goyon, *Rituels funéraires de l'ancienne Égypte* (Paris, 1972), 74.

[76] See also the figure in Lanzone (n. 53 above).

[77] Trans. T. G. Allen, *The Book of the Dead or Going Forth by Day* (Chicago, 1974), 160–1, T 2.

[78] Bonner (n. 56 above), 157.

[79] Cf. e.g. the *udjat*-eyes on scarabs in Hornung/Staehelin, *Skarabäen*, 338, no. 752, pl. 84; 340, no.

755, pl. 84. See also ibid. 170–1, the discussion of their symbolism.

[80] See e.g. Cairo Museum, CG 38817, 38819, 38830 (n. 65 above).

[81] See n. 41 above.

[82] Faience (h. 4.2 cm.), Cairo Museum, CG 38810; Daressy, *Statues*, 202: 'Words spoken by Isis who opens a good year and who causes the two lands to live.' On new year's wishes in general, see Hornung/Staehelin, *Skarabäen*, 181–2.

To sum up, Ptah-Pataikoi manifest conceptions normally found elsewhere in a variety of gods.[83] These focus on solar regeneration and rejuvenation, and hence on protection, traditionally expressed by Ptah and Amun as creator-gods, Horus and Khons as youthful gods, Min as a fertility god, Sokar and Osiris as forms of the nocturnal sun-god.[84]

As on scarabs, their attributes may be freely blended.[85] Thus a Ptah-Pataikos in Fribourg wears a moon-disk, but is called Ptah,[86] while a statuette in Cairo joins back to back the hawk's face of Sokar with a normal human head, both crowned with a moon-disk;[87] the picture published by Lanzone has the attributes of Min (scourge and erect phallus), but is called 'Ptah-Sokar'.[88]

The possibility that Ptah-Pataikoi display a privileged relationship between dwarfs and Ptah as craftsmen cannot be excluded, although the only evidence for such a link dates to the Old Kingdom, and is indirect. In the Late Period, Hellenistic or Phoenician cults may have influenced Egyptian beliefs, but at present this association of dwarfs with crafts, and with metallurgy in particular, remains hypothetical.

---

[83] On this characteristic of minor deities, see Hornung, *Conceptions*, 69.

[84] For Morenz (n. 4 above), 283–4, Ptah-Pataikoi represent Ptah only, but as an 'Allgott'.

[85] Cf. Hornung/Staehelin, *Skarabäen*, 13–15 (on ram-headed scarabs, pushing a lotus flower instead of the sun-disk).

[86] See nn. 13 and 41 above.

[87] Cairo Museum, CG 38818 (n. 13 above).

[88] See n. 53 above.

8

Physical Minorities

We may begin to understand the position of dwarfs in Egyptian society through a survey of Egyptian attitudes towards other categories of physical deviance, at birth and later in life.

There is little evidence on the occurrence of congenital anomalies in ancient Egypt. Monstrous births were not recorded in long lists as in the ancient Mesopotamian omen series *šumma izbu* 'if a foetus . . .'.[1] This may be due to the fact that Egyptians generally did not fear abnormal births as inauspicious. As evil was usually banished from language, they may also have been reluctant to report misfortunes.

Only an oracular amuletic decree dating to the Third Intermediate Period shows some concern about anomalous births. The god promises to protect the woman against three accidents: 'We shall save her from a Horus birth(?), from miscarrying(?) (and) from giving birth to twins.'[2] As Edwards notes, the term *ḏꜣj.t*, here translated by 'miscarriage', usually denotes an 'irregularity', a 'failure',[3] and thus could refer to a physical abnormality. Twins, *ḥtr*, appear as the other cause of anxiety. As Baines has demonstrated, twins were not welcome, but there is no evidence that one or both were suppressed at birth, as happens in many cultures.[4] The probable existence of a taboo on twins may account for the scarcity of extant examples, as well as for the lack of a specific word for twins before the Third Intermediate Period.[5]

There is no archaeological or literary evidence for the practice of exposure. Medical texts report that the newborn child was washed with cold water and that its first

---

[1] On birth omens in Mesopotamia, see e.g. E. Leichty, 'Teratological Omens', in *La Divination en Mésopotamie ancienne et dans les régions voisines* (Paris, 1966), 131–9; J. V. Kinnier Wilson, 'Organic Diseases of Ancient Mesopotamia', in *Diseases in Antiquity*, 203–4; M. Leibovici, 'Génies et démons en Babylonie', in *Génies, anges et démons* (Paris, 1971) (SourcesOr 8), 102.

[2] I. E. S. Edwards, *Oracular Amuletic Decrees of the Late New Kingdom* (London, 1960) (Hier. Pap. in the BM, 4th ser.), i. 66–7, (110)–(115); trans. rev. by J. Baines.

[3] *Wb* v. 518.5–8; Edwards, ibid., n. 68.

[4] J. Baines, 'Egyptian Twins', *Orientalia*, 54 (1985), 461–82; id., *LÄ* vi (1986), s.v. Zwilling, 1436. Cf. E. A. W. Budge, *Osiris*, ii (New York, 1973) (1st edn. 1911), 224–8, on the different attitudes towards twins in Africa (as a lucky or an unnatural event in Sudan and Benin).

[5] Baines (n. 4 above), identifies only three possible cases of human twins from the Old to the New Kingdom.

reactions (position of head, tone of screaming) were carefully observed,[6] but there is no allusion to criteria which would determine its elimination. Diodorus explains that this custom was unknown in Egypt because food was so easy to find that all children could be reared.[7] The earliest mentions of exposure (especially of girls) date to Graeco-Roman times, and they come from the Greek population.[8] Malformed newborn babies may, however, have been eliminated in dynastic Egypt, but there is no evidence of a state regulation; it probably remained a private decision, whose circumstances are conjectural.[9]

Little is known about the status of the malformed in Egyptian society. Sayings in wisdom texts give a positive image of attitudes towards human limits. Works like the New Kingdom *Instruction of Amenemope* teach that care for the old, sick, and malformed is a moral duty, because 'Man is clay and straw, the god is his builder' (24. 8–14).[10] As human life is determined by God, the 'wise man' should respect people afflicted by reversals of fortune. These precepts are repeated in demotic instructions, like that of Papyrus Insinger:

> The blind one whom the god blesses, his way is open.
> The lame one whose heart is on the way of the god, his way is smooth. (11. 24–12. 1).[11]

Workmen were generally not driven too hard.[12] Janssen has shown that the men working in the royal tombs at Thebes in the New Kingdom were easily exempted from work because of an illness, such as eye-disease, or an accident, such as the sting of a scorpion, or just because of weakness.[13] Sick workers were attended by a doctor who seems to have been attached to the necropolis workmen. A few documents reveal that maimed workers could receive some financial support from their master, but this may

---

[6] See H. von Deines *et al.*, *Grundriss der Medizin der alten Ägypter*, iv. 1, *Übersetzung der medizinischen Texte* (Berlin, 1958), 291–5.

[7] Diod. Sic. 1. 80. 3–6.

[8] Cf. the letter of the Greek soldier Hilarion instructing his wife to expose his child if it was a girl; B. P. Grenfell and A. S. Hunt (eds.), *Oxyrhynchus Papyri*, iv (London, 1904), 243–4, no. 744, ll. 8–10. For further evidence, see R. Taubenschlag, *The Law of Greco-Roman Egypt in the Light of the Papyri (332 BC–AD 640)* (New York, 1944), 103–4, 114, 139 n. 26, and L. R. F. Germain, 'Aspects du droit d'exposition en Grèce', *Revue historique de droit français et étranger*, 47 (1969), 187–8, 194–5.

[9] Cf. C. M. Turnbull, 'Survival Factors among Mbuti and other Hunters of the Equatorial Rain Forest', in *African Pygmies*, 116: 'Children born deformed are simply smothered at birth and not referred to again, and there is no mourning.'

[10] See full text below 150. Cf. also 2. 4–6: 'Beware of robbing a wretch, of attacking a cripple.

Don't stretch out your hand to touch an old man' (trans. Lichtheim, *Literature*, ii. 150).

[11] Trans. Lichtheim, *Literature*, iii. 194. For parallels in Jewish wisdom texts, see. M. Lichtheim, *Late Egyptian Wisdom Literature in the International Context* (Fribourg and Göttingen, 1983) (OBO 52), esp. 140 ff. Cf. also the conventional catalogues of virtues in tombs, e.g. in the Old Kingdom tomb of Nefer-Seshem-Re, called Sheshi; Lichtheim, *Literature*, i. 17. For the Late Period, see e.g. the biography of Djedkhonsefankh from Luxor; Lichtheim, *Literature*, iii. 17 (3).

[12] But see the description of the harsh working conditions of convicts in Nubian gold mines, as reported by Diod. Sic. 3. 13. 3.

[13] J. J. Janssen, 'Absence from Work by the Necropolis Workmen of Thebes', *SAK* 8 (1980), 127–52. See also M. Helbing, *Der altägyptische Augenkranke, sein Arzt und seine Götter* (Zurich, 1980), 64–5.

not have been the rule.[14] No legal text mentions provisions for persons disabled at birth, by accident or at war, as in archaic and classical Greece. Social help seems to have relied on private initiative alone, but extant evidence is so sparse that one cannot draw any definite conclusion.[15]

In iconography, ideal proportions reflect social status. Except in the Amarna period, major figures usually conform to the traditional idea of beauty: they have young and well-formed bodies, sometimes slightly corpulent to exhibit their prosperity.[16] Physical irregularities are shown in minor figures only, as displayed in a group of large Old Kingdom tombs at Saqqara; outdoor workers (boatmen, fishermen, bird-netters, herdsmen) are often bald, with deformed lower limbs (genu recurvatum), umbilical hernia or genital hypertrophy.[17] These abnormalities appear as an attribute of their humble role and status.

In literature, physical malformations are similarly associated with low-ranking professional activities. The best-known example is the ludicrous enumeration in the Middle Kingdom *Satire of the Trades*.[18] Each profession is characterized by unpleasant physical attributes: the carpenter has cramps, the potter is crushed, the gardener has a bulging neck, the fingers of the farmer are swollen. These malformations, which are not congenital, appear clearly as the marks of inferior status, and are ridiculed.

Physical irregularities, however, do not seem to have disqualified people from fulfilling public and priestly offices.[19] A lame man, for example, could perform divine service. A man with a thin and atrophied leg (result of poliomyelitis?) is depicted offering to Astarte on a stela dating to the nineteenth dynasty.[20] There are also cases of

[14] A.-P. Leca, *La Médecine égyptienne au temps des pharaons* (Paris, 1971), 401–2; Helbing (n. 13 above), 65–6. On labour conditions, see also H. Sigerist, *A History of Medicine*, i, *Primitive and Archaic Medicine* (New York, 1967), 253–63; H. Buess, 'Sozialmedizinisches aus dem alten Aegypten', *Praxis*, 46 (1957), 1009–11; J. Pirenne, *La Religion et la morale dans l'Égypte antique* (Neuchâtel, 1965), esp. 110–18.

[15] See in general H. Bolkestein, *Wohltätigkeit und Armenpflege im vorchristlichen Altertum* (Utrecht, 1939), 1–33. For the social assistance procured by pious foundations, see e.g. the inscription of the nomarch Karapepinefer at Edfu; K. Sethe, *Urkunden des aegyptischen Altertums*, i, *Urkunden des alten Reiches*² (Leipzig, 1933), 254, 13 ff.; Pirenne (n. 14 above), 71–2.

[16] On the Egyptian canon of proportions, see e.g. H. Schäfer, *Principles of Egyptian Art*⁵, trans. and rev. by J. Baines (Oxford, 1986), 16–19; K. R. Weeks, 'Art, Word and the Egyptian World View', in K. R. Weeks (ed.), *Egyptology and the Social Sciences* (Cairo, 1979), 69–71.

[17] For discussions of these representations (mid-5th/6th dyn.), see Weeks, *Anatomical Knowledge*, 137 (chart), and 139–43; id. (n. 16 above), 72; H. G. Fischer, 'The Ancient Egyptian Attitude towards the Monstrous', in A. E. Farkas *et al.* (eds.), *Monsters and Demons in the Ancient and Medieval Worlds. Papers Presented in Honor of Edith Porada* (Mainz am Rhein, 1987), 22. See also P. Ghalioungui, 'Some Body Swellings Illustrated in Two Tombs of the Ancient Empire and their Possible Relation to âaâ', *ZÄS* 87 (1962), 108–14 (on possible representations of bilharzia).

[18] Lichtheim, *Literature*, i. 184–92; see also the New Kingdom Papyrus Lansing, ibid. ii. 169–72.

[19] Cf. *Lev.* 21.20 for an enumeration of infirmities (lameness, blindness, etc.) preventing a man from being a priest.

[20] Copenhagen, Ny Carlsberg Glyptothek, AEIN 134; O. Koefoed-Petersen, *Les Stèles égyptiennes* (Copenhagen, 1948), 35–6, no. 44; Leca (n. 14 above) 279, pl. IX. See also the Old Kingdom depiction of a court official with a hump(?); Copenhagen,

lame kings, like King Siptah (1204–1198 BC), whose mummy shows that he suffered from club-foot.[21] Attendants with club-feet or humps also seem to have been appreciated by the Old and Middle Kingdom élite, who often paired them with dwarfs (E 18, 29b, 54c, 67, 68; pls. 19.1a, b, 20.2, fig. 9.21).

Some anomalies were valued positively thanks to divine associations and seem to have enjoyed a special status. Thus twins were assimilated to divine pairs, like Seth and Horus. They had a partly fused social identity, and joint burials; they could hold important offices, like the Old Kingdom 'overseers of manicurists' Niᶜankhkhnum and Khnumhotpe and the New Kingdom architects Suty and Hor.[22] In the Graeco-Roman Period, twin Egyptian girls, Thaues and Taous, seem to have performed the roles of Isis and Nephthys at the burial of the Apis bull at Saqqara.[23]

The blind had a special status, perhaps because a sudden deprivation of sight could be explained by a religious transgression.[24] They seem to have been traditionally qualified as choir singers in temple and funerary feasts, and, from the Middle Kingdom on, as harpists and singers in the household of the élite.[25] Blackman suggests that they were particularly appreciated by noblemen because they were less likely to seduce the women they entertained.[26] In some depictions, the blindness of the musician may be symbolic; it expresses the skill or the piety of the musician, as in the New Kingdom tomb of Raᶜia at Saqqara, where the tomb-owner is depicted as a blind man when he plays before two deities but is shown with normal eyesight in the other scenes.[27] Similarly, some deprivation of sight mentioned in the texts on votive stelae denotes not real, but spiritual blindness.[28] There is no certain evidence about blind kings and high officials, apart from a few late mentions in Greek authors.[29] Blind people were also given lower-ranking roles, such as door-keeper or herdsman, as in the New Kingdom *Tale of Truth and*

Ny Carlsberg Glyptothek, AEIN 942; O. Koefoed-Petersen, *Catalogue des bas-reliefs et peintures égyptiens* (Copenhagen, 1956), 20–1, no. 12, pl. 18; Leca (n. 14 above), 241, fig. 53.

[21] Leca (n. 14 above), 279, fig. 75; A. T. Sandison, 'Diseases in Ancient Egypt', in A. and E. Cockburn (eds.), *Mummies, Disease and Ancient Culture* (Cambridge, 1983), 32, fig. 2.1.

[22] On Suty and Hor, see Baines (n. 4 above), 461–3. On Niᶜankhkhnum and Khnumhotpe, see ibid. 463–70.

[23] Baines, ibid. 472–3.

[24] F. Jonkheere, 'Le Monde des malades dans les textes non médicaux', *CdE* 25/50 (1950), 213–32; H. de Meulenaere, 'La Légende de Phéros', *CdE* 28/56 (1953), 248–60; H. Brunner, *LÄ* i (1975), s.v. Blindheit, 830–1.

[25] Brunner, ibid. 829; S. Schott, *Das schöne Fest vom Wüstentale. Festbräuche einer Totenstadt* (Wiesbaden, 1953), 78–80; Helbing (n. 13 above), 66–9. For a choir of four blind men, see e.g. N. de G. Davies, *The Tomb of Two Sculptors at Thebes* (New York, 1925) (RPTMS 4), 31 n. 4, pl. v. See also the blind harpist and singer in the tomb of Wekhhotpe; A. M. Blackman, *The Rock Tombs of Meir*, ii (London, 1915) (ASE 23), pl. xxi, 2.

[26] On blind musicians and male jealousy, see Blackman, ibid. 12–13.

[27] G. T. Martin, *The Tomb-Chapel of Paser and Raᶜia at Saqqara* (London, 1985) (MEES 52), 12 n. 7, frontispiece, pls. i and 22; L. Manniche, 'Symbolic Blindness', *CdE* 53/105 (1978), 13–21.

[28] Cf. the New Kingdom stelae in M. Tosi and A. Roccati, *Stele e altre epigrafi di Deir el Medina* (Turin, 1972), 81 (50046), 85 (50050), 86 (50051).

[29] Hdt. 2. 111 (King Pheros; 12th dyn.?); 2. 137, 140 (King Anysis; 24th dyn.?); Diod. Sic. 1. 58 (King Sesoosis; 12th dyn.?).

Falsehood.[30] The crucial function of land measurement after the inundation is sometimes shown performed by blind people, perhaps because one could not question their impartiality. This attribution may also be satirical.[31]

Some anomalies had negative associations. The insane, defined as men possessed by the soul of a dead man, a god or a demon, seem to have been prevented from entering temples. This prohibition occurs in a text in the Roman Period temple at Esna which describes a feast of Khnum. Sauneron suggested that this measure may have been taken to exclude people who used an alleged mental or epileptic trouble to impress pilgrims and earn a living as prophets or seers.[32]

Giants, on the other hand, do not seem to have attracted much attention. Grapow observes that they are not mentioned in medical texts.[33] In literature, they represent mainly foreigners, especially fierce enemies. They are not shown in iconography, perhaps because they could not be easily differentiated from major figures within Egyptian conventions, as Fischer notes.[34]

To sum up, the Egyptian attitude to the malformed is ambivalent. It hovers between respect, veneration, and rejection according to the divine associations attached to specific types of malformation. The unusual appearance of dwarfs inspired similarly complex reactions, from forms of rejection to appreciation, as I now show.

[30] Lichtheim, *Literature*, ii. 211–14, 4.1; 5.8; 10.7. See also Brunner (n. 24 above), 829, nos. 18 and 22.

[31] See e.g. the representation of a blind land-surveyor in the tomb of Menna at Sheikh'Abd el-Qurna; PM i². 135, I; *Atlas*, i. 232.

[32] S. Sauneron, 'Les Possédés', *BIFAO* 60 (1960), 111–15.

[33] H. Grapow, *Grundriss der Medizin der alten Ägypter*, iii, *Kranker, Krankheiten und Arzt* (Berlin, 1956), 38.

[34] Fischer (n. 17 above), 24.

9

Human Dwarfs

Our information on the position of short-statured people in daily life is based essentially upon iconography. Most representations come from funerary monuments—tomb reliefs and stelae, sarcophagi, statuary—which were meant to recreate for the deceased the conditions of his daily life on earth. Few depictions have been found in settlements (statuary) and cult contexts (temple reliefs, votive stelae, and figure vases).

The chronological distribution of these pictures is uneven. Most reliefs date from the Old Kingdom, while most statuettes are pre- and early dynastic or come from Middle Kingdom sites. No firm conclusion can be drawn from the relative paucity of evidence belonging to specific periods or contexts. Very few town sites have been excavated, and there may be gaps because of the chances of discovery; changes in the iconographic repertory may also reflect changes in aesthetics and decorum rather than social transformations.

The relative abundance of pictorial representations contrasts with the scarcity of written evidence. Apart from four Old Kingdom inscriptions, two New Kingdom texts and the fourth-century biography of Djeho, it consists mostly of captions which give the dwarfs' names and titles.

I now examine the iconographic, epigraphic, and literary material available for each major period of Egyptian history in separate sections.

THE PREDYNASTIC PERIOD (to *c.*3200 BC)

The earliest representations of dwarfs are small ivory statuettes found in the predynastic cemeteries of Upper Egypt, at el-Ballas and presumably also Naqada (E 88–95). They depict male and female dwarfs in the same pose as fully grown human figures: standing, naked, with shaven heads and protruding ears, their arms hanging down along the body. The pubic area of female dwarfs is marked by a series of holes, as in statuettes of normal-sized women (E 92; pl. 27.1).[1] Pathological abnormality is clearly indicated by very short bent limbs, sometimes severely malformed (coxa vara). An anatomical detail

---

[1] Cf. the statuettes of full-sized figures in J. Capart, *Primitive Art in Egypt* (London, 1905), 166–7, figs. 128–9.

which may be revealing is that one of the male dwarfs has no penis-sheath, unlike full-sized males; this may indicate that he did not have a full adult status (E 94; pl. 27.2). No conclusion can be drawn from the other male figure, which is badly preserved (E 93).

The age and sex of the burials in which the figurines were found is unknown, but the funerary context suggests that dwarfs were not rejected as disturbing, frightening creatures; they were valued highly enough to accompany the deceased in the next world.

The function of the figurines has been interpreted variously. For most scholars, they represent the favourite attendants of the deceased.[2] Some objects were perhaps made for the living. One statuette is perforated from the back to the front, and may have been part of a mechanical toy (E 93);[3] another has a peg at the base, and was fastened to a stand or belonged to a piece of furniture (E 94; pl. 27.2). These statuettes may also have been early images of fertility and family guardians. A male figurine holds a child (E 93), an activity recalling Middle Kingdom statuettes of female dwarf nurses (E 123–d 125, 146; pls. 30.3–4, 34.2) and the protective role of Bes.

The scene depicted on a limestone cylinder may also transmit early beliefs in the supernatural powers of dwarfs (E 10; fig. 9.1 see p. 106). It depicts the mastering of dangerous animals. Two full-sized ithyphallic men stand among three crocodiles, a lizard, a scorpion, and possibly also a bee. They are fighting; one holds a weapon in the right hand, while the other raises his right arm. Behind this group, four bandy-legged dwarfs, seen in front, stand in a row. They may represent apotropaic beings: their power is suggested by their unusual pose, tête-bêche, recalling magical groupings.[4] They are also associated with a lizard, a reptile occasionally assimilated to snakes, but also a symbol of regeneration, which is sometimes held by dwarf gods in later periods.[5] The scene may thus combine different manifestations of apotropaic powers: dwarfs, warriors, and benevolent animals (lizard, bee) opposed to evil, symbolized by scorpions and crocodiles.[6]

Quibell published a decorated jar from Naqada depicting four dwarfs in a scene with boats;[7] the vessel is genuine, but the painting is a forgery, as Brunton demonstrated.[8]

---

[2] See e.g. E. Naville, 'Figurines égyptiennes de l'époque archaïque', *RecTrav* 22 (1900), 67; R. D. Barnett, *Ancient Ivories in the Middle East* (Jerusalem, 1982), 16.

[3] See R. H. Randall, *Masterpieces of Ivory from the Walters Art Gallery* (New York, 1985), 35, 42.

[4] For magical tête-bêche groupings on scarabs see e.g. the motif of two crocodiles in Hornung/Staehelin, *Skarabäen*, 123 and nos. 797–9, 901, B 51, B 85, D 21. For a discussion of its symbolic meaning, see Sourdive, *La Main*, 472–3.

[5] On the symbolism of lizards, see Hornung/Staehelin, *Skarabäen*, 109–10, and Raven, *Pataekos*,

13. See e.g. the Bes-god on a Middle Kingdom magical knife; F. Legge, 'The Magic Ivories of the Middle Empire', *PSBA* 27 (1905), 141, pl. VII, fig. 11.

[6] On the ambivalence of crocodiles, see also Hornung/Staehelin, *Skarabäen*, 122–6.

[7] Cairo Museum, CG 11557 (h. 16 cm.); J. E. Quibell, *Archaic Objects*, i (Cairo, 1905) (CGC), 116, pl. XXII. Discussed as genuine by G. D. Hornblower, 'Funerary Designs on Predynastic Jars', *JEA* 16 (1930), 14, figs. 1–2; Rupp, 'Zwerg', 299 (1).

[8] G. Brunton, 'Modern Painting on Predynastic Pots,' *ASAE* 34 (1934), fig. 5, 149–56, esp. 152 ff.

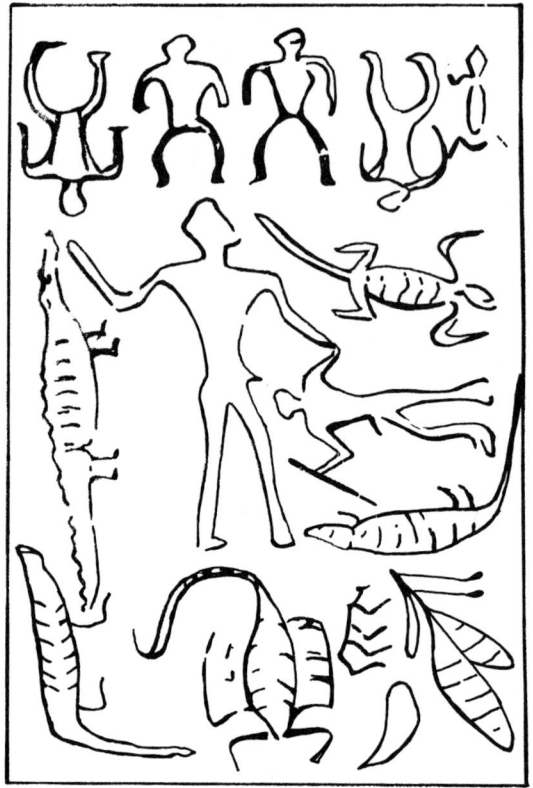

FIGURE 9.1. (E 10) Cylinder, limestone (w. 5.5 cm.). Cairo Museum, CG 14518

The Early Dynastic Court (2920–2770 BC)

A larger range of monuments (statuettes, graves, funerary stelae) comes from the early dynastic court. This material is generally similar in type and meaning to that from the Old Kingdom.

The best information comes from the dwarfs' burials in the royal tombs at Abydos and those of high officials at Saqqara. They were found in the rows of small subsidiary graves belonging to minor members of the royal entourage, including pet dogs, who were probably killed to accompany their master in the hereafter.[9] The dwarfs are identified by skeletal remains or by depictions on funerary stelae, inscribed with names, which were set up for the deceased (E 1, 3–9; pl. 17.2–3). As the superstructures above the royal tombs are now completely destroyed, most stelae were found scattered, or else

[9] On human sacrifice, see M. A. Hoffman, *Egypt before the Pharaohs* (London, 1980), 275–9, esp. 276, table XII; B. G. Trigger, 'The Rise of Egyptian Civilization', in *Ancient Egypt*, 52.

their original position is not recorded; it is possible to associate them with specific skeletons only in exceptional cases (E 1 and S 5; E 8 and S 7). As with dogs, graves of dwarfs were found only in the tomb complexes of kings of the first dynasty.[10] Their number suggests that short-statured people were much favoured companions; at least three dwarfs seem to have belonged to the entourage of King Djer (S d 4–6; E 1–3; pl. 17.1), one to that of King Wadj or of a high official (S 3; pl. 2.2), four to that of King Den (E d 4–d 7; pl. 17.2), two to that of King Semerkhet (E 8–9; S 7; pl. 17.3), and two to that of King Qaʿa (S d 8–9).

Like normal-sized retainers, dwarfs were buried in wooden coffins (S 3–5; pl. 2.2); their funerary equipment consisted of pottery (S 3; pl. 2.2) or stone vessels, linen (S 5), personal ornaments, such as bracelets of flint, and pieces of furniture (S d 4).

Their depiction on the funerary stelae shows a marked interest in their unusual morphology. As Rupp notes, craftsmen gave them a pose which exhibits their most typical features. While full-sized adults are depicted in a seated pose which is a mark of status, dwarfs stand with feet apart, so as to display their massive trunk, a slight back deformity and plump abdomen (E 3, 6, 8; pl. 17.2–3).[11] The standard logogram or determinative for dwarf, which uses the same conventions, was probably introduced in this period.

Kaplony observes that the stelae of dwarfs and dogs from the cemetery of King Den are of a better quality and workmanship than those of full-sized retainers (E d 4–d 7; pl. 17.2).[12] This confirms that the short men had a special place in the household, and, as markers of prestige, may have had a similar position to that of favourite dogs.

Some of the dwarfs' names seem to designate functions. Wediwesekh, 'he who places the collar', thus may indicate that the short man was involved in jewellery-making or that he supervised dress, a role often performed by dwarfs in the Old Kingdom (E 6; pl. 17.2). Another name, Simanetjer, 'the one who gladdens the god' (E d 7) is reminiscent of the jb3 dancing of Old Kingdom pygmies, which had comparable qualities; it may refer to a similar performance.[13] The remaining names do not have intelligible meanings, except Setinepu, 'whom Anubis begot', which expresses a common wish (E d 5), and Nofret, 'the good one', a very common name which may have no specific meaning in relation to the dwarf (E 8–9; pl. 17.3).

The location of the burials in the tomb complexes is also revealing. Kaplony has shown that the chamber of the dwarf Shedi is situated on the north side of the complex of King Djer, next to store-rooms containing chests full of linen (E 1; S 5). This location, together with the fact that remains of fine linen were found in his grave, may imply that the short man was keeper of the wardrobe.[14] In the eastern part of the same

[10] Stelae of dogs seem to occur only during the reign of King Den; Kaplony, *Inschriften*, i. 375.

[11] Rupp, 'Zwerg', 272–4.

[12] Kaplony, *Inschriften*, i. 374–5.

[13] Kaplony, *Inschriften*, i. 375. See also Hornung, *Conceptions*, 46.

[14] E. Amélineau, *Les Nouvelles Fouilles d'Abydos, 1897–1898*, i (Paris, 1904), 229–30, esp. no. 8; Kaplony, *Inschriften*, i. 217 and 375.

complex, another dwarf's burial is surrounded by graves of women containing fragmentary pieces of furniture, which might allude to his role as personal attendant (S *d* 4).[15]

A dwarf is depicted on a stone vessel from the cemetery of King Djer (E 2; pl. 17.1). His dress and attributes are those of a major figure. He wears a kilt and a broad collar, and holds in his right hand a short object no longer visible on recent photographs. This item, which is similar to the strip of cloth commonly held by officials and by dwarfs in the Old and Middle Kingdoms (E 48, 66; figs. 9.8, 9.20), may be an attribute of his privileged position.[16]

Dwarf statuettes, made of various materials (glazed ware, stone, ivory), were discovered at several sites in temple areas or deposits (E *d* 98–*d* 110; pls. 27.3–28.1) and in graves (E 97).[17] Most figurines are very crude (e.g. E *d* 102; pl. 27.3), but some are very high quality pieces which must have belonged in the inner élite (E 106–8, *d* 111; pls. 27.4, 28.1).

Their precise meaning is uncertain. For Dreyer, these figures may represent the physical characteristics of dwarf donors,[18] which would imply that the dwarfs living at the court had the privilege of making offerings in state sanctuaries. Three ivory statuettes found in the great temple deposit at Hierakonpolis suggest that these offerings were connected with the protection of birth (E 106–108; pls. 27.4, 28.1).[19] They are relatively large (11.9–15.3 cm.), with a tenon at the base, and were probably fixed to stands or to pieces of furniture. They depict female dwarfs with unusual pose and attire. Two wear long dresses and heavy wigs, hanging in a strand over each shoulder and down the back, which give them a dignified appearance (E 106–7; pl. 27.4). In the Middle and New Kingdoms, similar large wigs are associated with Hathor and characterize Middle Kingdom fertility figurines, occasionally shown as dwarfs.[20] The third figure also has a wig, but is naked; her belly is slightly protruding, perhaps indicating pregnancy (E 108; pl. 28.1); her right hand, and probably also her left, now missing, is laid over her chest, a pose best paralleled by New Kingdom depictions of pregnant women.[21] This interpretation is supported by a fourth ivory figurine of

---

[15] Amélineau, ibid. 103–4; Kaplony, *Inschriften*, i. 215–16.

[16] On the meaning of this piece of cloth, see E. Staehelin, *Untersuchungen zur ägyptischen Tracht im alten Reich* (Berlin, 1966) (MÄS 8), 162–3; H. G. Fischer, 'An Elusive Shape within the Fisted Hands of Egyptian Statues', in *Ancient Egypt in the Metropolitan Museum Journal 1–11 (1968–76)* (New York, 1977), 143–55; A. Fehlig, 'Das sogenannte Taschentuch in den ägyptischen Darstellungen des Alten Reiches', *SAK* 13 (1986), 55–94, esp. 69–71.

[17] Cf. also the grotesque male figure from the early temple area at Abydos (h. 7.6 cm.); Brooklyn

Museum, Charles Edwin Wilbour Fund, 1958, 58.32.1; W. Needler, *Predynastic and Archaic Egypt in the Brooklyn Museum* (New York, 1984), 349–50, no. 278.

[18] G. Dreyer, *Elephantine VIII, Der Tempel der Satet, Die Funde der Frühzeit und des alten Reiches* (Mainz am Rhein, 1986) (AV 39), 60.

[19] For the general context of the finds, see B. Adams, *Ancient Hierakonpolis, Supplement, the F. W. Green MSS* (Warminster, 1974), 8 ff., 113.

[20] See below 140–1.

[21] For a similar identification, see Raven, *Pataekos*, 11.

unknown provenance (E *d* 111), depicting a similarly naked female dwarf, with long hair hanging down her back, who holds a child like Middle Kingdom dwarf nurses (e.g. E 123–4, 146; pls. 30.3–4, 34.2).

These figurines could have been dedicated to a goddess associated with birth, such as Hathor, by short-statured women, who have great difficulty in childbirth.[22] Another possibility is that the statuettes symbolized normal-sized pregnant women, assimilated to dwarfs as good-luck figures. It is also conceivable that female dwarfs acted as helpers of birth in human society, as skilled small-handed midwives or as nurses.

OLD KINGDOM (2575–2134 BC)

Distribution

In the Old Kingdom, dwarfs appear as familiar members of the households of noblemen. They are depicted in more than fifty tombs, mostly of the fifth and sixth dynasties. This liking for dwarfs was shared by élite men and their wives, like Queen Meresᶜankh (E 11a; fig. 9.2), Queen Khentkaus I (E 25), Queen Nebet (E 35) and Waᶜtetkhethor (E 52c; pl. 23a, b), who also had dwarf attendants, usually female.

In a single tomb, dwarfs often are shown performing different functions, in two places, as in the tomb of Inti (E 59a, b; fig. 9.13a, b) or in three, as in the tomb of Serfka (E 61a–c; fig. 9.15a, b).[23] Occasionally they appear in large numbers, like the six goldsmiths in the tomb of ᶜAnkhmaᶜhor (E 49a), or the twelve bearers in the chapel of Waᶜtetkhethor (E 52c; pl. 23a). This profusion of dwarf retainers may not correspond to reality. For artistic reasons, the figures may be multiplied to fill the space, rather as rows of cattle or offering-bearers are extended.[24] This development in composition, which is characteristic of Saqqara tombs of the early sixth dynasty, may reflect the aspiration of the deceased to own an elaborate household appropriate to his rank.

Some pictures, however, may be more than just artistic motifs. Dwarf retainers were probably kept in families throughout generations, and the same individual might be shown in an idealized youthful form in chapels belonging to several members of a family. For example, the dwarf depicted in the chapel of Meryteti (E 52d; fig. 9.9) may be one of those shown in the chapel of his father Mereruka (E 52a, b; pl. 22.1, fig. 9.11). Some families may have had a tradition of keeping dwarfs; neighbouring tombs at Giza and Saqqara, such as the group north of Teti pyramid contain similar scenes with dwarf servants which may reflect such a family institution (E 48–54). It is, however, very difficult to establish if there is any family relationship between these tomb-owners. The

---

[22] For a possible relation to an early form of the cult of Hathor, see G. D. Hornblower, 'Predynastic Figures of Women and their Successors', *JEA* 15 (1929), 29–47, esp. 36–9.

[23] In two places: E 11, 20, 29, 36, 38, 49, 52, 53, 59, 62. In three: see also E 54.

[24] See e.g. the procession of servants with cattle in the tomb of ᶜAnkhmaᶜhor; A. Badawi, *Nyhetep-Ptah and ᶜAnkhmaᶜhor* (Berkeley, Calif., 1978), figs. 48–9. See also the extended row of servants in the tomb of Mereruka; P. Duell, *The Mastaba of Mereruka*, i (Chicago, 1938) (OIP 31), pls. 74–6.

only certain examples are the scenes in the chapels of Mereruka, Waʿtetkhethor, and Meryteti, who were husband, wife, and son (E 52a–d; pls. 22.1, 23; figs. 9.9, 9.11), and in the tombs of Meresʿankh and Nebemakhet, who were mother and son (E 11a, 12; fig. 9.2).[25] Similar representations of dwarfs in neighbouring tombs may also be due to the fact that the same artists worked in a particular area, or that they were inspired by the decoration of a nearby or well-known tomb, whose owner was unrelated.

Household Offices

Dwarfs are depicted performing a limited range of functions in the household. Four main offices are distinguished: personal attendant, animal-tender, jeweller, and entertainer. These functions were usually held by normal-sized persons, and one person could perform several of them. Dwarfs also take a part in a few outdoor activities, such as bird-catching and boating.

Personal Attendants

Sixteen reliefs and two statuettes show male and female dwarfs bringing objects related to clothing and bodily care. In most examples they carry a box of varying size on their shoulders or on top of their heads.[26]

Small boxes perhaps contained jewellery, as suggested by a relief in the tomb of Inti (E 59a; fig. 9.13a). A dwarf stands in a line of bearers before two dwarfs who are making jewellery, just below a damaged scene of metal-working.[27] The short man holds a broad collar which the small goldsmiths may have just completed, and carries on his head a box which could be filled with other costly items. A similar box is set before the overseer of dwarf jewellers in the tomb of Wepemnefert (E 24; fig. 9.14). The dwarf and hunchbacked attendants of Seshemnufer I may also bring the master's jewel-boxes (E 18; pl. 19.1a, b); both figures are themselves graced with various ornaments, a pectoral, bracelets, and anklets.

Fine linen was usually kept in soft bags, like that brought by the dwarf in the tomb of Niʿankhnesut (E 57; pl. 24.1) and the limestone figurine from the tomb of Nikauinpu (E 115; pl. 29.1),[28] but it may also have been placed in larger boxes or chests. Dwarfs carrying a box or a bag of linen may stand outdoors, under the litter carrying the tomb owner (E 37; pl. 22.2), or in a punting scene (E 50, 57; pl. 24.1). Junker explained this detail as showing that they had to supply their lord with fresh clothes or an extra

---

[25] D. Dunham and W. K. Simpson, *The Mastaba of Queen Mersyankh III* (Boston, 1974), 25. On kinship and similarities in tomb decoration, see Y. Harpur, *Decoration in Egyptian Tombs of the Old Kingdom* (London and New York, 1987), 13 ff.

[26] See E 17, 18, 37, 39, 52c, 59a; pls. 19.1, 22.2, 23, figs. 9.3, 9.6, 9.13a.

[27] For a reconstruction of the two registers, see R. Drenkhahn, *Die Handwerker und ihre Tätigkeiten im alten Ägypten* (Wiesbaden, 1976), 21, no. XVI.

[28] Cf. the full-sized servants carrying linen in the tomb of Ty (E 29b; pl. 20.2), Nikauhor (E 31) and Inti (E 59a; fig. 9.13a). See also the chart in Vandier, *Manuel*, iv, 177, fig. 33, nos. 27–9.

FIGURE 9.2. (E 11a) Giza, tomb of Queen Meres^cankh III

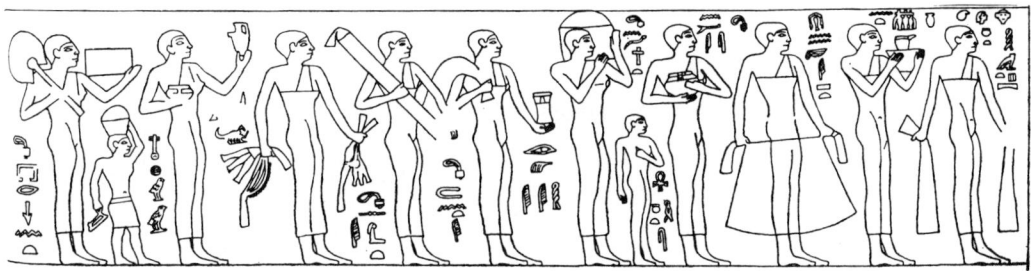

FIGURE 9.3. (E 17) Tomb of Wehemka. Hildesheim, Pelizaeus-Museum, 2970

garment if the weather cooled down.[29] Some bring other accessories for outdoor excursions, such as sandals and a walking-staff.[30]

In the chapel of Wa^ctetkhethor, a procession of female dwarfs carry items related to attire, such as fans and ointment jars (E 52c; pl. 23a, b); captions above two large chests reveal that they contained precious perfumes (myrrh) and pots of oil.[31] In the tomb of Ibi, the dwarf holds a mirror (E 62b), while a limestone statuette depicts a small man weighed down by four pots, perhaps ointment jars (E 114). The head-rest brought by a dwarf in the tomb of Nesutnufer suggests more specifically attendance in the bedroom (E 19a; fig. 9.4), like the unpublished fragment of relief from the funerary temple of Neuserre^c which depicts the curved legs of a dwarf beside a table (E d 27).

[29] H. Junker, *Giza*, xi (Vienna, 1953), 65.
[30] Sandals: E 17, 19b, 25, 31, 35, 57; pl. 24.1, figs. 9.3, 9.4. Walking-staff: E 19b; fig. 9.4.
[31] Similar inscriptions are carved above a chest on a relief from the tomb of Akhtihotpe now in the Louvre; *Atlas*, iii. 181, i, pl. 88. On the different types of chests, see Vandier, *Manuel*, iv. 142, fig. 45.

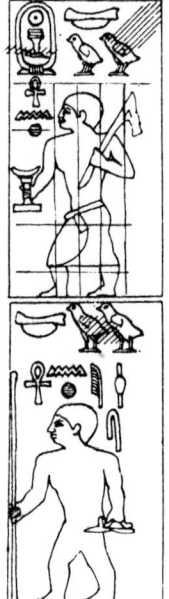

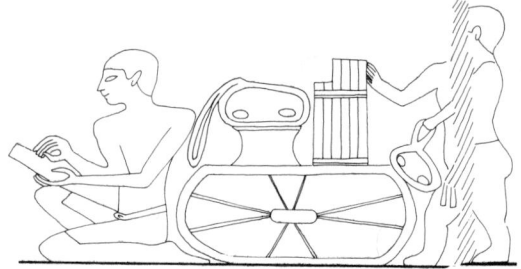

FIGURE 9.5. (E 36b) Saqqara, tomb of Nufer

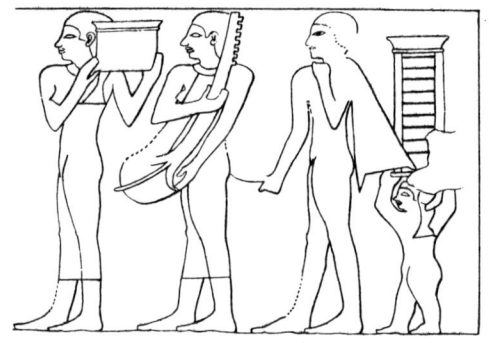

FIGURE 9.4. (E 19) Giza, tomb of Nesutnufer

FIGURE 9.6. (E 39) Hagarsa, tomb of Kaemnefert

FIGURE 9.7. (E 43) Giza, tomb of Khentkaus

In many respects, these dwarfs seem to have the same status as full-sized attendants. Usually, they stand in a row of servants carrying similar items, as in the tombs of Wehemka (E 17; fig. 9.3) and Nikauhor (E 31). Apart from the naked dwarfs in the tombs of Nesutnufer, Nufer and Niᶜankhkhnum/Khnumhotpe (E 19b, 36b, 37; pl. 22.2, figs. 9.4–5), males commonly wear a plain short kilt, like normal-sized servants (e.g. E 57, 59a–b; pl. 24.1, fig. 9.13a, b). Most female dwarfs wear the ordinary long dress with two wide shoulder straps, sometimes with a necklace as in the tomb of Nebet (E 11a, 35, 52c; pl. 23, fig. 9.2); a few seem to be naked, but wear ornaments, like the female dwarf in the tomb of Seshemnufer I (E 11b, 18, 39, 58; pl. 19.1b, fig. 9.6).

Dwarfs also hold the same titles as fully grown servants. They are captioned *jmj-r3 ššr* 'overseer of linen', *jmj-r3-pr* 'overseer of the house' or 'steward', and *šms(w)* 'retainer'.[32] They may have supervised the weaving of cloth, like the dwarf Seneb, who was both *ḫrp ššrw* 'overseer of the dwarfs in charge of linen' (E 113, b 2, c 1), and *jmj-r3 mr pr ʿ3* 'overseer of weaving in the palace' (E 113, c 3). The title *sd3wtj* 'seal-bearer/treasurer' held by several dwarfs may refer to care for precious articles, such as oils, perfumes and jewellery.[33] The names of the dwarfs are often indicated, such as Neferkhuu (E 17; fig. 9.3), Radjedefankh, and Ankhiwedjsu (E 19a, b; fig. 9.4).[34]

A few iconographic details, however, differentiate dwarf attendants from the rest of the household. Several wear an elaborate pointed kilt, sometimes reaching below the knees (E 19a; fig. 9.4), a garment denoting a higher-ranking position.[35] The piece of linen in the clenched hand of the dwarf Ankhef in the tomb of Khentika: Ikhekhi has a similar positive meaning (E 48; fig. 9.8). This strip of cloth is usually held by major figures, probably as an attribute of their leisured status; the fact that a dwarf holds it may indicate that he led a similarly unburdened life.[36] This hypothesis is supported by the depictions of dwarfs in pointed kilts in the tombs of Meryteti (E 52d; fig. 9.9) and Nikauisesi (E 54a), who stand with empty hands, as if their value did not depend on any action they might perform.

Composition and captions also single dwarfs out. Some are shown in a privileged relationship to their master. In the tomb of Khentkaus, the dwarf is set apart from the servants in the same register (E 43; fig. 9.7); he is turned towards the couple seated on a chair, waving his fly-whisk before their legs.[37] In the tomb of Khentika: Ikhekhi, the master is seated and extends his hand to take the box held ceremoniously by his dwarf (E 48; fig. 9.8). The short man deals directly with his lord, like an important functionary; a caption gives his name, Ankhef, and title, overseer of linen. The dwarf in the tomb of Mereri is in the same pose, but his master is boating and faces away from him (E 50); he also has a name, Redji, and the title of overseer of linen; in addition, he is a

---

[32] *Jmj-r3 ššr*: E d 46, 50, 117; pl. 29.2. *Jmj-r3-pr*: E 48, 54a; fig. 9.8. *Šms(w)*: E 31, 33; fig. 9.10.

[33] See the dwarfs in the tombs of Waʿtetkhethor (E 52c; pl. 23), Niʿankhnesut (E 57; pl. 24.1), and Nikauisesi (E 54c).

[34] See also E 37 (Nisuqed), 48 (Ankhef), 50 (Redji), 54a (Irienptah), 54c (Iti), 57 (Hotep), 58 (Sʿankhu-Hathor(?).

[35] See also E 43, 48, 50, 53a–b, 54a–b, 62b; figs.

9.7–8. See E. Staehelin, *LÄ* vi (1985), s.v. Schurz, 729.

[36] See n. 16 above.

[37] Cf. the similar whisk made of three strips of animal skin held by the lord in the tomb of Nufer at Saqqara: A. M. Moussa and H. Altenmüller, *The Tomb of Nefer and Ka-hay* (Mainz am Rhein, 1971), pl. 24.

ka-servant, a revealing title which was granted only to higher-ranking members of the household.[38]

Yet this special status is ambivalent. In a few scenes, dwarfs are assimilated to children or to women. In the tomb of Wehemka, a male dwarf and a male child appear in the same line of female bearers (E 17; fig. 9.3). Their presence among women is intriguing since rows of servants usually comprise only attendants of the same sex; it may imply that, like the child, the dwarf could be associated with females because he was not regarded as a full man. In the tomb of Nufer at Saqqara, a male dwarf performs the role of a child (E 36b; fig. 9.5); he stands by a table with writing implements behind two seated scribes writing on rolls of papyrus, and holds a palette in his right hand. His function is paralleled by that of the naked child in the tomb of Kaemrehu, who similarly holds a palette for a scribe standing and writing;[39] unlike the child, the dwarf is not naked, but wears a belt with strips, a low-status garment. Female dwarfs, however, appear with adult women only (E 11a–b, 18, 35, 39; pl. 19.1, figs. 9.2, 9.6).

Animal-Tenders

On twenty-two reliefs, dwarfs look after household pets, especially dogs and monkeys. This office was performed by male servants, and was assigned to male dwarfs only.[40] This function was often associated with that of overseer of linen. Thus, the dwarf leading a dog in the tomb of Neferirtenef is captioned *šmsw* 'retainer' (E 33; fig. 9.10), and the one in the tomb of Seshemnufer: Tjeti *jmj-r3 sšr* 'overseer of the linen' (E d 46); another animal-tender carries a bag of linen in the tomb of Ni'ankhnesut (E 57; pl. 24.1).[41] Dwarf animal-keepers may also be paired with normal-sized servants carrying bundles of linen.[42]

Animal-keeping scenes show a similar pattern: usually the dwarf accompanies his lord going out in a litter to supervise his estate. The animals follow in a row (e.g. E 52b; fig. 9.11) or are distributed in two sub-registers (E 29b, 33; pl. 20.2, fig. 9.10). The dwarf may stand near his master as he surveys agricultural pursuits (E 33, 57; pl. 24.1, fig. 9.10) or beneath his chair (E 44, 60, 61a; fig. 9.15a).

Most often the pets are led by a leash, sometimes attached to a buckle as in the tomb

[38] See S. Allam, 'Le *ḥm-k3* était-il exclusivement prêtre funéraire?', *RdE* 36 (1985), 1–15. See also U. Schweitzer, *Das Wesen des Ka im Diesseits und Jenseits der alten Ägypter* (Glückstadt etc., 1956), 86. For Middle Kingdom examples of contracts with ka-priests, see G. A. Reisner, 'The tomb of Hepzefa, Nomarch of Siût', *JEA* 5 (1918), 79–98.

[39] *Atlas*, i, pls. 402–3.

[40] On animal pets in general, see E. Brunner-Traut, *LÄ* iii (1980), s.v. Lieblingstier, 1054–6. There is an exception in the tomb of Meres'ankh, where a woman carrying a chest leads a monkey;

Dunham/Simpson (n. 25 above), fig. 8, second register from the bottom (= E 11a; fig. 9.2). A female animal-tender may be shown here because this is a woman's tomb.

[41] For full-sized servants leading pets and carrying bags of linen, see e.g. the tomb of Pepi'ankh at Meir; A. M. Blackman, *The Rock-Tombs of Meir*, v (London, 1953) (ASE 28), pl. XXXI; Vandier d'Abbadie, 'Singes', 163–4, fig. 19.

[42] See e.g. the tombs of Ty (E 29a–b; pl. 20.1–2), Kanufer (E 15; pl. 18.2) and Nikauhor (E 31).

FIGURE 9.8. (E 48) Saqqara, tomb of Khentika: Ikhekhi

FIGURE 9.9. (E 52d) Saqqara, tomb of Mereruka, chapel of Meryteti

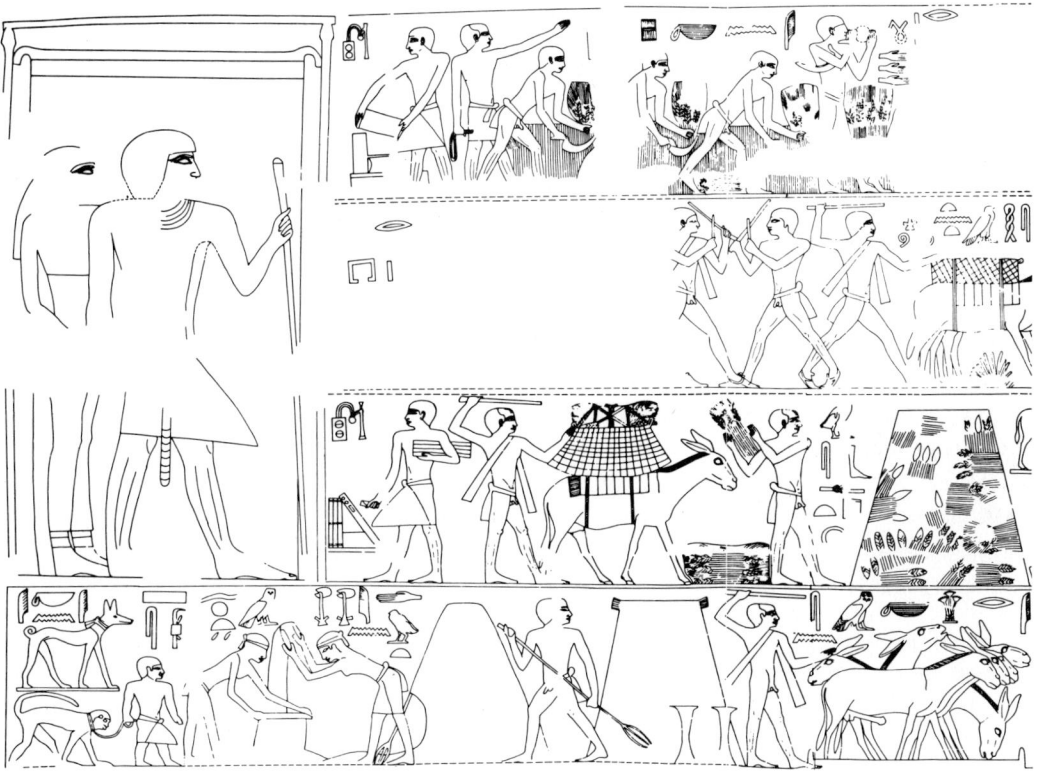

FIGURE 9.10. (E 33) Tomb of Neferirtenef. Brussels, Musées Royaux, E 2465

FIGURE 9.11. (E 52b) Saqqara, tomb of Mereruka

of Mereruka (E 52b; fig. 9.11). Dogs generally wear leather ties around the neck, but they may be free, as in the tombs of Ty (E 29a; pl. 20.1), Neferirtenef (E 33; fig. 9.10) and Mereruka (E 52b; fig. 9.11).

The dogs are tall hounds, _tzm_, which were used for hunting in the desert.[43] Their graceful silhouettes contrast humorously with their short plump minders. Naturalistic details may be added, such as foam running from the mouth, possibly to indicate the impatience of the animal to be unfastened (E 51). Dogs' names are sometimes indicated, as in the tombs of Neferirtenef (E 33; fig. 9.10) and Serfka (E 61c; fig. 9.15b).[44]

The monkeys are slender cercopitheci, not baboons (cynocephali), probably because the latter are very strong and could be dangerous.[45] In the tomb of Itisen, a dwarf appears exceptionally in a register above a baboon, but the animal is led by a full-sized servant (E 22). Unlike dogs, no ape has a name. When depicted with dogs, the monkeys seem to be very docile. On their own, however, they exhibit their natural inclination to play the fool. They may behave very disrespectfully with normal keepers, but are quite gentle with the dwarfs.[46] The monkey may be seated on the shoulders of the dwarf, playing with his leash as in the relief of Nufer: Idu (E 44), or it may hold a paw to its mouth, probably eating fruit stolen from the basket carried by the short man (E 15, 49b; pl. 18.2). In the tomb of Serfka, the animal is seated before the dwarf, and eats some food from a vessel set on the ground (E 61a; fig. 9.15a). As monkeys easily shin up trees, they may have been employed as tame fruit-pickers, a useful aid for a short-statured man.[47]

[43] H. G. Fischer, _LÄ_ iii (1980), s.v. Hunde, 77–81. For their use in desert hunting, see e.g. C. Aldred, _Egyptian Art_ (London, 1980), 88, fig. 48 (tomb of Ptahhotpe).

[44] On dogs' names, see J. M. A. Janssen, 'Über Hundenamen im pharaonischen Ägypten', _MDAI(K)_ 16 (1958), 176–8, esp. 179, no. 7 (_jknj_: tomb of Neferirtenef) no. 16 ('_bjw_?: tomb of Serfka). See also H. G. Fischer, 'A Supplement to Janssen's List of Dogs' Names', _JEA_ 47 (1961), 152–3, and H. Brunner, 'Zur Hundeinschrift des AR', _ZÄS_ 95 (1969), 72.

[45] A baboon is thus the determinative for _qnd_ 'to be furious'; Gardiner, _Grammar_, sign-list, E 32. See e.g. the baboon attacking a man in the Old Kingdom tomb of Djadjaemankh; Vandier d'Abbadie, 'Singes', 154, fig. 5.

[46] Cf. the monkey pinching the nose of an attendant in the tomb of Pepiꜥankh at Meir; Blackman (n. 41 above), pl. XVII; Vandier d'Abbadie, 'Singes', 163–4, fig. 20.

[47] I am grateful to H. Whitehouse for this suggestion. Cf. the monkey in a tree in the tomb of Nefermaꜥet; W. M. F. Petrie, _Medum_ (London, 1892), pl. XVIII; Vandier d'Abbadie, 'Singes', 171, fig. 35. For monkeys as fruit-pickers, see L. Keimer,

Some members of the élite had a liking for more prestigious pets. In the tomb of Ni^{c}ankhnesut, a dwarf leads a tame leopard which seems to pay no attention to the accompanying animals, two dogs and a monkey (E 57; pl. 24.1). The animal was probably trained to hunt with the hounds, as depicted in the tomb of Neferma^{c}et, where a leopard walks before a kneeling huntsman above a scene with a hound attacking three jackals. As Petrie observed, the leopard does not seem to be wild, but is prowling like a tame hunting-animal.[48] There is no other evidence of this practice before the New Kingdom,[49] but other felidae could be tamed, such as the lion which accompanied Ramses II on the battlefield and slept in front of his tent.[50]

Dwarf keepers often carry with dignity the characteristic attribute of animal-tenders, a short baton (E 28), which is sometimes spatulate (E 22) or ends in a hand (E 29a–b, 52b, 60; pls. 20.1–2, fig. 9.11).[51] Sourdive notes that these batons are large in relation to the size of the dwarfs, which suggests that the same object was used by short-statured and normal attendants; only the dwarf Pepy in the tomb of Ty seems to have a miniature baton suited to his size (E 29a; pl. 20.1).[52] Like the leash, this attribute probably displayed the authority of the short keeper over the animal and, more generally, his status.[53]

No Old Kingdom title is known to describe this function as animal-tender, except perhaps for *jwḥw* which occurs in the tomb of Ty (E 29a; pl. 20.1) and in a title of the dwarf Seneb (E 113c 2).[54]

The attire and posture of the dwarf keepers enhance their special status. As higher-ranking attendants, they often wear pointed kilts, while normal-sized animal-tenders only have plain kilts.[55] They usually stand very close to their lord, below his litter, his chair, or in the register below his figure.[56] The composition of the scene in the tomb of Neferirtenef shows this special intimacy between dwarfs and major figures (E 33; fig. 9.10). A dwarf stands with pets in the register below his lord who oversees work in the fields; the dwarf faces the same way as his lord, towards the agricultural scenes. Like the scribes in the fields, he wears a pointed kilt contrasting with the modest attire of the

'Pavian und Dum-Palme', *MDAI(K)* 8 (1939), 42–5; I. Wallert, *Die Palmen im alten Ägypten* (Berlin, 1962) (MÄS 1), 99.

[48] Tomb of Neferma^{c}et, Cairo Museum, JdE 43809; Petrie, ibid. 25, pl. XVII; *Musée égyptien du Caire*, no. 25a.

[49] For material, see O. Keller, *Die antike Tierwelt*, i (Leipzig, 1909), 86–7. See e.g. the leopard led by Nubians in the tomb of Rekhmire; N. de G. Davies, *The Tomb of Rekh-mi-Re^c at Thebes* (New York, 1943), pl. XIX. For tame leopards in Greece, see G. Koch-Harnack, *Knabenliebe und Tiergeschenke* (Berlin, 1983), 105–19.

[50] See the depictions on the rock-cut temple at el-Derr in Lepsius, *Denkmaeler*, 3. 183b, 184a. See

also the tame lion standing by the side of Tutankhamun in a fowling scene on the golden shrine; Cairo Museum, JdE 61481; I. E. S. Edwards, *The Treasures of Tutankhamun* (Harmondsworth, 1972), no. 25, left side, lower register.

[51] For a list of depictions of this baton, see Sourdive, *La Main*, 73 (table).

[52] Sourdive, *La Main*, 25, 85, and 94–5.

[53] Vandier d'Abbadie, 'Singes', 159; Sourdive, *La Main*, 19 ff., 96–7.

[54] For a discussion of this term, see above 31–2.

[55] See E 28, 29a–b, 33, 43, 48, 50–1, 52d, 53a–b, 54a–b, 55–6, 60, 62a–b; pls. 20.1–2, figs. 9.7–10.

[56] Below the litter: E 20a, 28, 29a, d 40, d 46, 51, 52b, 54b; pl. 20.1, fig. 9.11. Below the chair: E 44,

FIGURE 9.12. (E 23) Giza, tomb of Ireru

workers, who are naked or wear a belt with strips. He is singled out by his title *šmsw*, 'retainer'.

Funerary practices also show the favour which familiar dwarfs could enjoy. Thus the dog-keeper of Ipy was buried in the rock-tomb of his master (E 60; S 14). Garstang found his coffin in a subsidiary shaft of the inner chamber; the burial contained four calcite vases and a head-rest placed under the head of the deceased. A second burial was found in the same shaft, probably that of another attendant, but normal-sized and with no funerary equipment.[57]

Dwarfs are also shown leading a larger type of animal, the 'young oxen' which symbolized a wealthy estate.[58] On four reliefs, dwarfs stand for the normal servants, probably because they added a touch of prestige to the parade. In the tombs of Ireru (E 23; fig. 9.12) and Kakhent (E 38a), they are conspicuously malformed, but elsewhere they are only reduced in size (E *d* 42, *d* 47), which leaves room for doubt as to whether they are dwarfs. Usually the procession of cattle is led up to scribes recording property, or to the master seated before an offering-table; often a scene of slaughter follows. The presentation of desert animals seems to have been reserved for fully grown servants, who may carry the animals on their shoulders or even in their arms.[59]

Jewellers

On twelve reliefs, dwarfs are making jewellery. This trade was assigned to male members of the estate, and was similarly confined to male dwarfs. Since jewels are garments, Drenkhahn plausibly suggests that this office may have been connected with

60 61a; fig. 9.15a. Below the master: E 33, 57; pl. 24.1, fig. 9.10.

[57] J. Garstang, *The Burial Customs of Ancient Egypt* (London, 1907), 40–1, fig. 28.

[58] On these processions, see Vandier, *Manuel*, v. 13–17. On slaughtering scenes, see A. Eggebrecht,

Schlachtungsgebräuche im alten Ägypten und ihre Wiedergabe im Flachbild bis zum Ende des Mittleren Reiches (Munich, 1973).

[59] Cf. Badawi (n. 24 above), figs. 35 (bottom register) and 36 (top and bottom registers). On these scenes in general, see Vandier, *Manuel*, v. 27–45.

dwarfs' special care for clothing.[60] This relationship is clearly displayed in the tomb of Inti (E 59a; fig. 9.13a). Three dwarf jewellers are placed between attendants associated with dress: a man carrying a bag of linen on the left, and a group of sandal-makers on the right. Similarly, in the tomb of Ptahhotpe II the dwarfs appear in a grooming scene (E 34; pl. 21); attendants (hairdressers, manicurists, overseers of linen) are busy performing the toilet of the tomb-owner who listens to music and faces four dwarfs fastening large collars.

Dwarf craftsmen seem to have had very specific tasks. They are not depicted melting or hammering cast metal, which might be too difficult for them—although gold is not hard to hammer. Usually, they string beads, as in the tombs of Wepemnefert (E 24; pl. 19.2, fig. 9.14) and Mereruka (E 52a; pl. 22.1), or fasten a locket to the ends of a collar (E 34, 36a; pl. 21), a delicate task appropriate to their small hands.[61] In the tomb of Kaemrehu (E 30), the dwarfs each hold upright two pieces of metal (?), which may indicate that they are stretching or gilding; they may also be simply threading beads.[62] Often completed articles are displayed on a shelf above their heads.[63]

In most reliefs, dwarfs work in pairs, standing or seated, both holding a larger collar (E 24, 34, 36a; pls. 19.2, 21, fig. 9.14), a pectoral (E 52a; pl. 22.1) or a counterpoise (E 49a). There may be two, three, four, or even six dwarfs, as in the tomb of ᶜAnkhmaᶜhor (E 49a).[64] Some dwarfs had furniture made to fit their size. In several tombs, they are seated on very low stools, so that their feet reach the ground (E 12, 34, 49a, 52a; pls. 21, 22.1), or they work at small tables (E 36a, 62a). Their specific needs, however, were not always respected, and some dwarf goldsmiths are perched uncomfortably on standard large stools (E 16, 24, 55; pl. 19.2, fig. 9.14).

Two scenes in the tombs of Wepemnefert and Mereruka are enlivened by dialogues; these may be 'socially prescribed utterances', as Weeks suggests.[65] Both stress the goldsmiths' desire to work quickly. In the tomb of Mereruka, one of the first pair of dwarfs says: 'It is very good, my companion' (or 'The man who is with me is very good'), while a dwarf of the second group says: 'Hurry up, get it done' (E 52a; pl. 22.1). In the tomb of Wepemnefert, a dwarf similarly enjoins: 'Make haste with this necklace in order that it may be finished', while his companion replies 'As surely as Ptah loves you, I should like to finish it today' (E 24; fig. 9.14).

[60] Drenkhahn (n. 27 above), 45.

[61] On this activity, 'threading of gold necklaces' (stj nbw), see Drenkhahn (n. 27 above), 18–21, 45–6.

[62] Gilding: M. E. Vernier, La Bijouterie et la joaillerie égyptienne (Cairo, 1907), 134; P. Naster, 'Die Zwerge als Arbeiterklasse in bestimmten Berufen im alten Ägypten', in D. O. Edzard (ed.), Gesellschaftsklassen im alten Zweistromland und in den angrenzenden Gebieten, XVIII. Rencontre assyriologique internationale, München, 1970 (Munich, 1972), 141. Melting: P. Montet, Les Scènes de la vie privée dans les tombeaux égyptiens de l'Ancien Empire (Strasburg,

1925), 283. Threading beads: L. Borchardt, Denkmäler des alten Reiches (ausser den Statuen) im Museum von Kairo, i (Berlin, 1937) (CGC), 235; Atlas, i, pl. 404; Drenkhahn (n. 27 above), 18, IV.

[63] See E 12, 34, 59a, 62a; pl. 21, fig. 9.13a.

[64] Two dwarf jewellers: E 30, 59a; fig. 9.13a. Three: E 12. Four: E 16, 24, 34, 36a, 52a, 55, 61b, 62a; pls. 19.2, 21, 22.1, figs. 9.14, 9.15a.

[65] K. Weeks, 'Art, Word, and the Egyptian World View', in K. Weeks (ed.), Egyptology and Social Sciences (Cairo, 1979), 74.

(a)

(b)

FIGURE 9.13a. (E 59a) Dishasha, tomb of Inti
b. (E 59b) Ibid.

FIGURE 9.14. (E 24) Giza, tomb of Wepemnefert

The reference to the god Ptah, the patron of craftsmen, is intriguing, but need not imply that Old Kingdom dwarfs already had a particular relation to Ptah; they could invoke the god as goldsmiths, not because they are dwarfs.[66] Similarly, the small man in the tomb of Nikauisesi has a name compounded with Ptah (E 54a), but such names are common at Saqqara in the fifth and sixth dynasties.

Dwarf jewellers do not seem to have been a special class of craftsmen. None of them is naked; they wear plain kilts, sometimes pointed, like their normal-sized companions. Usually they work among craftsmen, often goldsmiths, as in the tombs of Nebemakhet (E 12), ꜥAnkhmaꜥhor (E 49a), and Mereruka (E 52a; pl. 22.1), or metalworkers and sculptors, as in the tombs of Kaemrehu (E 30) and Kairer (E 55).

In a few reliefs they work together with fully grown men. In the tomb of Nebemakhet, a dwarf holds a necklace together with a full-sized goldsmith (E 12); in the tomb of Serfka, he brings a necklace to an overseer kneeling behind the dwarfs (E 61b; fig. 9.15a); in the tomb of Wepemnefert, the normal-sized supervisor comments on the work with encouraging words: 'You will please its possessor! I have seen that (ornament)' (E 24; fig. 9.14).

The dwarfs are differentiated only by their physical disproportion, which is, however, not emphasized to create a grotesque or spectacular effect. Rupp suggests that some designers may have tried to soften the abnormal shortness of the dwarfs by showing them standing, as in the tombs of Ibi (E 62a) and of Ptahhotpe II (E 34; pl. 21),[67] but this treatment could be chosen for compositional reasons. Naster notes that these dwarfs are never shown presenting necklaces to their master, except in the tomb of Inti (E 59a; fig. 9.13a), because that function was reserved for attendants of a higher rank.[68] But, as seen above, jewels may have been kept in the boxes brought by dwarfs. Montet made the humorous—and rather implausible—suggestion that dwarfs were chosen as goldsmiths out of caution: they would be easily captured if they tried to run away with precious items.[69]

These pictures of short jewellers differ profoundly from the image of Indo-European dwarf goldsmiths. While Indo-European dwarf smiths are powerful sorcerers who master the mysterious fusion of metal or the transmutation of primordial elements and make magical objects, Old Kingdom goldsmiths prosaically thread beads, or fasten together necklaces produced by other people, sometimes in the same relief. This reflects the fact that metalworking never developed a mythology of its own in Egypt; this activity, and mining in particular, had a low status and is almost absent from texts.[70] There is hardly any allusion to dwarf jewellers in other eastern countries,[71] except in the

[66] See Rupp, 'Zwerg', 277 n. 1; Drenkhahn (n. 27 above) 46 n. 24.

[67] Rupp, 'Zwerg', 280.

[68] Naster (n. 62 above), 141.

[69] Montet (n. 62 above) 276; id., 'Ptah patèque et les orfèvres', *Revue Archéologique*, 40 (1952), 2.

[70] See e.g. the description of the 'stoker' in the *Satire of the Trades*; Lichtheim, *Literature*, i. 188.

[71] So for Naster (n. 62 above), 142.

FIGURE 9.15a. (E 61a, b) El-Sheikh Sa'id, tomb of Serfka
b. (E 61c) Ibid.

Ugaritic Keret epic, where dwarf craftsmen may be identified as the makers of the horse's metal bit.[72]

Dwarf Entertainers

Three reliefs and two figurines depict male and female dwarf entertainers, dancing, singing, or playing the music (E 13, 32, 45, 116, 118; pls. 18.1, 29.3, 30.1, figs. 9.16–18).

---

[72] W. F. Albright, 'Dwarf Craftsmen in the Keret Epic and Elsewhere in North-West Semitic Mythology', *IEJ* 4 (1954), 1–4.

FIGURE 9.16. (E 13) Giza, tomb of Debheni

FIGURE 9.17. (E 45) Giza, tomb of Nunetjer

FIGURE 9.18. (E 32) Tomb of Ka‘aper

Dwarf dancers appear in two Giza tombs among normal-sized performers. In the tomb of Debheni, seven women execute a variant of the secular *jb3* dance before the deceased (E 13; pl. 18.1, fig. 9.16). They are divided into two groups: on the left, four women dressed in short skirts perform a step; they raise their right arms and legs, left hand on hip. They are followed by three women in long dresses who clap the rhythm.[73] The female dwarf is treated separately. She stands behind the group of clappers, and is set apart by a tall cylindrical box ending in a head of feline shape; she is naked, with a shaven head, while the full-sized dancers are dressed and wear floral crowns. She seems to imitate the step of the first group of women; her right foot is stretched forward, and her left hand hangs by her side, but her right arm is not fully stretched because of her stiff joints. Is she a member of the troop of dancers, or an attendant in charge of the performers' accessories? Tall boxes similar to that set before her often occur in scenes of dancing; they are usually placed behind the group of clappers, as in the tombs of Serfka and Inti,[74] or among objects related to dress, as in the tomb of Meres'ankh (E 11a; fig. 9.2).[75] They may have contained linen, but not exclusively; in the tomb of Neferbauptah, a man takes a baton out of a similar box.[76]

In the tomb of Nunetjer, the dwarf (probably female) appears in the middle of two groups of women performing a lively dance (E 45; fig. 9.17).[77] In the left group, the dancers raise their right arms and legs, shaking a sistrum in the right hand and grasping a curved stick in the left; on the right, the women lift their left leg and arm, right hand on hip. The dwarf seems to belong to the first group of dancers; like them, she holds a sistrum and copies their movements, but unsuccessfully because she cannot lift her arms as high as her companions. She is also distinguished by special dress; she is crowned with a wreath of flowers, like the group of women clappers squatting on the left, but wears a plain belt with strips, unlike the surrounding dancers in short kilts with crossed shoulder-straps.

In both tombs, the performance of the dwarfs has no clear religious meaning, but adds a humorous touch to the scene.[78] The short women interchange with monkeys, which appear in musical contexts among human performers. In the tomb of Serfka (E 61; fig. 9.15a), a monkey joins in a dance similar to that executed by the dwarf in the tomb of Nunetjer (E 45; fig. 9.17), standing before women clappers and imitating the

---

[73] For a discussion of the steps of this variant, see Brunner-Traut, *Tanz*, 19–20. Cf. ibid. 15, fig. 3, similar full-sized *jb3* dancers in the tomb of Ty, and 20, fig. 5 in the tomb of Hetepherakhti.

[74] For the tomb of Serfka, see E 61; fig. 9.15a. For the tomb of Inti, see W. M. F. Petrie, *Deshasheh* (London, 1898) (MEEF 15), pl. XII.

[75] Dunham/Simpson (n. 25 above), 16, reg. 5 (3), fig. 8 ('a clothes bag').

[76] Sourdive, *La Main*, 6, pl. II, fig. 2.

[77] On this dance, see Brunner-Traut, *Tanz*, 27–30, esp. 28 ('Der lebhafte "regellose" Gruppentanz mit Sistren und Hölzern').

[78] H. Junker, *Giza*, x (Vienna, 1951), 136; W. Guglielmi, 'Humor in Wort und Bild auf altägyptischen Grabdarstellungen', in H. Brunner, *et al.* (eds.), *Wort und Bild* (Munich, 1979), 184.

movements of the dancers. The figure is badly preserved, and Davies suggested at first that the thin, small silhouette could be that of a short-statured person.[79]

The substitution dwarf/monkey is more conspicuous in an unusual scene in the tomb of Ka^{c}aper at Saqqara (E 32; fig. 9.18). A dwarf and a monkey are depicted in two sub-registers before a large scale figure of the deceased embracing his wife. The dwarf stands before a man whose upper part is lost, probably a flute-player. Unlike full-sized singers, who are usually seated on the floor with their elbows resting on their knees, the dwarf is standing, perhaps to lessen the difference in size with him and the musician, or to preserve the unity of the composition. He raises his left arm and makes a special gesture with his right hand, touching the little finger with his right thumb. Hickmann identified similar gestures as chironomic; they could be either rhythmical signs to direct a performance or melodic signs indicating the tone chosen for the singer.[80] In the register below, a monkey in the same pose substitutes for the short man, but it faces a harpist.

An unusual ivory figurine in Baltimore may also represent a dwarf entertainer (E 118; pl. 30.1). The man is naked, with short hair. His slightly flexed legs are suggestive of dancing; his hands are clasped over his belly, and may be clapping to beat the rhythm. Randall notes that there is a dowel hole in the head and that the hands are pierced, perhaps to make the statuette twist. The feet are missing and may have been set on a peg.[81] Like the el-Lisht toy (E 122; pls. 30.2, 31), the dwarf was perhaps mechanical and could move. But while the el-Lisht dwarfs may represent pygmies, the appearance of the Baltimore dwarf is not exotic and conforms to Old Kingdom conventions.

The only depiction of a dwarf musician is that of a harpist from the tomb of Nikauinpu at Giza (E 116; pl. 29.3a, b); again the small man interchanges with a tame ape. Monkeys are often associated with harpists, as in the tomb of Serfka, where a monkey is seated behind the musician (E 61; fig. 9.15a). Middle Kingdom faience models also depict monkeys playing the harp.[82] However, unlike even a well-trained monkey, the dwarf could have been a talented musician. From the Middle Kingdom on, another physical deficiency, blindness, characterized harpists. Is this a coincidence? Blackman suggested that malformed musicians were preferred because they would not

[79] N. de G. Davies, *The Rock-Tombs of Sheikh Saïd* (London, 1901) (ASE 10), 13. *Contra*: Weeks, *Anatomical Knowledge*, 177, D 4 (monkey). For material on dancing monkeys, see Brunner-Traut, *Tanz*, 33–4. Cf. R. A. Caminos, *Late-Egyptian Miscellanies* (London, 1954), 83: 'Apes are taught to dance' (Pap. Anastasi iii. 4. 1).

[80] H. Hickmann, 'La Chironomie dans l'Égypte pharaonique', *ZÄS* 83 (1958), 96–127, esp. 106–7, fig. 18.

[81] R. H. Randall, *Masterpieces of Ivory from the*

Walters Art Gallery (New York, 1985), 35, 42.

[82] For material, see B. J. Kemp and R. S. Merillees, *Minoan Pottery in Second Millenium Egypt*, (Mainz am Rhein, 1980), 146, pls. 10, 11, 17. See also C. Ziegler, *Catalogue des instruments de musique égyptiens, Musée du Louvre* (Paris, 1979), 87 (playing the double-flute), 117 (playing the lyre). Cf. Ael. *NA* 6. 10: 'Under the Ptolemies Egyptians taught baboons their letters, how to dance, how to play the flute and the harp' (trans. A. F. Scholfield (Loeb 1959)).

bewitch the ladies of the household.[83] It is unlikely that Nikauinpu chose a dwarf as a harpist for such reasons, because his tomb yielded two other models of harpists, but full-sized, male and female.[84]

Outdoor Scenes

Short people are very rarely involved in outdoor activities, probably because they were regarded as too precious for performing heavy tasks. Their high-pitched voices, however, may have been employed in bird-catching, as is suggested by two similar scenes in the tombs of Hesy, Hapi (E 53a) and Akhtihotpe (E 56). Beaters stand around a tree partly covered with a net, while the dwarf raises his left hand to his mouth, probably to shout, and waves his right hand to frighten the birds.[85]

In three scenes, dwarfs are associated with boats. They stand at the stern of the ship (E 59b; fig. 9.13b), at the prow (E d 38b), or on top of the cabin (E 64; pl. 24.2).[86] In each scene, the dwarf holds a club-shaped object or a long sceptre in his raised right hand, probably to direct the navigation of the ship.[87] Light weight could have qualified dwarfs as pilots. They also interchange with monkeys. Apes are found standing at the same place, holding the rudder, as on the Hanover fragment (E 64; pl. 24.2), or holding a sceptre, as in the tomb of Nufer.[88] Like monkeys, dwarfs may have been used as entertaining, good-luck companions.

The context of the scene in the tomb of Kakhent (E d 38b) is exceptional; the short man is not among fishermen, but in a skiff carrying the tomb-owner's wife. He has no obvious malformation, and his identification as a dwarf is very uncertain. He could also be a young man; he stands in the place usually taken by the son of the family, and bears the title of scribe, as do all the rowers in the boat.[89]

Integration in Society: Seneb

Although many dwarfs seem to have been regarded as human pets, some achieved the rank of court officials. The best known example is of the dwarf Seneb, who was buried with his wife in a mastaba at Giza (E 41, 113; frontispiece, pl. 28.1, fig. 9.19a–d).[90] Like

---

[83] A. M. Blackman, *The Rock Tombs of Meir*, ii (London, 1915) (ASE 23), 12–13.

[84] See W. S. Smith, *A History of Egyptian Sculpture and Painting in the Old Kingdom* (London, 1946), 101, pl. 27e.

[85] See Vandier, *Manuel*, iv. 313–18.

[86] For similar scenes, see Vandier, *Manuel*, v. 718, 722, and C. Boreux, *L'Art de la navigation en Egypte jusqu'à la fin de l'Ancien Empire* (Cairo, 1925) (MIFAO 50), 472–3.

[87] On these batons, see Boreux, ibid. 404–7, 411–14, fig. 174c ('élingue de corde') and 476, fig. 186. See also Sourdive, *La Main*, 70–2.

[88] Moussa/Altenmüller (n. 37 above), pl. 23; Sourdive, *La Main*, 71, pl. xxi, fig. 2.

[89] Cf. the similar scene in the tomb of Mereruka; *Atlas*, iii. 289–90, pl. 115B.

[90] On the different possible datings of his tomb, see N. Cherpion, 'De quand date la tombe du nain Seneb?', *BIFAO* 84 (1984), 34–54 (4th dyn.), and esp. 34–5 n. 3 on previous attempts. See also e.g. H. Junker, *Giza*, v, *Die Mastabas des Snb und die umliegenden Gräber* (Vienna, 1941), 3–6 (end of Old Kingdom); K. Baer, *Rank and Title in the Old Kingdom* (Chicago, 1960), 123–4, no. 441 (mid-6th dyn. or later).

most Giza tombs, his burial was plundered and his body has disappeared, but his biography and depictions reveal the main stages of his career.

Twenty titles are inscribed on Seneb's false door and on the plinths of his statues (E 113b, c). They give his offices in no clear chronological or ranking order.[91] The meaning of some of them is obscure. Several common titles are purely honorific, such as *smr* 'companion' (c 14, *smr pr* 'friend of the House (the palace)' (c 15), or the epithet *mrjj nb.f* 'beloved of his Lord (the king)' (b 5; c 18).

His functions at court are comparable with those of ordinary dwarfs. Seneb cared for clothing; he was *jmj-r3 mr pr*3 'overseer of weaving in the palace' (c 3),[92] and he controlled the work of the dwarfs in charge of the linen as *ḥrp* 𓈗 *sšrw* (b 2; c 1), which suggests the presence of other dwarfs in the palace. The title *ḥrp ḥwwt nt mw* 'overseer of the administration of *mw*' (c 6) might also be related to the previous ones; Junker suggested that *mw* could be an archaic word denoting some kind of linen.[93] Seneb was also *jmj-r3 jwḥw* 'overseer of the *jwḥw*', probably animal-tenders like the dwarf Pepy (E 29a; pl. 20.1).[94] These titles may indicate that Seneb started his career as a dwarf in charge of linen and perhaps also of pets, and later achieved a higher rank; his earliest functions would have been too low-ranking to be mentioned. Another possibility is that Seneb was born into a high-ranking family and was soon charged with supervising activities commonly performed by dwarfs.

As an official, he was *wr-ʿj* 'great one of the litter' (b 1; c 4), and held several offices related to the care of ceremonial boats. He was *ḥrp ʿprw kzw* 'overseer of the crew of the *kz* ships' (c 7), boats that were probably reserved for royal or cult uses,[95] and *sd3wtj-ntr Wn-ḥr-b3w* 'keeper of the God's seal of the *Wn-ḥr-b3w* boat' (c 8), a papyrus bark used in specific festivals.[96]

Various priesthoods contributed to his income; he was attached to the funerary service of Kings Khufu and Raʿdjedef (c 9; 10). As a sign of respect, he gave his three children names referring to them: his son was called Radjedefankh 'May Raʿdjedef live', his elder daughter Awibkhufu 'Happy is Khufu', while his younger daughter was named Smeretradjedef 'Companion of Raʿdjedef'.[97] He was also *ḥm ntr w3djt* 'priest of Wadjet' (b 3; c 11), priest of 'the large bull which is at the head of *Stpt*' (c 12) and of the bull *Mrḥw* (c 13). These two latter titles are especially interesting since they suggest an early association of dwarfs with sacred bulls. Seneb's marriage to a woman of high rank completed his status; she held priesthoods of Hathor and of Neith.[98]

The location of Seneb's burial may be related to his specific role at the court; he was

[91] Junker, ibid. 12–18 (transcription, trans. and comm.); Baer, ibid. 123–4, no. 441.

[92] J. Pirenne, *Histoire des institutions et du droit privé de l'ancienne Egypte*, ii (Brussels, 1934), 423, no. 98 reads 𓈘 *š*, 'garden pool', in accordance with Gardiner, *Grammar*, sign-list, N 39.

[93] Junker (n. 90 above), 14–15, no. 6; *Wb* ii. 53.

Contra: Pirenne (n. 92 above), who reads 'maître des châteaux de l'eau'.

[94] For a discussion of the word, see above 31–2.

[95] Junker (n. 90 above), 15, no. 7.

[96] Ibid. 15–16, no. 8.

[97] Ibid. 19–20.

[98] Ibid. 18.

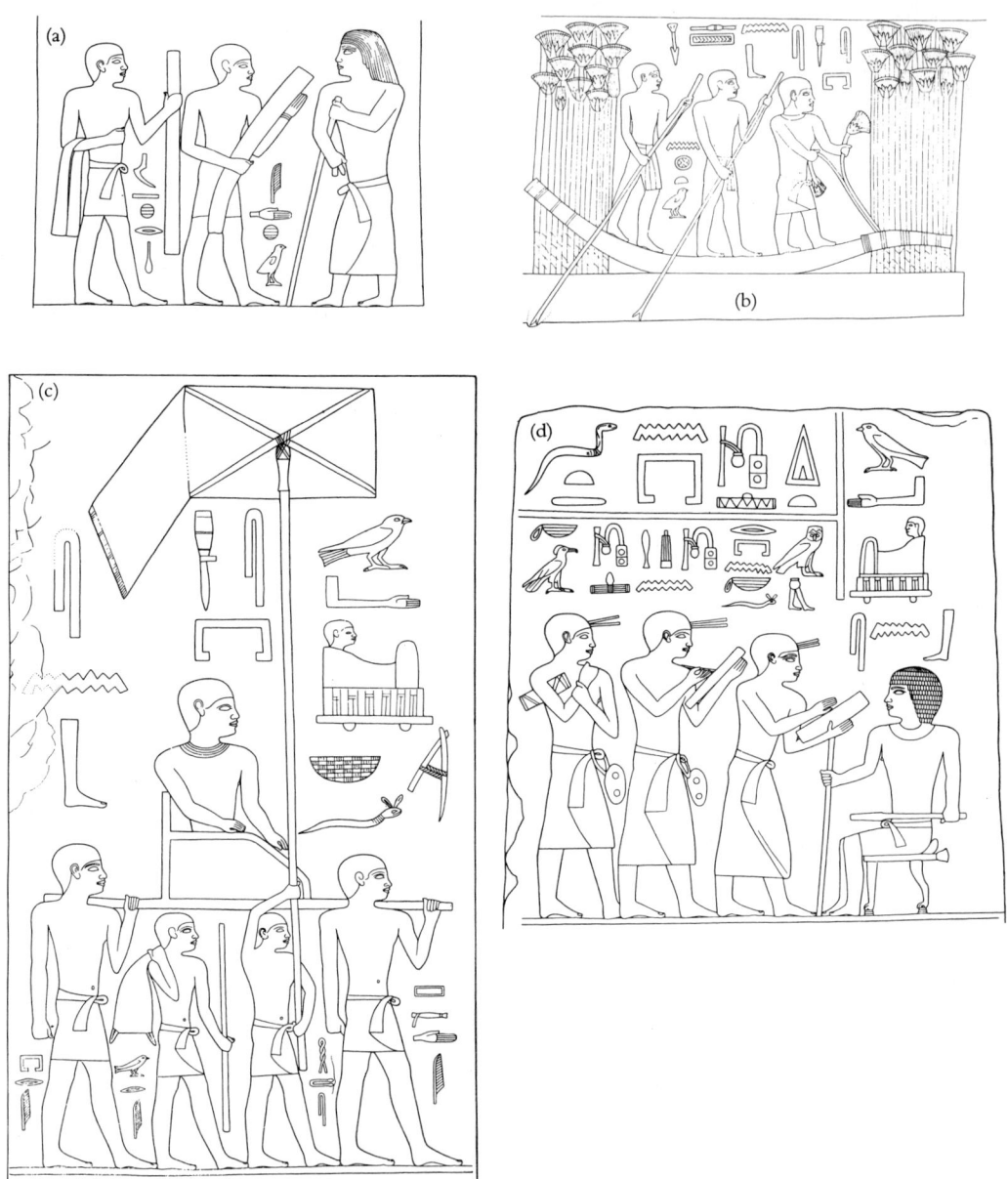

FIGURE 9.19. (E 41b, e, g, h) Tomb of Seneb. Cairo Museum, JdE 51297

buried next to an official named Ankhu who held two similar titles: *wr-ˁj* 'great one of
the litter' and *jmj-r3 mr prˁ3* 'overseer of weaving in the palace'.[99] The fine limestone
false door of Seneb is divided into registers depicting Seneb engaged in the standard
activities of a wealthy courtier: being censed by his son (E 41a), inspecting linen (E 41b;
fig. 9.19a) and cattle (E 41f), receiving accounts (E 41h; fig. 9.19d), going out in a litter
(E 41g; fig. 9.19c) and on various boats in the marshes (E 41c, d, e; fig. 9.19b).

His clothing and attributes conform to his rank. He wears a short curly wig (E 41a, h;
fig. 9.19d), a long wig (E 41b; fig. 9.19a), various kilts, especially a long one reaching
below the knee (E 41b; fig. 9.19a), a priestly panther-skin garment (E 41a),[100] and broad
collars (E 41d, e, g, h; fig. 9.19b–d). He holds a sceptre (E 41h; fig. 9.19d), a long
walking staff (E 41a, d, f, h; fig. 9.19d),[101] small pieces of cloth (41b, f; fig. 9.19a) or a
larger one slung over the shoulder (E 41a).[102] As a high official, he also owned two pet
dogs captioned with names (E 41f).[103]

Several elements in the decoration of the tomb are unusual and show the desire to give
Seneb a very dignified image. As already noted, the designer compromised between two
conventions to indicate both rank and physical characteristics: he depicted Seneb on a
large scale, as tall as his fully grown servants, but rendered his dwarfish proportions
which were essential to his social and religious role.

In most scenes, the composition softens the enlargement of the figure. Junker
observed that on the *š3bt* skiff, Seneb only half-kneels, while paddling sailors squat on
both legs in order to lessen their size (E 41c). Similarly, the scribe rendering accounts of
cattle is depicted at a slightly smaller scale than the figures in the register below; he is
not placed directly before Seneb, but stands at the entrance of the pavilion, separated by
two dogs, which preserves the eminence of the personage (E 41f). In another scene,
three scribes diminish in size towards the dwarf in order to minimize the difference (E
41h; fig. 9.19d).[104]

Another peculiarity is that Seneb is never depicted with his wife in the same relief.
One would expect to find her squatting beside her husband in the papyrus skiff (E 41e;
fig. 9.19b), or seated together with him in the scene of rendering accounts (E 41h; fig.
9.19d). She is probably omitted because she would have disturbed the compromise
found by the designer. As a major figure, she would have to be shown larger than her
servants, but she would then be larger than her husband. Seneb could not be seated
beside her without being in a demeaning position, with his feet not reaching the ground;
the solution found for the statue-group is impracticable in two-dimensional side view

[99] Ibid. 22.

[100] E. Staehelin, *Untersuchungen zur ägyptischen Tracht im Alten Reich* (Berlin, 1966) (MÄS 8), 48 ff.; Cherpion (n. 90 above), 37 and 52, table B for parallels.

[101] For a discussion of these staffs, see Cherpion (n. 90 above), 37 (*mdw* staff), 38 (*sḫm* sceptre).

[102] On these pieces of cloth, see n. 16 above.

[103] For a discussion of their names, see Junker (n. 90 above), 74–6.

[104] Ibid. 35–6.

(E 113c; frontispiece). Seneb's wife therefore appears alone with her children, but in the register just below the scene of rendering accounts.[105]

Some poses of Seneb also reveal his unusual status. His short arms, for example, do not allow him to adopt the same pose as fully grown men. In the censing scene, he holds a long stick which is not set away from the body, as is the rule, but seems to lean against his legs (E 41a).[106] For the same reason, when performing the $z\check{s}\check{s}$ $w3\underline{d}$ ceremony, Seneb is not shown in a male stance, raising a papyrus stem, or wielding a throwing-stick, but gently pulls papyri with both hands, like a woman (E 41e; fig. 9.19b).[107] Yet he might have liked to exhibit his physical strength; in the sailing scene, he holds the rigging of the boat with both hands, as if to show that he could perform this hard work—or that he specially enjoyed sailing (E 41d).[108] Several items of furniture were probably made to order, like staffs, low stools (E 41a, f, h; fig. 9.19d), and the short litter, with a low back and large side panels which hide his malformed legs (E 41g; fig. 9.19c).[109]

The tomb contained three statues of Seneb, one in wood (E 113a), one in granite (E 113b) and one in limestone (E 113c; frontispiece, pl. 28.2). These were sealed in stone boxes, but only the limestone one was found complete. The wooden one had disintegrated, but Junker was able to observe that it was about 30 cm. high and depicted Seneb standing, holding a walking-staff in his left hand and a sceptre in his right.[110]

The limestone group shows Seneb seated with his wife and children, like a normal courtier. The treatment of the group is exceptional. As in reliefs, Seneb's malformation is not hidden, but softened: his short legs do not hang down the seat without reaching the ground, but are bent under him. Two of his children occupy the position of his legs; this preserves the harmony of the composition, and creates an appearance of normality without disguising Seneb's physique. The children are not shown as inheriting his disorder.[111]

The implications of the case of Seneb should not be overstressed. His high position may be due to unknown propitious circumstances, such as being perhaps the son of a powerful father, and his career may have been quite exceptional. Yet the fact that he married a normal wife and had children demonstrates that his abnormality was not a taboo creating insurmountable obstacles to social and perhaps also conjugal achievement.[112]

[105] Junker (n. 90 above), 35, 86–91, fig. 23.

[106] Ibid. 37.

[107] Junker, ibid. 66; Y. Harpur, $z\check{s}\check{s}$ $w3\underline{d}$ scenes of the Old Kingdom, GM 38 (1980), 53–60, esp. 57.

[108] Junker (n. 90 above), 65.

[109] Junker, ibid. 35, 83; Vandier, Manuel, iv. 334.

[110] Junker (n. 90 above), 119. Cf. the wooden statues of Ka^{c}aper (h. 112 cm.), Cairo Museum, CG 34, and that of Merire-Hashetef (h. 73 cm.), Cairo Museum, JdE 46992, in the same pose, holding similar sceptre and staff; Musée égyptien du Caire, no. 40 and 64.

[111] See e.g. the discussion of E. L. B. Terrace and H. G. Fischer, Treasures of the Cairo Museum (London, 1970), 65–8; C. Aldred, 'Statuaire', in J. Leclant (ed.), Le Temps des pyramides (Paris, 1978), 200–1; id. (n. 43 above), 77; Musée égyptien du Caire, no. 39.

[112] The tomb of another dwarf official, Pernankh, has been discovered in the western field at Giza in 1989–90; for bibliog., see Introd. n. 8.

Other dwarfs were buried in their own mastabas, like the dwarf Petpennesut at Giza (E *d* 112), and probably also Khnumhotpe at Saqqara (E 117; pl. 29.2). The figure of Petpennesut is carved only with his name, while the modest career of Khnumhotpe is alluded to by two titles inscribed on the plinth: overseer of linen and ka-priest, like Redji, the dwarf of Mereri (E 50). These titles suggest that Khnumhotpe belonged to the household of a high official. His priesthood may indicate that he was particularly appreciated by his lord, who may have offered his burial, as happened to the dwarf Djeho in the Late Period (E 84; pl. 26.2). But the office of ka-priest was also transmitted within a family, and Khnumhotpe could have inherited it along with other privileges.[113]

Religious Role

The priesthoods of Seneb show that dwarfism did not preclude holding religious offices. There is, however, no certain Old Kingdom representation of a dwarf performing a religious function. In the tombs of Kaemnefert (E 39; fig. 9.6) and Meres^cankh (E 11a; fig. 9.2), a female dwarf carries on top of her head a tall, narrow box, which is perhaps a ritual object. This item may be compared with the shrine brought by a pair of fully grown women in the same line of bearers in the relief of Meres^cankh (E 11a; fig. 9.2), which resembles the hieroglyphic depiction of the divine Booth of Anubis.[114] This lack of evidence may be related to the general sparseness of such material in Old Kingdom iconography.

Written evidence supplements our knowledge. Two inscriptions from Giza and Saqqara confirm that dwarfs may have belonged to temple staff. The Saqqara inscription is carved on a limestone offering basin; it says that the official Shedi held the priestly titles of 'Overseer of the *s* ∬ dwarfs of the god's palace', and of 'One who belongs to the *s* ∬ dwarfs'.[115] The latter title was inherited by his eldest son Nebeminenet.[116]

A stela from Giza similarly reveals that Ma, among other priestly titles, such as 'scribe of the temple' and 'possessor of reverence before the great god', was 'Overseer of the *s* ∬ dwarfs'.[117] This function was held by his forefathers who are listed in the far left column under the heading title of 'He who belong(s) to the dwarfs of the god's palace'.

In both inscriptions the charge of overseeing dwarfs is thus a significant honorary office. It is the only title attributed to Shedi and his son in Saqqara, and to the ancestors of the Giza scribe Ma.

---

[113] On ka-priests, see n. 38 above.

[114] See A. M. Abu Bakr, *Excavations at Giza, 1949–1950* (Cairo, 1953), 33, fig. 27; I am very grateful to Y. Harpur for this information. Compare with the usual shrine shapes in M. Müller, *LÄ* v (1984), s.v. Schrein, 709–12.

[115] Cairo Museum,-; H. G. Fischer, 'Five Inscriptions of the Old Kingdom', *ZÄS* 105 (1978),

47–52 (end of 6th dyn.?).

[116] Trans. Fischer, ibid. 48 (right side and bottom of the basin).

[117] Cairo Museum, CG 1652; L. Borchardt, *Denkmäler des alten Reiches (ausser den Statuen)*, ii (Cairo, 1964) (CGC), 113; H. G. Fischer, 'Chroniques, Monuments of the Old Kingdom in the Cairo Museum', *CdE* 43/86 (1968), 310–12.

Little is known about the cult in Old Kingdom temples, but these titles suggest a possible role for dwarfs in them. They may not have held a priestly function, but could have belonged to the same category of auxiliaries as musicians and singers.[118] They perhaps cared for the linen kept in the temple. Fischer has shown that the sign *s* which accompanies the dwarf determinative in both inscriptions, may be a substitute for the word *sšr*, linen; as a derived adjective, it could mean 'clothier' or 'overseer of the linen'.[119] Like Old Kingdom lay priests, dwarfs may have held this function on a part-time basis only, and alternated it with their office in the households of officials. These entertaining attendants would contribute to making the temple an attractive residence for the god.[120]

On both the Giza and Saqqara monuments the name of the deity is not given, but Fischer observes that in the tomb of Debheni *ꜥḥ nṯr* 'palace of the god' denotes the location of a festival of Apis.[121] Dwarfs are found attached to the cult of sacred bulls from the Old Kingdom on, and it is tempting to see here another possible early indication of this association.[122] This suggestion, however, is very hypothetical since the term 'god's palace of Upper Egypt' also describes temples of Nekhbet, Wadjet, Re, and of all gods in general on the Palermo Stone, as Fischer notes.[123]

In addition two texts, the letter of King Pepy II to Harkhuf, and a spell in the Pyramid Texts (§ 1189), mention another type of dwarf, called *dng*, who may be identified as a pygmy, and who is intimately associated with ritual performances.

We have seen that the *dng* described by King Pepy II was appreciated as an exotic luxury, more precious than 'the gifts of the mine lands and of Punt' (l. 21).[124] The rarity of the small man seems to have qualified him for a specific function, the performance of *jb3w nṯr*, dances of the god (ll. 7 and 17), which would delight the heart of the king.

The circumstances of the performance of Harkhuf's *dng* are unknown. Was it executed in the king's palace, when the products brought back from the South were presented to the king? The youthful Pepy II may have been delighted by a dwarf's dance to enliven this solemn ceremony. Or was it performed in a cult context?

---

[118] Cf. the records of staff on duty in the funerary temple of Neferirkare Kakai; P. Posener-Kriéger, *Les Archives du temple funéraire de Néferirkarê-Kakaï* (Cairo and Paris, 1976), 565–609 (musicians, craftsmen, butchers, etc.).

[119] Fischer (n. 115 above), 48, figs. 5a and b. On normal-sized 'overseers of linen' in funerary temples, see e.g. Posener-Kriéger (n. 118 above), 599–600, and on the different types of linen kept in the temple, ibid. 341–67.

[120] On this aspect of cult, see Hornung, *Conceptions*, 229; J. Assmann, *Ägypten, Theologie und Frömmigkeit einer frühen Hochkultur* (Stuttgart, 1984), 59–60.

[121] S. Hassan, *Giza*, iv (Cairo, 1943), 168, fig. 118, col. 15; Fischer (n. 115 above), 50. For Old Kingdom references to the cult of Apis in Memphis, see E. Otto, *Beiträge zur Geschichte der Stierkulte in Aegypten* (Leipzig, 1938), 11–15.

[122] Cf. the titles of the dwarf Seneb (E 113c 12–13), priest of 'the large bull which is at the head of *Stpt*' and of the bull *Mrhw*.

[123] Fischer (n. 115 above), 50. Cf. Seneb's title *ḥm nṯr w3djt* (E 113b 3, c 11).

[124] See text above 25–6.

The formulation of the text throws some light on the meaning of this dance. Goedicke notes that it was performed for the king in his official and cult role (*njswt-bjtj*, 'King of Upper and Lower Egypt'), and not in his everyday person (*ḥm*), which suggests that it may have taken place in a religious context.[125]

The term *nṯr* probably designates the sun-god here. In New Kingdom and Late Period liturgical texts, the word *jbꜣ* describes a dance celebrating Re at his rising.[126] On a mythical level this dance was performed by baboons from south-eastern deserts, who also greet the newborn sun by shouting and singing.[127] In the Pyramid Texts (§ 608; § 1437), these baboons are said to be sons of Re.

Pygmies' dancing, like baboons' rejoicing, may have been associated with solar beliefs. They gladdened the heart of the king, because they celebrated his power as son of Re.[128] On a ritual level, pygmies could suitably have taken on the role of legendary baboons. Both types of beings have the same southern provenance and similar physical proportions. These similarities have often led to questioning the status of pygmies as full humans. From antiquity on, many travellers described pygmies as agile, hairy, ape-like creatures, sometimes with a tail.[129] This ambiguous status still appears in modern Tibesti vocabulary: Wolff noted that *dunku* (related to *dng*?) means 'baboon', and that the animal is regarded as a bewitched man and is taboo.[130]

The spell in the Pyramid Texts (§ 1189) fits in such a solar ritual context.[131] The deceased king is also described in his divine capacity as *njswt-bjtj*; he performs an *jbꜣ* dance before the throne of the sun-god, perhaps to celebrate solar renewal, and hence his own regeneration in the afterlife. He appears in the guise of a pygmy symbolizing his special solar affinities. The dancing of Harkhuf's *dng* could also have formed part of a special occasion, such as the *sed*-festival, as later evidence suggests (E 73; E 87).

[125] H. Goedicke, *Die Stellung des Königs im Alten Reich* (Wiesbaden, 1960), 9–10; D. Silverman, 'Pygmies and Dwarves in the Old Kingdom', *Serapis*, 1 (1969), 53.

[126] Brunner-Traut, *Tanz*, 77.

[127] For depictions of baboons adoring the rising sun-god, see e.g. the vignettes from funerary papyri in R. O. Faulkner, *The Ancient Egyptian Book of the Dead* (London, 1985), 42 (London, BM 10479/11) and 43 (London, BM 10472/1). In relief, see e.g. the solar altar with four baboons from Abu Simbel; Cairo Museum, JdE 42.955; C. Desroches Noblecourt, *Le Grand Pharaon Ramsès II et son temps* (Montréal, 1985), no. 2.

[128] On this quality of the king, see J. Assmann, *Der König als Sonnenpriester* (Glückstadt, 1970), (ADAIK 7), 60–5.

[129] For antiquity, see e.g. Ctesias, *FGrH*, 688, F 45

= Photius, *Bibl.* 72. 46a–b; Nonnosus in Photius, *Bibl.* 3. 3a. 25–6. Cf. G. Schweinfurth, *The Heart of Africa*, ii (London, 1873), 137–8, who reported that he had been told about the existence of little men with tails, covered with long hair, and with long beards, but added that he never observed that characteristic. Yet H. M. Stanley, *In Darkest Africa*, ii (London, 1890), 40, noted that 'the fell over the body [of a pygmy] was almost furry, being nearly half an inch in length.' For further references, see R. Hartmann, *Die Nigritier*, i (Berlin, 1876), 493–502; D. MacRitchie, *Encyclopaedia of Religion and Ethics*, v (1912), s.v. Dwarfs and Pygmies, 123; H. F. Wolff, 'Die kultische Rolle des Zwerges im alten Ägypten', *Anthropos*, 33 (1938), 464–6.

[130] Wolff, ibid. 467. Cf. the ape-like Greek Cercopes below 190–1.

[131] See text above 28.

FIRST INTERMEDIATE PERIOD (2134–1970? BC)

Relatively few decorated monuments survive from the unstable First Intermediate Period. Local rulers and high officials had their tombs carved and painted with scenes of everyday life, but the quality of the representations is often very poor.

Only the tomb of Meru at Nagᶜ el-Deir shows a dwarf, clad in a kilt (E 65). The short man is very clumsily rendered with broad shoulders, long arms, and skinny short legs. He leads a gazelle by the leash before his master, and lifts his right hand in a gesture of respect.[132] This motif may derive from Old Kingdom scenes of presentation of desert animals, which are normally led by full-sized servants.

MIDDLE KINGDOM (1970?–1640? BC)

Middle Kingdom archaeological evidence is essentially funerary. Relief or painted decoration in private tombs becomes rarer and is limited to the essential, but this paucity is counterbalanced by the mass-production of wooden or glazed models of servants, animals, and food products which are deposited in the burial chamber. Some evidence comes from a domestic context, principally from el-Lahun (Kahun). Written sources, on the other hand, are absent. Apart from captions with names or titles, no text or inscription mentions human dwarfs.

Household Offices

As in the Old Kingdom, dwarfs appear as prestige attributes of provincial notables. They perform a similar range of offices in the household, but with a few changes. Most are personal attendants or nurses, while only a few are entertainers or animal-tenders. There are no more dwarf jewellers, bird-catchers, and pilots. A new motif occurs, that of dwarf-shaped offering stands for the domestic cult. Some of these representations also seem to have functioned as fertility figures and may reflect the emergence of dwarf gods as protectors of childbirth and good-luck charms.

In two-dimensional representations, short-statured attendants are found on six funerary monuments from Middle and Upper Egypt. On one of the stelae of Wepwawetaa from Abydos, a dwarf in a long kilt, with arms reaching below the knees, stands before a row of offering bearers (E 72). He is shown at a minuscule scale (a third of the size of the other figures), probably in order to fit the small space left free below the plants carried by the first servant; his figure may also be an afterthought in the composition. He holds no offering and appears as a luxury servant.[133]

On another stela from Abydos, the dwarf stands under the chair of the deceased (E 70; pl. 24.3).[134] He is the only member of the household depicted on the monument. He is

---

[132] For the identification of the animal, see Vandier, *Manuel*, v. 6 ff., esp. 8.

[133] Cf. the similar pose of the Old Kingdom

dwarfs in the tombs of Meryteti (E 52d; fig. 9.9) and Nikauisesi (E 54a).

[134] Cf. the similar location of Old Kingdom

dressed in a pointed kilt and stands in a dignified pose, his left hand laid on his chest, in an attitude which may express deference.[135] He is captioned with his name, Amenemhetsonbe.

The governors of the Oryx nome had a special liking for unusual attendants. Several dwarfs appear in their tombs at Beni Hasan. In an unpublished relief in the tomb of Baqt II, a dwarf follows his master who oversees agricultural and hunting pursuits (E d 69). Amenemhet also owned a short-statured woman with an abnormally big head, larger than those of fully grown servants, and rough facial features (E 66; fig. 9.20).[136] She stands before two large chests and carries a square fan, perhaps to indicate that she was in charge of linen and toilet objects.[137] She is not distinguished by special attire: like normal-sized servants, she wears a long wig and a dress with wide shoulder-straps. In her right hand she holds a piece of cloth which may be an attribute of her higher-ranking position.[138]

As in the Old Kingdom, dwarfs were occasionally paired with persons with physical malformations. In the tomb of Khety a dwarf and a club-footed man accompany the deceased in an outdoor excursion (E 67). Captions above their heads describe their afflictions: *nmw* 'dwarf' and *ḏnb* 'crooked'.[139] These men seem to have been held in great favour: they wear pointed kilts, and stand just behind the tomb-owner; they are also at a larger scale than the full-sized servants in the upper registers (except for the man carrying a parasol). In the tomb of Baqt I, there is a hunchback behind the dwarf and the club-footed man (E 68; fig. 9.21); the figures also wear pointed kilts, and are captioned with words referring to their abnormalities (*nmw*, *ḏnb*, *jw*). In this case, they are not at a larger scale than the other servants. In both these tombs, the malformed attendants bring no accessories and are not involved in a specific activity. This lack of employment enhances their prestige status.

In statuary, dwarf attendants appear mostly as small figurines, usually made of glazed ware, but also of wood, ivory, or lead. The majority of the statuettes come from the el-Lisht cemeteries (E 122–134; pls. 30.2–33.2), but one was also found at Beni Hasan (E 119), two at Deir el-Bersha (E 120–1; fig. 9.22), one at el-Haraga (E 144; pl. 33.3), one

dwarfs in the tombs of Nufer: Idu (E 44), Ipi (E 60), Serfka (E 61a; fig. 9.15a) and Ibi (E 62b).

[135] The servant leading pet animals in the Old Kingdom tomb of Ptahhotpe II makes a similar gesture with his right hand (E 34; pl. 21). See also the similar attitude of the club-footed man in the tomb of Baqt I (E 68; fig. 9.21). For further examples, see H. Müller, 'Darstellungen von Gebärden auf Denkmälern des alten Reiches', *MDAI(K)* 7 (1937), 100–8, esp. figs. 31–2, 37.

[136] In the painting, the total height of her body measures 28.5 cm. and her head 6.5 cm. (ratio: 1/4.3), while a full-sized servant in the register below measures 34.5 cm. with a head of 6 cm. (ratio: 1/5.7).

[137] A woman carries a similar fan in the tomb of Djehutihotpe at Deir el-Bersha; P. E. Newberry, *El Bersheh*, i (London, 1892) (ASAE 3), 37, pl. xxx.

[138] Cf. the dwarf in the tomb of King Djer (E 2; pl. 17.1), Khentika: Ikhekhi (E 48; fig. 9.8) and Seneb (E 41b, f; fig. 9.19a). For a discussion of this attribute, see above n. 16.

[139] On these terms, see above 30–2.

FIGURE 9.20. (E 66) Beni Hasan, tomb of Amenemhet

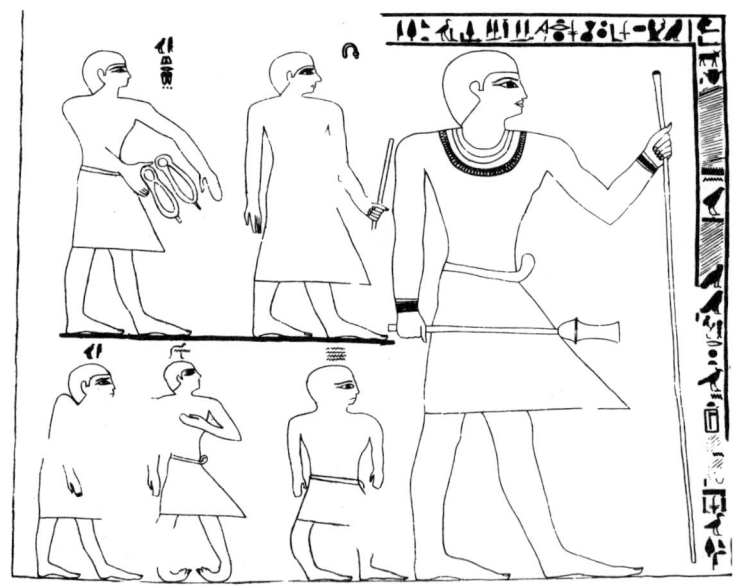

FIGURE 9.21. (E 68) Beni Hasan, tomb of Baqt I

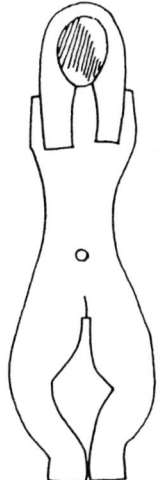
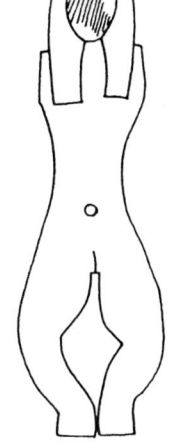

FIGURE 9.22. (E 121) Wood (h. 9.2 cm.). Cairo Museum, JdE 34.299

FIGURE 9.23. (E *d* 191) Glazed ware (h. 5.2 cm.). Boston, MFA 11.1524

at el-Badari (E *d* 145), two at Abydos (146–7; pl. 34.2), and one at Thebes (E *d* 148);[140] outweighing all this material numerically is a group of over forty statuettes which was found in the votive deposit of the Temple of the Obelisks at Byblos in Lebanon (E 150–*d* 190).[141] Nine are of uncertain or unknown provenance (E 149, *d* 191–6; pls. 33.4, 35.1, fig. 9.23).

In most examples, the dwarfs are naked, with shaven heads, sometimes with black dots rendering their short stubbly hair (E 134, 192, *d* 196; pl. 33.2). Accessories, such as belts, necklaces, and fingernails, are painted in black (E 134, 192–3; pl. 33.2, 4). As Bourriau notes, abnormal physical features are generally exaggerated: the head is overlarge, with a strikingly flat top, and the limbs are extremely short and stubby, with protruding stomach and buttocks, and an oversized phallus.[142] This grotesque appearance contrasts with the realism of a few ivory and wooden figurines, such as of the dwarf now in the British Museum (E 195; pl. 35.1a, b) and the female dwarf in Liverpool (E 146; pl. 34.2), both of which are rendered without caricature.

[140] The material from the cemetery around the North Pyramid at el-Lisht is mostly unpublished; it is being studied by J. Bourriau, to whom I am very grateful for the information she has provided on these objects.

[141] The Byblos statuettes are all deposited in the National Museum in Beirut and are not accessible for study. They are mostly unpublished, and their descriptions in my catalogue are based upon those of M. Dunand, *Fouilles de Byblos, II, 1933–1938* (Paris, 1958), 759–64.

[142] J. Bourriau, *Pharaohs and Mortals: Egyptian Art in the Middle Kingdom* (Cambridge, 1988), 122. Oversized genitals are particularly conspicuous among the Byblos statuettes (E 150–1, 156, 168, *d* 190).

Most figurines are standing, the arms on the belly (e.g. E 147), or hanging down their sides (e.g. E 192), occasionally with open palms (E 120, 133–4; pls. 32.4, 33.2). Some figurines have informal attitudes; they touch their ears or head with one or both arms (E 168, 170–2, *d* 175, 193; pl. 33.4), they hold the right hand to their mouth, like children (E 128, 147, 153; pl. 32.1), or lie on their back (E *d* 187–8, *d* 190).[143] Others squat or kneel and seem to drink from a vessel (e.g. E *d* 157–8, *d* 163, *d* 178, *d* 196); these figurines have no conspicuous physical malformations, and they could represent boys.

Several figurines repeat the Old Kingdom theme of the dwarf bearer; they carry jars on their shoulders (e.g. E 129; pl. 32.2), or hold a box or a vessel before their belly (e.g. E *d* 130–1; pl. 32.3).

The dwarf in the British Museum holds his right hand on his chest (E 195; pl. 35.1a), like the dwarf on the Abydos stela (E 70; pl. 24.3).[144] Together with the delicacy of the carving and the quality of the material (ivory), this pose suggests that the figurine was not a conventional genre piece; it could have been made for a wealthy official who was fond of his personal attendant.

Four statuettes depict dwarf nurses. The best preserved is a wooden figurine from Abydos (E 146; pl. 34.2). The woman is naked, and her hair is divided into three plaited locks; in her left arm, she carries a child who clings tightly to her body; her pierced right hand held an object that is now missing. Three faience nursing-figurines from el-Lisht are rather cruder. All three wear puffed wigs, but their pose and attire vary. One is standing, holding the child within her cloak (E 123; pl. 30.3), another, in a long dress, seems to suckle the child (E 124; pl. 30.4), while the third squats with the child across her knees (E *d* 125). This motif of nursing may continue a more ancient tradition, known only from predynastic and early dynastic statuettes (E 93, *d* 111). There are no Old Kingdom depictions of dwarfs nursing children, but such scenes were not part of the iconographic repertoire of the period.

Very few Middle Kingdom monuments show dwarf animal-tenders, although dogs, cats, and monkeys often appear in tomb reliefs and statuary. In relief, only a stela from Abydos associates a dwarf with an animal in a scene of presenting cattle (E *d* 71). Like Old Kingdom dwarfs, the small man leads an ox before a line of attendants; he is very crudely rendered, with a large paunch and short spindle-legs. In statuary, two Byblos figurines may depict dwarfs with a monkey seated on their shoulders (E 159, *d* 179). Other statuettes show dwarfs, standing or kneeling, carrying a quadruped in their arms or on their shoulders (E *d* 148, *d* 174, *d* 176–7, 183, *d* 191; fig. 9.23). Raven suggests that these quadrupeds could be calves representing the newborn Re,[145] which would imply

---

[143] For similar gestures, see e.g. the figurines of children from Elephantine (early dynastic–Old Kingdom); Dreyer (n. 18 above), figs. 42–71 (right hand to mouth), 72, 88–9 (left hand over the ear).

[144] A Middle Kingdom ivory statuette of an official in a long pleated kilt has a similar pose, but with the left arm to the chest; Baltimore, WAG 71.509 (h. 21.3 cm.); Smith, *Art and Architecture*, 184, fig. 177.

[145] Raven, *Pataekos*, 12 n. 62 (about E *d* 148 and *d* 191). On the symbolism of calves, see P. Behrens, *LÄ* iii (1980), s.v. Kalb, 296–7.

that the statuettes alluded to the role of dwarfs as defenders of the sun-god, and hence of newborn children assimilated to Re. The archaeological context of one of the statuettes (E *d* 148) supports this interpretation: the figurine comes from the tomb in the Ramesseum which contained various medico-magical items for the protection of childbirth, including a female Bes statuette.[146]

A few objects depict dwarf entertainers. The best known is the mechanism made of ivory, probably a toy, which was found in the blocking of the burial chamber of a young girl at el-Lisht. It is composed of a long ivory board, on which three male dwarfs stand, each set on a drilled spool-like peg (E 122; pl. 31a, b); a fourth figure, smaller than the other three, stands on a square tenon to be inserted in a larger base, perhaps made of wood (pl. 30.2). The entire object probably fitted in a larger wooden case, as is suggested by the remains of rotted wood found along its sides. The mechanism was very delicate and complex, as Lansing explained:

The underside of the base is peculiar in that a long channel is cut halfway between one edge and the holes cut to receive the 'spools'. Horizontal holes are drilled from the channel to the spools . . . Threads were tied to the spools and passed through the holes and out to the end of the base by way of the channel. The figures were then twisted so as to wind up a certain amount of thread on the 'spools'. By pulling the threads the three figures could be made to turn, a slight jerk producing a change in position, a stronger one a full pirouette.[147]

Lansing added that the fourth figure probably did not move; the two holes drilled through its rectangular base may have been for pegs to fasten it to mortises. Yet a thread could have passed through these holes too, possibly to make the dwarf slide laterally. All the dwarfs are naked, with slightly flexed legs, and wear large bands or shawls across their chests; two have bead necklaces. As seen above, they may represent dancing pygmies.[148] The figures on the ivory board have uplifted hands, while the detached figure claps his hands against his breast, perhaps to beat the rhythm. Lansing noted that one figure has a 'pursed' mouth, which may indicate that he is whistling or singing.[149] This ingenious toy was made to entertain the living, and had been used: Lansing observed scratches around the holes on the board and traces of mending.[150]

Dwarf musicians may also be seen in a few fragmentary glazed figurines. The upper part of a flautist, with a flat-topped skull and achondroplastic (or negroid) facial features, was found at el-Lahun (E *d* 141). Another is a surface find from el-Lisht showing a naked man beating a tambourine (E *d* 132; pl. 33.1). In side view, his kneeling pose

---

[146] On the material from that tomb, see also above 70.

[147] A. Lansing, 'The Egyptian Expedition 1933–1934', *BMMA* 29 (1934), part ii, 32. Modern toys still use the same principles; see e.g. the booklet by G. Brouillart, *Utilisons les déchets et vieux objets sans emploi* (Paris, 1942), 12–13, fig. 11: 'Une danseuse

dont le tutu prendra son joli bouffant lorsqu'on tirera sur le bout de la ficelle qui aboutit à un élastique après avoir fait deux fois le tour de la bobine supportant la silhouette de ce rustique automate.'

[148] See above 43.

[149] Lansing (n. 147 above), 31.

[150] Lansing, ibid. 32.

resembles that of New Kingdom amuletic Bes musicians (pl. 4.1), but his identification as a dwarf is very uncertain; the head is missing, and his proportions, with no conspicuous malformation, could be those of a boy.[151]

Scholars usually regard this group of small statuettes as toy servants which were meant to create the same humorous effect as models of animals, in particular monkeys, imitating humans.[152] However, t'eir poses, attributes, and find-contexts suggest that they also had a symbolism related to beliefs in dwarfs as guardians of family and fertility.

Most statuettes have the same physical appearance as New Kingdom amuletic Ptah-Pataikoi: they have squat bodies with large, flat-topped heads and snub-noses. These similarities have led some scholars to identify them as early forms of Ptah-Pataikoi.[153] The poses of a few figurines reinforce this association; some hold the palms of their hands open, either with the arms raised (E 144; pl. 33.3) or along the body (E 120, 133–4; pls. 32.4, 33.2). These gestures may express submission, praise or prayer, and thus imply a relation to the supernatural world.[154]

The corpulent belly, sometimes grossly protuberant, of both male and female dwarfs (e.g. E 144; pl. 33.3) recalls the appearance of the goddess Taweret, the guardian of childbirth and motherhood; it evokes also the bulging paunches of fecundity figures. This physical feature may have had a similar symbolism, evoking the protection of pregnancy, and more generally of fertility.[155]

Female dwarfs, in particular, share several characteristics with the so-called 'concubine' figurines depicting women, usually naked, graced with various ornaments (jewellery, body painting), with no legs shown below the knees; one of their functions was to ensure the fertility of the deceased in the afterlife.[156] Like them, female dwarfs are usually naked and have similar hairstyles: they wear typical wigs of long hair or have three long plaits.[157] Belts and necklaces also adorn dwarf figures, male and female (E 134, 147, 193; pls. 33.2, 4). Like 'concubine' figurines, four dwarf maids nurse small children (E 123–d 125, 146; pls. 30.3–4, 34.2).

These iconographic associations suggest that statuettes of male and female dwarfs may

---

[151] See also the Byblos figurines E d 161, d 165, 180 (holding a tambourine?).

[152] So for Hayes, *Scepter*, i. 221–3; A. M. Badawi, 'Le Grotesque: invention égyptienne', *Gazette des Beaux-Arts*, 66 (1965), 189–98. On monkeys acting like humans, see B. J. Kemp and R. S. Merillees, *Minoan Pottery in Second Millennium Egypt* (Mainz am Rhein, 1980), 146–7 and 162.

[153] See e.g. G. Brunton, *Qau and Badari*, i (London, 1927) (BSAE 44), 64: 'The first is apparently a figure of Ptah-Sokar.' See also R. Engelbach, *Harageh* (London, 1923) (BSAE 28), 12.

[154] E. Brunner-Traut, *LÄ* ii (1977), s.v. Gesten, 577–8, 1c.

[155] On the enlarged bellies of fecundity figures and Taweret, see Baines, *Fecundity Figures*, 95–7, 127.

[156] See in general C. Desroches Noblecourt, '"Concubines du mort" et mères de famille au Moyen Empire', *BIFAO* 53 (1953), 7–47, and G. Pinch, 'Childbirth and Female Figurines at Deir el-Medina and el-'Amarna', *Orientalia*, 52 (1983), 405–14; id., *Votive Offerings to Hathor in New Kingdom Temples*, D.Phil. thesis (University of Oxford, 1984, pub. forthcoming).

[157] Puffed wigs: E 121, 123–d 125, 133, 170–2, d 185?; pls. 30.3–4, 32.4, fig. 9.22. Three plaited locks: E 120, 146, d 182, d 184; pl. 34.2.

have served as apotropaia for women and children. Find contexts reinforce this hypothesis. The female dwarfs from Deir el-Bersha come from the tomb of a woman (E 120–1; fig. 9.22), while the figurine from el-Badari was found in the tomb of a very young child (E d 145). As seen above, the el-Lisht toy belonged to a young girl (E 122; pls. 30.2, 31); the object was discovered with four 'concubine' figurines and a statuette of Taweret.[158] This group of objects was perhaps meant to ensure the protection of the child in the afterlife; the toy could have referred to the notion of solar regeneration associated with pygmies' dancing. Other graves contained small faience models of animals (cats, hedgehogs, lions, baboons, hippopotami) together with the dwarf figurines (E 123–d 125, 147; pls. 30.3–4).[159] These animals recall those depicted with Bes-gods on magic wands, and they may have had the same protective function. The find context of the Byblos figurines is analogous (E 150–d 190); the dwarfs were associated with figurines of 'concubines' and of animals, such as rams, lions, crocodiles, and frogs.[160]

The meaning and function of these statuettes seem thus to be many-sided. On an everyday level, they represent servants performing the office they held in the household of the deceased. But they seem also to have functioned as fertility figures, and more generally as apotropaic beings, like the contemporary Bes-gods. Their connection with pregnancy and delivery may have a funerary meaning; like Bes, these dwarfs could have attended the rebirth of the deceased in the afterlife.

In a domestic context, dwarfs seem to have played a similar protective role. At el-Lahun, Petrie found several large stands, made of clay or limestone, which were carved in the shape of naked dwarfs, sometimes very crudely modelled. The figures, usually male, have large flat heads with protruding ears, pendulous breasts, and short bent legs. They derive from the motif of the dwarf bearer: some support a dish or a cup with both arms (E d 137, 143; fig. 9.24), or the vessel is placed on top of the head (E d 138–d 140; pl. 34.1); occasionally two dwarfs stand back to back (E d 136, d 142).

Petrie discovered similar stands with dishes on top, but in the form of columns with plain or lotus capitals, which 'whenever they are found charged, have a cake of dough stuck in the dish'; he suggested that those objects served for the 'household offerings of daily bread' in the domestic cult.[161] The dwarf-shaped stands may have contained similar offerings, or supported dishes or lamps. Petrie also found two crude figures of Taweret which probably belonged in this ritual context.[162] A similar range of items related to the domestic cult comes from the New Kingdom settlement of Deir el-Medina.[163] Large jars in the form of Bes, or with Bes images (fig. 6.5a), recall the form

---

[158] Lansing (n. 147 above), 30 and 35, fig. 29.

[159] Kemp/Merillees (n. 152 above), 112, 115–58, 160–2 (E 147); the tomb also contained models of food products, servants and entertainers (wrestlers, musicians). Ibid. 167 (E 123–d 125).

[160] See Dunand (n. 141 above), 741–67.

[161] W. M. F. Petrie, *Illahun, Kahun and Gurob* (London, 1891), 11. See also id., *Kahun, Gurob, and Hawara* (London, 1890), 26.

[162] Id. (1891), 11.

[163] See Bruyère, *Deir el Médineh*, 102–4, esp. fig. 35 (Bes water-jars), 204–11, esp. 205 (lamp-stands).

FIGURE 9.24. (E 143) Limestone. Present location unknown

of the dwarf-shaped stands of el-Lahun, and may have had a similar apotropaic function.[164]

The protective role of the el-Lahun stands is also suggested by their resemblance to a stand in Leiden which Raven attributes to the Middle Kingdom. It represents not a human dwarf, but a demonic being, probably female, with a protruding belly, holding in either hand a snake.[165] The top of the head is carved with a shallow circular cavity, implying its use as a stand. These attributes suggest that it was an apotropaic figure in a domestic context; like the el-Lahun stands, it probably carried bread offerings or a basin of water in a private shrine.

Integration in Society

There is some evidence that dwarfs could have a normal position in Middle Kingdom society. Two statues come from private tombs. A limestone statuette found in the el-Riqqa cemetery may render the abnormal proportions of a dwarf (short trunk, long arms), but in a stylized way (E d 135). His name, Nihor, is carved on the back pillar. The owner was probably a person of some rank, as his attire (wig, long kilt) shows.

A wooden statue of unknown provenance depicts a dwarf dressed in a plain kilt, his left hand closed, his right palm turned outward possibly to express his wish to receive offerings (E 197; pl. 34.3a, b).[166] The figure is very finely carved; the dwarf has a stocky body with short legs, a large head with a snub nose and a flat, elongated skull.

---

[164] Raven, *Pataekos*, 14–15.
[165] Ibid, fig. 1 and pl. 1; see also above 90.
[166] On this attitude, see E. Delange, *Catalogue des statues égyptiennes du Moyen Empire, 2060–1560 av. J.C.* (Paris, 1987), 172 n. 1. See e.g. the Middle Kingdom statue of Ii with both palms opened (h.

The plinth, which does not belong originally to the statue, bears a fragmentary inscription. The text describes 'an offering which the king gives to Ptah and Sokar (that they may provide) an offering of bread and beer, foods and nourishment, ointment and clothing for the Lady of the House, Itasenbet(?) Mereryt, the justified'.[167] Itasenbet cannot be the male owner of the statue which was probably fitted to the base in antiquity or later.

The tombs of two dwarfs with skeletal material are recorded. The tombs themselves do not differ from those of fully grown people. At Beni Hasan, a proportionately short woman, Seneb (S *d* 15), was buried with the standard funerary equipment of an élite lady (jewels, toilet objects, models of servants, musical instruments).[168] In a more modest burial at Asyut, a reused sarcophagus contained the remains of a male dwarf, possibly achondroplastic, wearing a necklace of silver beads (S 16); as Chassinat and Palanque noted, this ornament indicates that the deceased was not an unimportant figure.[169]

NEW KINGDOM AND THIRD INTERMEDIATE PERIOD, EXCLUDING EL-ᶜAMARNA (1539–750? BC)

New Kingdom iconography differs significantly in style, composition and subject-matter from that of earlier periods. Dwarfs are no longer shown in tombs of officials, except at el-ᶜAmarna, but they are represented in a few temple reliefs and in statuary. This rarity contrasts with the widespread occurrence of the dwarf gods Bes and Ptah-Pataikos, who appear in a variety of forms in the households of the élite. In literature, human dwarfs are mentioned only in two texts.

The Amarna period, which is characterized by a radical departure from the traditional iconographic canons, is analysed in a separate section.

Household Offices

Dwarfs do not appear in New Kingdom tomb paintings and reliefs, though scenes which earlier included them continue to occur.[170] At meals, the owner and his guests often are entertained by dancers and musicians, but no dwarf joins in the troop, although monkeys do.[171]

25 cm.), Cairo Museum, CG 65842; D. Wildung, *L'Âge d'or de l'Égypte, Le Moyen Empire* (Fribourg, 1984), 105, fig. 94.

[167] Delange (n. 166 above), 171.

[168] J. Garstang, *The Burial Customs of Ancient Egypt* (London, 1907), 41, 113–14, 226, figs. 102–4, pl. v (tomb contents).

[169] E. Chassinat and Ch. Palanque, *Une Campagne de fouilles dans la nécropole d'Assiout* (Cairo, 1911) (MIFAO 24), 15, and n. 2.

[170] Davies (n. 49 above), pls. LII–LV.

[171] For New Kingdom representations of dancing monkeys, see e.g. the group of musicians consisting of a blind harpist, a lute player, a dancer, and a monkey in T. Säve-Södebergh, 'Eine Gastmahlsszene im Grabe des Schatzhausvorstehers Djehuti', *MDAI(K)* 16 (1958), 285, fig. 3. See also the ostraca with monkeys playing the double flute before dancing Nubians; E. Brunner-Traut, *Die altägyptischen Scherbenbilder (Bildostraka) der deutschen Museen und Sammlungen* (Wiesbaden, 1956), 98–100, nos. 100–1, pls. III and XXXV.

One terracotta figure vase depicts a naked dwarf carrying a pot on his left shoulder (E 204; pl. 37.3). The statuette repeats the motif of the dwarf bearer, but with new associations. Its physical appearance is ambiguous: the figurine has male genitals and breasts, but also a protruding belly and large hips. This modelling recalls the figure vases, usually in calcite or terracotta, depicting pregnant women, which are similarly naked, with dwarfish proportions, bulging abdomen, large hips, and elongated breasts, like fecundity figures and Taweret.[172] Usually they hold both hands on their paunches, perhaps to evoke the rubbing of the unguents which the vessels contained, or to emphasize the size of their bellies. The resemblance of the male dwarf bearer with these figurines may reflect the relationship of dwarfs with childbirth and motherhood, which is expressed so strongly in contemporary magical spells; the vessel might also have contained some medicine for pregnant women.

Two calcite figure vases show female musicians, a girdle round the hips, holding a lute (E d 201, 205; pl. 37.2). Their proportions are similar to those of the pregnant women mentioned above, and the figurines may have similar associations with fertility and sexuality.

It does not follow from this relative lack of material that dwarfs were no longer kept in the households of officials. They may have ceased to be shown in tomb decoration because they were replaced by new prestige motifs, such as scenes of investiture or funerary banquets; banquet and temple musicians are then characterized by another physical defect, blindness.[173]

Dwarf attendants occur in less formal iconographic contexts, such as the satiric-erotic papyrus in Turin, which depicts the debauchery of a priest and a singer of Hathor (E 81; fig. 9.25). A short man appears underneath the chariot drawn by two girls on which the protagonists couple. He seems to be the attendant of the singer: he holds a kind of bag, which may contain the lady's effects, and probably watches over the monkey playing in the horses' reins above his head. He is treated in the same scurrilous way as the other figures: sparse tufts of hair garnish the back of his bald skull, and his short kilt reveals a large erect phallus. He is the only male figure without a partner; he waves his left hand in vain to attract the attention of the pair of girls. A miniature figure occurs in another scene involving the same partners; Omlin identified it as a dwarf, but as it is well-formed, and without conspicuous genitals, it could be a child.[174]

The presence of a dwarf near a singer of Hathor may allude humorously to the association of the goddess with the dwarf god Bes. This document foreshadows the

---

[172] For material, see E. Brunner-Traut, 'Gravidenflasche, das Salben des Mutterleibes', in A. Kuschke and E. Kutsch (eds.), *Archäologie und Altes Testament, Festschrift für K. Galling* (Tübingen, 1970), 35–48; id., 'Nachlese zu zwei Arzneigefässen', *WdO* 6 (1970–1), 4–6. See e.g. the terracotta figure vase from Abydos (h. 21 cm.), Cairo Museum, JdE 34403;

D. Randall-MacIver and A. C. Mace, *El Amrah and Abydos* (London, 1902) (EES 23), 72–5, pl. L; *Nofret* i. 24, no. 7.

[173] On blind musicians, see above 102.

[174] J. A. Omlin, *Der Papyrus 55001 und seine satyrisch-erotischen Zeichnungen und Inschriften* (Turin, 1973), 50.

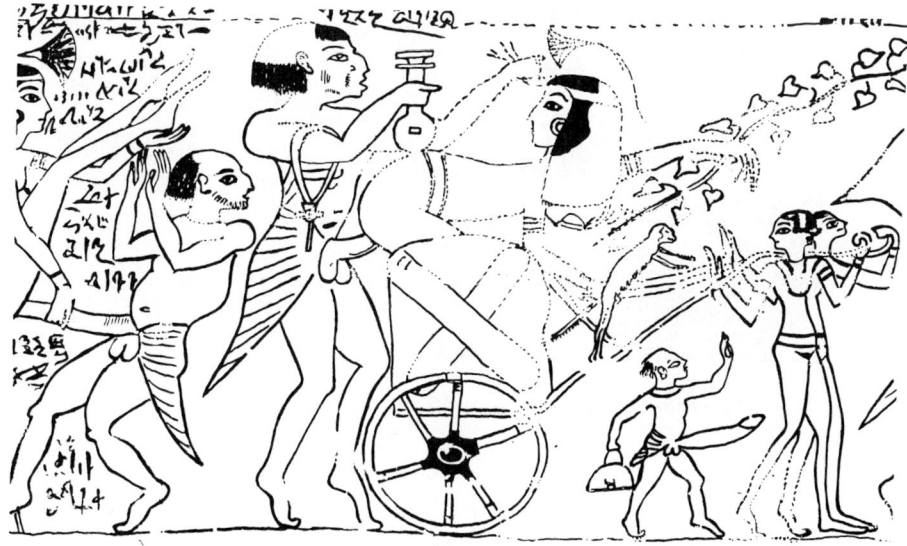

FIGURE 9.25. (E 81) Papyrus. Turin, Egyptian Museum, 55001

Hellenic treatment of dwarfs, turned into ithyphallic creatures rejected by women, sometimes ludicrous, but still associated with religious symbols.[175]

Dwarfs might also be seen in a few scenes of the birth arbour depicted on ostraca and model beds. On ostraca, the woman is usually seated and holds or suckles a newborn child; on model beds, she is reclining with the child lying at her side. There is sometimes a small black figure near the woman, on the lower corner of model beds, or behind the seated woman on ostraca; it performs a lively dance with uplifted arms, sometimes holding an object in its hand. Pinch suggests that this may be a dwarf attendant enacting the role of Bes and dancing to ensure the protection of the newborn child; yet, as no example has clearly malformed proportions or iconography recalling Bes, it could also be a child, perhaps a Nubian.[176]

Religious Role

The rarity of dwarfs in private iconography is partly counterbalanced by their occurrence in religious contexts. They are depicted on two reliefs, one in the temple of Amenophis III in Soleb (E 73), the other from the temple of Osorkon III at Tell Basta

[175] See below 218–19, and esp. 236–7 (G 18; pl. 52).

[176] For material, see Pinch (n. 156 above), 409–10. See e.g. the model bed in Bruyère, *Deir el Médineh*, 141–2, pl. XLIV, and the ostracon from Deir el-Medina in J. Vandier d'Abbadie, *Catalogue des ostraca figurés de Deir el Médineh*, i (Cairo, 1937), 71, VA 2344, pl. LIII. Cf. the dancing boy on the ostracon ibid. ii (Cairo, 1959), 187, VA 2858, pl. CXX.

FIGURE 9.26. (E 82) Festival hall of Osorkon III. Philadelphia, University Museum, E 226

(E 82; fig. 9.26). Both reliefs are associated with one of the most important Egyptian religious ceremonies, the *sed* or Jubilee festival, which took place theoretically after thirty years of reign to renew the power of the king.[177]

In the temple of Amenophis III (E 73), the dwarf, proportionately short, but with a beard, performs a rejoicing *jhb* dance with four normal-sized figures.[178] The caption states that the dancers are 'from Punt', which may mean either that a specific pygmy is shown, or that the part of a pygmy was enacted by a dwarf. This combination of exotic provenance and short stature recalls the Old Kingdom *jb3* dance of Harkhuf's *dng* which took place in a related context.[179] The presence of a dwarf, ethnic or pathological, in a *sed*-festival may have reinforced the regeneration symbolism of the ritual of royal renewal.

The fragment of relief from the *sed* festival hall of Osorkon III depicts three well-proportioned dwarfs in kilts holding long staffs (E 82; fig. 9.26); the elbow of a fourth one appears at the edge of the block. They stand in a row before the temple priests, and are captioned with titles describing their office. may mean *s'š3*, 'guards',[180] or 'chiefs (*ḥ3t*) of the numerous ('*š3*) *s* dwarfs', 'dwarf clothiers'.[181] The latter reading would imply the continuity of the tradition of dwarf keepers of the wardrobe of the god's statue, a function suggested by two Old Kingdom incriptions.[182] Their specific role in the *sed*-festival is unclear.

The calcite figure vase found in the temple of Hathor at Serabit el-Khadim also suggests a presence of dwarfs in temples (E 203; pl. 37.1). It depicts a dwarf carrying a jar that probably contained unguents to anoint the cult statue, like the calcite vessels

[177] Material in E. Hornung, E. Staehelin, *Studien zum Sedfest* (Basle and Geneva, 1974) (ÄH 1), 16–42; K. Martin, *LÄ* v (1984), s.v. Sedfest, 782–90.

[178] On *jhb* dances, see Brunner-Traut, *Tanz*, 80–1.

[179] See above 132–3.

[180] *Wb* iv. 55. 14–18.

[181] Fischer (n. 115 above), 49.

[182] See nn. 115 and 117 above.

FIGURE 9.27. (E 75a) El-ᶜAmarna, tomb of Panehesi

found at the same site, which depict Bes and the cow of Hathor.[183] Another calcite figure vase of a dwarf bearing a vessel may come from a similar context (E 206; pl. 37.4); two holes are drilled below the abdomen, and a phallus may have been set in one of them, enhancing the fertility associations of the figure.[184]

EL-ᶜAMARNA (1353–1336 BC)

At el-ᶜAmarna, the capital of Akhenaten, dwarfs occur in several tomb reliefs and in statuary (E 74–9, 198–d 200; pls. 25, 35.2–3, fig. 9.27).

Household Office

The same two dwarfs, suffering from severe limb deformities (coxa vara) with club-feet, are depicted in six tombs of high officials (E 74–9; pl. 25, fig. 9.27); they usually accompany the sister of the queen, the princess Mutnedjmet. They wear ordinary dresses with wide shoulder-straps, probably indicating that they are female.[185] A very damaged Amarna relief in Cairo may depict a male figure in a kilt (or a belt with strips?), and locks of hair (E 79).

As Hanke notes, the dwarfs are differentiated from other courtiers, who are mostly anonymous, by captions giving their names and titles.[186] One is called: 'the vizier of the queen', 'the Sun', and the other 'the vizier of his mother', 'For ever' (E 75a, 76, 77a; pl. 25, fig. 9.27). These titles are ironical; the vizier is the highest position after the king,

[183] On calcite vessels as votive offerings to Hathor, see Pinch (n. 156 above), 535–8. See also the calcite unguent-vases from the tomb of Tutankhamun; Edwards (n. 50 above), nos. 3–4.

[184] On the presentation of votive phalli to Hathor shrines, see G. Hornblower, 'Phallic Offerings to Hathor', *Man*, 52 (1926), 81–2; id. 'Further Notes on Phallicism in Ancient Egypt', *Man*, 53 (1927), 150–3;

Pinch (n. 156 above), 434–8, 444–9.

[185] So for N. de G. Davies, *The Rock Tombs of el Amarna*, ii (London, 1905) (ASE 14), 13–14. *Contra*: Weeks, *Anatomical Knowledge*, 174.

[186] R. Hanke, *Amarna Reliefs aus Hermopolis. Neue Veröffentlichungen und Studien* (Hildesheim, 1978), 21–2, no. 14.

and the queen, or his mother, would not have one. The humorous contrast between
these prestigious titles and the grotesque appearance of the dwarfs suggests that they had
a jester-like function.[187] This hypothesis may be supported by a word play in their titles;
Wolff pointed out that the word for vizier (*ṯȝty*) is close to that for nestling (*ṯȝ*), and thus
may have a derogatory connotation.[188] These titles might also allude to the role of
dwarfs as counsellors, a function often fulfilled by jesters in later historical periods.[189]

However, these dwarfs may also have been credited with divine affinities. The names
'the Sun' and 'For ever' relate them to cosmic powers, and in particular revive the
traditional solar associations of dwarfs and pygmies. These beliefs fit with the cult of the
sun-disk and could have been assimilated by Akhenaten's religion. The unusual physical
appearance of the Amarna dwarfs stress their solar symbolism. Their severely contracted
limbs evoke those of the scarab-beetle Khepri; they also have strange flat skull-tops
which seem reminiscent of those of Ptah-Pataikoi, crowned with the sacred insect.[190]
Like other traditional motifs, dwarfs may have been used by Akhenaten to serve his
ideological innovations. The dwarfs accompany the royal family in scenes of worship of
the Aten disk, and they may have appeared as exterior marks of the court's devotion to
the sun-god (E 75a; fig. 9.27).[191]

The dwarfs' faces are destroyed on all these reliefs. Davies related this to their
notoriety under the reign of Akhenaten, but this is unlikely, since the heads of other
figures at the same level are also defaced.[192]

In statuary, three figurines depict dwarfs carrying on their shoulders jars which are
hollowed to contain cosmetics. Their poses are similar, but their dresses vary; one has an
elaborate pleated kilt (E *d* 200; pl. 35.3), another a plain one with the belt placed low
below the navel to emphasize his belly (E 198; pl. 35.2), and the third only a sash (E *d*
199). One of them has royal associations: he carries a vessel carved with the cartouches
of Akhenaten and Nefertiti (E 198; pl. 35.2)[193]

The presence of a female dwarf on a calcite boat from the tomb of Tutankhamun, the
second successor of Akhenaten, may be a continuation of this Amarna fashion (E 202;
pl. 36a–c). The piece is very finely worked; details are incised and filled with coloured
pigments and gold. The dwarf is naked with armlets, and wears a curled wig with a
sidelock of hair; her achondroplastic facial features are very naturalistically rendered; like
the Amarna dwarfs, she has bowed legs (coxa vara) with club-feet. She stands steering

---

[187] See Guglielmi (n. 78 above), 185; H. G.
Fischer, 'The Ancient Egyptian Attitude towards the
Monstrous', in A. E. Farkas *et al.* (eds.), *Monsters and
Demons in the Ancient and Medieval Worlds. Papers
Presented in Honor of Edith Porada* (Mainz am Rhein,
1987), 23.

[188] Wolff (n. 129 above), 451.

[189] For material, see E. Tietze-Conrat, *Dwarfs and
Jesters in Art* (London, 1957).

[190] On the iconographic assimilation of dwarfs and
scarabs, see above 50, 53–4.

[191] On Akhenaten's manipulations, see in general
D. O'Connor, 'New Kingdom and Third
Intermediate Period', in *Ancient Egypt*, 220–1.

[192] N. de G. Davies, *The Rock Tombs of el Amarna*,
vi (London, 1908) (ASE 18), 21 n. 2.

[193] B. v. Bothmer, 'The Dwarf as Bearer', *BMFA*
47/267 (1949), 10, fig. 4.

with a pole (now missing) at the stern of the boat; at the prow a squatting woman in the same attire holds a lotus flower. The boat is placed on a small pedestal in a pylon-shaped tank to be filled with water or perfume.[194]

The meaning of the boat is unclear. Was it a palace ornament, as Carter suggested, or was it related to some specific custom? There is an open catafalque or chest in the centre, which is surrounded by four papyrus columns, as on a funerary boat.[195] But stem and stern end in the heads of gazelles, animals which may be associated with bridal customs, as Aldred suggests.[196] A boat depicted on a Third Intermediate Period relief may provide a revealing iconographic parallel. It carries a shrine with the goddess Bastet, which is surrounded by two papyrus columns with Hathor capitals, a column with the falcon Horus and a standard with a lion; the prow ends in a cobra, and the stern in a gazelle's head.[197] Horus, the lion, and the cobra refer to solar powers, while Bastet, Hathor capitals, and the gazelle's head allude to the protection of women. On the Tutankhamun piece, the gazelles' heads, together with the dwarf, might similarly refer to the female sphere.

Religious Context

Apart from a scene of worshipping of the sun-disk (E 75a; fig. 9.27), human dwarfs are not found in religious contexts at el-ʿAmarna. I mention only a doubtful representation in a *sed*-festival scene on a talatat of Amenophis IV (E *d* 80). Two figures in long kilts, with outstretched arms, stand before two kneeling bald men; they wear large wigs with horns or ears which may be feline masks. Several elements relate them to dwarfs. They are conspicuously smaller than the kneeling men, they appear in a *sed* festival, like the slightly earlier Soleb dwarf (E 73), and their wigs or felid masks recall the dwarf god Bes. Men wearing feline masks also occur in the tomb of Kheruef at Thebes and in the temple of Osorkon III at Tell Basta, both in a *sed*-festival context.[198] In the tomb of Kheruef, the kinship of these men with Bes and fecundity figures in general is stressed by their pendulous breasts and fat bellies.[199]

Masked figures seem thus to have been a regular feature of *sed*-festivals, but it is difficult to define more closely their relation to human or divine dwarfs. The personages depicted on the talatat have no physical malformation, and their small size could be due to compositional constraints.

[194] See H. Carter, *The Tomb of Tut-ankh-Amen*, iii (London, 1933), 127–8 (water and flowers); Weeks, *Anatomical Knowledge*, 175, C2 (scent).

[195] Cf. one of the wooden models of boats found in the treasury; Edwards (n. 50 above), no. 2.

[196] See C. Aldred, *Jewels of the Pharaohs* (London, 1971), 119, figs. 47–9 (gazelles on 18th-dyn. diadems of ladies of royal blood).

[197] A. el-Sawi, 'Preliminary Reports on Tell Basta Excavations', *ZÄS* 104 (1977), 129, fig. 4. I am grateful to D. Jones for this reference.

[198] Tomb of Kheruef: PM i². 298; Epigraphic Survey, *The Tomb of Kheruef* (Chicago, 1980), pls. 39–40; Sourdive, *La Main*, 56–9, pl. XIX. Temple of Osorkon III: PM iv. 28; E. Naville, *The Festival Hall of Osorkon II in the Great Temple of Bubastis* (London, 1892) (MEEF 10), pl. XV, 5; Sourdive, *La Main*, 59–64, pl. XX.

[199] Sourdive, *La Main*, 126.

New Kingdom texts express an ambivalent view of dwarfs and pygmies. In two texts, dwarfism appears as a disability which provokes uneasy feelings.

The *Instruction of Amenemope* says:

> Do not laugh at a blind man,
> Nor tease a dwarf (*nmw*),
> Nor cause hardship for the lame.
> Do not tease a man who is in the hand of the god,
> Nor be angry with him for his failings. (24. 8–12)[200]

The injunction shows that dwarfs had to be respected. On the other hand, it suggests that short-statured people were exposed to bad treatment as much as other deformed persons. No other literary or didactic text uses the figure of a dwarf to entertain or instruct the reader, although their activities, such as animal-tending, would have provided easy subject matter for satire.

In the Chester Beatty dreambook, the appearance of a dwarf creates anxiety. An omen states: 'Seeing a dwarf (*nmj*): bad; the taking away of half of his life' (iii. 8.13).[201] This negative reference should not, however, be taken literally. The interpretation is probably based upon an association of ideas, as in other omens: because the dwarf is half-sized, the life of the dreamer will be shortened by half.[202] It may also express the assimilation of dwarf gods with the sun who appears during twelve hours a day (half-day), as el-Aguizy notes.[203]

LATE PERIOD (750?–332 BC)

While the iconography of the dwarf gods Bes and Ptah-Pataikos becomes very diverse in the Late Period, only three depictions of human dwarfs are preserved, all in strongly religious contexts.

The most important is the sarcophagus of the dwarf Djeho, who was buried in the same pit as his master Tjaiharpta at Saqqara (E 84; pl. 26.2). It is a rare example of joint burial of master and servant. The sarcophagus is made of granite, an expensive material, and is of very fine workmanship. The outside of the lid is carved with a depiction of the dwarf in an unusual pose, in full profile, both body and face. He is naked, and exhibits typical achondroplastic features; the figure may be at life-size (120 cm.).

The lid and the sides of the sarcophagus are carved with a biographical text which reveals the key elements of his life. Djeho is first introduced as: 'The dwarf Djeho,

[200] London, BM 10.474; trans. Lichtheim, *Literature*, ii. 160. H. O. Lange, *Das Weisheitsbuch des Amenemope aus dem Papyrus 10.474 des British Museum* (Copenhagen, 1925), 121.

[201] London, BM 10.683; trans. A. H. Gardiner, *Chester Beatty Gift*, i (London, 1935) (Hieratic Pap. in the BM 3rd ser.), 17.

[202] On this literature, see S. Sauneron, 'Les Songes et leur interprétation dans l'Égypte ancienne', in *Les Songes et leurs interprétations* (Paris, 1959) (SourcesOr 2), 32–8.

[203] El-Aguizy, 'Dwarfs', 57.

possessor of reverence, son of Petekhons, true of voice, born of Tawenshe who is called Tenthapu, true of voice.' The names of his parents have no foreign connotations, which suggests that he was a native Egyptian.

Djeho also states that he belonged to the household of the lord Tjaiharpta. His master is described as a high official, 'close to the king, who was the chief financial officer of Upper Egypt'. As Spiegelberg pointed out, his parents are referred to only by their names and have no titles.[204] This 'homo novus' adopted what was probably the custom of the élite and took a dwarf companion. Djeho expresses his loyalty in praying to Osiris-Apis to let him continue to attend the deceased in the afterlife: 'May my body be exalted beside him in his tomb . . . May you cause, that I remain beside him . . . in exchange for what he has done for me.' This last sentence suggests that Tjaiharpta paid for the sarcophagus of Djeho, a costly present which reveals the regard in which he held his dwarf.

But the main element of Djeho's life appears to be dancing. He twice presents himself as a sacred dancer performing a *ḥbj* dance for the burials of the Apis and Mnevis bulls: 'I am the dwarf (*nmw*) who danced in Kem and in Shenqebeh at the festival of everlasting' (sides), and 'the dwarf who danced in Kem on the day of the interment of Apis-Osiris, the great god, king of the gods; who danced in Shenqebeh on the day of the festival of everlasting of the Osiris of Mnevis, the great god' (lid).

Since the earliest times, sacred bulls had been regarded as the living exemplars of major gods, especially the Apis bull for Ptah,[205] the Mnevis bull for Re-Atum,[206] and both after their deaths for Osiris. In the Late and Ptolemaic periods, the cults of these sacred bulls became very popular. Their burials were major public religious events, which were celebrated in the two places mentioned by Djeho: at Saqqara (Kem) for Apis bulls,[207] and at Heliopolis (Shenqebeh) for Mnevis bulls.[208]

The different stages of the burial of Apis are quite well documented, and it is possible to guess at what moment Djeho's performance took place. He probably danced when the mummified Apis was carried along the way leading to the Serapeum, in a procession enlivened by the shouting, singing, and dancing of the priests; the atmosphere was very 'dionysiac', as Plutarch reports: 'but as for what the priests openly do in the burial of the

[204] W. Spiegelberg, 'Das Grab eines Grossen und seines Zwerges aus der Zeit des Nektanebes', *ZÄS* 64 (1929), 79. For the titulary carved on the sarcophagus of Tjahorpta, see ibid. 76–9, and J. Baines, 'Merit by Proxy: The Biographies of the Dwarf Djeho and his patron Tjaiharpta', in forthcoming *DEA* 78 (1992).

[205] *Wb* i. 344.14 (*wḥm n Ptḥ*); Plut. *De Is. et Os.* 20, 29, 43. E. Otto, *Beiträge zur Geschichte der Stierkulte in Aegypten* (Leipzig, 1938), 23–34; E. Kiessling, 'Die Götter von Memphis in griechisch-römischer Zeit', *AfP* 15 (1953), 26–7. Apis is also occasionally

associated with Re-Atum and Horus.

[206] *Wb* i. 344.15 (*wḥm n rꜥ*); Otto (n. 205 above), 38–40; L. Kákosy, *LÄ* iv (1982), s.v. Mnevis, 165–7.

[207] *Km* could be Athribis, but it probably refers to the Serapeum here; see the discussion by Spiegelberg (n. 204 above), 80 n. 1; Otto (n. 205 above), 19–21. On the locations of these tombs at Saqqara, see J. Vercoutter, *LÄ* v (1984), s.v. Serapeum, 809.

[208] Spiegelberg (n. 204 above), 81 n. 2; Otto (n. 205 above), 36; Kákosy (n. 206 above), 165.

Apis when they transport its carcass on a raft, this in no way falls short of Bacchic revelry, for they wear fawn-skins and carry thyrsus-rods and produce shouts and movements, as do the ecstatic celebrants of the Dionysiac rites.'[209]

The presence of a dwarf in these rituals reflects very ancient symbolic affinities between dwarfs and sacred bulls. Like dwarf gods, bulls were closely associated with fertility, and represented the same divine powers, Re and Ptah; Apis and Mnevis also acquired a cosmic dimension in the course of time.[210] The earliest evidence for the association of dwarfs with the cult of bulls is in the titulary of Seneb, priest of 'the great bull which is at the head of *Stpt*' and of the bull *Mrhw* (E 113c 12–13). In the Graeco-Roman period, it is illustrated by terracottas depicting Bes holding or crowned with a naos containing a bull (pl. 11.2), or the bull appearing behind the head-dress of the god.[211] This motif may have been transmitted to Greece along with the belief in the protective power of dwarfs.[212]

The relationship between dwarfs and Apis can be defined more precisely. At Memphis, the sacred bull lived in a building in the precinct of the temple of Ptah, and the clergy of Ptah organized its burial ceremony.[213] As Ptah was worshipped in the form of a dwarf in this temple, it is not unlikely that the priests chose a dwarf dancer to take on the role of the god. As dwarfs were strongly associated with birth and regeneration, this dance may also have celebrated the renewal of the Apis and his second birth in the afterlife. The pictorial rendering of Djeho seems to reflect his religious role. His full profile pose stresses his resemblance to Ptah-Pataikoi figurines: he has the same flat shaven head, with a small button nose, and the same half smile. Thus, besides being the attendant of a wealthy high official, Djeho was essentially a sacred dancer. The emphasis on his religious function is very significant, revealing that this position was the main constituent of his social identity; it gave a positive value to his abnormality.

Did Djeho also execute dances at private burials? There is a passage in the story of Sinuhe where the author says: 'the dance (*hbj*) of the *mww* dancers is done at the door of your tomb.'[214] These *mww* dances occur in literature and iconography from the Old Kingdom on; they were performed by two, three or four men characterized by tall crowns made of reeds, who welcomed the deceased at the entrance of his tomb, and, on

[209] Plut. *De Is. et Os.* 35 (trans. Griffiths). On the burial ritual, see Otto (n. 205 above), 19–23; J. Vercoutter, *LÄ* i (1975), s.v. Apis, 339–42.

[210] On bulls and fertility, see Otto (n. 205 above), 1–4, 11–13; E. Hornung, 'Die Bedeutung des Tieres im alten Ägypten', *Studium Generale*, 20/2 (1967), 74. This notion of fertility is also expressed in the myth of the cosmic cow; E. Hornung, *Der ägyptische Mythos von der Himmelskuh* (Fribourg and Göttingen, 1982) (OBO 46), 96–101. Apis and Mnevis stand for

the sun and the moon in a Coptic text; see Kákosy (n. 206 above), 166 n. 40.

[211] For material, see above 82.

[212] See e.g. the terracotta group of a dwarf standing on the back of a bull or cow from Aegina (G 125; fig. 13.6); discussion below 201.

[213] See Hdt. 2. 153; 3. 37. See also Vercoutter (n. 209 above), 339–42.

[214] Lichtheim, *Literature*, i. 229, ll. 194–5.

a mythical level, watched over the regeneration of the sun-god in the primeval waters of Nun.[215]

An altered version of the Sinuhe motif occurs on a New Kingdom stela from Saqqara and a Ptolemaic stela from Akhmim. The term *mww* is replaced by *nmw* 'dwarf'; the word is written with a dwarf determinative in the New Kingdom inscription, and with a dwarf logogram in the Ptolemaic one.[216]

Do these writings imply that dwarfs, like Djeho, also performed dances at private funerals?[217] Or are they due to a misunderstanding? A third document, a Middle Kingdom manuscript, has another, but meaningless, variant: *mww* is corrupted into *nnyw* 'the weary ones';[218] this suggests that the writings of *nmw* are also corruptions, but probably not innocent ones. As Altenmüller demonstrates, *mww* dancers share several symbolic characteristics with dwarfs, which may explain the confusion. Both types of being are famous for their ritual dancing, performed before the king or Re, and both are described as god's dancers; on a mythical level, both are liminal figures related to the protection of the sun-god. These similarities probably influenced the reinterpretation implicit in this writing.

Two pairs of round tambourine membranes now in the Ashmolean and Cairo museums depict human dwarfs engaged in rituals involving various deities, especially female and Bes-gods (E *d* 85, 86; fig. 9.28).

The Cairo pair is better preserved (E 86; fig. 9.28a, b). The membranes are detached from their frame, but their similar size and motifs suggest that they belonged to the same object.[219] On one membrane, a female musician beats a round tambourine before Isis who is seated and lifts a cup, probably filled with wine, in her left hand. Between these figures, a female dwarf in a dress dances, her right hand raised, her left hand hanging down, on a kind of pedestal patterned after a lotus flower; she is proportionately short, but with very long arms, like the woman on the Brooklyn obelisk (E *d* 83; pl. 26.1).

On the second membrane, a Bes-god, clad in a sashed kilt, substitutes for the dwarf. He has the same pose as amuletic figurines of Bes in side view, the right hand laid on the

[215] See G. Jéquier, 'A propos de la danse des Mouaou', *REA* 1 (1927), 144–51; Brunner-Traut, *Tanz*, 43, 53–9; H. Altenmüller, 'Zur Frage der *Mww*', *SAK* 2 (1975), 1–37.

[216] New Kingdom stela, Cairo Museum, JdE 3299; PM iii². 737; A. Mariette-Pacha, *Monuments divers recueillis en Egypte et en Nubie*, i (Paris, 1872), pl. 61; J. Berlandini, 'Varia Memphitica VI', *BIFAO* 85 (1985), 41–62. Ptolemaic stela, Cairo Museum, CG 22054; K. Piehl, *Inscriptions hiéroglyphiques recueillies en Europe et en Egypte*, i (Stockholm and Leipzig, 1886), pl. LXXIII; ibid. ii (Leipzig, 1888), 66–7; A. Bey Kamal, *Stèles ptolémaïques et romaines*, i (Cairo, 1905), 51–4, esp. 53, l. 15; ibid. ii (Cairo, 1904), pl. XVII.

[217] See e.g. R. el-Sayed, *La Déesse Neith de Sais*, i (Cairo, 1982) (BdE 86), 130–1; Berlandini (n. 216 above), 46–8; El-Aguizy, 'Dwarfs', 59; Bourriau (n. 142 above), 122.

[218] See A. H. Gardiner, *Notes on the Story of Sinuhe* (Paris, 1916), 70, v. 194–5; J. Settgast, *Untersuchungen zu altägyptischen Bestattungsdarstellungen* (Glückstadt, 1963), 43–4; Altenmüller (n. 215 above), 34–6.

[219] L. Borchardt, 'Die Rahmentrommel im Museum zu Kairo', *Mélanges Maspero*, i (Cairo, 1935–8) (MIFAO 66), 2. For reconstruction and parallels, see L. Manniche, 'Rare Fragments of a Round Tambourine in the Ashmolean Museum, Oxford', *AcOr* 35 (1973), 33–4.

FIGURE 9.28. (E 86a, b) Tambourine (diam. 25 cm.). Cairo Museum, CG 69351 and 69352

chest, the left holding his tail and the legs slightly flexed. Both the dwarf and Bes are marked as foreign. The short woman is black, which contrasts with the light skins of the musician and the goddess; she performs a step belonging to a Nubian dance, as Wild demonstrates.[220] The skin of Bes is light coloured, but his attire is exotic. Isis also has a tall feather head-dress which gives her a southern touch and may assimilate her to Hathor, or possibly Anukis. The two membranes thus show the same scene, but with dancers on two different levels, divine and human; the dwarf and Bes seem to be interchangeable.

Similar motifs are found on the Ashmolean tambourine (E d 85). On one membrane, the decoration is divided into four registers. The first and fourth registers are badly damaged, but the second and third registers depict rows of figures playing the round tambourine. In the second register, the musicians are deities, male (Osiris, Thoth) and female, some with tall feather head-dresses like Isis on the Cairo membrane, some with horns and sun-disk. In the third register, female dancers clad in short kilts, with tall feather head-dresses, beat tambourines.

The other membrane is very fragmentary. Two gods are standing, one with the double-crown, followed by a figure with outstretched wigs, horns and sun-disk, holding the *ankh*-sign (Isis?). Between them a small female figure appears, perhaps a dwarf, as on the Cairo membrane.[221]

[220] H. Wild, 'Une Danse nubienne d'époque pharaonique', *Kush*, 7 (1959), 81–3.

[221] On these identifications, see Manniche (n. 219 above), 30–1.

As Manniche demonstrates, the decoration of both the Cairo and Ashmolean tambourines seem to allude to rituals connected with pregnancy and delivery. From the New Kingdom on, the round tambourine is played by female and Bes-figures in contexts related to birth, especially in birth-houses.[222] The prominence of Isis, the presence of Bes-like personages and of female figures playing the tambourine combine to suggest that both scenes show a celebration of birth and a ritual rejoicing performed in honour of the gods, in particular Isis, who ensured a safe delivery. On both objects, human figures—a female dwarf on the Cairo membrane, and young girls with feather head-dresses on the Ashmolean one—seem to play the part of the protective Bes-gods.[223]

GRAECO-ROMAN PERIOD (332 BC–AD 395)

In Hellenistic Egypt, human dwarfs appear only in objects in the Greek tradition. Some of these representations reflect Egyptian conceptions of dwarfism, but in a new iconography. I mention here the widespread motif of the dwarf dancer, often with an overlarge phallus, which expresses the traditional association of dwarfs with fertility and regeneration.[224]

The only representation in native Egyptian style is a relief of the reign of Ptolemy IX Soter II in the birth-house at Edfu (E 87). Behind the king, a dwarf in a dress (or cloak) stands on a sub-register above a niche; he lifts his left arm, his right crossed over the chest; this pose recalls the *ḥnw* jubilation dance 𓀜 ,[225] or, more generally, a gesture which is part of a ritual. The scene reveals the continuity of the practice of keeping dwarfs at court, which goes back to the early dynastic times.

---

[222] See also above 78.

[223] Manniche (n. 219 above), 31–2.

[224] For material, see W. Weber, *Ägyptisch-griechische Terrakotten* (Berlin, 1914); A. Ippel, *Der Bronzefund von Galjûb* (Berlin, 1922), 45–7; A. Adriani, 'Microasiatici o Alessandrini i grotteschi di Mahdìà?', *MDAI(R)* 70 (1963), 80–92; W. E. Stevenson, *The Pathological Grotesque Representation in Greek and Roman art*, Ph.D. thesis (University of Pennsylvania, 1975), esp. 46 ff.; N. Himmelmann-Wildschütz, *Alexandria und der Realismus in der griechischen Kunst* (Tübingen, 1983), 70 ff.; P. Blome, 'Affen im Antikenmuseum', in M. Schmidt (ed.), *Kanon. Festschrift Ernst Berger* (Basle, 1988) (AK Beiheft 15), 205–10, esp. 208 ff.

[225] Gardiner, *Grammar*, sign-list, A 8; it resembles also the determinative for dancing 𓀜 .

10

Conclusion

The image of short-statured people in Egypt is essentially positive. Their physical anomaly was not only tolerated, but accepted and valued as a divine mark for its religious associations.

On a mythical level, dwarfs were closely associated with solar and rejuvenating powers. Their abnormal bodies with short bent limbs were equated with that of the sacred scarab-beetle Khepri, one of the main hypostases of the sun-god Re. They stand on the side of humans, especially of children and women, in the crucial times of transition between life and death, repelling evil and bringing regeneration. Yet these benevolent deities remained marginal to the conventional pantheon and to systems of representation. Ptah-Pataikoi appear mostly in amuletic forms, while Bes is not shown in the main areas of temple relief, but in more specialized contexts associated with Hathor.

In daily life, the position of short-statured people is more ambiguous. Throughout the dynastic period, they appear in inferior roles, as members of the retinue of the élite. They were excluded from a number of arduous or menial activities, such as those of smiths, cooks, or outdoor workers, which were not compatible with their role as markers of their masters' prestige. The symbolic associations of dwarf gods seem to have influenced their status. A number of corresponding functions can be observed;[1] like their divine counterparts, dwarfs watched over bodily care, clothing, and jewellery; as auspicious and protective beings, they looked after young children,[2] and small-handed female dwarfs may have served as helpers during delivery.

Liminal status is suggested by their frequent association with other malformed people (hunchbacks or club-footed),[3] or with exotic people from Nubia (E 19; fig. 9.4)[4] or Punt (E 73). They also relate in an ambiguous way to animal pets, as though they were themselves regarded as human 'pets' or curiosities. In Old and Middle Kingdom reliefs,

---

[1] Cf. *contra* K.-J. Seyfried, *LÄ* vi (1986), s.v. Zwerg, 1433: 'Ein direktes Wechselverhältnis zwischen diesen "profanen" Zwerg-Vorkommen und den sog. zwerghaften Gottheiten . . . ist nicht offensichtlich.'

[2] As was later customary in European courts. For iconographic material, see E. Tietze-Conrat, *Dwarfs and Jesters in Art* (London, 1957).

[3] E 18, 29b, 54, 67, 68; pls. 19.1, 20.2, fig. 9.21.

[4] H. Junker, *Giza*, iii (Vienna, 1938), 179, nos. 12–13.

they often appear in the small spaces reserved for pets, such as below the lord's chair, and they seem to be interchangeable with monkeys as humorous entertainers, as in the Old Kingdom tomb of Ka'aper (E 32; fig. 9.18). Some poses and attributes imply rejection, especially those assimilating male dwarfs to children or women, as though they were not seen as full men. They are commonly naked, like children; some have no penis-sheath (E 94; pl. 27.2), or wear a sidelock (E 202; pl. 36c); they may adopt youthful or female poses, such as putting a hand to the mouth (e.g. E 128; pl. 32.1), or pulling papyri with both hands (E 41e; fig. 9.19b).

Acceptance is shown by a few examples of dwarfs in high-ranking positions. Some short-statured people belonged to the élite and owned a tomb, like Petpennesut (E d 112), Seneb (E 113; frontispiece, pl. 28.2), Itasenbet (E 197; pl. 34.3), and perhaps also Khnumhotpe (E 117; pl. 29.2). They seem to have been regarded as persons in the full legal sense, who could marry and inherit civil and religious functions, such as that of ka-servant (E 50). Some dwarfs were literate, like Seneb and perhaps also the scribe Karesy (E d 38). They received the same type of burial as ordinary Egyptians. They were not distinguished by special names. Most of their names express an apotropaic wish, such as Seneb 'may he be healthy' (E 113),[5] or place the newborn child under the protection of a god or a king, such as Radjedefankh 'King Redjedef lives' (E 19a; fig. 9.4).[6] Others, like Mereri 'the beloved one' (E 44), may refer to the lord's good will towards his dwarf, but only if the name was acquired some time after birth.[7]

In the religious sphere, their recurrent role as dancers in ritual contexts may reflect their continuing affinity with solar and rejuvenating powers. Throughout the dynastic period, they perform dances such as the *jb3w ntr* of Harkhuf's pygmy, the *jhb* dance of the Soleb dwarf (E 73), and the *hbj* dance of Djeho for the burials of the Apis and Mnevis bulls (E 84; pl. 26.2). This institution of dancing dwarfs may go back to the beginning of dynastic times, as is suggested by the name *sjm3 ntr* 'the one who gladdens the god' of a dwarf from the retinue of King Qa'a (E d 7). It is likely that these dances were executed by indigenous dwarfs as well as by pygmies (or black dwarfs), when the latter were available, but the two groups were not fully interchangeable; the southern provenance of an authentic pygmy, endowed as it was with solar affinities, must have added a special value to the performance.

A few texts and monuments show dwarfs on duty in sanctuaries, either as auxiliaries belonging to the temple staff, like the dwarfs mentioned in two Old Kingdom inscriptions and in the temple of Osorkon III (E 82; fig. 9.26), or as a priest attached to the funerary cult of a king, like Seneb (E 113; frontispiece, pl. 28.2). Dwarfs wearing Bes-masks may have impersonated dwarf gods in specific rituals. Several

---

[5] See also Ankhiwedjsu 'the living one ordained him' (E 19b; fig. 9.4) and Ankhef 'may he live' (E 48; fig. 9.8).

[6] See also Neferkhuu 'the good one protects' (E

17; fig. 9.3) and S'ankhu-Hathor (?) 'the one whom Hathor keeps alive' or 'brings to life' (E 58).

[7] See also possibly Nofret 'the good one' (E 8, 9; pl. 17.3).

representations, such as the reliefs in the Graeco-Roman temple of Hathor at Philae (pl. 9.1a, b), may be lively illustrations of such performances, where real dwarfs could add greater realism to the ceremony;[8] this type of religious travesty is well documented in Greece for satyrs.[9]

It must, however, be borne in mind that our information is limited by chances of discovery, changes in decorum, and the social provenance of the documents. Most representations of dwarfs were made for the élite; they show only the short-statured people who lived in the company of high officials. We have no information on those who lived in country villages. It is conceivable that dwarfs were brought to the court by their parents or officials, given that they were sought after for their special qualities. It also remains unclear whether dwarf retainers were free men or some kind of slaves. The status of dependants in the Old Kingdom is uncertain, but some categories of servants might have been under a kind of bondage. Junker suggested that the title *jsww* of the dwarf attendants in the tomb of Nesutnufer (E 19a, b; fig. 9.4) could describe such a subordinate status, since the term derives from the root *jsw*, which means exchange, payment, but such etymological argument is very problematic.[10] The identification of pygmies and the understanding of their position is also limited by the paucity of the evidence, such as the absence of extant burials or skeletal remains.

Some gaps are revealing. No medical text, for example, takes note of dwarfs. Disturbances related to dwarfism, such as weakness of the legs, short sight, and deafness, occur in medical papyri, but they are never connected to a pathologically short stature.[11] This absence may be due to the relatively small quantity of extant medical texts, and to the rare occurrence of this disorder. Egyptian physicians may also have preferred to describe those diseases which they thought possible to cure, but their curiosity might have been stimulated by this unusual disorder. They could have attempted to prescribe some medicine or spells to help a short patient grow,[12] or they could simply have tried to analyse the physical constituents of the phenomenon.

The silence of medical texts may be related to other gaps in the textual evidence. Only four Old Kingdom and two New Kingdom non-religious texts and inscriptions mention short people, pathological or ethnic. This scarcity contrasts with the numerous allusions to dwarf gods in religious hymns and magical spells from the Middle Kingdom on. No source gives the prayers of a dwarf who wished to become taller or to be relieved from

---

[8] Cf. F. Daumas, 'Les Propylées du temple d'Hathor à Philae et le culte de la déesse', *ZÄS* 95 (1968), 17 ('Que des offrandes à la déesse aient eu lieu et que par endroits, des prêtres travestis en singes ou en Bès aient pris part aux rites, c'est ce qui est infiniment probable à Dendara et sûr à Philae').

[9] On the kinship between Bes and satyrs, see esp. F. Jesi, 'Bes e sileno', *Aegyptus*, 42 (1962), 257–75; H. Stricker, 'Bes de danser', *OMRO* 37 (1956), 35–48.

[10] H. Junker, *Giza*, v (Vienna, 1941), 8–9. Cf. Gardiner, *Grammar*, sign-list, F 44: *jsw* 'exchange', and 132: *m-jsw* 'as a payment'. See also W. Helck, *LÄ* v (1984), s.v. Sklaven, 985.

[11] See e.g. the various diseases of the lower limbs; G. Lefebvre, *Essai sur la médecine égyptienne de l'époque pharaonique* (Paris, 1956), 156–60.

[12] Cf. the illusory treatments for cataracts in Lefebvre, ibid. 81–3, and for the rejuvenation of an old man, ibid. 173.

some physical disorder. Not the slightest allusion can be discerned in the conventional formulae, and no veiled term seems to have expressed the secret hope of a small person. Restricted growth is also not mentioned in the oracular amuletic decrees which aimed at the magical prevention of birth-irregularities.

These gaps in the evidence have a positive meaning. They suggest that ancient Egyptians welcomed short statured people. They considered dwarfism neither as a disease to be cured, nor as the result of a religious transgression to be countered. As it was not feared as a prejudicial event, pregnant women omitted to ask for protection against it, and, on the contrary, invoked Bes and other small gods as protectors during delivery. This malformation was closely related to the religious sphere, and therefore appeared only in religious terms in text, such as in the Harris Magical Papyrus.[13] This special status may partly account for the silence of the other textual sources.

[13] Spell U 8–V 9.13; H. O. Lange, *Der magische Papyrus Harris* (Copenhagen, 1927), 72 and 74, ll. 1–4; G. Roeder, *Der Ausklang der ägyptischen Religion* (Zurich, 1961), 175.

III

GREECE

II

Terminology

Two Greek words define a person of abnormally short stature: πυγμαῖος and νᾶνος. Πυγμαῖος is the earliest known term. It occurs first in the *Iliad* where it designates little men living on the banks of the river Ocean who are attacked by migrating cranes.[1] The word derives from πυγμή, 'fist', 'boxing', which also denotes a measure of length, the cubit 'from the elbow to the knuckle', equivalent to about 18 δάκτυλοι (*c*.35 cm.), which suggests that pygmies were imagined as extremely small persons with rather pugnacious impulses.[2] Νᾶνος first occurs in Aristophanes: in a lost play, the Ὁλκάδες or *Merchantships*, and in the *Pax* in the form νανοφυής 'dwarfish', describing small crab-like dancers.[3]

As in Egypt, the two terms do not seem to have defined a specific pathology; they were interchangeable and could both refer to ethnic or pathological short stature, proportionate or disproportionate. In the works of Aristotle, for example, πυγμαῖοι is found describing pathological dwarfs deformed in their parts, proportionately reduced in size, and African pygmies;[4] when analysing the origin of stunted growth in mules, γίννοι, and pigs, μετάχοιρα, Aristotle names human dwarfs νᾶνοι in the *Historia Animalium*, and πυγμαῖοι in the *De Generatione Animalium*.[5] The two terms do not seem to have acquired a specific meaning with the course of time. Later writers, such as Longinus and Isidore of Seville, employ both words as if they were synonymous.[6]

Archaeological finds confirm that the two terms had a similar meaning. Herodotus calls 'pygmies' figures which the Persian King Cambyses saw in the temple of Ptah-Hephaistos at Memphis.[7] It is quite likely that these statuettes represented Ptah-Pataikoi, Egyptian dwarf gods who were very popular the time of Cambyses' invasion (525–522 BC). They depict achondroplastic dwarfs with large heads, long trunks, and short,

---

[1] *Il.* 3. 3 ff.

[2] See P. Chantraine, *Dictionnaire étymologique de la langue grecque* (Paris, 1968), s.v. πύξ, 955; LSJ, s.v. πυγμή, 1550; H. Frisk, *Griechisches etymologisches Wörterbuch*, ii (Heidelberg, 1970), s.v. πυγμή, 619–20.

[3] Aristoph. *Pax*, 790; Gell. *NA* 19. 13. 3 (Kassel/Austin iii (2), fr. 441).

[4] Arist. *Gen. An.* 2. 8. 749ᵃ4–6 (malformed

dwarfs); *HA* 8. 12. 597ᵃ (African pygmies); *Pr.* 10. 12. 892ᵃ18 (proportionate dwarfs).

[5] *HA* 6. 24. 577ᵇ; *Gen. An.* 2. 8. 749ᵃ5.

[6] Longinus, *Subl.* 44. 5: οἱ Πυγμαῖοι, καλούμενοι δὲ νᾶνοι; Isid. *Etym.* 11. 3. 7: *ut nani, vel quos Graeci Pygmaeos vocant.*

[7] Hdt. 3. 37 (πυγμαίου ἀνδρὸς μίμησίς ἐστι); see full text above 84.

slightly bowed limbs.[8] Herodotus therefore obviously understood the word πυγμαῖος as a synonym for a pathological dwarf.

Μικρός, an epithet denoting shortness in general, is also found describing dwarfs. Pygmies are called μικροί ἄνδρες by Basilis in his *History of India*, and γένος μικρόν by Aristotle.[9] Similar epithets for smallness and thinness, such as λεπτός and βραχύς, might have described a dwarf on occasion, but no instance is known to me.

---

[8] Cf. pls. 12–16.

[9] Basilis, *FGrH* 718, F 1 = Ath. 9. 390b; Arist.

HA 8. 12. 597ᵃ. See also Ctesias, *FGrH* 688, F 45 = Photius, *Bibl.* 72. 46a (μικροί).

12

Iconographic Conventions

Iconography reflects the sensitivity of the Greeks to the human body, its proportions, its integrity. Greek artists, like Egyptian, had little interest in showing human physical anomalies. Monsters are usually composed of human and animal elements otherwise normal when looked at separately. The hundred eyes of Argos are neatly designed and scattered over a well-proportioned body, Geryon has three normal upper bodies, and apart from his single eye, Polyphemos has no conspicuous malformation; even the Giants are muscular naked men, seldom strikingly tall, only sometimes distinguished by a peculiar bushy hairstyle.[1] Hybrid creatures, such as Gorgo, sirens, and centaurs, are similarly composed of healthy human and animal parts;[2] only satyrs may have slightly unusual bodies, obese, acromegalic or hunchbacked.[3] A few legendary figures are characterized by physical anomalies, such as the decrepit Geras or the lame Hephaistos.[4] Their deformity is, however, never emphasized and even tends to disappear with the course of time. Hephaistos is commonly lame in one of two feet in archaic vase-painting (fig. 13.5), but has a normal constitution in classical pictures, where his lameness is suggested only by his riding.[5] Similarly, ordinary humans are

---

[1] See e.g. K. Schefold, *Götter- und Heldensagen der Griechen in der spätarchaischen Kunst* (Munich, 1978), fig. 78 (Tityos); figs. 141–7 (Geryon); figs. 185–7 (Alkyoneus); figs. 353, 356 (Polyphemos); id., *Die Göttersage in der klassischen und hellenistischen Kunst* (Munich, 1981), figs. 124, 126, 131–2 (Giants); figs. 173–4 (Argos).

[2] A possible exception is a geometric centaur from Lefkandi with six fingers, which may emphasize his supernatural character, or may be a simple error; J. Boardman, *Greek Sculpture: The Archaic Period* (London, 1978), fig. 4; M. R. Popham *et al.* (eds.), *Lefkandi*, i (London, 1979), pls. 251–2; ibid. (1980), 168–70.

[3] See e.g. the acromegalic satyr in a cup, Florence, Museo Etrusco, 4211; *ARV* 121.22; Boardman, *ARFH* i, fig. 116. See also the fat satyr playing the flute on a cup, New York, private (once Castle Ashby 193); *ARV* 172.1, *Add²* 184; Boardman, *ARFH* i, fig. 83.

[4] On Geras, see G. Q. Giglioli, 'Una pelike attica da Cerveteri nel Museo di Villa Giulia a Roma con Herakles e Geras', in G. E. Mylonas and D. Raymond (eds.), *Studies presented to D. M. Robinson*, ii (St Louis, 1953), 111–13, pls. 36a, 37d–e; H. A. Shapiro, 'Notes on Greek Dwarfs', *AJA* 88 (1984), 391–2.

[5] For the iconography of Hephaistos' deformity, see below 198–9.

very rarely individualized by physical deformities, apart from marginal figures, such as highwaymen or foreigners.[6]

Greek artists also disliked showing mutilated bodies. Visual scenes of violence never compete with the realistic crudity of literary descriptions. The grim details of the cannibal dinner of the Cyclops, for example, as narrated in the Odyssey, are only suggested in art.[7] In war scenes, bodies are pierced through, blood flows from wounds; the heroes die but are not crippled.[8] The few extant scenes of cruel slaughter show abnormally wild behaviour due to madness, such as when enthusiastic maenads tear to pieces little animals, or dismember Pentheus.[9] I know of only one representation of a severely disabled person, a seventh-century terracotta from Sicily, which depicts a man deprived of both legs and an arm. His condition is probably due not to an accident but rather to a congenital disorder, because he would probably not have survived such drastic amputations; he seems to have been trained to use his right arm to move, as suggested by the muscular development of his right pectoral and the pose of his right arm, with the palm on the ground.[10] This sensitivity towards mutilation is also present in the conventions of war: Greek prisoners may be killed, but their bodies are never disfigured.[11]

Depictions of dwarfs thus constitute a significant exception to the traditional stock figures of vase-painting and statuary. This suggests that dwarfism was not seen as an irreducible monstrosity, to be hidden, but as an acceptable physical anomaly which could be shown on drinking-vessels. The fact that their bodies are complete, although shrunken or distorted, may partly explain this exception to the rule.

IDENTIFICATION

As was the case in Egypt, the representation of dwarfs conformed to specific artistic conventions. A review of the stock figures presenting similar physical features first lets us define how the condition was stereotyped in Greek art. Secondly, an attempt will be made to identify the different types of dwarfism depicted.

[6] See in general W. Binsfeld, *Grylloi. Ein Beitrag zur Geschichte der antiken Karikatur*, Diss. (Cologne, 1956), 31–5; V. Zinserling, 'Physiognomische Studien in der spät-archaischen und klassischen Vasenmalerei', *WZRostock*, 16 (1967), 570–5; ead., 'Die Anfänge griechischer Porträtkunst als gesellschaftliches Problem', *AAntHung* 15 (1967), 283–5; D. Metzler, *Porträt und Gesellschaft* (Berlin, 1971), esp. 81–128.

[7] *Od.* 9. 288–96; see also Eur. *Cyc.* 398–404. In iconography, see e.g. the Laconian vase in Paris, Cabinet des Médailles, 190; C. M. Stibbe, *Lakonische Vasenmalerei* (Amsterdam and London, 1972), pl.

94.1 (giant seated holding two legs).

[8] On representations of scars, see J. Boardman, 'An Anatomical Puzzle', *AA* (1978), 330–3.

[9] See e.g. the bleeding trunk of Pentheus on a psykter, Boston, MFA 10.221; *ARV* 16.14, *Add²* 153; Boardman, *ARFH* i, fig. 28.

[10] J. Dörig, *Art antique. Collections privées de Suisse Romande* (Geneva, 1975), no. 143; *Médecine antique, IVᵉ colloque international hippocratique* (Lausanne, 1981), 91–2, no. 61. For amputation, see G. Penso, *La médecine romaine* (Paris, 1984), fig. 146.

[11] P. Ducrey, *Le Traitement des prisonniers dans la Grèce antique* (Paris, 1968), 206–8, 313–32.

Miniatures

Miniature men can seldom be positively identified as pathological dwarfs. An extreme smallness usually depicts a low-ranking figure, whose subordinate role is stressed by his pose and activity. Thus servants are often shown kneeling or squatting, a position not in accordance with the dignity of a free man, unless he is in some abnormal state.[12] The short bearded man on a hydria in the Villa Giulia, for example, could be a familiar servant (G 48; pl. 61.1b): he stands in the middle of revellers, and holds a pot in which one of them relieves himself.[13] Similarly, the small bearded man on an amphora in Bonn has no conspicuous pathological deformity; he holds out a liver to be inspected by a departing warrior, and substitutes for the young servant who carries the viscera in similar scenes of hepatoscopy.[14]

Miniature figures filling a small space left free under the handles of the vase or below a motif may be due to compositional need. For example, the small bearded man on a terracotta plaque by Exekias cannot be identified as a dwarf.[15] He is conspicuously shorter than his companions, and has a slightly abnormal pose, with flexed legs, which enhances his paunchy belly and fat buttocks. These features may be associated with a pathological short stature, but the limbs are not significantly shortened. Exekias may have reduced the size of the figure in order to fit him before the mules of the funerary cart and indicate his inferior status, probably as a groom.[16] The smallness of winged homunculi, Erotes or feeble εἴδωλα, souls of the deceased, likewise does not imply dwarfism. These airy supernatural creatures are represented as slender young men with well-built bodies.[17] No depiction of the soul of a deceased dwarf has so far been found.

Dwarfs must therefore have conspicuous physical malformations to be identified with certainty. As naturalistic features are usually not rendered in archaic vase-painting, there are very few representations of human dwarfs on black-figure vases. The miniature man on a pyxis in the Louvre may be identified as a pathological dwarf because his miniature

---

[12] On these criteria, see N. Himmelmann, 'Archäologisches zum Problem der griechischen Sklaverei', *Abh Mainz*, 13 (1971), 614–57.

[13] For men relieving themselves in a pot, see G. Vorberg, *Glossarium eroticum* (Stuttgart, 1932), figs. 335, 380, 382. For young slaves helping their nauseous masters, see e.g. the cup, Berlin, SM (West) 2309; *ARV* 373.46, *Add²* 226; Rühfel, *Kinderleben*, 66, fig. 38; 70 n. 166 (refs. to similar scenes).

[14] Amphora fr., Bonn, University, 945.47; Himmelmann (n. 12 above), 635–6, fig. 35. See twelve similar scenes in J.-L. Durand and F. Lissarague, 'Les Entrailles de la cité', *Hephaistos*, 1 (1979), 92–108. See also the miniature bearded servant on the amphora in Boston, MFA 99.517; *ABV* 241.25, *Add²* 61; *CVA* USA 14, Boston, III H,

pl. 22.1–2.

[15] Funerary plaque fr., Berlin, SM (West) 1814; *ABV* 146.22, *Add²* 41; Himmelmann (n. 12 above), 626–7, fig. 14; Boardman, *ABFH*, fig. 105.2.

[16] But grooms are normally younger. Cf. the young man placed below the head of horses on a psykter in Naples, Museo Nazionale, N Sant 38; *ABV* 29.3; *CVA* Italy 20, Naples 1, III H e, pl. 9.3.

[17] For small Erotes, see e.g. the cup in Berlin, SM (West) 2291; *ARV* 459.4, *Add²* 244; A. Greifenhagen, *Griechische Eroten* (Berlin, 1957), 67, fig. 51; Boardman, *ARFH* i, fig. 310. For εἴδωλα, see e.g. the lekythos in London, BM B 639; Haspels, *ABL* 227.8; Boardman, *ABFH*, fig. 261, and the lekythos in Athens, NM 1926; *ARV* 846.193, *Add²* 297; C. Bérard, 'L'Ordre des femmes', in C. Bérard *et al.*, *La Cité des images* (Lausanne and Paris, 1984), fig. 150.

size does not seem to be due to composition (G *d* 1; pl. 41b). He has a beard and wears a wreath around the neck, which suggests that he has a full adult status like his normal-sized companions.[18] The small man digging in a hole on a Corinthian pinax has peculiar proportions (long trunk, short limbs), but this could also be due to his location in a narrow space (G *d* 2; fig. 15.1). In scenes of myth, dwarfs can be recognized from the context. In black-figure, pygmies are usually depicted as miniature humans, but they are identified by their fight against cranes; the size of their opponents highlights their abnormal smallness (e.g. G 40; pl. 58). Similarly, the Cercopes are shown as well-built men, but small in proportion and with beards (G 104, 106–9; pls. 71.2, 72.1, figs. 13.2–3). Only the two bearded gnomes on a skyphos in Paris have clearly malformed bodies (G 113; pl. 75a, b).

Children

The physical appearance of a dwarf may resemble that of a child. On red-figure vases, children often display an ambiguous disproportion between their large heads and the fullness of their short limbs. These plump babies seem to illustrate the theory of Aristotle on the dwarfishness of children,[19] but they can easily be recognized as infants. Babies often wear amulets hanging on a string across their chest (*crepundia*) to protect them against the evil eye.[20] They have childish interests; they push a toy cart, wave a rattle, play with pet animals or crawl assiduously towards a table, or another piece of furniture. In addition, pictures of children are mostly found on a special type of vessel, the choes, miniature jugs which were made for them on festal occasions, such as the Anthesteria.[21]

Occasionally the attributes characterizing children and dwarfs blend. Children may have occupations unusual for their age, like the two young athletes boxing on a chous in Boston, but no sign of physical maturity indicates that they are pathologically short adults.[22] On the pelike in Laon, however, two plump beardless figures are probably dwarfs (G 12; pl. 48.2). They have the corpulence of children, associated with features that betray their adult age; the man standing on the left is half-bald, and has sparse side-whiskers; both have adult sexual organs. Dwarfs may also be depicted on choes. A small man in Dresden has a thin, proportionate body, but with a beard and normal sexual organs, which reveal his greater age (G 14; pl. 49.2); the painter may have chosen to picture him on this kind of jug to stress his resemblance to a child.

[18] Cf. also the small, but young servant with a slightly unusual body on an amphora in Cambridge, Fitzwilliam Museum, GR 27.1864 (48); *ABV* 259.17, *Add²* 67; *CVA* Great Britain 6, Cambridge 1, III H, pl. 23.2.

[19] Arist. *Part. An.* 4. 10. 686ᵇ12: 'all children are dwarfs'.

[20] See e.g. the chous in Athens, NM 14527; Rühfel, *Kinderleben*, 136, fig. 74. See also the chous in Athens, NM 15875; G. van Hoorn, *Choes and Anthesteria* (Leiden, 1951), no. 11, fig. 24.

[21] Rühfel, *Kinderleben*, 125–31.

[22] Chous, Boston, MFA 95.53; Van Hoorn (n. 20 above), no. 368, fig. 132; Rühfel, *Kinderleben*, 147, fig. 83. See also the children acting on a chous in Paris, Louvre, CA 2938; Rühfel, *Kinderleben*, 152, fig. 86.

Caricatures

A few red-figure vases depict personages with over-large heads contrasted with meagre bodies. Where is the limit between caricature and portrait? The painter describes in both cases a deviation from the physical norm, when the model is a person physically malformed. Some scholars simplify the problem by considering all pictures of dwarfs as caricatures of normal Athenians.[23] The depiction of a pathological case must, however, show a non-fatal malformation, while an imaginary figure may have an impossible combination of physical deformities.

Thus an askos in the Louvre depicts a man combining a miniature body with an over-sized head (pl. 38.1).[24] The striking hypertrophy of his head is a clear biological impossibility; the man is probably a caricature influenced by a theatrical stock figure, such as Hermonios, an old man whose mask is similarly characterized by a long pointed beard, a bald head, and piercing glance.[25] The deformity of the naked man depicted on a fragment of plate in Athens is also unrealistic (pl. 38.2).[26] He is squatting and exhibits large dangling genitals. His skinny arms, like those of the man on a pyxis in Boston (pl. 38.3),[27] seem to indicate rickets rather than dwarfism. His obscene pose, emphasized by his pinched nose, confirms the comic intention of the painter.[28] The man on a cup in the Vatican, caught in the middle of a conversation with a talkative fox, offers a more ambiguous appearance (pl. 38.4):[29] the disproportion between his head and his trunk is striking, but more naturalistic than in the previous examples; his size seems relatively normal when compared to that of the fox. The painter perhaps intended to caricature the fabulist Aesop, as Schefold and others propose.[30] Ancient authors mention that Aesop was very ugly and malformed, but they do not specify that he was abnormally short

---

[23] On the dwarf in Erlangen (G 9; pl. 47.1), see e.g. G. Lippold, 'Zu den Imagines Illustrium', *MDAI(R)* 52 (1937), 44–7. On the dwarf in Dresden (G 14; pl. 49.2), see P. Herrmann, 'Erwerbungsbericht der Dresdener Skulpturensammlung 1899–1901', *AA* 3 (1902), 117, no. 35, and Deubner, *Att. Feste*, 244.

[24] Paris, Louvre, G 610; G. E. Rizzo, 'Caricature antiche', *Dedalo*, 7 (1926), 405–6 (with fig.); Zinserling (n. 6 above), 573, pl. 128, 5; Metzler (n. 6 above), 101, fig. 11; P. Ghiron-Bistagne, *Recherches sur les acteurs dans la Grèce antique* (Paris, 1976), 151, fig. 61.

[25] Ghiron-Bistagne, ibid. 149–50. A plastic vase made by Sotades shows the face of a similar old man in Ferrara, Museo Nazionale, 20401; *ARV* 766.5, *Add²* 286; N. Alfieri, *Spina, Museo Archeologico Nazionale di Ferrara*, i (Bologna, 1979), 48, no. 108.

[26] Plate fr., Athens, NM ACR 1073; B. Graef, F. Langlotz, *Die antiken Vasen von der grossen Akropolis zur Athen*, ii. 2 (Berlin, 1931), no. 1073, pl. 83;

Zinserling (n. 6 above), 572, pl. 128, 1; Metzler (n. 6 above), 87, fig. 3.

[27] Pyxis fr., Boston, MFA 10.216; *ARV* 81; Rizzo (n. 24 above), 405–6 (with fig.); Zinserling (n. 6 above), 572, pl. 128, 3; Metzler (n. 6 above), 97, fig. 10.

[28] A man is squatting in a similar position on a cup in Paris, Louvre, G 5; *ARV* 71.14, *Add²* 167; Vorberg (n. 13 above), 382, 1. See also the satyr on a rhyton in Baltimore, WAG 48.2050; *ARV* 765.15, *Add²* 286; H. Hoffmann, *Attic Red-figured Rhyta* (Mainz am Rhein, 1962), pl. IX. 1.

[29] Cup, Vatican, 16552; *ARV* 916.183, *Add²* 304; Rizzo (n. 24 above), 405–7 (fig.); Zinserling (n. 6 above), 572–3, pl. 128, 4; Metzler (n. 6 above), 94–5, fig. 7.

[30] K. Schefold, *Griechische Dichterbildnisse* (Zurich, 1965), 89, pl. 3b. Cf. Phil. *Imag.* 1. 3, about the fox which leads the chorus of the actors of Aesop's fables.

too. Iconography may transmit a lost oral tradition; a Roman copy now in the Villa Albani shows a man with a very short stunted trunk who may represent the fabulist.[31] The man depicted on the Ferrara cup has strikingly abnormal proportions too (G 7; pl. 46.1). He has short thin limbs and an over-large head which seems to bend under its own weight; his smallness is indicated by the size of the child who walks before him. I include him in the category of human dwarfs, but as an example of a clumsy rendering of the pathology.

Actors

In a theatrical context, figures with odd morphologies are also matters for doubt. Did the painter depict actors mimicking a deformity or persons suffering from a real physical defect? On a skyphos in Boston, a short man with meagre limbs kneels before a chorus of men riding on ostriches (pl. 39.1).[32] His stunted silhouette evokes that of a dwarf, but his bearded mask could cover a beardless young man; when upright, his size could be normal. If he is a dwarf, he is too skilfully disguised and cannot be identified.

This ambiguity is particularly striking on vases from the sanctuary of the Kabeirion near Thebes, and on those from South Italy which illustrate the Phlyax comedy.[33] In both series of pictures, counterfeit people with big heads, thin limbs, and over-sized genitals are shown. Their unrealistic physical form may be inspired by real dwarfs, or by Egypto-Phoenician dwarf gods, as in the case of the Kabeirion vessels. These dwarfish figures are so numerous and so similar that their deformity must be seen as a convention showing the parts of normal actors.

To sum up, in black-figure vase-painting, leaving aside mythical figures, such as pygmies and Cercopes, pathological dwarfs cannot be recognized with certainty because they only have miniature bodies. In red-figure vase-painting, however, they are characterized by the combination of the following features: First, a marked physical malformation. Most dwarfs have large heads, long trunks, and short limbs. Their abnormal anatomy is often stressed by their nakedness, which contrasts with the clothing of their companions. In Athenian vase-painting, only one dwarf is fully dressed in a himation (G 13; pl. 49.1), and five wear a chlamys on the shoulder (G 4, 7, 8; pls. 43a–c, 46.1–2). It is probably not a coincidence if the only draped dwarf is proportionately short (G 13; pl. 49.1): the detailed rendering of his miniature body would not have added much to

---

[31] Rome, Villa Albani, 964 (h. 56 cm.); P. Richer, *Le Nu dans l'art*, ii (Paris, 1926), 343, figs. 473–4; M. D. Grmek, 'Les Affections de la colonne vertébrale dans l'iconographie médicale et les arts antiques', *Dossiers Histoire et Archéologie*, 123 (1988), fig. on 57; R. Bol in P. C. Bol (ed.), *Forschungen zur Villa Albani. Katalog der antiken Bildwerke*, i (Berlin, 1989), cat. no. 75, 227–31, pls. 126–9.

[32] Skyphos, Boston, MFA 20.18; Ghiron-Bistagne

(n. 24 above), 259, fig. 110.

[33] For the Kabeirion vases, see the material collected by P. Wolters and G. Bruns, *Das Kabirenheiligtum bei Theben*, i (Berlin, 1940), and K. Braun and T. E. Haevernick, *Das Kabirenheiligtum bei Theben*, iv, *Bemalte Keramik und Glas* (Berlin, 1981). For the Phlyax comedy, see e.g. Trendall, *PhV²*, and M. Bieber, *The History of the Greek and Roman Theater²* (Princeton, 1961), 129–46.

the characterization of his condition. Secondly, the indication of physical maturity. Most figures have beards (e.g. G 3, 5, 6; pls. 42, 44–5), or a moustache (G 13; pl. 49.1), often associated with incipient baldness, which emphasizes their bulging heads, sometimes with wrinkled foreheads (e.g. G 3, 5, 8, 9; pls. 42, 44b, 46.2, 47.1); they have mature sexual organs (e.g. G 8, 18–19; pls. 46.2, 52, 53.1).

MEDICAL DIAGNOSIS

In red-figure vase-painting, the rendering of anatomy is often very accurate, and a diagnosis may be attempted on several depictions.[34]

A disproportionate short stature, especially achondroplasia (pl. 1.1–2; fig. 1.1b), is the commonest condition. Thus, on the Peytel aryballos, the short man holding a hare presents the typical skeletal malformations associated with achondroplasia (G 5; pl. 44b). He has a long broad trunk, thick thighs, and short legs. His bulging forehead, enhanced by incipient baldness, is associated with a strong lower jaw revealed by sparse side-whiskers spread along his cheeks; the depression at the root of the nose is indicated by a large hook drawn at its wing. His hairy chest is muscular, as shown by the fullness of the inferior line of the chest muscle (pectoralis major) drawn in brown lines on his abdomen. Typical folds of skin appear on the thighs below the buttocks, and are emphasized by the shortness of the lower leg (tibia).[35]

Although the Clinic Painter and the Followers of Makron had an interest in rendering unusual anatomical details and attitudes,[36] no figure made by the group resembles this dwarf figure, except satyrs, similarly naked, with snub noses, bushy brows, and hairy chests. Two satyrs in particular are very small,[37] and two others infibulated.[38] But their beards are usually longer, with tufts sprouting out downwards.

The achondroplastic dwarfs depicted on an askos in Hamburg (G 76; pl. 66.2) and on a cup in Leipzig (G 77; pl. 67.1) have similar muscular formations. The biceps of the dwarf in Leipzig is marked by a curved brown line on his right arm, as is the calf muscle (gastrocnemius) of his left leg. When the abdomen is seen in profile, the lumbar lordosis, associated with prominent buttocks and a swelling belly, is clearly indicated.[39] This paunchy abdomen may denote muscular hypotonia, as suggested by the accessory line in diluted glaze parallel to the belly of the pygmy in Compiègne (G 67; pl. 64b).

[34] For anatomical terminology, see D. C. Kurtz, *The Berlin Painter* (Oxford, 1983), esp. 18–36, figs. 6, 8, 10.

[35] Cf. Kunze/Nippert, *Genetics*, 16, fig. 17 (achondroplasia).

[36] See e.g. the man turning his head backward while running on a cup in Havanna, Lagunillas; *ARV* 811.51; unpub. (Beazley archive).

[37] Small satyrs: cup, London, BM E 66; *ARV* 808.2, *Add²* 291; Boardman, *ARFH* i, fig. 376. Cup,

Berlin, SM (West) 2534; *ARV* 826.25, *Add²* 294; *CVA* Germany 21, Berlin 2, pl. 100.

[38] Infibulated satyrs: cup, Boston, MFA 10.572; *ARV* 821.5, *Add²* 293; Caskey/Beazley, iii, pl. 87. Cup, Paris, Cabinet des Médailles, 812; A. De Ridder, *Catalogue des vases peints de la Bibliothèque Nationale* (Paris, 1902), pl. 21.

[39] See e.g. G 8, 11, 22, 67, 74, 75; pls. 46.2, 48.1, 54, 64, 65.2, 66.1.

Three dwarfs seem to be affected by a severe shortening of the lower extremities (mesomelic dwarfism). The dwarf on a pelike in Boston has especially short lower legs (tibia and fibula) and forearm (humerus) (G 8; pl. 46.2); his thick prominent lips and his large snub nose recall the features of a satyr drawn also by the Dwarf Painter, except for the nose which is flatter.[40] The dwarfs dancing on a stamnos in Erlangen (G 9; pl. 47.1) and on an oinochoe in Oxford (G 18; pl. 52) also have very short extremities. The portrait of the dwarf in Erlangen is especially delicately rendered. He has a small button nose, a large lower lip revealed by his moustache and beard, a small chin and mandible contrasting with his large cranium. His bulging forehead is subtly indicated by a fine line towards the eyes which are surmounted by thick brows giving him a thoughtful expression. Only one satyr by the Peleus Painter has similar facial features, again with a more shaggy beard.[41]

Cases of hypochondroplasia (pl. 1.3) may also have been depicted. The dwarf dancing on a krater in Zurich (G 23; pl. 55.1) has a normal profile with a straight forehead–nose line, associated with a disproportionate body, like the dwarf on a cup once in Munich (G 11; pl. 48.1) and the two gluttons in Laon (G 12; pl. 48.2).

Proportionate short stature (pl. 1.4) is recognizable in a few depictions. On a cup in Athens (G 13; pl. 49.1), a tiny man with slender limbs stands beside a seated woman; he is so small that he hardly reaches the level of her seat. He has a moustache, painted in diluted glaze, which indicates his adult age, and a strikingly large head, more than a quarter of his size, with a curiously elongated shape. He is probably affected by a pituitary disorder, as is the pygmy fighting on an amphora in Brussels who shows similar diminutive proportions (G 70; pl. 65.1). The dwarf holding a skyphos on a chous in Dresden is also a human miniature (G 14; pl. 49.2), but his features are very simplified, and no real diagnosis can be made.

Several figures show more stylized depictions of dwarfs. Restricted growth is recognizable, but an accurate identification is not possible. Thus, the two dwarfs jumping and dancing wildly on a skyphos in Paris associate quite normally proportioned bodies with achondroplastic snub noses and large foreheads (G 3; pl. 42). The dwarfs depicted by the Sabouroff Painter (G 6; pl. 45) and by the Washing Painter (G 10, 11; pls. 47.2, 48.1) show realistic elements of bodily disproportion (long trunks, short limbs), but have excessively large heads, with heavy facial features, which evoke faintly those of achondroplastic dwarfs.

The intriguing picture of a female dwarf in Munich may be the only case of short trunk dwarfism (G 16; pl. 51a, fig. 1.1d). The woman has a very short neck, a stunted trunk with a protruding thoracic cage, and her back is deformed by a pronounced lumbar lordosis inducing prominent buttocks. Her arms appear of relatively normal length, as do her legs, if the painter intended them to be imagined at the bottom of the

---

[40] Amphora, Edinburgh, Royal Scottish Museum, 1872.23.10; *ARV* 1011.12; unpub. (Beazley archive).

[41] Krater frr., Syracuse, Museo Nazionale, 24114; *ARV* 1041.1; unpub. (Beazley archive).

pot, below the ornament. These features may indicate that she is affected by metatropic dwarfism, or by mucopolysaccharidoses, a condition inducing metabolic disorders which would explain her negroid features. This interpretation relies, however, on the addition of the missing lower legs. Her deformity could also be due to caricature. The woman's appearance of health is not entirely consistent with that of persons suffering from the severe complications derived from this disorder, such as limitation of mobility of the joints and paraplegia.

As in Egyptian art, these pictures are not fully naturalistic. Three errors in particular may reflect folk-beliefs about dwarfs. First, most red-figure dwarfs and pygmies belong to the disproportionate type, and more particularly to the achondroplastic. They have snub noses, associated with beards and balding heads. This combination of features occurs in reality, but not with such high frequency, since achondroplasia has the same incidence as hypochondroplasia. It is thus not a medical fact, but an iconographic convention which suggests the affinities of dwarfs with the world of Dionysos. Satyrs have very similar facial features: thick eyebrows, depressed nasal bridge, thick lips and beard, associated with incipient baldness. I note, for example, the similarity between the face of the dwarf in Erlangen (G 9; pl. 47.1) with that of the satyr on the Taranto cup.[42] The two groups of figures are subtly distinguished: the noses of the dwarfs are usually neatly button-shaped, while those of satyrs may be very irregular and stumpy like a boxer's. The beards of dwarfs also are not as large and bushy as those of satyrs, but small and pointed with a few sparse tufts of hair on the cheek.[43] This special hairstyle sometimes characterizes marginal figures such as foreigners,[44] and some Athenian revellers.[45]

The second non-medical feature is that dwarfs often have larger or more conspicuous genitals than normal-sized Athenians.[46] It was seen as an inelegant attribute, as Dover demonstrates,[47] and may have stressed their affinities with satyrs.

Thirdly, no woman is shown, apart from the one depicted on the skyphos in Munich (G 16; pl. 51a), which suggests that the rendering of female deformity was under some

[42] Cup, Taranto, Museo Nazionale; *ARV* 860.3, 1672, *Add²* 298; Boardman, *ARFH* ii, fig. 65.

[43] See e.g. G 5, 8, 9, 14; pls. 44b, 46.2, 47.1, 49.2.

[44] See e.g. the skyphos in Zurich, private; H. J. Blösch (ed.), *Greek Vases from the Hirschmann Collection* (Zurich, 1982), 80–1, no. 39. On representations of foreigners in general, see W. Raeck, *Zum Barbarenbild in der Kunst Athens* (Bonn, 1981).

[45] See e.g. the man with a thin pointed beard in an erotic scene on a cup, once Munich, Arndt; *ARV* 339.55, *Add²* 218; Boardman, *ARFH* i, fig. 241. For revellers with unusual hairstyle, see also: cup, Paris, Louvre, G 25; *ARV* 316.5, *Add²* 214; P. Hartwig, *Die griechischen Meisterschalen* (Stuttgart, 1893), pl. 9. Cup, Berlin, SM 3198; *ARV* 402.13; H. Licht, *Sittengeschichte Griechenlands* ii (Dresden and Zurich, 1926), 205. Cup, Basle, Cahn coll. 63; *ARV* 421.77; unpub. (Beazley archive).

[46] See e.g. G 8, 11, 14–15, 74, 77–9; pls. 46.2, 48.1, 49.2, 50.1, 65.2, 67.1–2, 68.1.

[47] K. J. Dover, *Greek Homosexuality* (London, 1978), 125–9.

kind of taboo, while the notion of male abnormality was less disturbing, perhaps because of its Dionysiac connotation. The absence of depictions of the more severe malformations (e.g. pseudo-achondroplasia, hypothyroidism) may indicate a similar unease towards crippling diseases, perhaps mingled with compassion.

13

Dwarfs in Myth

Four categories of dwarfs or dwarfish beings are found in myth. Two are well documented in both literary and iconographic sources: the exotic pygmies, and the mischievous Cercopes. The two others belong to a category of deities which 'resist the clear shapes of Greek mythology':[1] demons associated with Hephaistos and smithcraft (the Kabeiroi, the Telchines, and the Daktyloi), and anonymous kourotrophic demons.

PYGMIES: WRITTEN SOURCES

From Ethnology to Myth

Homer said that, in the autumn, migrating cranes waged war on a population of dwarfs living by the stream of Ocean.[2] This story was repeated throughout antiquity, not as a myth created by a powerful poet, but as an ethnological reality.[3] In the sixth century, Hecataeus of Miletus described them in his *Periegesis*, or 'Journey around the World', a work which aimed to present a rational catalogue of existing lands and races of men, where pygmies were no legend.[4] Aristotle too certified that they were no fable.[5] Of all classical authors, a few only dared to raise doubts against their reality, 'for no man worthy of belief professes to have seen them', as Strabo states.[6]

As suggested by their name, a striking shortness was the salient physical characteristic

---

[1] Burkert, *GrRel*, 284.

[2] *Il*. 3. 2–6.

[3] The basic discussions remain E. Wüst, *RE* xxiii. 2 (1959), s.v. Pygmaioi, 2064–74; O. Waser, Roscher, iii (2), 1902–9, s.v. Pygmaien, 3283–317; P. Janni, *Etnografia e mito. La storia dei pigmei* (Rome, 1978) (review by H. Wölke, *Gnomon*, 55 (1983), 97–9), with earlier bibliog. For an ethnological and historical perspective, see principally P. Monceaux, 'La Légende des pygmées et les nains de l'Afrique équatoriale', *RH* 47 (1891), 1–64; C. Préaux, 'Les Grecs à la découverte de l'Afrique par l'Égypte, *CdE* 32/64 (1957), 284–312; M. Gusinde, *Kenntnisse und Urteile über Pygmäen in Antike und Mittelalter* (Leipzig,

1962) (Nova Acta Leopoldina, 162); L. L. Cavalli-Sforza, 'Evaluation of the State of Research', in *African Pygmies*, 363–7.

[4] Hecat. *FGrH* 1, F 328a, b. Cf. the opening sentence of his *Genealogies, FGrH* 1, F 1a: 'I write what I believe to be the truth, for the Greeks have many stories which, it seems to me, are absurd' (trans. L. Pearson, *The Oxford Classical Dictionary* (Oxford, 1970), s.v. Hecataeus, 490).

[5] Arist. *HA* 8. 12. 597ᵃ.

[6] Strab. 17. 2. 1 (trans. H. L. Jones (Loeb 1932)); ibid. 1. 2. 28; 2. 1. 9; 7. 3. 6. See also Philostr. *Vita Apol*. 3. 47; Gell. *NA* 4. 6; Rut. Nam. 1. 291–2.

of pygmies. Ctesias reports that they measure between one cubit and a half (69.3 cm.) and two cubits (92.4 cm.), while Basilis says that they are so minuscule as to ride on partridges;[7] this extreme smallness became a ludicrous commonplace in Hellenistic poetry.[8] Only Ctesias describes more precisely their physical appearance, but as a model of Greek ugliness: pygmies are snub-nosed, σιμοί, they dress in their long hair, and have over-large genitals, hanging down to the ankles.[9] Some authors added a few picturesque details on their habits. After Hecataeus, pygmies are traditionally imagined as farmers who cultivate stalks of wheats as tall as trees,[10] and live among miniature horses and cattle.[11] Aristotle and Philostratus say that they dwell in caves or under earth, while Pliny reports that they live in huts made of mud, feathers, and eggshells.[12] To repel the cranes they were believed to ride on rams, goats, and even partridges, with primitive weapons, arrows, and also castanets to frighten the birds;[13] Pliny adds that each year they organized an expedition to the nesting place of their enemies to destroy their eggs and kill their young.[14]

Most historians and geographers located the short men in Africa, along the shore of Ocean, or at the sources of the Nile,[15] while from the time of Alexander secondary traditions distributed them in other distant parts of the world, such as Caria, India, Thrace, and Thule, where cranes are seen migrating.[16] Herodotus twice mentions African short people. In his second book, he reports that five Nasamones from Libya had crossed the Sahara and reached a place with fruit-trees, where they were kidnapped by small black people who took them across swamps to their town. This town was along a river with crocodiles which Herodotus identifies as the Nile.[17] Scholars usually think that it was the Niger, or possibly an affluent of the Nile, the Bahr el-Ghazal.[18] In his fourth book, Herodotus speaks about the travels of a Persian explorer, Sataspes, who sailed along the African coast and reached after many months the coast of modern Guinea; he discovered short people, ἀνθρώπους σμικροὺς, dressed with palm-leaves, who

[7] Ctesias, FGrH 688, F 45 = Phot. Bibl. 72. 46a; Basilis, FGrH 718, F 1 = Ath. 9. 390b. Cf. also Pliny, NH 7. 26 ('Trispithames', i.e. c.66 cm. tall). Eust. Il. 3. 6, adds that some geographers even raised their size to five spans (c.1.15 m.), which is closer to reality.

[8] See e.g. the epigrams by Lucillius in Anth. Pal. II. 95. 265.

[9] Ctesias, FGrH 688, F 45 = Phot. Bibl. 72. 46b. For Greek aesthetic criteria of male genitals, see K. J. Dover, Greek Homosexuality (London, 1978), 125–9.

[10] Hecat. FGrH 1, F 328b; Philostr. Imag. 2. 22. 30–3.

[11] Ctesias, FGrH 688, F 45 = Phot. Bibl. 72. 46b; Arist. HA 8. 12. 597ᵃ; Philostr. Imag. 2. 22. 1.

[12] Arist. HA 8. 12. 597ᵃ; Pliny, NH 7. 26; Philostr. Imag. 2. 22. 26–7; Philostr. Vita Apoll. 3. 47.

[13] Hecat. FGrH 1, F 328a–b; Basilis, FGrH 718, F 1 = Ath. 9. 390b; Pliny, NH 7. 26.

[14] Pliny, NH 7. 26.

[15] Hes. fr. 150. 17–18 (R. Merkelbach and M. L. West, Fragmenta Hesiodea (Oxford, 1967)); Hecat. FGrH 1, F 328a–b; Arist., HA 8. 12. 596ᵇ; Strab. 1. 2. 28.

[16] In India: Ctesias, FGrH 688, F 45 = Phot. Bibl. 72. 46a–b. In Thule: Eust. Il. 3. 6. Pliny mentions several locations: Caria (5. 109), Thrace (4. 44), India (6. 70), Ethiopia (6. 188).

[17] Hdt. 2. 31–2.

[18] Monceaux (n. 3 above), 27–9 (Bahr el-Ghazal, or near Lake Chad); Préaux (n. 3 above), 293–4 (possibly Bahr-el-Ghazal); R. Carpenter, 'A Trans-Saharan Caravan Route in Herodotus', AJA 60 (1956), 231–42, esp. 239–40 (lake Chad).

fled to the mountains when the mariners landed.[19] Herodotus does not, however, call these short people 'pygmies', perhaps because he reserved the term for fabulous dwarfs only, as Janni suggests.[20] In the time of Justinian, Nonnosus gives another precise location; he reports that he saw on an island along the east African coast a tribe of timid, black, and hairy people of a very short stature. Like Herodotus, he does not call them 'pygmies'.[21]

How much do these descriptions owe to real facts? It seems likely today that pygmies once lived in the swamps of the White Nile, where they might have been directly or indirectly in contact with Egyptians, Greeks, and Romans.[22] Details in the stories may express some distorted real knowledge. Thus the huts made of eggshells reported by Pliny evoke the actual low hemispherical huts, made of saplings and tiled with leaves, of pygmies;[23] travellers perhaps described this form as egg-like, and hence as made of eggshells. The legend of the fight between cranes and small men might also have translated into a mythical language the fact that pygmies have hostile relationships with their tall black neighbours, the Nuer, Dinkas, and Chillouks, who are used to stand on one leg like birds. Yet, apart from shortness and egg-like huts, the image of pygmies transmitted by ancient sources contains many fabulous elements: pygmies live in tropical forests where no cranes enter, they are not farmers but nomads and hunters, they build temporary huts made of branches, and do not dwell in caves.

The role of the cranes is likewise a mixture of actual and false data. It is true that cranes are migrating birds.[24] Greeks indeed observed flocks of cranes flying north in the spring and south in the autumn.[25] They probably knew that the birds fly across Egypt down to Nubia and Abyssinia, and they may have caught some picking up seeds in their fields. It is also true that cranes may be very tall (above a metre, and up to 1.56 m. on average). Some cranes may have nested in the north of Greece where peasants could notice their agression against those who approached their young. Greeks may also have

---

[19] Hdt. 4. 43.

[20] Janni (n. 3 above), 31.

[21] Nonnosus, *FGrH* iv, 180 = Phot. *Bibl.* 3. 3a. 21–38. For the possible identification of this island with that of Nu'man on the coast of Arabia, see A. Nibbi, 'Punt and Pygmies in the Northern Red Sea', *DE* 2 (1985), 27–36, esp. n. 19. See also above 26–8.

[22] R. Hennig, 'Der kulturhistorische Hintergrund der Geschichte vom Kampf zwischen Pygmäen und Kranichen', *RhM* 81 (1932), 20–4; Préaux (n. 3 above); Cavalli-Sforza (n. 3 above), 364–7. Cf. the great surge of enthusiasm which followed the discovery of pygmy tribes by the end of the 19th cent. See G. Schweinfurth, *The Heart of Africa*, ii (London, 1873), 122 ff.; H. M. Stanley, *In Darkest Africa*, ii (London, 1890), 40–4, 92–6. As Monceaux (n. 3 above), 1–2, said: 'Il suffit de soulever la broderie pour apercevoir la trame de la légende.'

[23] On these huts, see L. L. Cavalli-Sforza, 'Demographic Data', in *African Pygmies*, 31–4.

[24] See e.g. F. Hüe and R. D. Etchecopar, *Les Oiseaux du Proche- et du Moyen-Orient* (Paris, 1970), 233 ('Grus cinerea), 236–7 ('Demoiselle de Numidie' or 'Anthropoides virgo'); H. Heinzel *et al.*, *Oiseaux d'Europe, d'Afrique du Nord et du Moyen Orient* (Neuchâtel, 1985), 110–11; S. Keith and J. Gooders, *BLV Vogelführer* (Munich etc., 1982), figs. 164, 166, 451–2.

[25] For ancient sources, see N. Douglas, *Birds and Beasts of the Greek Anthology* (London, 1928), 99–101; D'Arcy Wentworth Thompson, *A Glossary of Greek Birds*[2] (London, 1936), 68–75; J. Pollard, *Birds in Greek life and Myth* (London, 1977), 83–4. See e.g. Hes. *Op.* 448–51; Eur. *Hel.* 1479–94; Ael. *NA* 2. 1; Strab. 1. 2. 28.

heard about their impressive duels in nuptial parades when two tall males rise fighting in the air. These observations may have led to the idea that cranes could attack men, although they never do.

The possible influence of some direct or indirect knowledge of pygmies upon the formation of the story cannot be rejected totally. It does not, however, explain the function of the story, nor the meaning of its specific constituents in the Graeco–Roman tradition; it does not make clear, for example, why pygmies were seen as farmers, and cranes as man-killers. These imaginary elements reflect Greek preoccupations, which may be divided into three main themes: the representation of the end of the inhabited world, the conflict between mankind and the animal world, and the nightmare of farmers.

The End of the World: Ethnology and Teratology

Ancient Greek geographers and historians conventionally placed in distant countries, especially southern ones, people distinguished by some oddities, physical or mental. The land of the pygmies was thus always located at the end of the inhabited world near other marvellous peoples: some one-legged, others one-eyed; without lips, tongue, or neck; or with eyes on their shoulders, or with heads of dogs and a bark.[26] Pliny explained this marvellous production of monstrosities by the stronger capacity of the 'mobile element of fire to mould their bodies and carve their outlines' in southern countries.[27]

These descriptions of human curiosities were a necessary and reassuring act. They are still fascinating because they correspond to the archetypal unease created by the sight of human malformations. It is striking to note that most of these fabulous anomalies describe real genetic malformations. An unambiguous example is given by Ctesias asserting that 'in a distant land called Albania, men are born whose hair turns white in childhood and who see better by night than in the daytime'.[28]

'Singularity best characterises the monster' observes Gonzales-Crussi, who suggests that the imaginary multiplication of a physical anomaly into a race may have helped to stabilize the fright caused by its exceptional nature.[29] Abnormal people were located in conveniently remote countries where they could be no danger to normal men; the trouble induced by their difference could be debated freely, but in ethnological terms. Accounts of marvellous races did not provide an explanation, but gave the means of expressing a reaction of rejection, symbolized by geographic distance and emphasized by fancy descriptions of unusual habits.

Thus stories of pygmies may have emerged to account for the presence of

---

[26] See esp. the imaginative catalogues by Hellenistic historians and geographers (Ctesias, Onesicritus, Megasthenes, and others) reported by Gell. *NA* 4. 6 and 9. 4. 2 ff., and by Pliny, *NH* 7. 21–32 (India, Ethiopia).

[27] Pliny, *NH* 6. 187 (trans. H. Rackham (Loeb,

1942)). Cf. also Paus. 8. 29. 4 (India as the favoured birth place of monsters).

[28] Ctesias in Gell. *NA* 9. 4. 6 (trans. J. C. Rolfe (Loeb, 1927)). See also Pliny, *NH* 7. 23 (India).

[29] F. Gonzales-Crussi, *Notes of an Anatomist* (London, 1985), 91–102.

pathologically short people in Greek cities. This may explain why pygmies' ethnic features (colour of the skin, physiognomy) are almost never mentioned in ancient accounts, as if superfluous, while their abnormal smallness is always a subject of comment.

These stories aimed also to demonstrate that short people were perfectly innocuous beings. Pygmies are described resisting their enemies with prodigious, but useless, efforts, and they are always discredited with scorn as a race of 'puny' feeble men.[30] As Ballabriga notes, Hesiod and Oppian describe this weakness with adjectives like ἀμενηνός and ὀλιγοδρανής which usually denote figures with faint material substance (souls of the dead, dreams, dying warriors), as if to suggest that pygmies were hardly alive.[31] Short men seem to incorporate the evanescence of human life so often referred to in Greek literature.[32] Eustathius reports that their lives were very short, and he draws a parallel between their smallness and the briefness of their life.[33] Ballabriga adds that pygmies seem to have affinities with other unhappy inhabitants of the border of the Ocean, the spectral Cimmerians who live also in the presence of death, near the land of Hades, in a gloomy country of mist on which the sun never shines.[34]

It may be noted that the need to explain the emergence of abnormality is universal. Stories about *races* of short people are found in many parts of the world, even in North America where no pygmies ever existed.[35] This need is still present; the limits of our known world have now expanded to space and galaxies whose furthest inhabitants are imagined as *small* green men . . .

Mankind and the Animal World

In the fifth-century evolutionary perspective, pygmies represent mankind at its early stage of civilization, not yet clearly detached from the animal world. For Hesiod, they are born of Gaia, the Earth, as are all monstrous beings, and of Poseidon, as are the Laestrygones (giants dwelling in the earth), Ethiopians, and Black people;[36] Philostratus adds that they are brothers of the Giant Antaeus.[37]

The primitive nature of the pygmies is suggested by their habitat: like their

---

[30] Opp. *Hal.* 1. 622–3.

[31] Ballabriga, 'Nains', 57–9. See eg. *Od.* 19. 562 (for dreams); *Il.* 5. 887 (for a wounded god).

[32] See e.g. Pind. *Pyth.* 8. 95–6: 'We are creatures of a day: what thing is man, or what thing is he not. Man is but the shadow of a dream' (trans. L. R. Farnell, *The Works of Pindar* (London, 1930)).

[33] Eust. *Il.* 3. 6: Βραχύσωμοι καὶ αὐτοὶ καὶ ὀλιγοχρόνιοι ἐς παντελές.

[34] Ballabriga, 'Nains', 58–9; he quotes *Od.* 11. 14–19 (Odysseus finds there the entrance to Hades).

[35] See W. R. Halliday, 'Pygmies and Cranes', *CR* 35 (1921), 27 (Cherokee myth); R. Dangel, 'La Lutte contre les Pygmées chez les Indiens d'Amérique du Nord', *SMSR* 7 (1931), 128–35; A. Scobie, 'The Battle of the Pygmies and the Cranes in Chinese, Arab, and North American Indian Sources', *Folklore*, 86 (1975), 122–32; id., ibid. 88 (1977), 86–7. For further references, see Thompson, *Motif Index*, esp. F 451.1 ff. (on Chinese, Indian, and Finnish legends).

[36] Hes. fr. 150. 17–18 (Merkelbach/West); M. L. West, *The Hesiodic Catalogue of Women* (Oxford, 1985), 85 (trans.), and genealogical table on 178.

[37] Philostr. *Imag.* 2. 22. 1.

neighbours, the Κατουδαῖοι,[38] 'they dwell in the earth just like ants' says Philostratus.[39] They resemble the miserable state of mankind, as described by Aeschylus, before it was taught civilization by Prometheus: men were inconsistent, 'like figures in a dream', they dwelt beneath the ground 'like scurrying ants in sunless caves', and acted like beasts, irrationally, at random.[40] Greek ethnography knew other earth-dwellers, also characterized as not really full humans. Thus the Ethiopian Troglodytes mentioned by Herodotus have no speech but scream like bats; they are hunted like animals by their neighbours, the Garamantes.[41]

This primitive community is also marked by impiety, as mythographers under the Empire explain:[42] their fate was caused by their hybris, their non-observance of divine power. Once upon a time, the marvellous beauty of a pygmy woman, called Gerana or Oinoe, induced her people to worship her as a goddess. To punish this offence, Hera transformed her into a crane, 'a most hideous bird' says Aelian.[43] When the crane tried to come back to her people, the small men, frightened, repelled her brutally, and thereafter cranes waged war on them.

The motif of the battle against birds evokes the early times when animals chased off men. Pausanias reports that in a heroic past man-eating birds used to live in Greece at Stymphalos, in Arkadia.[44] These birds were the size of a crane and, adds Pausanias, were probably an Arabian breed. They plundered the country round about and could tear a man in pieces with their strong beaks. As his sixth labour, Heracles cleansed the lake of them; he frightened them away with bronze crotala or shot them,[45] and since then no bird was ever recorded as a threat to normal-sized men.[46] In the primitive land of pygmies, this legendary past is an eternal present. Like the Stymphalian birds, cranes plunder fields and kill farmers. Although pygmies use crotala like Heracles, and fight like a miniature army, their war never has a definite outcome. The primitive pygmies are opposed to another collectivity of the same force and skill. Several Greek anecdotes demonstrate that cranes were intelligent and knew a social life with laws. It was reported that their impressive flock was highly organized, with a leader and sentinels; in flight they formed an acute-angled triangle and it was thought that the younger birds were placed in the middle, while the older ones, who knew the journey, led the flight in

---

[38] Cf. Hes. fr. 150. 18 (Merkelbach/West).

[39] Philostr. *Imag.* 2. 22. 1 (trans. A. Fairbanks (Loeb, 1931)). See also id., *Vita Apoll.* 3. 47; Arist. *HA* 8. 12. 597ᵃ.

[40] Aesch. *PV* 441–53. About Greek notions of human progress see e.g. the commentary by M. Griffith, *Aeschylus, Prometheus Bound* (Cambridge, 1983), 164–8 (with earlier bibliog.).

[41] Hdt. 4. 183.

[42] Ael. *NA* 15. 29; Ath. 9. 393e–f; Ant. Lib. *Met.* 16; Ov. *Met.* 6. 90–2.

[43] Ael. *NA* 15. 29 (trans. A. F. Scholfield (Loeb, 1958)). See also Boio in Ath. 9. 393 f.

[44] Paus. 8. 22. 4–6; see also Diod. Sic. 4. 13. 2. Cf. P. Borgeaud, *Recherches sur le dieu Pan* (Geneva, 1979), 36.

[45] Apollod. *Bibl.* 2. 6; Diod. Sioc. 4. 13. 2; Paus. 8. 22. 4; Ap. Rhod. *Argon.* 2. 1052–7.

[46] Cf. the 'ravening birds' in the island of Ares, which attack men with darting feathers in Ap. Rhod. *Argon.* 2. 382 and 1033–89.

front.[47] Their proverbial vigilance even 'taught men the rules of government', adds Aelian.[48] Yet, little impressed by these qualities, Greeks and Romans used to eat cranes as a delicacy and their brains were believed aphrodisiac.[49] The birds were not only captured but even crammed like geese and swans.[50] We are never told about pygmies roasting their enemies; on the contrary, they are described meeting a shameful and restless death, mutilated by birds, an end usually reserved to cowards and enemies.[51]

The stories of the battle of the pygmies may have belonged to the category of comic poems, παίγνια, attributed to Homer, which caricatured the deeds of Homeric heroes by describing small innocuous animals at war, like frogs and mice in the *Batrachomyomachia*.[52] This parodic dimension is especially acute in the description by Philostratus of their assault against Heracles. Like a real army, the small men display phalanx, bowmen, slingers, and engines to attack the hero asleep 'on the soft sand' of Libya. Soldiers assail each part of his enormous body as if it were a citadel; they lay siege to his hands and feet, surround his head with 'fire for his hair, a mattock for his eyes, doors of a sort for his mouth . . . , gates to fasten on his nose, so that Heracles may not breathe when his head has been captured'.[53] The hero stands up and throws all of these miniature warriors in his lion's skin as if they were small animals, caricatures of the dangerous monsters captured in the course of his earlier labours.[54]

A Farmer's Nightmare

Greek peasants feared birds, for they plundered seeds. The birds of Aristophanes are very conscious of this power, and do not shrink from threatening to reduce mankind to famine by assaulting the fields and plucking out the eyes of cattle, as the birds of Stymphalos did.[55] It is likely that Greek farmers normally managed to protect their fields, but these stories may express an archetypal anguish. The miserable fate of the

[47] See e.g. Arist. *HA* 9. 10. 614[b]; Ael. *NA* 2. 1; 3. 13; Pliny, *NH* 10. 58–60. Cf. also Thompson (n. 25 above), 71–2.

[48] Ael. *NA* 3. 14. The birds even once served human justice: according to a tradition reported by Plutarch and other poets, a flock of cranes which had seen Ibycus being killed by robbers, brought to justice the two murderers. See e.g. Plut. *De garr.* 14, *Mor.* 509 f; *Anth. Pal.* 7. 745 (Antipater of Sidon); *Suda* s.v. Ἴβυκος. For futher references, see Thompson (n. 25 above), 73–4.

[49] Ael. *NA* 1. 44: 'Their brain possesses some kind of spell that leads women to grant sexual favours—if those who observed the fact are sufficient guarantee' (trans. Scholfield).

[50] Plut. *De esu carnium*, 2. 2, *Mor.* 997a. Further references in Thompson (n. 25 above), 74.

[51] See e.g. *Il.* 1. 1–5; 2. 391–3. Ballabriga, 'Nains', 61; A. Schnaufer, *Frühgriechischer Totenglaube*

(Hildesheim, 1970) (Spudasmata 20), 148–51.

[52] *Suda* s.v. Ὅμηρος; Procl. *Vita Homeri*, 76–7 (ed. A. Severyns (Paris, 1963)). See W. Schmid and O. Stählin, *Geschichte der griechischen Literatur* (Munich, 1929) (HdA VII. 1. 1), 226–31, and H. Ahlborn (ed.), *Pseudo-Homer. Der Froschmäusekrieg, der Katzenmäusekrieg* (Berlin, 1968). On similar, but Egyptian animal parodies (New Kingdom), see E. Brunner-Traut, *Altägyptische Tiergeschichte und Fabel* (Darmstadt, 1968).

[53] Philostr. *Imag.* 2. 22. 3 (trans. Fairbanks).

[54] Cf. Amm. Marc. 22. 12. 4: 'But though they kept up this agitation long and persistently, it was in vain that they barked around a man as unmoved by secret insults, as was Hercules by those of the pygmies' (trans. J. C. Rolfe (Loeb, 1940)).

[55] See e.g. Aristoph. *Aves*, 578–84. On the birds of Stymphalos, see nn. 44–6 above.

pygmies perhaps emerged in the nightmare of an overworked farmer, tormented by the idea of his fields ravaged by a flock of voracious birds, and his family dying of hunger.

Cranes in particular were traditionally called robbers of seed,[56] although a positive role was also recognized; in the autumn, their passage over Greece indicated the sowing-time.[57] Their greed was believed to be stronger in southern countries, like Egypt, where they find both a 'generous table', according to Aelian,[58] and weak opponents, as a fable of Babrius suggests: a farmer chases away cranes which are overrunning his fields; hit by stones, the birds cry out: 'Let us flee to the land of the pygmies. This man, it seems, is no longer trying to frighten us; already he is beginning to do something!'[59] Pomponius Mela states that the birds even managed to extinguish the race of pygmies by starving them.[60]

The legend of the geranomachia also offered to farmers the means to relieve another, more concrete, fear, that of ceaseless military campaigns. If the destruction of fields by cranes was imaginary, the damage caused by flocks of warriors was real, and the plays of Aristophanes resound with complaints of Greek peasants about these ravages.[61] Homer twice compares the warriors of the Trojan and Achaean armies to flocks of cranes.[62] The comparison can be reversed: the image of cranes destroying the harvest of pygmies may have evoked the ravages of warriors burning the fields of peasants. Before the periodical threat of war, Greek and pygmy farmers thus appear as powerless brothers. But to exorcize these fears, pygmies were presented as ludicrous small beings, as living caricatures of mankind.

PYGMIES: ICONOGRAPHY

In iconography, only the merciless battle between pygmies and cranes is depicted. The feminine figure of Gerana or Oinoe is not shown, though she had caused the primordial conflict according to authors of the Roman period. The battle is reduced to its two basic elements: tall birds opposed to short men.

In vase-painting, these pictures decorate essentially wine-vessels: cups (9), kantharoi (8), skyphoi (6), rhyta (5), plastic vases (5), kraters (3), amphorae (3), askoi (3), one oinochoe, one mug, and vases used by athletes and women: pelikai (9), hydriae (4), lekythoi (2), one pyxis and one aryballos. They are also found on a golden diadem from

---

[56] Antipater of Sidon in *Anth. Pal.* 7. 172. 1–4. The scholion on Aristoph. *Aves*, 232, similarly states: σπερμολόγων. οἱ γέρανοι (ed. J. W. White (Boston, 1914), 58).

[57] Hes. *Op.* 448–51.

[58] Ael. *NA* 2. 1 (trans. Scholfield).

[59] Babrius 26 (trans. B. E. Perry (Loeb, 1965)). See also ibid. 13, the beseeching stork: 'I'm not a crane, I don't destroy the seed, I'm a stork.'

[60] Pomponius Mela, 3. 81–2: *Fuere interius Pygmaei, minutum genus et quod pro satis frugibus contra grues dimicando defecit.*

[61] Ballabriga, 'Nains', 68; see e.g. Aristoph. *Ach.* 1023–5.

[62] See e.g. *Il.* 2. 459–69.

Rhodes (G 58), and a Corinthian portable altar (G 53; pl. 61.2). The earliest certain representations date to the sixth century (G 40, 57, 58; pl. 58).[63]

The geranomachy conforms to a schema which exhibits but few variations during the archaic and classical period. These variations follow the transformations of Greek narrative methods. In the archaic period, craftsmen often chose to show simultaneous actions in a scene; a frieze depicts combatants fighting in different places at different moments (e.g. G 40–2, 47–8; pls. 58–9, 60.3, 61.1a); in the classical period, they preferred to render a spectacular moment, rich in dramatic tension, like duels (e.g. G 70–1, 75, 78; pls. 65.1, 3, 66.1, 67.2).[64] In plastic vases, the pygmy is carrying a dead crane (e.g. G 63, 66, 103; pls. 63.3, 70.2).

The minimal combination of attributes which denotes a pygmy with certainty appears in the schema which opposes a dwarf and a crane. In some scenes, this traditional schema is not recognizable, and the identification of the geranomachy uncertain. Thus, elements of the scene depicted on the Northampton amphora allude to the legend of the pygmies (G d 56; pl. 62.1): minute men, holding weapons, are associated with tall birds. However, they are not fighting, but riding the birds, and they seem to threaten each other with clubs, so the scene cannot be identified as a traditional geranomachy. For Beazley, it may depict a new episode of the legend: 'the cranes have finally been conquered; the war is now civil war, and the cranes must serve their masters as steeds',[65] or it may be a decorative motif, which was only influenced by the legend of the pygmies.[66] Similarly, the small armed men depicted on three archaic sherds, facing long-necked birds, cannot be identified positively as pygmies (G d 51–2, d 54); they are involved in obscure actions, and the birds are also difficult to identify (ostriches or cranes?). As Dunbabin suggests, the scenes may illustrate other imaginary contests against birds, or unwritten variants of the tale.[67]

[63] For possible Mycenean forerunners on vessels from Enkomi and Ras Shamra, see C. Schaeffer, 'Sur un cratère mycénien de Ras Shamra', *BSA* 37 (1936–7), 212; V. Karageorghis, *La Civilisation préhistorique de Chypre* (Athens, 1976), 165, no. 123 (with earlier bibliog.). On representations of myth in general in Mycenean art, see id., 'Myth and Epic in Mycenean Vase-Painting', *AJA* 62 (1958), 383–7.

[64] See C. Robert, *Archaeologische Hermeneutik* (Berlin, 1919), esp. 180–6; N. Himmelmann-Wildschütz, 'Erzählung und Figur in der archaischen Kunst', *Abh. Mainz.* 2 (1967), 73–101, and the review by J. M. Hemelrijk, *Gnomon*, 42 (1970), 166–71; V. Dasen, 'Autour du dinos de Néarchos: essai sur la bande dessinée chez les Anciens', *Études de Lettres* 4 (1983), 55–73, esp. 64 ff.; A. M. Snodgrass, 'La Naissance du récit dans l'art grec', in C. Bérard et al., *Images et société en Grèce ancienne* (Lausanne, 1987)

(Cahiers d'Archéol. Rom. 36), 11–18.

[65] Cf. J. D. Beazley, 'Notes on the Vases in Castle Ashby', *PBSR* 11 (1929), 2; *CVA* Great Britain 15, Castle Ashby, 2.

[66] See B. Freyer-Schauenburg, 'Die Geranomachie in der archaischen Vasenmalerei', in *Wandlungen. Studien zur antiken und neueren Kunst, E. Homann-Wedeking gewidmet* (Waldsassen, 1975), 77. For further uncertain examples, see e.g. ibid. 76–7, pl. 14a–b, the man running before a water-bird on a Boeotian kantharos (Kiel B 49). See also the grotesque with knife facing two geese on a skyphos in Berlin, SM 3179; P. Wolters and G. Bruns, *Das Kabirenheiligtum bei Theben*, i (Berlin, 1940), 99, pl. 29. 2.

[67] T. J. Dunbabin, in T. J. Dunbabin et al., *Perachora: The Sanctuaries of Hera Akraia and Limenia*, ii (Oxford, 1962), 96–7.

On three red-figure vases, the painter isolated an element of the conventional schema; naked dwarfs brandish a club or a bow, but their opponents are not shown (G 74, 76–7; pls. 65.2, 66.2, 67.1). Do these dwarfs belong to a mythical or to a human world? Since the morphologies of pygmies and pathological dwarfs are similar, their identification is debatable. The dwarf on the Athens pyxis wears an animal skin on his arm and brandishes a club (G 74; pl. 65.2), as does the pygmy on the Brussels amphora (G 70; pl. 65.1). No crane is present, but the clubs and the animal skins, combined with the action of the figures, define them clearly as pygmies. The identification of the dwarf on the Leipzig cup is more uncertain (G 77; pl. 67.1). He too runs with a club, which suggests that he has bellicose intentions, but he wears a headband, which is awkward for a pygmy. I include this picture in the corpus of pygmies because it conforms to the traditional schema, although I admit the ambiguity of the headband; Greeks might have imagined pygmies with headbands, but they might also have depicted pathological dwarfs running with clubs (as hunters, for example). On the third vessel, an askos in Hamburg, a naked dwarf is bending a bow (G 76; pl. 66.2). On the other side appears his enemy, a winged horse, probably Pegasus. As a winged creature, the horse may have been a substitute for the traditional crane, which is depicted on a similar askos in Prague (G 79; pl. 68.1b).[68]

These pictures reflect two themes which are expressed in the literary tradition: the confusion between ethnology and pathology, and the closeness between mankind and the animal world. Two graphic themes are developed: the parody of the heroic world, and the caricature of Heracles.

Ethnology, Teratology, and Wildness

How did archaic and classical craftsmen render this fabulous remote country where dwarfs are the norm and birds attack men? Unlike writers, painters had little sense of the picturesque. That taste emerged only in Hellenistic art.[69] They were not inspired by the fancy descriptions of travellers or geographers, and the scenery is very plain. The South is not indicated by exotic animals or vegetation, such as camels, crocodiles, or palm-trees, though these are shown on contemporary vases;[70] the fight of a pygmy against a crane is depicted on a plastic vase at Ruvo in the form of a black man eaten by a crocodile-like fish (G 71; pl. 65.3a),[71] but the geranomachy itself has no additional

[68] On an askos in Paris, Louvre, G 447, Pegasus confronts Chimaera, another fantastic creature with foreign associations, which may, symbolically, be a substitute for pygmies; H. Hoffmann, *Sexual and Asexual Pursuit* (London, 1977), 13, no. 134, pl. XI. 4.

[69] See e.g. W. Brooks McDaniel, A Fresco Picturing Pygmies, *AJA* 36 (1932), 259–71, pl. IX; H. Whitehouse, 'In Praedis Iuliae Felicis: The Provenance of Some Fragments of Wall-Painting in the Museo Nazionale, Naples', *PBSR* 45 (1977), 52–68.

[70] See e.g. the camel on a pelike at St Petersburg, Hermitage, 614 (St 1603); *ARV* 288.11, *Add*² 209 (Argos Painter); Boardman, *ARFH* i, fig. 183. For palm-trees, see e.g. the suicide of Ajax on the amphora in Boulogne Museum 558; *ABV* 145.18, *Add*² 40; Boardman, *ABFH*, fig. 101. On palm-trees in Crete and Delos, see H. Baumann, *Die griechische Pflanzenwelt*² (Munich, 1986), 58–9.

[71] For other examples of funny Greek rendering of fish-like crocodiles, see E. Buschor, 'Das Krokodil des Sotades', *MJBK* 11, 1/2 (1919), 1–8, figs. 1–12.

FIGURE 13.1. (G 97) Kantharos (h. 20.9 cm.). Berlin, SM 3159

fantastic element. The few plants in the scene indicate an open space only; they are not African, but Greek, as on the rhyton in St Petersburg (G 60; pl. 62.2a).

Pygmies themselves are no ethnic curiosities. On archaic vases, pygmies are rendered as slender diminutive Greeks. Only on a few objects do they show paunchy bellies (G 45, 58; pl. 60.1). In red-figure, artists replaced ethnic characteristics with physical abnormality. Apart from a fragment of a hydria in Athens (G 62; pl. 63.2), pygmies are short-limbed dwarfs, with the typical 'negroid' features of achondroplasia (snub nose and thick lips). Pygmies are painted in black only on Attic and Italiote plastic vases, such as the Bonn rhyton (G 66; pl. 63.3), but without curly hair, apart from the Basle figure (G 64). This lack of ethnological imagination is intriguing since painters mastered the representation of negroes.[72] Only the Sotades Painter, who displays great interest in unusual figures and depicted a number of Blacks, renders pygmies with rather woolly hair (G 67–8; pl. 64). Fourth-century pygmies usually have proportionate, full-sized bodies (e.g. G 85; pl. 69.2), apart from those caricatured on Kabeirion vases (e.g. G 97; fig. 13.1).

Greek craftsmen also eschewed the exotic elements of pygmies' equipment mentioned in travellers' descriptions.[73] They rather keep the attributes which mark a primitive world, where man and animals are equal. Usually pygmies are naked, like wild satyrs,

[72] Cf. W. Raeck, *Zum Barbarenbild in der Kunst Athens* (Bonn, 1981), 164–213, and catalogue on 330–1; F. M. Snowden, *Blacks in Antiquity* (Cambridge Mass. and London, 1970). See e.g. the black warrior on a red-figure cup in Paris, Louvre, G 93; *ARV* 225.4, *Add²* 198; Snowden, ibid. 47, fig. 17.

[73] See e.g. the skirts made of palm-leaves mentioned by Hdt. 4. 43.

and use an animal skin as shield (e.g. G 70–1, 74, 79; pls. 65.1–3, 68.1); they wave knotty clubs, curved sticks, and carry slings.[74] Some wear a chiton (G 46, d 49, d 50; pl. 60.2). Only a few wear Oriental clothes, such as the Scythian soft pointed hats of the pygmies on the St Petersburg rhyton (G 60; pl. 62.2), and have foreign weapons, such as bows (G d 54, 60, 76; pls. 62.2, 66.2), and peltae (e.g. G 85, 92; pls. 69.2, 70.1), which stress their marginality.[75] Some pygmies caricature Greek heroic warriors; on the Brummer oinochoe (G 82; pl. 69.1), the pygmy is naked, with a crested helmet and round shield, and boldly brandishes his spear against a bird; others brandish swords or spears, and protect themselves with round shields.[76]

It is the cranes alone, which, as migrating birds, serve as space indicators; they locate the scene at an indeterminate distance, where the climate is hot, and where cranes go in winter. It is also the place where their behaviour is transformed. In Greece they are depicted as domesticated animals, which are fed by women,[77] while in Africa they attack men. Some are characterized as European cranes (Anthropoides virgo) with tufts of feathers at the tail (G 62; pl. 63.2), but most of them are not very realistically rendered, and resemble storks or herons (see e.g. the heron on G 79; pl. 68.1b).[78] For painters, the legendary cranes need only to be shown as tall birds—as tall as the pygmies- with impressively large wings and long necks.

The exotic provenance of pygmies may, however, have been suggested by other means. For example, I am tempted to see references to far-away countries in the personages accompanying the pygmies on the aryballos by Nearchos (G 41; pl. 59.1). The painter placed the frieze of pygmy warriors around the edge of the vase, and added on the handles figures associated with remote lands. On one handle he depicted Perseus, a hero who not only visited fabulous peoples in his quest for Medusa, but found love in Ethiopia in the attractive form of Andromeda, and on the other Hermes, the messenger of the gods; at the back of the handle three satyrs evoking a feeling of wild spaces are added.[79]

The fact that pygmies are often shown on rhyta, vessels of foreign origin, may also reflect exotic associations. The tale seems to have been particularly appreciated in Greek settlements of the Bosporus, perhaps because later forms of the myth located the pygmies in Scythia instead of Africa, as suggested by Metzger;[80] the popularity of the

[74] Clubs: see e.g. G 41, 46, 53, 61, 70–1, 75, 77–9, 80; pls. 59.1, 60.2, 61.2, 63.1, 65.1, 3, 66.1, 67.1–2, 68.1–2. Curved sticks and slings: G 40; pl. 58a–d.

[75] Raeck (n. 72 above), 204.

[76] Swords: e.g. G 48, 60, 92; pls. 61.1, 62.2b, 70.1. Spears: G d 51, 68, 82, 94, 97–8, d 101; pl. 69.1, fig. 13.1. Round shields: G 62, 82, 90; pls. 63.2, 69.1.

[77] See e.g. the two women and the man feeding cranes on a hydria, Baltimore, Robinson coll., CVA USA 6, Robinson coll. 2, III, I, pl. 34, and the crane

behind a chair on a hydria, Brussels, Musées Royaux, A 3098; ARV 493.2, Add² 249; CVA Belgium 3, Brussels 3, III, I, d, pl. 15. 1 and 16. 2e.

[78] Cf. the figures in Heinzel et al. (n. 24 above), 110–11, 35 (herons) and 43 (storks).

[79] This may also allude to the fact that satyrs are the sons of Hermes. See T. Carpenter, Dionysian Imagery in Archaic Greek Art (Oxford, 1986), 78–9 n. 13.

[80] Cf. G 84, 87–92; pl. 70.1; H. Metzger, Les Représentations dans la céramique attique du IVe siècle

myth of the fight of the Arimaspi against the Griffins, which took also place in Scythia, and conforms to a similar iconographic schema, may also have influenced the fourth-century iconography of the geranomachy.[81]

Parody of the Heroic World

Despite their smallness and their primitive weapons, pygmies act like normal-sized heroes and fight with great determination. No picture shows them as the 'feeble', shadowy creatures whom ancient authors suggested.[82] On the contrary, on the François vase they ride with a commanding appearance on the back of goats (G 40; pl. 58a, d), and some of them have the equipment of Greek heroic warriors.[83] On red-figure vases, the infibulation of some pygmies may stress their physical strength and assimilate them to athletes (G 67, 73, 75, 78; pls. 64, 66.1, 67.2).

Despite their weapons, the short men do not, however, resist the attacks of the cranes, and die in the same proportion. As Minto notes, the hopeless efforts of the pygmies are emphasized by the humorous contrast between their corpulence and the graceful slimness of the birds.[84] Often they have dangling genitals (e.g. G 71, 74, 79; pls. 65.2–3, 68.1), which the birds may maliciously attack, as depicted on a Kabeirion kantharos (G 97; fig. 13.1). A few pictures show dead or dying pygmies lying on the ground, mutilated by birds (G 40, 42–3, d 55, 58; pls. 58a, b, 59.2); these pictures may refer to a darker side of heroic life, to the horrifying death of warriors eaten by animals. This unpleasant scene is very rarely shown with normal human figures,[85] but could perhaps feature pygmies because they were not considered as full humans.

The comic aspect of the fight is partly due to reference to the figure of Heracles. Like pygmies, he is usually armed with a club, and uses an animal skin as a shield. On black-figure paintings, the hero fights the Stymphalian birds with a sling as do pygmies on the François vase (G 40; pl. 58d).[86] But the hero is involved in a fight against dangerous monsters which threaten the civilized world, while pygmies fight familiar birds. Small and malformed, they are the real monsters, while cranes, tall, thin and elegant seem to refer, like Heracles, to the Hellenic order.

By way of a joke, pygmies often are represented in the same attitude as the hero. In black-figure, when the pygmy is depicted alone, and as a human miniature, this resemblance may even make the distinction between the small man and Heracles impossible to draw (G d 49).[87] The comical effect of this substitution is more

---

(Paris, 1951) (BEFAR 172), 326–7.

[81] See e.g. the pelike in Madrid, Museo Arq. Nacional; K. Schefold, *Untersuchungen zu den Kertscher Vasen* (Berlin and Leipzig, 1934), no. 516, pl. 23.

[82] See nn. 30–3 above.

[83] See n. 76 above.

[84] A. Minto, *Il vaso François* (Florence, 1960) (AAT 6), 151.

[85] See e.g. the pithos, Athens, NM 2495; J. Schäfer, *Studien zu den griechischen Reliefpithoi des 8.-6. Jahrhunderts v. Chr. aus Kreta, Rhodos, Tenos und Boiotien* (Kallmünz Opf, 1957), 67–8, pl. x. 2.

[86] Brommer, *VL* 207–9; S. Woodford, *LIMC* v (1990) s.v. Herakles, iv, F, 54–7, nos. 2241–83.

[87] For Heracles: J.-J. Maffre, 'Collection Paul Canellopoulos (VIII)', *BCH* 99 (1975), 437–8; Woodford, ibid. 56–7, nos. 2262, 2276, 2278.

conspicuous in red-figure. The dwarf on the Brussels amphora (G 70; pl. 65.1) has, for example, the same pose as Heracles on an amphora in Munich.[88] Yet, while Heracles appears invincible against a crowd of small birds,[89] pygmies succumb to the assault of tall enemies. In fifth-century art, Heracles becomes less explicitly present, and no extant red-figure vase records his deeds against the Stymphalian birds, but the frequent representations of pygmies may, by virtue of their association with the Stymphalian birds, allude to a continued popularity of the hero.

To sum up, it is not impossible that stories about pygmies transmit some knowledge of African small people. The essential function of the myth was not, however, to account for the hypothetical existence of ethnic short men. As suggested by its parodic tone, it answered a number of archetypal fears within Greek society.

First, anxiety before the geographic and genetic unknown. The myth demonstrates that remote countries contain no danger; small, ephemeral, unhappy, and impious, pygmies emerge as the opposite of their tall, long-lived, beautiful, and pious southern neighbours, the Ethiopians.[90] The myth also accounts for the presence of pathological dwarfs at Athens. Physical abnormality becomes an exotic feature. Dwarfs appear thus as liminal, wild, but inoffensive beings, like powerless pygmies. Forms of rejection are expressed in terms of geographic distance and racial difference.

On a more general level, the fate of the small men may have expressed the eternal apprehension of farmers, worried about all that may destroy their wealth, whether birds or warriors. In this perspective, it is revealing that no author, apart from Pomponius Mela,[91] ascribed an end to the battle between pygmies and cranes.

CERCOPES: WRITTEN SOURCES

The ambiguous proportions of dwarfs, between childhood and adulthood, describe other marginal figures, such as the Cercopes, two brothers famed as a pair of wily and treacherous rascals.[92] Their name is often found in Old and Middle Attic Comedy,[93] and

---

[88] Amphora, Munich, Antikensammlungen, 2316; *ARV* 183.12, *Add²* 187; *CVA* Germany 20, Munich 5, pl. 209.3. Cf. also Heracles on an amphora in Paris, Louvre, F 384; Brommer, *VL*, 208.8; Haspels, *ABL* 238 (132); *Hommes, dieux et héros de la Grèce* (Rouen, 1982), 222–3, fig. 90.

[89] See the amphora in London, BM B 163; *ABV* 134. 28, *Add²* 36; Boardman, *ABFH*, fig. 95. See also the lekythos in Munich, Antikensammlungen, 1842; Haspels, *ABL* 195.8, pl. 9.2.

[90] On 'blameless Ethiopians', see e.g. *Il.* 1. 423–4; Hdt. 3. 17–25. See the full discussion by Ballabriga, 'Nains', esp. 74.

[91] See n. 60 above.

[92] For literary sources, see Preller/Robert, *GrMyth*, ii (2), 504–6. See in particular C. A. Lobeck, *Aglaophamus*, ii (Königsberg, 1829), 1296–308; K. Seeliger, Roscher, ii (1), 1890–4, s.v. Kerkopen, 1166–73; A. Adler, *RE* xi. 1 (1921), s.v. Kerkopen, 309–13.

[93] See e.g. Cratinus, Ἀρχίλοχοι, fr. 13 (Kassel/Austin iv); Hermippus, Κέρκωπες, frr. 36–41 (Kassel/Austin v); Eubulus, Κέρκωπες, frr. 52–3 (Kassel/Austin v); Menippus?, Κέρκωπες (*CAF* iii. 383). Adler (n. 92 above), 313, notes that only Eubulus' play shows Heracles and may have narrated the tale.

their story seems to have been very popular in Athens. It was probably a very ancient tale which is said to have inspired Homer to write a comic poem.[94] Yet only some later sources report it.[95]

Although modern scholars often describe the Cercopes as two 'gnomes', 'two mischievous dwarfs',[96] no ancient text clearly says that the Cercopes were especially short, apart from those which report that they were changed into small animals, monkeys.

Five different names are kept for the pair; some denote their character such as Sillos ('wag') and Tryballos ('sycophant'), or Passalos ('peg') and Akmon ('anvil', a name also given to an Idaean Daktylos) or Kandulos and Atlas, which might suggest ironically that they are small.[97] On a fragmentary black-figure cup in New York, they are confused with the monstrous twin sons of Poseidon, and are erroneously captioned Moliones.[98]

The story of the Cercopes has several affinities with that of the pygmies: it describes the adventures of marginal people, this time thieves, who are also confronted with the Greek champion of order, Heracles. Their human status is uncertain and, as a race, they change finally into exotic animals.

Highwaymen

The Cercopes were two malefactors who committed many evil deeds, some say in Boeotia, others at Thermopylae or in Lydia (Ephesus).[99] According to Nonnus, their mother, Theia, daughter of Okeanos, had warned them to keep off any man showing a black bottom, for a μελάμπυγος would master them.[100] Heracles was then serving Omphale and was cleansing the country of highwaymen, and had fallen asleep under a tree when the two brothers attempted to steal his weapons. The hero was soon wakened, easily caught both robbers and hung them upside down from a stick, face to face, like two young animals. When the two brothers saw his black hairy bottom, they realized that he was their famous μελάμπυγος. They began joking at his hairiness with

[94] *Suda* and Harpocration, s.v. Κέρκωπες; Procl. *Vita Homeri*, 77 (ed. A. Severyns (Paris, 1963)). Lobeck (n. 92 above), 1296–7, suggests that it may have been an appendix to the poem celebrating the conquest of Oichalia.

[95] Diod. Sic. 4. 31. 7; Apollod. *Bibl.* 2. 6. 3; Ov. *Met.* 14. 88. The fullest version of the story is preserved by Nonnus, *Narrationes ad Gregorium*, 39. 10–24 (ed. A. Westermann, *Mythographoi* (Braunschweig, 1843)).

[96] Thus, wrongly, J. M. Edmonds, *Fragments of Attic Comedy*, i (Leiden, 1957), 27, Cratinus, Ἀρχίλοχοι, fr. 13, n. f; id., ibid. ii (Leiden, 1959), 105, Eubulus, Κέρκωπες fr. 53, n. c; LSJ 943; P. Chantraine, *Dictionnaire étymologique de la langue*

grecque (Paris, 1968), s.v. Κέρκωπες, 520; H. Frisk, *Griechisches etymologisches Wörterbuch*, i (Heidelberg, 1960), s.v. Kobolde, 830.

[97] See Lobeck (n. 92 above), 1305–8; Adler (n. 92 above), 312–13; Seeliger (n. 92 above), 1170–1.

[98] D. v. Bothmer (personal communication). See id. and E. Böhr, 'Der Schaukelmaler', *AJA* 88 (1984), 82. On the Moliones, see R. Hampe, *LIMC* i (1981), s.v. Aktorione, 472–6.

[99] Boeotia: sch. on Lucian, *Alex.* 42. 4. 1. Thermopylae: Hdt. 7. 216. Ephesus: Apollod. *Bibl.* 2. 6. 3.

[100] Nonnus (n. 95 above). See also *Suda* and Photius, *Lexicon*, s.v. Μελαμπύγου τύχοις.

such wit that the hero burst out laughing and let them go.[101] Diodorus says that he killed some—there were more than two of them—and offered the others to Omphale,[102] while Zeus finally turns them into stones in other versions.[103]

Always qualified as liars, deceivers, their name became synonymous with robbers and cheats;[104] it gave rise to the word κερκωπία for 'trickiness' and the verb κερκωπίζω 'to play the ape'.[105] There was also at Athens a Κερκώπων ἀγορά or knaves-market for receivers of stolen goods according to late sources.[106] 'Practical' reasons could support the alleged smallness of the Cercopes. The word τοιχωρύχος 'the one who digs through the wall of a house', describes a burglar, which suggests that smallness was a significant quality of professional thieves.[107] The use of the word Pataikos as nickname for thieves may also allude to that quality.[108] This aspect of the Cercopes as short, perhaps also dwarf, rascals, is paralleled in other cultures; for example, the etymology of the German word for dwarf 'Zwerg', may derive from the Indo-European 'dhuerg-, dhreugh-, dhrugh-, dhreu', which means 'deceit'.[109]

Mankind and Wildness

Thus Heracles treated the Cercopes like small animals brought to the market. Their name evokes some resemblance to animals: it contains the word κέρκος 'the tail' and is found describing a long-tailed ape.[110] For Liddell–Scott–Jones it may be translated by 'man-monkey'.[111] There is also a possible association with κέρκος as denoting male organs, which would fit the common belief about dwarfs' sexual potency.[112]

In late versions of the tale, the Cercopes finally return to wildness, a miserable fate due to their impiety. According to Ovid, Jupiter was so exasperated by the continuous

[101] The proverb μὴ σύγε μελαμπύγου τύχοις, which is repeated by several authors in the *Corp. Paroem. Graec.* (e.g. Zenob. 5. 10), was already known to Archilochus fr. 211 (eds. F. Lasserre and A. Bonnard, (Paris 1958), 61–2). This blackness might also denote the dark complexion of Heracles and thus his physical strength; cf. *Heraion*, ii. 189 n. 2.

[102] Diod. Sic. 4. 31. 7.

[103] See e.g. *Suda*, s.v. Κέρκωπες; sch. on Lucian, *Alex.* 42. 4. 6.

[104] See e.g. Aeschin. *De falsa legatione*, 2. 40; Plut. *Quomodo adul.* 18, *Mor.* 60c, where the name denotes hypocrisy.

[105] Chantraine (n. 96 above), s.v. Κέρκωπες, 520; LSJ, s.v. κερκωπίζω, 943.

[106] See e.g. Diog. Laert. 9. 114; *Suda*, Harpocration and Hesychius, s.v. ἀγορὰ Κερκώπων sch. on Lucian, *Alex.* 42. 4. 5, and various authors in the *Corp. Paroem. Graec.* (e.g. Zenob. 1. 5; Diogen. 1. 3). See also W. Judeich, *Topographie von Athen²* (Munich, 1931) (HdA III, 2, 2), 309.

[107] I am very grateful to J. Boardman for this suggestion.

[108] Cf. below 215.

[109] See F. Kuge, *Etymologisches Wörterbuch der deutschen Sprache*[11] (Berlin and Leipzig, 1934), s.v. dhreu, 721; id.[21] (Berlin and New York, 1975), s.v. dhreu, 895 ('nuire en trompant'). Cf. also A. Walde and J. Pokorny, *Vergleichendes Wörterbuch der indogermanischen Sprachen*, i (Berlin and Leipzig, 1930), s.v. dhuerh, 871; s.v. dhreugh, 874.

[110] Chantraine (n. 96 above), s.v. Κέρκωπες, 520. See e.g. Manilius, *Astronomica*, 4. 668 (long-tailed ape).

[111] LSJ, s.v. κέρκος, 943.

[112] Frisk (n. 96 above), s.v. Κέρκος, 830–1. See e.g. Hesychius, s.v. κέρκος. ἀνδρεῖον αἰδοῖον, and s.v. οὐρά. ἡ κέρκος, καὶ τὸ αἰδοῖόν. The word has this specific meaning in e.g. Aristoph. *Thesm.* 239 and Herod. 5. 45. Yet Adler (n. 92 above), 313, rejects this association. Cf. also J. N. Adams, *The Latin Sexual Vocabulary* (London, 1982), 35–7. On dwarfs' sexual

misbehaviour of the Cercopes (described as a 'treacherous race'), that he transformed them into apes, living caricatures of man:

[He] changed the men to ugly animals in such a way that they might be unlike human shape and yet seem like them. He shortened their limbs, blunted and turned back their noses, and furrowed their faces with deep wrinkles as of age . . . He took from them the power of speech, the use of tongues born from vile perjuries, leaving them only the utterance of complaint in hoarse, grating tones.[113]

They were transferred to the islands of Procida and Ischia near the bay of Naples, which were then called Pithecussae 'ape islands' after them.[114] This connection between apes and dwarfs is also expressed by Diodorus reporting that: 'The animals which bear the name cynocephali are in body like misshapen men, and they make a sound like the whimpering of human beings.'[115]

CERCOPES: ICONOGRAPHY

The characteristics of the Cercopes suggest some kinship with Indo-European dwarfs,[116] but physical abnormality is never clearly expressed in texts; only iconography shows the Cercopes as small wild men.

The Cercopes are mostly depicted in archaic art since the first quarter of the sixth century.[117] The pictures conform to a similar schema; painters selected the moment when the Cercopes were carried by Heracles, hanging upside down from a pole. It is the high point of the story for the public who knew its outcome, a climactically farcical point which probably gave rise to improvised renderings of the dialogue between the hero and the two robbers. On seven vases, Heracles turns his head backward, as if reacting to a specific joke (cf. G 107, 109–10; pls. 71.2, 72.1–2);[118] yet the hero does not exhibit hairy private parts, but wears a chiton and a lion's skin.

On most depictions, the brothers are adolescents, almost as tall as the hero who carries them. Usually they hang with a stiff dignity, or wave their legs and arms with

---

potency, see also below 218–19, 236–7.

[113] Ov. *Met.* 14. 90–100 (trans. F. J. Miller (Loeb, 1916)). See also *Suda* and Harpocration, s.v. Κέρκωπες.

[114] Ov. ibid. J. G. Frazer mentioned that the tale may be 'a reminiscence of Phoenician traders bringing apes to Greek markets' (Apollod. *Bibl.* 2. 6. 3 (Loeb, 1921), 243 n. 3).

[115] Diod. Sic. 3. 35. 5 (trans. C. H. Oldfather (Loeb, 1935)).

[116] Cf. J. Grimm, *Deutsche Mythologie*³ i (Göttingen, 1854), 434–5 ('alle Zwerge und Elbe sind diebisch'); Thompson, *Motif Index* F 451.5.2.2. (dwarfs steal from humans), F 451.5.2.3 (dwarfs steal

magic objects), F 451.5.2.4 (dwarfs steal food and drink); F 451.3.2.1 (dwarfs turn to stone); F 451.4.12 (live in stones).

[117] For material, see *Heraion*, ii. 185–95; Brommer, *VL* 98–9; M. Pipili, *Laconian Iconography of the Sixth Century BC* (Oxford, 1987), 10.

[118] See also: Lekythos, Palermo, Museo Nazionale, 2635; Haspels, *ABL* 50; *Heraion*, ii. 191, fig. 43. Amphora, Madrid, Museo Arq. Nacional, 10917; *ABV* 308.78; *Heraion*, ii. 188, fig. 38. Lekythos, Oxford, Ashmolean Museum, 249; Haspels, *ABL* 49; *Heraion*, ii. 190, fig. 42. Lekythos, Agrigento, Museo Civico, Ex Giudice 67; Haspels, *ABL* 205.2; *Heraion*, ii. 192, fig. 44.

FIGURE 13.2. (G 104) Amphora (h. 44 cm.). Boulogne, Musée des Beaux-Arts, 413

FIGURE 13.3. (G 108) Pinax fr. (h. 6 cm.). Berlin, SM (West), F 767

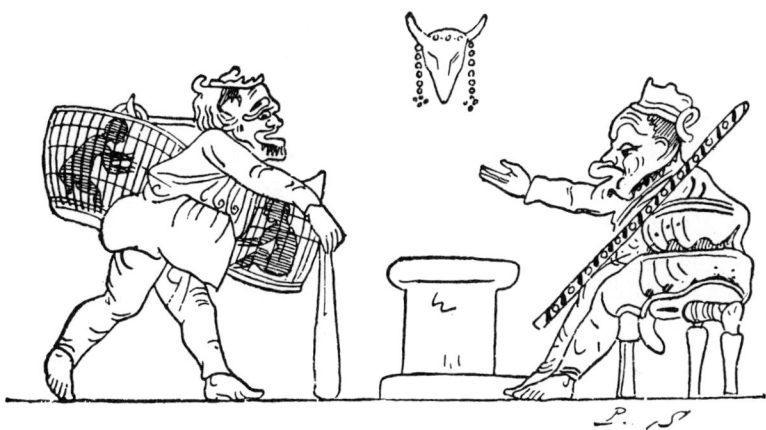

FIGURE 13.4. (G 111) Krater. Catania, Museo Civico, Biscari coll., 735

despair, as on a shield ornament in the Olympia Museum;[119] their wildness is suggested by their nakedness, and emphasized by their long hair (cf. G 105–7, pls. 71.1–2). On nine pictures, their liminality is increased by a physical peculiarity; the Cercopes are dwarfs, not short-limbed, but proportionate, like archaic pygmies. Usually they have a short beard (G 104, 107–110; pls. 71.2, 72.1–2, figs. 13.2–3), or no beards, but very short arms and rather bulky bodies, as on the Florence amphora (G 105; pl. 71.1). The red-figure scene on the krater in Munich reproduces this archaizing schema (G 110; pl. 72.2): the robbers are not achondroplastic dwarfs, but miniature adults with beards.

These pictures stress the similarities between pygmies and Cercopes observed in literary sources: the Cercopes appear as wild dwarfs mastered by Heracles, hanging from a pole like game brought home by a hunter, which questions their status as full humans.[120] On a fourth-century Lucanian pelike, the painter accentuated this ambiguous status (G 112; pls. 73–4). As Brommer notes, both brothers have marked animal features; one has pointed, satyr-like, ears, the other a monkey's face.[121] Both also have oversized genitals, which contrasts with the ideal ones of Heracles; this may refer to the association of dwarfs with the phallus, or to a possible play upon the use of κέρκος for both tail and male genitals.[122] An Italiote krater from Catania, probably related to phlyax comedy, may show the final transformation of the Cercopes into monkeys

[119] Shield, Olympia Museum, B 2198; K. Schefold, *Frühgriechische Sagenbilder* (Munich, 1964), 69, fig. 61a.

[120] See e.g. the hunter on a lekythos in Harvard, Fogg Art Museum, 1925.30.51; *ARV* 302.21, *Add²* 212; Boardman, *ARFH* i, fig. 210.

[121] F. Brommer, in J. Frel and S. Knudsen (eds.),

Greek Vases in the J. Paul Getty Museum, ii (Malibu, 1985), 203.

[122] See n. 112 above. Cf. the full-sized ithyphallic Cercopes on an olpe in Brussels, Musées Royaux, R 293; Brommer, *VL* 98.1; *Heraion*, ii. 189, fig. 41, 190, no. 5.

(G 111; fig. 13.4). Heracles brings to Eurystheus two small beings imprisoned in reed baskets hanging at both ends of his bow. They are so small that they could be monkeys.[123]

Representations of another episode of the story, the theft of the weapons of Heracles, show that the Cercopes could be assimilated to satyrs, likewise comical beings very close to the animal world. Painters occasionally replaced the two robbers by a troop of satyrs on Attic red-figure vases, and by Pans in South Italy.[124] On an Apulian chous, for example, two Pans are stealing weapons and food from Heracles, lying asleep near the remnants of his meal (pl. 39.2);[125] one Pan is carrying away the bow and the quiver of the god, the second holds a container, perhaps a hold-all or a cornucopia according to MacPhee.[126] Their misshapen appearance evokes the dwarfishness of the Cercopes, and enhances the comedy of the scene.

The Cercopes thus appear as marginal and exotic creatures, but not really inoffensive; they are highwaymen, and even their wit represents some threat. However, like pygmies, they are easily overcome by Heracles and finally by Zeus/Jupiter. The dwarfism of the brothers may bear witness to a lost version of the legend which associated their social marginality with a physical peculiarity. It is also tempting to discern in these figures the silhouettes of those mischievous gnomes sometimes very wicked, or very witty, but never fully human, who haunt Indo-European legends.

HEPHAISTOS AND SMALL DEMONS

A group of minor Greek divinities related to smithcraft, the Kabeiroi, the Telchines, and the Daktyloi, might originally have had the form of dwarfs; they are mentioned in vague or contradictory literary texts which date mostly to the Hellenistic and Roman periods, but derive from Greek sources going back as early as the fifth century BC.[127]

The mysterious Kabeiroi are the most well known.[128] These deities are closely associated with Hephaistos. One of their main sanctuaries was located on Lemnos, the island where it was said that Hephaistos fell down from heaven, and where he had his cult centre.[129] Acusilaus, Pherecydes, and Herodotus state that these gods are the sons or

[123] On this assimilation of dwarfs to monkeys in Hellenistic iconography, see P. Blome, 'Affen im Antikenmuseum', in M. Schmidt (ed.), Kanon. Festschrift Ernst Berger (Basle, 1988) (AK Beiheft 15), 205–10.

[124] See the catalogue established by I. MacPhee, 'An Apulian Oinochoe and the Robbery of Heracles', AK 22/1 (1979), 38–42 (7 Attic and one Apulian representations).

[125] Chous, Taranto, Ragusa coll. 126; Brommer, VL 193. D 5; MacPhee (n. 124 above), pl. 15. 1–3, 38 ff.

[126] MacPhee (n. 124 above), 39 n. 20.

[127] On this group of deities in general, see Hemberg, Kabiren, 300–1 (with full earlier bibliog.), with the review by J. Moreau, AC 20 (1951), 244–6, and H. J. Rose, 'Little Gods but very Wise', CR 65 (1951), 211–13.

[128] They are two, three, or four according to ancient sources. See in general O. Kern, RE x (1917), s.v. Kabeiros und Kabeiroi, 1399–450; Nilsson, GGR³ i. 670–2; Burkert, GrRel, 281–5.

[129] For sources, see Hemberg, Kabiren, 160–70; Burkert, GrRel, 281, esp. n. 4.

grandsons of Hephaistos.[130] Nonnus repeats that the Kabeiroi are the sons of Hephaistos, and adds that they are 'well skilled at the forge'.[131] 'Guilds of craftsmen, especially smith guilds with their special celebrations, may be seen in the background' of the mystery ceremonies which were performed in their sanctuaries, as Burkert notes.[132] Various sources allude to their short stature. Herodotus reports that they were depicted as dwarfs resembling 'the Phoenician Pataikoi' in a temple at Memphis.[133] Hesychius calls them crabs, καρκίνοι, a comparison which may refer to their dwarfish proportions: the large body, with short, bent legs of the crab could evoke the physical disproportion of the dwarf, as the scarab did for the Egyptians.[134] This relation of dwarfs to crabs is also suggested by Aristophanes in the *Pax* when he qualifies as dwarf-like the appearance of the sons of Carcinos, 'the crab'.[135] The image of the crab may also have symbolized smithcraft, especially its pincers, which evoke the smith's tools.[136]

The iconography of the wine vessels found in the Kabeirion near Thebes supports Herodotus' statement. Most vases are painted with grotesque pot-bellied figures, with distorted faces and large genitals, who caricature mythical and human characters.[137] A crab is depicted on a fragment.[138] This material has strong Dionysiac associations (scenes of symposion, satyrs, ivy decoration), which stresses the connection of the small gods with wine, and hence with Hephaistos through the episode of his return to Olympos.[139]

Etymology also suggests the relation of the Kabeiroi to dwarfs. Wackernagel first derived the name Kabeiroi from that of a Sanscrit god, *Kubera* or *Kuvera*, **Kabera*, which means 'the ill-shaped one'.[140] In Hinduism, this god rules over chthonic dwarf demons, the *Guhyakas* and the *Yakṣas*, 'spirits of darkness', who live in forests and mountains, and protect fertility and hidden treasures. In later traditions, *Kubera* himself becomes the god of buried treasure, and is represented as a pot-bellied dwarf.[141] Other

[130] Acusilaus, *FGrH* 2, F 48, and Pherecydes, *FGrH* 3, F 48 = Strab. 10. 3. 21; Hdt. 3. 37. They are depicted carrying a double-axe, like him; see e.g. M. C. Waites, 'The Deities of the Sacred Axe', *AJA* 27 (1923), 25–56, esp. 52, fig. 6.

[131] Nonn. *Dion.* 14. 19–22 (trans. W. H. D. Rouse (Loeb, 1940)).

[132] Burkert, *GrRel*, 281.

[133] Hdt. 3. 37. See full text above 84.

[134] Hesychius, s.v. Κάβειροι. On the similar symbolism of scarabs in Egypt, see above 50, 53–4.

[135] *Pax*, 790.

[136] M. Detienne, 'Le Phoque, le crabe et le forgeron', in R. Crahay, M. Derwa, and R. Joly (eds.), *Hommages à Marie Delcourt* (Brussels, 1970) (Latomus coll. 114), 230.

[137] See P. Wolters and G. Bruns, *Das Kabirenheiligtum bei Theben*, i (Berlin, 1940); M. Bieber, *The History of the Greek and Roman Theater*² (Princeton, 1961), 48–9, figs. 203–7; K. Braun and T. E. Haevernick, *Das Kabirenheiligtum bei Theben*,

iv, *Bemalte Keramik und Glas* (Berlin, 1981), esp. 21ff. See also the grotesque terracotta figurines found in the sanctuary: B. Schmaltz, *Das Kabirenheiligtum bei Theben*, v, *Terrakotten* (Berlin, 1974), 177–80, nos. 312–45, esp. nos. 312–13, pls. 24–5.

[138] Braun/Haevernick (n. 137 above), 48, no. 148, pl. 8. 1 (K 1510).

[139] Cf. Aesch. fr. 97 (*TrGF* 4) = Plut. *Quaest. Conv.* 2. 1, *Mor.* 632 f.

[140] J. Wackernagel, 'Über einige lateinische und griechische Ableitungen aus den Verwandtschaftswörtern', *ZVS* 41 (1907) 314–18. See also H. Frisk, *Griechisches etymologisches Wörterbuch*, i (Heidelberg, 1960), s.v. Κάβειροι, 750.

[141] G. Liebert, *Iconographic Dictionary of the Indian religions* (Leiden, 1976), s.v. Guhyaka, 98, s.v. Yakṣa, 143. See also H. W. Haussig (ed.), *Götter und Mythen des indischen Subkontinents* (Stuttgart, 1984) (Wörterbuch der Mythologie i. 5), s.v. Guhyakas, 82, s.v. Kuvera, 128–9, s.v. Yakṣas, 200–1.

derivations, however, are plausible, especially the Semitic *kabir*, 'great', but also Hittite *habiri* and Sumerian *kabar* for copper.[142]

Like the Kabeiroi, the Idaean Daktyls are related to Hephaistos and associated with mysteries.[143] Like the Olympian god, they are described as magicians, γόητες.[144] They are said to live in caves on Mount Ida, in Crete or Phrygia, and to have discovered the use of fire; they first worked copper and iron, and created all kind of useful objects.[145] Like the pygmies, named after πυγμή, a measure of length, their name, Daktyloi, may have implied that they were believed to be as small as a finger, but no ancient author clearly says it. Only Pausanias mentions that the oldest of them, Heracles, was depicted in the Arcadian sanctuary of Megalopolis as a small man about 'eighteen inches high' (45.7 cm.).[146] Pausanias attributes the same size to the bronze statue of Pan, another dwarfish being, which stood in an official building at Megalopolis.[147]

The Telchines from Rhodes or Crete, who are born from the sea, are also said to have invented smithcraft.[148] Like the Daktyloi, they first worked metal, created statues of gods and 'things which are useful for the life of mankind'; each of them was named after the metal that he invented.[149] Like Hephaistos, they made weapons for the gods, such as the sickle of the Titan Kronos and the trident of Poseidon, and remarkable jewels, such as the necklace of Harmonia.[150] Diodorus says that they kept their art secret, which again alludes to initiations and guilds of smiths.[151] Their name was derived from θέλγω, 'to bewitch',[152] and they are often referred to as magicians, γόητες, who possess redoubtable powers. Their malignant aspect prevails in ancient sources. The Telchines are referred to as slanderous, βάσκανοι, jealous, φθονεροί, wicked beings, πονηροί, who

---

[142] See the discussion by Hemberg, *Kabiren*, 318–25. For the Sumerian derivation, see e.g. Moreau (n. 127 above), 245–6; L. Deroy, 'L'Origine préhellénique des peuples méditerranéens,' in *Mélanges I. Lévy*, i (Brussels, 1953), 114–18.

[143] There are 5, 10, 20, or 100 of them, in different authors. For sources, see principally L. v. Sybel, Roscher, i (1884–6), s.v. Daktyloi, 940–1; O. Kern, *RE* iv (1901), s.v. Daktyloi, 2018–20; Hemberg, *Kabiren*, 346–9.

[144] Pherecydes, *FGrH* 3, F 47; Strab. 10. 3. 22.

[145] Hes. fr. 282 (Merkelbach/West); Sophocles fr. 366 (*TrGF* 4); Pherecydes, *FGrH* 3, F 47; Diod. Sic. 5. 64, 17. 7; Ov. *Met.* 4. 281. Cf. S. Marinatos, 'Zur Frage der Grotte von Arkalochori', *Kadmos*, 1 (1962), 86–94, on the finding of metallic objects (iron, silver, gold), esp. double-axes, in Cretan caves (votive deposits?). The Cretan Zeus, closely associated with the Daktyloi, is called *Welchanos*; see M. L. West, 'The Dictaean Hymn to the Kouros', *JHS* 85 (1965), 154; H. Verbruggen, *Le Zeus crétois* (Paris, 1981), 46.

[146] Paus. 8. 31. 3.

[147] Paus. 8. 30. 6.

[148] They are three, four, or nine according to ancient authors, and were also attached to Sicyon, Ceos, and Cyprus. The main source is Suet. Περὶ Βλασφημ, s.v. Τελχῖνες (ed. J. Taillardat (Paris, 1967), 92). For further sources, see principally Preller/Robert, *GrMyth*, i. 605–9; C. Blinkenberg, 'Rhodische Urvölker', *Hermes*, 50 (1915), 271–303; P. Friedländer, Roscher, v (1916–24), s.v. Telchinen, 234–43; H. Herter, *RE* v. 1 (1934), s.v. Telchin, 196–7; id., ibid. s.v. Telchinen, 197–224; Hemberg, *Kabiren*, 349–51.

[149] Diod. Sic. 5. 55. 2 (trans. C. H. Oldfather (Loeb, 1939)); Strab. 14. 2. 7 (iron and brass); Suet. Περὶ Βλασφημ.

[150] Callim. *Del.* 31 (Poseidon's trident); Suet. (sickle of Kronos); Stat. *Theb.* 2. 274 (Harmonia's necklace). Cf. Apollod. *Bibl.* 3. 4. 2 (necklace of Harmonia attributed to Hephaistos).

[151] Diod. Sic. 5. 55. 3.

[152] See e.g. *Suda* and *Etym. Magn.*, s.v. Θέλγει.

cast the evil eye.[153] They could 'summon clouds and rain and hail at their will and likewise could even bring snow', and they made Rhodes a desert by pouring the water of the Styx, mixed with sulphur, on plants and animals.[154] Stesichorus assimilates them to the Keres, spirits associated with disease and death.[155] Their name even became a nickname for 'spiteful backbiters'.[156] Suetonius, however, made a distinction between those who are craftsmen and benevolent, and those who are malignant.[157]

The physical appearance of the Telchines is indefinite and fearful. Suetonius says that they have gleaming eyes, dark eyebrows and a sharp glance; some have no hands, some no feet, or they have animal feet, with skin between the toes, like geese. They can also change their natural shape and transform into amphibian animals.[158] This description could evoke the image of a seal.[159] It recalls also the characteristics of the dwarfs in Indo-European traditions, as several scholars have remarked.[160] Like the Telchines, these gnomes are mostly smiths, they live in hills, caverns, or mountains, make magic objects and guard buried treasure.[161] They can bewitch people, change weather, destroy crops or other works of man.[162] They transform their shape at will, and may have bird or animal feet.[163]

Another point in common with Indo-European mythical dwarfs is that these Greek demonic beings are also described as belonging to an early race which once became extinct.[164] Thus the Daktyloi are said to have been the first inhabitants of Mount Ida, like the Telchines at Rhodes.[165] A Lemnian Kabeiros is called the first man; he is

---

[153] Strab. 14. 2. 7; Ov. *Met.* 7. 366: *Oculos ipso vitiantes omnia visu*; Photius and Hesychius, s.v. Τελχῖνες.

[154] Diod. Sic. 5. 55. 3 (trans. C. H. Oldfather (Loeb, 1939)); Strabo 14. 2. 7; *Suda*, s.v. Θέλγει; Zenob. 5. 41; Nonn. *Dion.* 14. 45–9.

[155] Stesichorus fr. 265 (ed. D. L. Page, *Poetae Melici Graeci* (Oxford, 1962)). On the Keres, see Nilsson, *GGR*³ i. 222–5; W. Burkert, *Homo Necans* (Berlin and New York, 1972), 250–5.

[156] LSJ, s.v. Τελχίν, 1774. See e.g. Callim. *Aet.* 1. 1 'In Telchinas'.

[157] Suet. Περὶ Βλασφημ.

[158] Diod. Sic. 5. 55. 3; Suet. Περὶ Βλασφημ.

[159] So for Detienne (n. 136 above), 219–33.

[160] See e.g. Blinkenberg (n. 148 above), 285–6; Friedländer (n. 148 above), 242–3.

[161] See in general J. Grimm, *Deutsche Mythologie*³ i (Göttingen, 1854), 415–40; H. W. Haussig (ed.), *Götter und Mythen im alten Europa* (Stuttgart, 1973) (Wörterbuch der Mythologie, i. 2), s.v. Zwerg, 97–8, s.v. Maahiset, 327–8, s.v. Nyrckes, 368–9, and C. Webster, 'Paracelsus and Demons: Science as a

Synthesis of Popular Beliefs', in *Scienze credenze occulte, livelli di cultura. Convegno internazionale di studi, Firenze, 1980* (Florence, 1982), 3–20. In particular, see Thompson, *Motif Index* F 451.3.4.2 (as smiths); F 451 4.1.11; 451.4.2.6; F 535.6.1 (live in hills and mountains); D 812.12; D 817.1.1 (make magic objects); F 451.7.1 (possess treasures); F 451.10.4 (make precious objects for the gods). Cf. also the discussion on the possible derivation of *Kobold* from Κόβαλος; Grimm, ibid. 468–70; A. Adler, *RE* xi. 1 (1921), s.v. Kobaloi; Hemberg, *Kabiren*, 326–7.

[162] Thompson, *Motif Index* F 451.3.3 (as magician); F 451.3.3.5.1 (bewitch people); F 451.5.2.5 (interfere with mortal's work).

[163] Thompson, *Motif Index* F 451.3.3.0.1 (can take what shape he wants); F 451.2.2 (feet of dwarfs).

[164] Thompson, *Motif Index* F 451.1.3; F 451.1.4 (origin of dwarfs); F 451.9.1 (emigration of dwarfs).

[165] Strab. 10. 3. 22 (Daktyloi). Diod. Sic. 5. 55. 1; Strab. 14. 2. 7 (Telchines). The Telchines were killed by Olympian gods; Ov. *Met.* 7. 367 (Jupiter); Suet. Περὶ Βλασφημ. (Apollo).

depicted on a Kabeirion vessel from Thebes with the name Pratolaos, looking at his parents Mitos and Krateia.[166]

Hephaistos has many connections with these small demons. Like them, he learned his craft in caves, located in the depth of Ocean, and could be heard at work with the Cyclopes underground in Mount Etna.[167] He is not only a talented smith, but a magician who fashioned the first woman, Pandora, and created enchanted weapons and jewels.[168] Scholia on Apollonius and Callimachus report a version which stress his resemblance with the gnomes of Indo-European folktales; they say that in the Lipari islands people used to leave pig-iron for Hephaistos with a fee, and got it back the following morning, forged in the shape of a sword, or of another object.[169]

His special qualities were marked by an unusual physical appearance: Hephaistos was lame in both feet, like his sons, Periphetes and Palaimonios.[170] Since antiquity, this physical peculiarity has given rise to various rationalistic and mythological explanations. Some modern authors, like Wilamowitz, and later Malten, suggested that his deficient lower limbs could reflect the original dwarf form of the god.[171] This hypothesis is supported by the fact that the deformity of his legs is commonly described by three epithets: κυλλός, κωλός, ἀμφιγυήεις, which may apply to crabs, and hence evoke an early dwarf form.[172]

In iconography, however, Hephaistos is hardly ever shown as a dwarf. In a few archaic vase paintings, Hephaistos seems to be slightly shorter than his companions, but this may be due to compositional needs (fig. 13.5).[173] Only an Apulian amphora in Foggia depicts Hephaistos as a plump, miniature man with a snub nose and slightly malformed limbs (G 114; pl. 76). The god has come to set free Hera who is bound to his magic throne; she is surrounded by Olympian, yet powerless, deities. Their dignified

[166] Page (n. 155 above), 985; Wolters/Bruns (n. 137 above), 96, pls. 5 and 44. 1; Braun/Haevernick (n. 137 above), 62, no. 302, pl. 22. 1. 2; Burkert, GrRel, 282.

[167] See e.g. Ap. Rhod. 4. 760–2; Callim. Dian. 46 ff., Del. 141–6; Nonn. Dion. 27. 97. Cf. Eur. Cyc. 20, 298, 599.

[168] Cf. Il. 8. 195; 15. 310; 18. 137 (weapons); Hes. Theog. 570–85; Op. 59–73 (Pandora); Paus. 1. 20. 2 (magical throne). For further sources, see Gruppe, GrMyth, ii. 1309–10.

[169] Sch. on Ap. Rhod. 4. 761; sch. on Callim. Dian. 46; Preller/Robert, GrMyth, i. 181. Cf. Thompson, Motif Index H 973.1, 973.3 (task performed by dwarfs).

[170] For sources, see Preller/Robert, GrMyth, i. 174–84; Gruppe, GrMyth, ii. 1304–18, esp. 1306–7; L. Malten, RE viii. 1 (1913), s.v. Hephaistos, 311–66, esp. 336–7.

[171] U. v. Wilamowitz-Moellendorff, 'Hephaistos',

NGG (1895), 241–4. Malten (n. 170 above), 336–7, first disagreed with Wilamowitz, but later supported his hypothesis: id. 'Hephaistos', JDAI 27 (1912), 256–60. Cf. also M. Delcourt, Hephaistos ou la légende du magicien (Paris, 1957), esp. 112–15.

[172] Cf. the description of the crab in Anth. Pal. 6. 196.

[173] Corinthian amphoriskos, Athens, NM 664; H. G. Payne, Necrocorinthia (Oxford, 1931), i. 118–24; ii. 314, no. 1073; T. Carpenter, Dionysian Imagery in Archaic Greek Art (Oxford, 1986), 15 ff., pl. 5; Kunze/Nippert, Genetics, 44, fig. 47 (diastrophic dysplasia, bilateral club-foot, traumatic club-foot). See also the Laconian cup, Rhodes Museum, 10711; M. Pipili, Laconian Iconography of the Sixth Century (Oxford, 1987), 54, fig. 80, nn. 541–3. On Hephaistos' deformity in iconography, see F. Brommer, Hephaistos. Der Schmiedegott in der antiken Kunst (Mainz am Rhein, 1978), 11, 16, nos. A 1, C 4, 5, 6?, 12, 15, 16, figs. 1, 8, pls. 11.1–3, 12.1, 13.3.

FIGURE 13.5. Corinthian amphoriskos. Athens, NM 664

attitudes contrast with the grotesque appearance of the magician smith, probably by way of humorous enhancement of his redoutable capacities.[174]

In archaic vase-painting, Hephaistos is also associated with dwarfs or dwarfish padded creatures in representations of his return to Olympos (G 113; pl. 75; fig. 13.5). On several Corinthian komos scenes, these padded figures may mimic foot deformities, but their relation to Hephaistos is unclear.[175]

To sum up, the qualities and functions of Hephaistos and minor deities may bear witness to pre-Hellenic, Indo-European beliefs in dwarf smiths. This possible origin, however, seems to have been completely ignored in Greece in historic times, which coincides with the general lack of interest of ancient authors in the dwarfs of myth.[176] The name of the Daktyloi was never related to the possible dwarfishness of the demons; it is usually explained as an image of their manual skill or of the hand of the nymph Anchiale, grasping and throwing the earth from which they are born.[177] The malformation of Hephaistos was usually accounted for by his fall from Olympos, or else

[174] Cf. the possible depictions of a dwarf Hephaistos on a cista in London, BM 745; H. B. Walters, *Catalogue of the Bronzes, Greek, Roman and Etruscan in the British Museum* (London, 1899), 132–4; G. Bordenache Battaglia and A. Emiliozzi, *Le ciste prenestine* (Florence, 1979), no. 34, pls. CLI–CLIII.

[175] See Carpenter (n. 173 above), 16 n. 19 (list of figures with twisted or misshapen feet). M. Azra Hincks, 'Le Kordax dans le culte de Dionysos', *RA* 17 (1911), 1, fig. 1. 1–5, publishes an aryballos where two dancers mimic lameness, while Dionysos

stands in an attitude which evokes strongly that of Bes: he is frontal, the panther's head of his animal skin hanging between his splayed legs.

[176] Cf. Rose (n. 127 above), 213, about this group of demonic beings: 'They lie outside the main stream of mythology, literature, and, generally speaking, of the interest of the higher and more articulate classes of society.'

[177] See Hemberg, *Kabiren*, 346, T 1. See e.g. Pherecydes, *FGrH* 3, F 47; Hellanicus, *FGrH* 4, F 89; Ap. Rhod. 1. 1129–31.

he is lame from birth,[178] while mythographers speculated on a possible symbolic reference to the imperfect nature of flame.[179] It is hardly possible to attribute his lameness to a single cause. Some modern authors, like Gruppe, related his misshapen feet, κυλλοποδίων, to the animal feet of satyrs, an interesting hypothesis which would stress the Dionysiac affinities of the god.[180] Others, like Rosner, suggested that the defect might arise out of a real disease among early blacksmiths, who may have suffered from poisoning due to high concentrations of arsenic in metal.[181] These explanations are perhaps all to some extent complementary.

KOUROTROPHIC DEMONS

Distribution

In the second half of the sixth century, small statuettes of dwarfs, usually made of terracotta, or sometimes of wood, are found all over the Greek world: on the Greek mainland (Argos, Delphi, Perachora, Tanagra), in the islands (Aegina, Amorgos, Cyprus, Delos, Melos, Paros, Rhodes, Samos), and in the colonies (Sicily, South Italy, North Africa, Asia Minor, Northern Greece, and the Black Sea).[182]

These objects were produced during some fifty years, between the first third of the sixth century and c. 520 BC. As Sinn notes, the size of the statuettes seems to have diminished with the course of time;[183] the earliest items in the series are relatively tall, such as, for example, the Kassel figurine, which dates to the second quarter of the sixth century (h. 19 cm.; G 154; pl. 78.3), while the most common ones, produced after 550 BC, are smaller (h. 7 cm. on the average).

The terracottas are made of a micaceous clay, pale orange (e.g. G 137–9, 154; pls. 77.1–3, 78.3) to reddish brown (e.g. G 145, 147; pl. 78.1). They are usually attributed to Rhodian, Samian, or other South Ionian workshops,[184] but some are also of Argive manufacture.[185]

[178] See e.g. *Il.* 18. 396 (lame by birth); Apollod. *Bibl.* 1. 3. 5 (fall from Olympos).

[179] Gruppe, *GrMyth*, ii. 1306. See e.g. Heracl. *Probl. Hom.* 26. 8–11.

[180] Gruppe, *GrMyth*, ii. 1306–7.

[181] E. Rosner, 'Die Lahmheit des Hephaistos', *Forschungen und Fortschritte*, 29 (1955), 362–3. See also Nilsson, *GGR*[3] i. 527 (blacksmiths have weak feet).

[182] See the map in Sinn, *Antidoron*, 88, fig. 1.

[183] Sinn, *Antidoron*, 87.

[184] For Rhodes see e.g. R. A. Higgins, *Catalogue of the Terracottas in the British Museum*, i (London, 1954), 56, no. 86; U. Liepmann, *Griechische Terrakotten, Bronzen, Skulpturen* (Hanover, 1975), 44, T 15; A. Caubet, 'Statuette en bois de démon égypto-grec d'époque archaïque', *RLouvre*, 19 (1969), 10–11. For Samos see e.g. A. Furtwängler, 'Zwei griechische Terrakotten', *ARW* 10 (1907), 322; P. C. Bol, *Liebieghaus, Ancient Art: Guide to the Collection* (Frankfurt am Main, 1981), 6–7; M. Bell, *Morgantina Studies*, i (Princeton, 1981), 15–16 (Ionian Aphrodite group); G. Kopcke, 'Neue Holzfunde von Samos', *MDAI(A)* 82 (1967), 110; Sinn, *Antidoron*, 87 nn. 8–12; id., 'Der sog. Tempel D im Heraion von Samos', *MDAI(A)* 100 (1985), 151–3.

[185] See e.g. figurines found on Aegina; U. Sinn, 'Der Kult der Aphaia auf Aegina', in R. Hägg *et al.* (eds.), *Early Greek Cult Practice. Proceedings of the Fifth International Symposium at the Swedish Institute at Athens, 26–29 June, 1986* (Stockholm, 1988), 152–3, esp. n. 27. See also F. Eckstein and A. Legner, *Antike Kleinkunst im Liebieghaus* (Frankfurt am Main, 1969), no. 35 (Boeotia).

Typology

The figurines are commonly divided in two types: statuettes (h. *c*.7 cm.), and plastic vases (h. 11.5–16.5 cm.). In both forms, the dwarfs are naked and stand on a plinth, with bent legs, the hands clasped over their protruding belly, often with creases suggesting fatness (e.g. G 154, 162, 183; pls. 78.3, 79.3, 80.1). The neckless head is very large, with long wavy hair hanging down the back, and the genitals are infantile. Some dwarfs wear either a petasos (e.g. G 146; pl. 78.2), or a pilos which resembles Scythian soft caps (e.g. G 139, 206; pls. 77.3, 80.3),[186] in addition to sandals (G 145; pl. 78.1). Some carry a shield (G 206; pl. 80.3), or a rectangular basket with offerings (G 156; pl. 79.2). Often a smaller figure, crudely made and attached after moulding, is seated on their left shoulder (e.g. G 138, 146, 183, 202; pls. 77.2, 78.2, 80.1, 80.2). A few authors identified these figures with monkeys,[187] but they more likely depict small children, as is suggested by their tightly convoluted legs.[188]

Origin and Function

The iconographic type of the statuettes derives from the Egyptian dwarf gods Bes and Ptah-Pataikos.[189] The pose of the figurines copies that of Ptah-Pataikos: like the Egyptian god, the dwarfs are standing, with clenched fists laid on the chest; they have human faces, as opposed to the animal features of Bes, and subdued smiles. In the Greek adaptation, the legs of the dwarfs are not bent sideways, but forward, and the symbolic attributes of Ptah-Pataikos (knives, feathers, snakes, scarabs) are abandoned.

Some attributes are borrowed from the iconography of Bes. Dwarfs carrying on their shoulder a small seated figure evoke representations of Bes nursing a smaller figure of himself or carrying a child (e.g. G 183; pl. 80.1; cf. pl. 7.3), while armed dwarfs may be influenced by representations of Bes as a warrior (e.g. G 206; pl. 80.3; cf. pl. 10.1–2). But the exotic attributes of Bes, such as the plumed head-dress, the animal skin, and the grimacing face with a protruding tongue, are missing.

A few isolated variants show other Egyptian influences. For example, a figurine once in Kassel represents a dwarf on top of a bull (G 125; fig. 13.6), which might refer to the special links between dwarfs and sacred bulls in Egypt.[190] The carapace of a scarab is depicted on the back of a Corinthian statuette (pl. 40.3b). The origin of the motif is clearly Egyptian: it derives from representations of dwarf gods with beetle-shaped chests

[186] Authors normally describe this cap as a pilos, except Sinn, *Antidoron*, 87 n. 19 (helmet), and Furtwängler (n. 184 above), 329 (Scythian soft cap).
[187] See e.g. Winter, *TK* i. 213, no. 6; Higgins (n. 184 above), 57, no. 93; Liepmann (n. 184 above), 44, T 15.
[188] So for Furtwängler (n. 184 above), 324; Bol (n. 184 above), 26; Sinn, *Antidoron*, 87.
[189] For Ptah-Pataikoi as the prototype: J. Boehlau,

Aus ionischen und italischen Nekropolen (Leipzig, 1898), 155–6; Furtwängler (n. 184 above), 322–4; Higgins (n. 184 above), 56; J. Boardman, *The Greeks Overseas*[2] (London, 1980), 147; Bell (n. 184 above), 16. For Bes as the prototype: Sinn, *Antidoron*, 88–9.
[190] For bull-vases in general, see e.g. Higgins (n. 184 above), 59, no. 98. On dwarfs and bulls in Egypt, see above 82, 152.

FIGURE 13.6. (G 125) Terracotta. Formerly Kassel, Habich coll. Present location unknown

(figs. 5.2–3). It evokes the assimilation of dwarfs to the sacred scarab-beetle Khepri, and hence to solar and regeneration concepts.[191] This motif does not seem to have inspired other Greek artistic works. The Corinthian attempt remained unique, probably because the symbolism attached to scarabs in Egypt had no analogy in Greek traditions.

The Hellenization of the Egyptian dwarf motif goes back to the seventh century. At that period, figurines of Bes and Ptah-Pataikos are found with other Egyptian votive offerings in several Greek tombs and sanctuaries;[192] their meaning seems to have been well understood by the Greek dedicators, who, for example in Samos, consecrated figurines of dwarf gods to the local cult of Hera as kourotrophos (pl. 7.2).[193] Egyptian dwarf gods may have become so popular that Greek craftsmen created their own iconographic type, probably by the end of the seventh century. The type may have originated in Samos, a place which had close contact with Egypt during the archaic period.[194] A possible seventh-century prototype, of Samian workmanship, was found in the Heraion (G 155; pl. 79.1).[195] The piece, made of wood, differs stylistically from the mass-produced terracotta dwarfs. It is closely but freely inspired from Egyptian models, especially Bes. The dwarf has a round head, with large eyes, a broad, flat nose, and thick lips which evoke the facial features of Bes; he also had a phallus, now missing, which was set in the hole below his abdomen.[196] His attitude is influenced by Bes too: he is

[191] See above 50, 53–4.

[192] On the relationships between Egypt and Greece in the archaic period, see in general Boardman (n. 189 above), 111 ff. See e.g. the amuletic Bes and Ptah-Pataikoi offered to Artemis in Paros; O. Rubensohn, *Das Delion von Paros* (Wiesbaden, 1962), 169, pl. 35, 1–4 (cf. G 140–1). For similar amulets in Rhodes, see e.g. G. Jacopi, *Clara Rhodos*, 6–7 (1933), 304–7, nos. 5–8 (cf. G 153), and in Cyprus, E. Gjerstad *et al.*, *The Swedish Cyprus Expedition*, ii (Stockholm, 1935), pl. CLIX, nos. 16, 23–5, the tomb of a woman with amuletic Bes (no. 18), Isis (no. 17) and Taweret (no. 19).

[193] See Sinn, *Antidoron*, 90, 92 and n. 67. See e.g.

the bronze figurine of Bes in U. Jantzen, *Samos*, viii (Bonn, 1972), 14, pl. 19 (B 1226).

[194] On the relationships between Egypt and Samos, see K. Parlasca, 'Zwei ägyptischen Bronzen aus dem Heraion von Samos', *MDAI(A)* 68 (1953), 127–36; W. M. Davis, 'Egypt, Samos, and the Archaic Style in Greek Sculpture', *JEA* 67 (1981), 61–81; Sinn (n. 184 above), 152–3.

[195] For its attribution to Samian workshop, see Kopcke (n. 184 above).

[196] On ithyphallic Bes, see e.g. Tran Tam Tinh, *Bes*, 102, no. 45, pl. 80. See also above 81–2, and the Graeco-Roman phallophorous procession on pl. 11.3.

accompanied by a small figure which is not seated on his shoulder, but held in his left arm, as in representations of nursing Bes (pl. 7.3); this small figure is naked, and wears a tight cap which evokes Egyptian headgear. The piece is so far unique, and does not seem to have directly influenced the conventional type of terracotta dwarfs. The creation of the standard type, however, may also be Samian, as suggested by the hairstyle of an early statuette in Kassel (G 154; pl. 78.3).[197]

The statuettes convey the Egyptian notion of dwarfs as family guardians. Greek dwarfs seem to have ensured the same protective function as their Egyptian models. A large number of terracottas were found in the sanctuaries and temene of female divinities, such as those of Hera at Argos (G 115),[198] Perachora (G 119), and Samos (G 155–6; pl. 79.1–2), that of Athena on Lindos (G 147–151), of Artemis on Paros (G 140–1) and at Ephesus (G 193), of Aphaia on Aegina (G 121–2), and of Demeter and Kore at Tocra (G 185–6), Catania (G 163–6), Selinus (G 175, 177–8) and Gela (G 167). Inscriptions and votive gifts reveal that these female deities were worshipped as kourotrophic in many places.[199] As Sinn notes, the dwarfs might represent their hypostases, and had a similar role.[200] They cared for young children, as suggested by the babies carried on their shoulders, and probably also for mothers. In particular, they guaranteed the welfare, εὐτροφία, of the child. An unusual variant from Samos depicts a dwarf carrying a rectangular basket with cakes on top of his head (G 156; pl. 79.2); as Sinn demonstrates, this food is not a votive offering, which is normally put on a kanoun, a round tray, but real food for the child.[201] This notion of εὐτροφία was perhaps also expressed by the fatness of the figures. Like their Egyptian counterparts, Greek dwarfs also guarded children in the afterlife. Several figurines come from tombs of children, as in Samos (G 154; pl. 78.3), Megara Hyblaea (G 168–70), Selinus (G 176) and Syracuse (G 180, 182); in composite pieces, the infant occasionally makes uncertain gestures, raising both arms to his head as on a statuette in Frankfurt (G 202; pl. 80.2), in a gesture which may express grief.[202]

The popularity of these kourotrophic dwarfs, however, lasted for a relatively short time. The production of the statuettes does not go beyond the general fashion for Eastern objects. Is it because the belief in benevolent dwarf gods was purely foreign, as

---

[197] Cf. the kouros in H. Kyrieleis, *Führer durch das Heraion von Samos* (Athens, 1981), 45, fig. 32. For further stylistic elements, see Sinn, *Antidoron*, 87 nn. 10–12.

[198] Kopcke (n. 184 above), 110 n. 25, mentions several unpublished figurines from that site.

[199] See Th. Hadzisteliou Price, *Kourotrophos: Cults and Representations of the Greek Nursing Deities* (Leiden, 1978), 90–100 (Cyprus), 122 (Aegina), 144–6 (Argos), 146 (Perachora), 149–50 (Paros), 150–2 (Mykono-Delos), 152–3 (Samos), 154–6 (Rhodes), 181–6 (Sicily). See also Sinn, *Antidoron*, 90–1 (Samos); Sinn (n. 185 above), 152 ff. (Aegina).

[200] Sinn, *Antidoron*, 91.

[201] On the kanoun, see in general J. Schelp, *Das Kanoun. Der griechische Opferkorb* (Würzburg, 1975); see e.g. models of offering trays in R. S. Stroud, 'The Sanctuary of Demeter and Kore on Acrocorinth', *Hesperia*, 34 (1965), 23, pl. 11; Sinn, *Antidoron*, 88, figs. 6a–b. Cf. Hdt. 3. 48 (on the offering of sesame and honey cakes to youths in the Samian cult of Artemis).

[202] Furtwängler (n. 184 above), 324. Cf. Bol (n. 184 above), 26: the child tries to 'shield him from his first glimpse of the underworld'.

Bell suggests?[203] Hellenic beliefs in dwarf demons may also be expressed by these kourotrophic figures.[204] The Telchines, Kabeiroi, and Daktyloi are described as protectors of divine children. The Kouretes, for example, who are assimilated to the Daktyloi, helped to conceal the newborn Zeus from the devouring instinct of his father Kronos by drowning his cries with the clashing of their weapons on their shields.[205] The weapons carried by some Greek dwarfs, like that in Munich (G 206; pl. 80.3), may refer to these legends.[206] In return Cretan myths might have influenced the image of Egyptian dwarf gods, in particular Late Period representations of Bes as a warrior, waving a sword and a shield to chase off evil powers (pls. 10.1–2, 11.2).

The figure of the dwarf as kourotrophos left some traces beyond the archaic period. It may later have influenced the creation of another Greek motif, that of Silenos, often ambiguously short, as pedagogue,[207] which has an Egyptian counterpart in the figure of Bes.[208] This positive image of dwarfs probably influenced their position in everyday life, as Attic vase-painting shows.

---

[203] Bell (n. 184 above), 16.

[204] For Telchines, see Blinkenberg, *Lindos*, i (1931), 560–1: 'Il en reste pourtant l'idée juste que ce type a dû représenter des êtres créés par les croyances populaires helléniques, et dont la dénomination a peut-être varié selon les endroits. Quant à l'île de Rhodes, j'ai proposé depuis longtemps le nom de Telchines'; id. *Clara Rhodos*, 8 (1936), 18. For Pataikoi and Kabeiroi, see Furtwängler (n. 184 above), 326–9; Hadzisteliou Price (n. 199 above), 76 (Kabeiroi). See also the discussion of these various hypotheses in Sinn, *Antidoron*, 87 n. 14.

[205] Strab. 10. 3. 11; Apollod. *Bibl.* 1. 1. 7. See also the Cretan hymn to Zeus (4th cent. BC) in M. L. West, 'The Dictaean Hymn to the Kouros', *JHS* 85 (1965) 149–59. Cf. Diod. Sic. 5. 55. 1 (Poseidon). For further lit. refs, see Preller/Robert, *GrMyth*, i. 653–5, esp. 654; O. Immisch, Roscher, ii (1), 1890–

4, s.v. Kureten und Korybanten, 1587–628, esp. 1600–1; F. Schwenn, *RE* xi. 2 (1922), s.v. Kureten, 2202–9, esp. 2206–7; Hadzisteliou Price (n. 199 above), 193.

[206] Cf. also the figurine seen in Gela in 1889 by Furtwängler (n. 184 above), 324, which depicts a dwarf bowman(?): 'er streckt den linken Arm vor und biegt den rechten so wie im Motiv des Bogenschiessens'.

[207] See e.g. F. A. G. Beck, *Album of Greek Education* (Sydney, 1975), pl. 12, fig. 67; pl. 13, figs. 71, 73, pl. 15, figs. 81–3. See also the terracotta figurine of Silenos with the infant Dionysos, in G. R. Davidson, *Corinth*, xii (Princeton, 1952), 51, no. 331, pl. 29.

[208] See e.g. the terracotta lamps with Bes pedagogues; Tran Tam Tinh, *Bes*, 104, nos. 74, 76.

14

Physical Minorities

How was a short-statured person accepted in Greek communities? Did a dwarf child receive the same education as a normal child? Once adult, did he possess and exert the same rights as his contemporaries? Could he be elected to a magistracy? If severely crippled did he benefit from private or public assistance?

No dwarf is mentioned in the sources describing Greek laws and customs. Information on the Greek attitude towards physical deviance must, as was the case for Egypt, be sought in secondary sources, such as accounts of the exposure of abnormal children and of legal provisions for the disabled. All these references apply in particular to Attica. I distinguish here between congenital handicaps (lameness, blindness, deafness) and disabilities due to human or natural violence (war, crime, accident). I consider three key moments in the life of a normal Athenian: his birth, his entry into the community, and his function in the politeia, in order to show how a citizen who was physically different passed these points.

THE EXPOSURE OF ABNORMAL BIRTHS

In Athens, a law attributed to Solon gave the father, as the head of the family, all power over the newborn infant. The father recognized his child officially at the feast of the Amphidromia which took place a few days after the birth (five, seven, or ten days according to different sources);[1] a sacrifice was made, and the child was carried around the domestic hearth. The ceremony marked symbolically the entry of the child into the familial group, and hence into the community. Only then was it legally born, and it probably received a name on the same occasion.[2] The existence of the practice suggests that the ceremony could also not happen. Within these critical first days after birth, it seems to have been admitted that an unwanted baby might be put away in a public place

---

[1] For ancient sources, see E. Saglio, *DA* i (1877), s.v. Amphidromia, 238–9; P. Stengel, *RE* i. 2 (1894), s.v. Amphidromia, 1901–2. See also M. Golden, 'Names and Naming at Athens: Three Studies', *Echos du Monde Classique*, 5 (1986), 252–6.

[2] For a discussion of naming ceremonies, see Golden, ibid.

where he might die or be rescued.[3] No Athenian law, however, is recorded. The decision seems to have depended entirely upon the will of the father, and very likely also upon that of the mother. The exposure was thus not a matter of public, but of private law.[4]

The only legal text mentioning the exposure of the newborn child in the fifth century comes from Gortyn in Crete; an inscription defines the status of the child of a divorced woman.[5] The law gives the mother the right to expose her baby if her previous husband does not recognize the child. If the divorced woman is a slave, she has to present her baby to the master of her previous husband, which suggests that the master also had as a rule full power over the children of his slave household.

The causes for exposure vary: in myths and in the plays of Middle and New Comedy, unwelcome children are mostly illegitimate or female.[6] A physical abnormality could probably also justify this practice, as suggested by Plutarch, Plato, and Aristotle.

Sparta

Plutarch assigns to Lycurgus, the legendary founder of Spartan institutions, a law which gave to the State the right to decide whether a newborn could be accepted in the community or must die:

> Offspring was (sic) not reared at the will of the father, but was taken and carried by him to a place called Lesche, where the elders of the tribes officially examined the infant, and if it was well-built and sturdy, they ordered the father to rear it, and assigned it one of nine thousand lots of land; but if it was ill-born and deformed, they sent it to the so-called Apothetae, a chasm-like place at the foot of Mount Taygetus, in the conviction that the life of that which nature had not well equipped at the very beginning for health and strength, was of no advantage either to itself or to the state.[7]

Unfortunately Plutarch does not specify what criteria defined the disability of a child. Were all types of handicap, physical and mental, even slight ones (club-foot, polydactylism, deafness) pointed out? The rejected newborn infant is described only as ἀγενής, ἄμορφος, as not belonging to a race. Plutarch adds that the women then washed

---

[3] The only direct mention of the practice is Plato, *Tht.* 160e–161a. See the material collected by G. Glotz, *DA* ii (1892), s.v. Expositio, 930–9; La Rue Van Hook, 'The Exposure of Infants at Athens', *TAPhA* 51 (1920), 134–45; E. Weiss, *RE* xi. 1 (1921), s.v. Kinderaussetzung, 463–71; W. Kroll, ibid. 471–2; H. Bolkestein, 'The Exposure of Children at Athens and the ἐγχυτρίστριαι, *CPh* 17 (1922), 222–39; A. Cameron, The Exposure of Children and Greek Ethics', *CR* 46 (1932), 105–14; J. Rudhardt, 'Sur quelques bûchers d'enfants découverts dans la ville d'Athènes', *MH* 20 (1963), 10–20, esp. 17 ff.; R. F. Germain, 'Aspects du droit d'exposition en Grèce', *Revue historique de droit français et étranger*, 47 (1969),

177–97 (with earlier bibliog.); A. W. Gomme and F. H. Sandbach, *Menander: A Commentary* (Oxford, 1973), 34–5.

[4] Germain (n. 3 above) 189–92.

[5] R. F. Willetts (ed.), *The Law Code of Gortyn* (Berlin, 1967) (Kadmos suppl. 1), 29, 41–2, col. iii, esp. ll. 44–52. See also Ael. *VH* 2.7, who reports that the practice was forbidden at Thebes.

[6] See e.g. the exposure of an illegitimate son (Ion) in Eur. *Ion*, 946–60. For further examples see e.g. Glotz (n. 3 above), and the discussion by Gomme/ Sandbach (n. 3 above).

[7] Plut. *Lyc.* 16. 1–2 (trans. B. Perrin (Loeb 1914)).

the babies with wine, instead of water, to test their constitution 'for it is said that epileptic and sickly infants are thrown into convulsions by the strong wine and lose their senses'.[8] Schmidt suggests that this test was performed before the elders and that the reactions of the baby determined their verdict.[9] This wine bath may also not have been a test but an initiatory rite which confirmed the acceptance of the child in the community, as den Boer notes.[10] The meaning of wine may have been related to the fact that it was an invigorating beverage, often assimilated to blood.[11] Plutarch may have lost the original meaning of the rite, and confused the wine bath with that of cold water which, for example, Soranus mentions in his *Gynaeceia* as a means of hardening babies among Teutons and Scyths.[12] Plutarch equally does not make clear how superfluous lives were suppressed in Sparta. He only mentions that rejected newborn infants were taken away. They were probably thrown down the escarpments of Taygetus, or simply abandoned there.

This measure aimed openly to preserve the superiority of the small, but powerful ruling class by keeping its members physically perfect. Those who were not citizens, Helots and Perioikoi, do not seem to have been submitted to such selection.

Plutarch describes the law as one of the former peculiarities of the Spartan constitution. As Schmidt suggests, this regulation may have disappeared at the end of the fifth century.[13] King Agesilaus of Sparta (447–360 BC) was, after all, lame in one foot; he was accepted by the Laconians despite a warning of the Delphic oracle which said:

> Boasting Sparta, be careful not to sprout
> a crippled king, you are sure-footed;
> unexpected troubles will overtake you,
> the lamentations of the storm of war
> destroyer of Mankind.[14]

The physical lameness of Agesilaus was found less disturbing than the uncertain provenance of the other contestant for the throne, Leotychidas, who was perhaps not the true son of Agis, the previous King of Sparta.[15] Plutarch reports, however, that

---

[8] Ibid. 3.

[9] M. Schmidt, 'Hephaistos lebt. Untersuchungen zur Frage der Behandlung behinderter Kinder in der Antike', *Hephaistos*, 5–6 (1983–4), 134.

[10] W. den Boer, *Laconian Studies* (Amsterdam, 1954), 234–41.

[11] See R. B. Onians, *The Origins of European Thought* (Cambridge, 1951), 217–22.

[12] Soranus, *Gynaeceia*, 2. 12 (81), 1–3 (ed. J. Ilberg, *Die Überlieferung der Gynäkologie des Soranos von Ephesos* (Leipzig, 1910)). For a full discussion of this practice, see N. Horsfall, 'Numanus Remulus:

Ethnography and Propaganda in Aen. ix. 598 f', *Latomus*, 30 (1971), 1110 n. 2. On the curative effect of the different sorts of wine see e.g. Hippocrates, *Reg.* 50–2.

[13] Schmidt (n. 9 above), 134.

[14] Paus. 3. 8. 9 (trans. P. Levi (Penguin 1971)). Cf. also Xen. *Hell.* 3. 3; Plut. *Ages.* 3. 3–5; *Lys.* 22. 4.

[15] See A. Brelich, *Gli eroi greci, un problema storico-religioso* (Rome, 1958), 245; J. Bremmer, 'Medon, the Case of the Bodily Blemished King', in *Perennitas, Studi in onore di A. Brelich* (Rome, 1980), 69–70.

King Archidamos (361–338 BC) was fined by the Spartans for marrying a very small wife who would supply them with kinglets.[16]

Athens

Similar legislation does not seem to have existed in Athens. Plato and Aristotle thus postulate that the ideal State should control the size of the human population following the Spartan model. They both demonstrate that mental and moral achievement is connected with physical strength; men must produce vigorous offspring to guarantee the survival of a strong élite. The ideal family must be small, if possible limited to a single son.

In the *Republic* of Plato, for example, the conventional family does not exist any longer. The State takes all decisions concerning the selection and education of the young who are brought up together. Severe eugenic measures are applied in order to keep the population of the Republic physically and mentally strong: all babies born from over-age parents (an age of 55 for the father and of more than 40 for the mother) or from inferior citizens must be put away.[17] As at Sparta, children born in the upper class of the citizens, the philosopher-guardians, are even more severely controlled: all newborn infants displaying deficiencies must be systematically suppressed, 'dispose(d) of in secret, so that no one will know what has become of them'.[18] The term defining these abnormal babies, ἀνάπηρα, is very vague and can designate physically as well as mentally handicapped children.

Plato does not fix a time limit for this selection, which was probably intended to happen within the first days of life. Nor does he explain the fate of the unwanted child clearly. He only states that the children are taken away to a remote place, that is ἀπόρρητος, ἄδηλος, and hidden, which is probably an euphemism for abandoning or killing them. This reserve suggests that his Athenian audience would not have welcomed so direct a mention of the killing of babies.

In the *Politics*, Aristotle prescribes similarly that 'measures must be taken to ensure that the children produced may have bodily frames suited to the wish of the lawgiver'.[19] Mating must be very strictly regulated (age of the partners, moment of intercourse, number of children). A law prohibits the rearing of deformed children, also vaguely defined by the term πεπηρωμένα.[20] Like Plato and Plutarch, Aristotle implies that abnormality is synonymous with physical weakness. He anticipates the reluctance of public opinion to accept such a regulation: if overpopulation cannot be resisted by means

---

[16] Plut. *De lib. educ.* 1. 2, *Mor.* 1d.

[17] Plato, *Rep.* 5. 9. 459d–461e. See also the measures prescribed in *Leg.* 6. 775. Cf. J. J. Mulhern, 'Population and Plato's Republic', *Arethusa*, 8/2 (1975), 265–81, and M. Moïssidès, 'Le Malthusianisme dans l'antiquité grecque', *Janus*, 36 (1932), 169–79.

[18] Plato, *Rep.* 5. 9. 460c (trans. P. Shorey (Loeb, 1930)).

[19] Arist. *Pol.* 7. 14. 2 (trans. H. Rackham (Loeb, 1932)).

[20] Ibid. 10: 'As to exposing or rearing the children born, let there be a law that no deformed (πεπηρωμένον) child shall be reared' (trans. ibid.).

of exposure because 'the regular customs hinder any of those being exposed', abortion shall be practised.[21]

Delcourt related the exposure of abnormal births in Greece to religious or superstitious necessity.[22] She thought that physical defects inspired a sacred horror, as in Mesopotamia, and were considered as evil omens, impurities which death alone could cancel.[23] There is, however, no text showing this connection in Greece. On the contrary, Plutarch, Aristotle, and Plato justify the exposure of deformed children only on demographic and eugenic grounds. Their laws aimed at ensuring the future of their community in creating a strong élite.[24]

These views must be taken cautiously because they come from philosophers. They were repeated by other Greek intellectuals of the Hellenistic period,[25] but they cannot be regarded as revealing for the popular opinion. No other document gives evidence for the practice of exposing abnormal newborn children in Greece. Very likely, exposure was a permitted custom, but not prevalent in Attica; indeed Athenians seem to have disapproved of it. Isocrates thus praises Athens on the ground that the exposure of infants, among other 'atrocities', is not practised there.[26] Children were loved, and Aristotle himself says in the *Rhetorica* that a man needs numerous children, male and female, to be happy.[27]

As Schmidt has shown, the myth of the lame god Hephaistos confirms this aversion to exposure. The newborn child was thrown down from Olympos by his mother Hera because of his 'disgraceful' deformity,[28] but this deed is described as a shameful act; Hera behaved as an abnormal mother, 'with a dog face',[29] deprived of feelings. The

---

[21] Ibid. See A. Preus, 'Biomedical Techniques for Influencing Human Reproduction in the Fourth Century BC', *Arethusa*, 8/2 (1975) 237–63.

[22] M. Delcourt, *Stérilités mystérieuses et naissances maléfiques dans l'antiquité classique* (Liège, 1938), 29–49. *Contra*: P. Roussel, 'L'exposition des enfants à Sparte', *REA* 45 (1943), 5–17.

[23] Cf. J. G. Février, 'Un Sacrifice d'enfant chez les Numides', *Mélanges Isidore Lévy*, i (Brussels, 1955), 161–70, for evidence of the sacrifice of a deaf child to Baal on a punic stela.

[24] For a discussion of the ideological use of these principles in modern totalitarian regimes, see Schmidt (n. 9 above), esp. 133 nn. 1–3, 151–2 for bibliog. (on the notion of 'lebensunwertes Leben' in the Third Reich).

[25] See e.g. Menander, fr. 656 K: 'There is nothing more wretched than a father, except another one who is father of more children' (trans. F. C. Allinson (Loeb, 1951)).

[26] Isoc. 12. 122–3. The characters of Middle and New Comedy also feel the need of justifying this practice by exceptional circumstances. Cf. La Rue

van Hook (n. 3 above), Bolkenstein (n. 3 above), and Schmidt (n. 9 above), 139–42.

[27] Arist. *Rh*. 1361ᵃ6. On the importance of children at Athens, see G. Raepsaet, 'Les Motivations de la natalité à Athènes aux Vᵉ et IVᵉ s. avant notre ère', *AC* 40 (1971), 80–110. For a demonstration of the demographic impossibility of a high rate of female infanticide, see D. Engels, 'The Problem of Female Infanticide in the Greco-Roman World', *CPh* 75 (1980), 112–20.

[28] *Hymn. Hom. Ap.* 314–16: 'But my son Hephaistos whom I bare was weakly among all the blessed gods and shrivelled of foot, a shame and a disgrace to me in heaven, whom I myself took in my hands and cast out so that he fell in the great sea' (trans. H. G. Evelyn-White (Loeb, 1914)); see also *Il.* 18. 396; Paus. 1. 20. 2. Cf. the similar fate of the ugly Priapus, who was rejected by his mother Aphrodite, as mentioned by the schol. on Ap. Rhod. *Argon*. 1. 932; see H. Herter, *RE* xxii. 2 (1954), s.v. Priapos, 1917.

[29] *Il.* 18. 396. Cf. Schmidt (n. 9 above), 151.

abandoned child survived. Hephaistos was reared by Thetis and the Nymphs, and had a successful destiny.[30]

The elimination of abnormal children probably happened naturally: the rate of infantile mortality was very high, and babies suffering from severe anomalies, such as hydrocephalic or acephalic births, siamese twins or cyclops, did not survive because of their delicate constitution. On the other hand, babies who are born with additional or missing limbs are stronger, and could have survived without special care; they were, however, probably exposed by their parents, as suggested by their absence from literary and medical descriptions. In this context, dwarfs benefited from a special positive circumstance: on the whole, growth disturbances cannot easily be detected at birth, but become progressively evident during infancy after the time allowed for exposure. Despite their slightly abnormal appearance, usually characterized by folded skin, overlarge skull, or very small size, newborn children so afflicted were accepted and reared as normal children. The community only realized with the passage of time that one of its members was physically different.

DISABLED CITIZENS

Education

Conjectures only can be made about the intellectual and physical education reserved for disabled children. A crippled child could not be trained in athletics as a normal child; he was perhaps set free to exercise by himself, or had a special tutor if his parents were rich. If his mental capacities were normal, he probably received the same intellectual education as other children (writing, arithmetic, and music). The case of the lame poet Tyrtaios, who is said to have been a schoolmaster at Athens, suggests that a malformation in the legs did not prevent a man from being educated.[31] However, Tyrtaios is also said to have suffered as a result of prejudice; Pausanias reports that he was 'not thought at all clever', and a scholion on Plato adds that he was despised by the Athenians, who sent him to the Laconians to evince their contempt for Sparta.[32] Thus his poetic talent was recognized only at Sparta.

Aristotle and Plato, for their part, often related physical abnormalities to mental deficiencies. For example, Aristotle expressed very negative judgements on the intellect of dwarfs: they do not, for example, reason properly, they have little memory, and need much sleep.[33] These views were related to the widespread ideal of καλοκἀγαθία, which associated physical beauty with moral and mental perfection. But this ideal did not prevent the recognition of true beauty; does not Plato himself describe Socrates as a

---

[30] For ancient sources on Hephaistos, see Preller/Robert, *GrMyth*, i. 174–84; Gruppe, *GrMyth*, ii. 1304–18; L. Malten, *RE* viii. 1 (1913), s.v. Hephaistos, 311–66.

[31] Paus. 4. 15. 6.
[32] Schol. on Plato, *Leg.* 1. 629 a.
[33] Arist. *Part. An.* 4. 10. 686ª30; *Mem.* 453ᵇ1; ibid. *Somn.* 457ª24. See the discussion below 217–19.

short, fat, but very wise man? In the *Protagoras*, he adds that no one should despise an ugly, short (σμικρός) or crippled man, for physical appearance is only due to luck.[34] In a society which paid such attention to physical abilities, it must, however, have been more difficult for abnormal persons to fulfil important functions in the polis.

Official Entry to the Community

At the age of 18, the young Athenian was officially registered in his deme and, says Aristotle, was enrolled in the ephebeia.[35] Very likely his physical abilities were examined by the Council, as were those of skirmishers, knights, and foot-soldiers fighting in the ranks of the cavalry.[36] This enrolment marked his entry into the community: the group of new ephebes took a solemn oath of fidelity to their country, and made an official circuit of all the temples of Athens. After two years of garrison duty, the ephebes could exercise their full rights as citizens. The only known condition for their registration was to have an Athenian father, and, after 451, an Athenian mother too.[37]

How did the disabled pass this key moment? If they had Athenian parents, they could be registered among their demesmen. But what happened if they could not be trained like the others because of a physical handicap? Were they exempted from the ephebeia, or did they perform another type of public service? Or were they seen as eternal children, or as women?

No Greek text mentions such a problem. However, exemption probably did exist. The only evidence is found on a papyrus from Egypt which dates from AD 220.[38] It lists sixty-six ephebes from Memphis and states that nineteen are exempted because of physical deficiencies: some are ὑπερμεγέθεις, i.e. out-sized, overgrown, the others are 'falling short of sight', i.e. suffering from impaired vision.[39] Despite their physical inability, these young men entered the ephebic corps officially, but probably did not take part in the agon.

A similar measure may have been applied to disabled ephebes at Athens. They were perhaps admitted into the group, but exempt from participation in the exercises. They might have been kept busy cleaning the weapons of their companions, cooking for them, or even entertaining them.

Participation in the Politeia

Could a disabled person hold an important public office? He could certainly attend the discussions in the Ecclesia, the public assembly, which was open to all citizens. All

---

[34] Plato, *Prt.* 13. 323d.
[35] Arist. *Ath. Pol.* 42.
[36] Ibid. 49. 1–2.
[37] Ibid. 26. 4. See e.g. D. M. MacDowell, *The Law in Classical Athens* (London, 1978), 67 ff.
[38] M. N. Tod, 'An Ephebic Inscription from Memphis', *JEA* 37 (1951), 86–99.
[39] Impaired vision is the most common infirmity justifying exemption from liturgy in Roman Egypt. Cf. N. Lewis, *The Compulsory Public Services of Roman Egypt* (Florence, 1982) (Papyrologica Florentina XI), 95 and 165–7.

magistrates, elected or drawn by lot, were submitted to a review, the δοκιμασία, by the Council.[40] Did this examination include a medical selection?

In his speech for an invalid, probably lame, Lysias mentions that specific functions required physical selection.[41] He reports that the Thesmothetai, the magistrates in charge of drawing lots for the archonship would disqualify the candidacy of an invalid. The *Etymologicum Magnum* specifies that those applying for religious offices, such as that of βασιλεύς, were physically examined.[42] The existence of an unwritten religious taboo is confirmed by the absence of any reference to disabled holders of sacred functions. No crippled magistrate is recorded at Athens, apart from the legendary Medon, the first Athenian archon, son of King Kodros, who was lame in one foot, as Pausanias reports.[43] The choice of King Kodros was, however, first violently opposed by his other son, Neleus, because of the lameness of Medon; the Delphic oracle finally supported Medon.[44] King Agesilaus of Sparta also was lame in one foot.[45]

Outside Attica, however, a few texts suggest that persons injured by accident or in war continued in office if their injuries were not too severe. For example, Herodotus tells us that the diviner Hegesistratos from Elis was practising at the battle of Plataea despite his artificial foot, made of wood, the result of a heroic escape from a Spartan prison.[46] In the third century, the Aetolians had a disabled strategos, Ariston, but he could not take part in campaigns any longer;[47] Philip of Macedon kept full power over his army, although he had lost an eye in war.

The respect of the Athenian community towards its maimed heroes can be inferred from a special provision created in the archaic period to sustain them financially when they returned home. Aristotle states that: 'There is a law which lays down that those who possess less than three minae and who are physically maimed, ἀδύνατοι, so as to be incapable of work, are to be examined by the council and to be given two obols a day for maintenance at public expense.'[48]

Plutarch assigns the law to Solon and adds that the provision was made specifically for soldiers injured in war.[49] The amount of the allowance seems to have changed with the course of inflation: in the first half of the fourth century, the orator Lysias claims for his client, an invalid, one obol a day;[50] in 336, Aristotle says that two must be given,[51]

---

[40] Arist. *Ath. Pol.* 45. 3, 55. 2–4, 56. 1, 59. 4, 60. 1. On the general procedure of the dokimasia, see MacDowell (n. 37 above), 167–9.

[41] Lysias, *Περὶ τοῦ Ἀδυνάτου*, 24. 13.

[42] *Etym. Magn.*, s.v. Ἀφελής.

[43] Paus. 7. 2. 1.

[44] On Indo-European parallels for bodily blemishes, see Bremmer (n. 15 above), 70–6.

[45] See n. 15 above.

[46] Hdt. 9. 37–8.

[47] Polyb. 4. 5. 1.

[48] Arist. *Ath. Pol.* 49. 4.

[49] Plut. *Sol.* 31. 2. See E. Ruschenbusch, Σόλωνος Νόμοι, *Die Fragmente des Solonischen Gesetzwerkes mit einer Text- und Überlieferungsgeschichte* (Wiesbaden, 1966), 124. Solon also made provision for war-orphans according to Diog. Laert. 1. 55. Cf. J. J. Buchanan, *Theorika. A study of Monetary Distributions to the Athenian Citizenry During the Fifth and Fourth Centuries B.C.* (New York, 1962), 1–3.

[50] Lysias, *Περὶ τοῦ Ἀδυνάτου*, 24. 8. 13.

[51] Arist. *Ath. Pol.* 49. 4.

while Philochorus, at the end of the fourth century, mentions five.[52] This law is so far unique in Greece, as Hands notes.[53]

Yet it should not give too positive a picture of the attitude of the Athenians towards their disabled compatriots. The regulation was not made to comfort the congenitally disabled, but to reward the sacrifice of valiant soldiers who had defended their polis. An anecdote concerning Alexander the Great, and reported by Diodorus, reveals that the return of maimed soldiers could be morally very difficult. Diodorus says that on his way to Persepolis, Alexander met a group of 800 Greeks who had been captured by the Persians years before.[54] Their appearance was horrifying. They were all mutilated by the cutting-off of nose, ears, hands, or feet. Alexander was deeply moved and promised to help them to return home. After deliberation, however, the Greeks decided to stay forever in Persia; they would not face, alone, the look of those who could not understand their distress: 'If they were brought back safely, they would be scattered in small groups, and would find their abuse at the hands of Fortune an object of reproach as they lived on in their cities. If however they continued living together, as companions in misfortune, they would find a solace for their mutilation in the similar mutilation of the others.'[55]

Thus, the place reserved for people suffering from congenital handicaps seems to have been regulated by unwritten laws which may have restricted their access to official positions. Some physical defects, however, were compensated by special powers. Since Homer, the blind, whether from birth or due to some transgression, were often associated with poetry, song, and prophecy. They appear as beings especially close to the gods, who have access to knowledge not available to the ordinary man.[56] Peculiar capacities were also granted to the lame: Hephaistos is a skilful smith and a magician.[57] Tyrtaios, lame and perhaps also one-eyed, was a gifted poet who moved the whole Spartan nation. Dwarfs were likewise credited with peculiar capacities to which literary sources only allude, as we shall see.

---

[52] Harpocration and *Suda* s.v. ἀδύνατοι.

[53] A. R. Hands, *Charities and Social Aid in Greece and Rome* (London, 1968), 100. See also H. Bolkenstein, *Wohltätigkeit und Armenpflege im vorchristlichen Altertum* (Utrecht, 1939), 273–4.

[54] Diod. Sic. 17. 69. 2. See also Curt. 5. 5–24; Just. *Epit.* 11. 14. 11–12.

[55] Diod. Sic. ibid. (trans. C. B. Welles (Loeb, 1963)).

[56] See A. Esser, *Das Antlitz der Blindheit in der Antike*[2] (Leiden, 1961) (*Janus* suppl. 4), esp. 96–103;

R. G. A. Buxton, 'Blindness and Limits: Sophokles and the Logic of Myth', *JHS* 100 (1981), 22–37, esp. 27–35. Cf. Dio Chrys. *Or.* 36. 10–11: 'Moreover all these poets are blind, and they do not believe it possible for anyone to become a poet otherwise' (trans. H. Lamar Crosby (Loeb, 1940)).

[57] Cf. *Il.* 8. 195, 15. 310, 18. 137 (weapons); Hes. *Theog.* 570–85, *Op.* 59–73 (Pandora); Paus. 1. 20. 2 (magical throne). For further sources, see Gruppe, *GrMyth*, ii. 1309–10, and above 198.

15

Human Dwarfs

WRITTEN EVIDENCE: WORKS OF FICTION

Literary sources on pathological short people go back to the fifth century BC. There are only a few allusions in works of fiction. Dwarfs do not occur as characters in the plots of Greek drama, not even in the comic works of Aristophanes, but this may be due to the chances of discovery. Aulus Gellius mentions that the word νᾶνος was used in one of the lost plays of Aristophanes, the Ὁλκάδες, or *Merchant-ships*, but he does not say in what context.[1] Two extant plays use physical shortness as a comic element. In the *Pax*, Aristophanes viciously lampoons the three sons of Carcinos who were talented dancers, but of small stature; the author portrays them with irony as 'home-bred quails, dwarfish dancers (νανοφυεῖς) with hedgehogs' necks, snippets of dungballs, hunters after gimmicks'.[2] In the *Vespae* he compares them to tiny animals, spiders, prawns, crabs, and wrens.[3] Pherecrates, in a fragment of οἱ Ἄγριοι or *Savages*, also insists on the smallness of the three brothers which contrasted with the size of their ambitions.[4] Their parts may have been played by young boys, as Sommerstein suggests,[5] but if available, real dwarfs would have fitted such roles perfectly, enhancing the buffoonery of the characters. However, no record of such actors is known.

No other literary text can be related to pathological dwarfs, but stories about people with markedly thin and small bodies may indicate the qualities and defects which were traditionally associated with a short size. Thus physical smallness provoked laughter and gave rise to many scornful jokes.[6] Athenaeus reports that the poet Philetas of Cos, tutor of Ptolemy II, was so thin, λεπτός, that 'he had to wear on his feet balls made of lead to keep him from being upset by the wind',[7] while the Athenian Philippides, compared

---

[1] Gell. *NA* 19. 13. 3 (Kassel/Austin iii (2), fr. 441).
[2] Aristoph. *Pax*, 790.
[3] Aristoph. *Vesp.* 1500–13.
[4] Sch. on Aristoph. *Vesp.* 1509; *CAF* i. 149, no. 14.
[5] A. H. Sommerstein (ed.), *Wasps* (Warminster, 1983), 246, v. 1501.

[6] On the use of physical defects in poetry, see F. J. Brecht, *Motiv- und Typengeschichte des griechischen Spottepigramms* (Leipzig, 1930) (Philologus suppl. 22, 2), 88–96, esp. 89–91.
[7] Ath. 12. 552b (trans. C. B. Gulick (Loeb 1933)); Ael. *VH* 9. 14. See the same motif in *Anth. Pal.* 11. 100 (Lucilius).

with a living corpse, or 'a plucked chicken' in the Middle and New Comedy, provoked the invention of a new expression, 'to be philippidized' for 'to be very very thin'.[8] The name Pataikos, which refers to Phoenician and Egyptian dwarf gods, may also transmit a negative image of dwarfs, associated with fraud and gluttony. Ancient authors derived it from ἀπατᾶν 'to cheat', or πατέομαι 'to feed on',[9] and it was used as a nickname for thieves and sycophants.[10]

Short stature, however, was not believed to detract from courage; as Brelich noted,[11] several Greek heroes, such as Odysseus, Ajax, son of Oïleus, and Tydeus, were relatively small, βραχεῖς, but brave and quick-footed fighters.[12] A well-developed sense of wit was also traditionally associated with physical smallness. The satirical poet Hipponax of Ephesus is said to have had an ugly, small, μικρός, and thin, λεπτός, body. This was compensated by his strong temper, ἀκρότονος, which could be redoubtable. He was caricatured in an insulting way by the sculptors Bupalos and Athenis, and as a revenge, he wrote such biting lampoons against his enemies that they committed suicide.[13] Socrates himself is described as a short, fat man, but unusually clever.[14] A fable attributed to Aesop illustrates this association of smartness and physical smallness:

After Zeus had fashioned men he told Hermes to put intelligence in them. Hermes made a vessel for measuring it and poured an equal quantity into every man. It was enough to fill the little men full, so that they became wise. But the dram was too small to percolate all through the bodies of the big men; so they turned out rather stupid.[15]

Very little is known about the life and the physical appearance of Aesop himself, but his wit seems also to have been associated on occasion with some physical peculiarity. Some late authorities report that he was malformed and extremely ugly, like Thersites,[16]

---

[8] Ath. 12. 552d–f. Cf. ibid. the story about the soothsayer Archestratus 'who was placed on a scale and found to have the weight of a penny'. See also *Anth. Pal.* 11. 88–94, 99–101, 104–9, 111.

[9] E.g. Orosius in *Etym. Magn.* s.v. πάταικος; cf. Hesychius, s.v. Γίγγρων and Εὐφράδης: πάταικος ἐπιτραπέζιος.

[10] See e.g. Aeschin. *In Ctes.* 3. 189; Dio Chrys. *Or.* 52. 9. Further references in W. Pape, *Wörterbuch griechischer Eigennamen* (Braunschweig, 1911), 1145, and P. M. Fraser and E. Matthews, *A Lexicon of Greek Personal Names*, i (Oxford, 1987), s.v. παταίκιον, παταικίων, πάταικος, 365.

[11] A. Brelich, *Gli eroi greci un problema storico-religioso* (Rome, 1958), 234–6. I am very grateful to N. Horsfall for this reference.

[12] See e.g. *Il.* 3. 193 (Odysseus), and N. Horsfall, 'Some Problems in the Aeneas Legend', *CQ* 29 (1979), 380–1, on Lycoph. *Alex.* 1243–4 (Odysseus,

νάνος); *Il.* 2. 527 (Ajax); 5. 800 (Tydeus); Eust. *Il.* 2. 511 (Mynias).

[13] Ath. 12. 552d; Ael. *VH* 10. 6; Pliny, *NH* 36. 11–14 (*foeditas voltus*). Cf. the English poet Alexander Pope (1688–1744), who was a very spirited short-statured man (slightly above 1.50 m.), and a hunchback.

[14] Plato, *Symp.* 215a–b.

[15] *Fab.* no. 153, 'Why giants are boobies' (trans. S. A. Handford (Penguin 1954)).

[16] Planudes, *Vit. Aesop.* (B. E. Perry, *Aesopica* (Urbana, Ill., 1952), 214–15, no. 2; 222–3, no. 30; 228, no. 56). For Thersites, see *Il.* 2. 210–20; 265–77. On these associations, often fictional, between physical appearance, character, and the nature of poetry, see M. R. Lefkowitz, *The Lives of the Greek Poets* (London, 1981), esp. 27 (Hipponax and Archilochus).

while the slight pictorial evidence suggest that he was seen as relatively short (pl. 38.4).[17]

This list of spirited short men may be longer. Plutarch reports that a man, called Pataikos, boasted that he had inherited Aesop's soul, which suggests to Flacelière that he was another fabulist.[18] His name, Pataikos, may either have described his size, or have stressed his association with the qualities of wit attributed to dwarfs.

MEDICAL TEXTS

The Hippocratic Text

A very small number of ancient medical texts describe congenital disorders.[19] The Hippocratic texts report only a few examples of congenital abnormalities, such as the birth of a defective foetus without a skeleton or of a child with the right arm adhering to his trunk.[20] No dwarf is mentioned, but a late fifth-century treatise, *De Genitura*, displays a theory of the causes of abnormal births which could apply to restricted growth. The author considers newborn infants with a small size (λεπτός), a weak constitution (ἀσθενής) and crippled (ἀνάπηρος).[21] This general terminology could be used for dwarfs; the verb πηρόω occurs in Aristotle's description of the process of reducing the embryo.[22] In this Hippocratic text, growth disturbances are related to the time of gestation. The embryo remains small and delicate if the womb is too narrow, or if the womb does not close well and loses the nutriment of the child.[23] This process is compared with the development of a cucumber placed in a vessel: the plant takes the shape and the size of its container during its growth.[24] An apparently normal woman can thus have many small and weak children if her womb is in fact too narrow. A newborn child can also be crippled if the mother is injured during pregnancy; the embryo is deformed where the womb has been damaged.[25] The author adds that malformed parents may either have normal children, or transmit their specific physical deformity; when a part of the body is defective, the humour which constitutes sperm is

---

[17] See above 169–70.

[18] Plut. *Sol.* 6; R. Flacelière (*Plutarque*, ii, Paris, 1961), 16 n. 1.

[19] See in general M. D. Grmek, *Les Maladies à l'aube de la civilisation occidentale* (Paris, 1983), 25 ff.; C. S. Bartsocas, 'An Introduction to Ancient Greek Genetics and Skeletal Dysplasias', in *Skeletal Dysplasias* (New York, 1982), 3–13; id. 'Goiters, Dwarfs, Giants and Hermaphrodites', in *Endocrine Genetics and Genetics of Growth* (New York, 1985), 1–18.

[20] *Epid.* 2. 19 (foetus carnosus); *Epid.* 5. 13 (right arm adhering to the trunk). Cf. also *Nat. Puer*, 31 (twins). For a few further examples of genetic dis-

orders in Hippocratic texts, see H. Haser, *Lehrbuch der Geschichte der Medicin und der epidemischen Krankheiten*, i, *Geschichte der Medicin im Alterthum und Mittelalter* (Jena, 1875) (repr. Hildesheim and New York, 1971), 204.

[21] *Genit.* 9–11. See the commentaries by I. M. Lonie, *The Hippocratic Treatises, 'On Generation', 'On the Nature of the Child', 'Diseases IV'* (Berlin and New York, 1981), 139–46.

[22] Arist. *Gen. An.* 2. 8. 749ᵃ.

[23] Hippoc. *Genit.* 9. 2, 10. 2.

[24] Ibid. 9. 3.

[25] Ibid. 10. 1.

weakened and can transmit the abnormality to the offspring.[26] The same reasoning is used by the author of *De Aere, Aquis, Locis* describing the inheritance of the Longheads, 'for the seed comes from all parts of the body, healthy seed from healthy parts, diseased seed from diseased parts'; thus 'bald parents have bald children, grey-eyed parents grey-eyed children, squinting parents squinting children, and so on with other physical peculiarities.'[27] According to this theory, one could infer that dwarfism was seen as a disorder either inherited from abnormal parents, or acquired during pregnancy. This conjecture is supported by similar statements in Aristotle's discussion of the causes of dwarfism.

Aristotle

In the *De Generatione Animalium*, Aristotle defines monstrosities as 'a class of things unlike the parents', and 'contrary to Nature, not any and every kind of Nature, but Nature in her usual operations'; these imperfect beings are characterized by the presence of superfluous parts as well as of mutilations, 'since both deficiency and excess are monstrous'.[28] Aristotle gives many examples of monstrous animals; he describes external abnormalities, such as a chicken with four legs and four wings, a two-headed snake, a goat with a horn upon its leg, and internal monstrosities, such as animals born with an incomplete liver or without a spleen.[29] He notes that these disorders seem to be commoner in species producing many young, and thus usually less frequent in man 'for he produces for the most part one young one and that perfect'.[30] In places like Egypt, however, where women give birth to many children, monstrosities are more common.[31] Aristotle does not give examples of human physical abnormalities, apart from dwarfism which is described in other passages. As in the Hippocratic treatises, he thinks that congenital and acquired physical defects can be transmitted to the offspring, but not always.[32]

In several treatises, Aristotle considers dwarfs, called νᾶνοι and πυγμαῖοι, as forming, with children and animals, a category of inferior beings, characterized by disproportionate bodies. In the *De Partibus Animalium*, the dwarf, νᾶνος, is defined as a person with an overlarge upper body, 'big in the trunk, or the portion from the head to the residual vent', and a small lower part, 'where the weight is supported and where locomotion is effected'.[33] Similarly, 'all children are dwarfs', and creep instead of walking till their lower limbs grow and their trunk becomes proportionally smaller; animals whose upper part is too heavy are thought to have developed forelegs or

---

[26] Ibid. 11. 1. See Lonie (n. 21 above), 144–5.

[27] *Aer.* 14 (trans. W. S. Jones (Loeb, 1923)).

[28] Arist. *Gen. An.* 4. 4. 770ᵇ (trans. J. A. Smith and W. D. Ross, *The Works of Aristotle*, v (Oxford, 1912)). See also ibid. 4. 3. 769ᵇ.

[29] Ibid. 4. 4. 770ᵃ–771ᵃ.

[30] Ibid. 770ᵃ (trans. Smith and Ross).

[31] Ibid. 770ᵃ; *HA* 7. 4. 584ᵇ (twins).

[32] *Gen. An.* 1. 17. 721ᵇ; 1. 18. 724ᵃ; *HA* 7. 6. 585ᵇ. See also Pliny, *NH* 7. 50.

[33] *Part. An.* 4. 10. 686ᵇ, 1–20 (trans. A. L. Peck (Loeb, 1937)).

wings.[34] In the *De Historia Animalium*, Aristotle adds that dwarfs have an abnormally large phallus, like small mules, γίννοι.[35] Only in the *Problemata*, a late compilation of Aristotelian treatises, the author defines clearly two types of dwarfs: disproportionate, who 'have limbs of children', and proportionate, 'small as a whole', who look like the human miniatures depicted on shops.[36]

According to Aristotle, these disproportions between the parts of the body induce specific metabolic disorders. In the *De Partibus Animalium*, he states that dwarf-like beings (animals, children, and dwarf adults) are less intelligent than normal-sized men, because the weight of a heavy upper part disturbs reasoning, 'for weight hampers the motion of the intellect and of general sense.'[37] Memory is also inferior: the heaviness of a large head impairs the impulses of thought, which are therefore erratic.[38] Like children and 'big-headed types', dwarfs also tend to sleep too much; the warmth provided by food is cooled by their large upper parts, and the resultant flowing back of the cold paralyses their system for a longer time than usual, inducing deep sleep.[39] In the *Problemata*, Aristotle adds that persons with ill-proportioned bodies are more easily affected by fatigue, which is not equally distributed in their limbs.[40] These intellectual deficiencies are counterbalanced by other qualities that the author mentions vaguely by the term δύναμις: 'Even among human beings, children, when compared with adults, and dwarf adults, when compared with others, may have some characteristics, δύναμιν, in which they are superior, but in intelligence, νοῦν, at any rate, they are inferior.'[41]

As in the Hippocratic texts, Aristotle relates the causes of growth disturbances to the time of gestation, and adds possible external causes. In *De Generatione Animalium*, Aristotle states that 'human dwarfs . . . become deformed in their parts and stunted in their size during the time of gestation, and thus are comparable with μετάχοιρα (malformed pigs) and γίννοι (mules)';[42] the deficiency was perhaps attributed to a too narrow uterus, which would conform to the Hippocratic theory. In the *De Historia Animalium*, dwarfism is more precisely attributed to a disease, νόσημα, during gestation, whose nature is not specified.[43] Aristotle may have meant a deficiency of the sperm, since he states in *De Generatione Animalium* that a lack or an excess of heat of the sperm can produce deformed offspring.[44] In the *Problemata*, he adds that growth disturbances may also occur after birth, during early infancy, because of a lack of food or space, which produces specific conditions: inadequate food causes 'imperfect' dwarfs, with a disproportionate body, like that of children, while a shortage of space produces proportionate dwarfs or 'pygmies'.[45] Aristotle illustrates this theory by describing how

---

[34] *Part. An.* 4. 10. 689[b] ff. (quadrupeds), 695[a] (birds).

[35] *HA* 6. 24. 577[b].

[36] *Pr.* 10. 12. 892[a].

[37] *Part. An.* 4. 10. 686[a–b] (trans. Peck).

[38] *Mem.* 453[b].

[39] *Somn.* 457[a].

[40] *Pr.* 5. 22. 883[a].

[41] *Part. An.* 4. 10. 686[b], 23–7 (trans. Peck).

[42] *Gen. An.* 2. 8. 749[a] (trans. A. L. Peck (Loeb, 1942)).

[43] *HA* 6. 24. 577[b].

[44] *Gen. An.* 2. 8. 749[a].

[45] *Pr.* 10. 12. 892[a].

some people 'try to reduce the size of animals after birth, for instance by bringing up puppies in quail cages',[46] where their limbs progressively crush and bend. Aristotle does not specify if such treatment was also applied to humans, but later Longinus reports that human dwarfs were kept in boxes, γλωττόκομα, in order to stunt their growth, which suggests that the practice did exist, if only in Roman times.[47]

Aristotle thus appears to be the first author who comments on restricted growth, and uses his observations in discussions of human physiology. His theory is, however, incomplete. It concerns essentially short-limbed dwarfs, and their description is simplified. Aristotle does not mention the special development of their facial features, nor the fact that their upper limbs are often shortened. This shows that Aristotle depended at least as much on theory as on observation; he was elaborating theoretical principles on the development of living beings, and included dwarfs mainly because of their strikingly large upper part.[48] This may partly explain why he does not mention other types of restricted growth, such as short-trunk dwarfism, which would not fit in his discussion. This absence may also be due to the fact that these disorders are less common, and that affected children died early. Yet his statements about the secondary disorders of dwarfs are significant. They reflect folk-beliefs which appear in myths involving dwarfs: short people are associated with children and animals, as in the myth of pygmies and Cercopes, they are less intelligent than normal-sized men, but they have peculiar sexual capacities. This last statement, based upon no physical reality, had a long life: it was repeated by scholiasts and lexicographers, and has been echoed in literature and in iconography up to modern times.[49]

Dwarfs do not occur in medical texts of later periods, not even in the works of Galen. No inscription mentions dwarfs, although votive inscriptions recording cures describe a wide range of diseases; at Epidauros, for example, no extant inscription refers to restricted growth, nor to related disorders.[50] This may be related to the fact that, as in Egypt, short-statured people were credited with special capacities which were valued positively; these capacities could not be accounted for in the traditional categories of classical literature, but could be shown in iconography.

On a more general level, the scarcity of medical accounts on genetic anomalies may betray a lack of interest in the pathology of children, as Grmek suggests.[51] Because of

[46] Ibid. (trans. W. S. Hett (Loeb, 1936)).

[47] Longinus, *Subl.* 44. 5: 'Not only do the cages in which they keep the pygmies or dwarfs, as they are called, stunt the growth of their prisoners, but their bodies even shrink in close confinement' (trans. W. H. Fyfe (Loeb, 1927)).

[48] Cf. G. E. R. Lloyd, *Early Greek Science: Thales to Aristotle* (London, 1970), 118 ff.

[49] See e.g. Hesychius, s.v. νᾶνος. ἐπὶ τῶν μικρῶν. ὡς νᾶνον καὶ αἰδοῖον ἔχοντα μέγα. οἱ γοῦν νᾶνοι μεγάλα ἔχουσιν αἰδοῖα. See also *Suda* and Photius, s.v. νάνος or νᾶνος. On the modern association of dwarfs with sexuality, see e.g. the personage of the dwarf in the short story by M. Tournier, 'Le Nain rouge', in *Le Coq de Bruyère* (Paris, 1978) (Folio), 103–21.

[50] See e.g. K. Herzog, *Die Wunderheilungen von Epidauros* (Leipzig, 1931) (Philologus suppl. 22. 3), esp. 98–105.

[51] Grmek (n. 19 above), 21–30.

the high rate of infantile mortality, newborn children with severe abnormalities could not be observed over a long period; most will have died very early, either because of metabolic disorders, or because they were abandoned, following the practice of exposure. It is also probable that parents kept such births secret, or did not mention the cause of the premature death of their baby.

ICONOGRAPHY: GREEK MAINLAND

Distribution

The earliest representations of human dwarfs may be identified in two sixth-century Attico-Boeotian and Corinthian paintings (G *d* 1–*d* 2; pl. 41, fig. 15.1), but most pictures are found on Athenian vessels of the second half of the fifth century (G 3–23; pls. 42–55). The higher frequency of the depictions after 450 BC conforms to an important change in the stock scenes of painters whose interest in scenes of everyday life and in the rendering of naturalistic poses and anatomy increased perceptibly at that period.

Short-statured people seem to have been familiar figures in all classes of the population. Their pictures are found on vessels of extremely varying quality. Besides expensive vases, finely decorated with elaborate drawings, like the bell krater in private hands (G 23; pl. 55.1) or the Peytel aryballos (G 5; pl. 44), we find vessels of more common shapes and of a lesser quality, such as the skyphoi in Munich, in Yale, and in the Louvre (G 3, 16, 22; pls. 42, 51, 54), or the pelike in St Petersburg and the chous in Dresden (G 10, 14; pls. 47.2, 49.2).

Unlike pygmies, dwarfs are not depicted on vases normally used by women, such as hydriae or pyxides. As a rule, they decorate wine-vessels, such as cups (7), skyphoi (3), choes (2), oinochoe (1), stamnos (1), krater (1), and rhyton (1), and vessels for oil, such as pelikai (4), and the Peytel aryballos, which probably contained some unguent (G 5; pl. 44).

From Stereotype to Portrait

These pictures of dwarfs seem to render a variety of bodies as though each figure referred to a different real individual. However, the living models who inspired the painters cannot have been many. Some dwarfs must have been very familiar in the potters' area. It may be possible to rediscover their silhouettes by comparing the pictures of human and mythical dwarfs made by contemporary painters, sometimes working in the same studio. Striking similarities appear when grouping the extant depictions of dwarfs and pygmies according to J. D. Beazley's attributions to workshops. Thus in 450, in the group of Polygnotos, the Epimedes Painter gave to his pygmy (G 70; pl. 65.1) facial features closely resembling those of the dwarf depicted by the Peleus Painter (G 9; pl. 47.1). It is likely that we have here two depictions of the same man. At the same period, the workshop of Sotades produced two rhyta and three figure-vases showing very similar achondroplastic pygmies, with very short limbs, distended abdomens and large heads (G 64–8; pls. 63.3, 64). These figures may render the anatomy of an individual who was also depicted on skyphoi in the Louvre and in Yale by painters working in the same studio (G 3, 22; pls. 42, 54). In 420, the Phiale Painter worked with the Dwarf Painter in the workshop of the Achilles Painter; they too may have shown the physical characteristics of the same short-limbed individual (G 8, 18; pls. 46.2, 52). At that period, four vases depict twin dwarfs in a scene, eating (G 12; pl. 48.2), dancing (G 15, 20; pls. 50.1, 53.2) and training (G 10; pl. 47.2) together. This peculiar feature might reflect reality, either the presence of two dwarf brothers, or of two dwarfs of the same age, at Athens. It may also be a playful scheme which aimed to emphasize the oddity of the scene; the workshop of the Washing Painter, which produced two twin figures of dwarfs (G 10, 12; pls. 47.2, 48.2), also depicted twin satyrs.[52]

Three pictures in particular may represent attempts at portraiture, within the limits of the iconographic conventions defined above.[53] The achondroplastic dwarf depicted on the Peytel aryballos probably shows a specific individual (G 5; pl. 44b). The scene is unique: it is the only known picture of a Greek clinic, indicated by cupping devices hanging on the wall, and by a row of patients waiting for their turn. The doctor, who is very young, is depicted in the middle of the scene, treating a patient (blood-letting?). The picture may have been ordered by the doctor himself as a container for some special ointment which he could leave on a shelf as a personalized decoration. The presence of the dwarf, who may be his servant, or another patient, adds a realistic touch to the scene. The short man dancing on a stamnos in Erlangen may also be a definite person (G 9; pl. 47.1). His achondroplastic physiognomy is very carefully studied, and he has a name, Hippokleides, which can be reconstructed above his head.[54] The dwarf depicted on a cup in Athens (G 13; pl. 49.1) is not individualized by a name, but his morphology differs strikingly from the dwarf stereotype. He has no beard, but a moustache alone, which is very uncommon, his limbs are extremely thin, and his skull inordinately elongated.

Caricature and Parody

Red-figure representations of dwarfs bear witness to the development of an increasing interest in the rendering of individuality and of caricature, its corollary. To depict a physiognomy in detail provides means of making fun of it, as Raeck notes.[55] Yet, despite the mastery of new drawing techniques, Greek painters do not seem to have

---

[52] See e.g. the two satyrs playing on a potter's wheel on an oinochoe, London, BM E 387; *ARV* 1134.10, *Add²* 333; L. Deubner, 'Spiele und Spielzeug der Griechen', *Die Antike*, 6 (1930), 166, fig. 11.

[53] See above 173–4.
[54] On this name, see below 224.
[55] Cf. W. Raeck, *Zum Barbarenbild in der Kunst Athens* (Bonn, 1981), 37–9, 218–19.

exploited the burlesque possibilities offered by the unusual anatomy of short people. No picture indulges in rendering dwarfs as ugly or ludicrous creatures. Vase-painters reveal, on the contrary, a very ingenious sense of humour: when a dwarf is funny, it is not only through his physical aspect, but by reference to existing genre scenes or well-known anecdotes. Painters played with images, like writers with words, and some composed parodies by replacing normal protagonists with unusual figures; the procedure is best known with satyrs or apes.[56] These visual references partly escape us because of the gaps in the material, but some allusions can be suggested. Thus, in four scenes dwarf figures may represent standard stock figures in a ludicrous way, as living caricatures. These examples of visual satires may be identified by comparing our pictures with extant similar scenes.

The scene on the St Petersburg pelike seems to ridicule the numerous red-figure pictures of training athletes (G 10; pl. 47.2). It is perhaps one of the first illustrations of rising criticism of the new training methods, which no longer aimed to produce harmonious bodies, but rather professional athletes.[57] The Washing Painter may have used the image of malformed athletes to satirize a specialization, the pankration, an art which mingled wrestling and boxing. One of the dwarfs is training on a punch-bag of the larger sort, as prescribed by Philostratus.[58] Their stocky bodies also evoke the ideal proportions of the pankratiast, 'small of build', bulky, and with an overdeveloped upper body, as for boxers,[59] and may caricature the physical type produced by too intense exercise. A meat diet, which was believed to give greater strength, was part of the new art of training. The coarse faces of the dwarfs, with stupid expressions, may be intended to render the disastrous effects of this regimen on intelligence.[60] The scene is very uncommon, since satyrs, and not mortals, are usually shown aping athletic activities.[61] It is paralleled only by a pelike in New York, where an unconventional pair of boxers seems to be ridiculed: the couple is ill-assorted in size (one athlete is very small), and they exercise in time to the music of pipes, heads up, with very inspired expressions.[62]

On the Peytel aryballos, the depiction of a dwarf chatting with a man may parody conventional homosexual courting scenes (G 5; pl. 44a, b). In the work of the Clinic

---

[56] See in general W. Binsfeld, *Grylloi. Ein Beitrag zur Geschichte der antiken Karikatur*, Diss. Cologne, 1956.

[57] On these criticisms see e.g. E. N. Gardiner, *Greek Athletic Sports and Festivals* (London, 1910), 122–45; M. I. Finley and H. W. Pleket, *The Olympic Games: The First Thousand Years* (London, 1976), 113–27. On athletes' deformities, see Lucilius in *Anth. Pal.* 11. 75–8, 81, 258.

[58] Philostr. *De Gymnastica*, 57.

[59] See Philostr. *De Gymnastica*, 36. See e.g. Melissos of Thebes in Pind. *Isthm.* 4. 49 ff. On boxers' similar deformity, see Xen. *Symp.* 2. 17–18.

[60] Cf. e.g. Xen. *Mem.* 1. 2. 4; Aristoph. *Pax*, 33–4.

[61] See e.g. the satyr athletes on a krater in Munich, Antikensammlungen, 2381; *ARV* 221.14, *Add²* 198; Boardman, *ARFH* i, fig. 163.

[62] Pelike, New York, MMA 49.11.1; *ABV* 384.19, *Add²* 101. Parody: D. v. Bothmer, 'Attic Black-Figured Pelikai', *JHS* 71 (1951), 41–2; merely a clumsy rendering: H. A. Harris, *Sport in Greece and Rome* (London, 1972), fig. 19. Cf. J. P. Cèbe, *La Caricature et la parodie dans le monde romain antique des origines à Juvénal* (Paris, 1966) (BEFAR 206), 350–1, on pygmies caricaturing athletes.

Painter, the older lover is commonly shown in the same pose as the man on the aryballos: he rests his weight on a walking stick, and converses with his Eromenos.[63] Usually he offers a living hare to his lover, but the animal may be dead, like that carried by the dwarf.[64] On the Peytel aryballos, a humorous effect is created by the presence of a stumpy hairy dwarf at the place of the expected handsome Eromenos. Moreover, the dwarf is in the same age-group as his Erastes, while in courting scenes one of the lovers is older. The unusual location of this meeting enhances the comedy of the parody; the two men are shown not in a palaestra, a place suited to the appraisal of naked bodies, but in a clinic, among sick or injured people.

Another parodic scene may be identified on a skyphos in Munich (G 16; pl. 51a, b). A female dwarf stands, naked but with earrings, a large headband, and a wreath around her bun. She is holding a skyphos in her hand as does, for example, another naked woman on a cup by Onesimos.[65] She is not involved in a simple komos, but takes part in a cult scene, as the scene depicted on the reverse suggests. A tall phallus, enlivened with wings and an eye, is crowned with an offering basket.[66] It is erected in honour of a fertility god, very likely out of doors, as indicated by the two branches sprouting out on each side. Before the phallus, a skyphos is set on the trapeza.[67] This combination of ritual objects characterizes Dionysiac rites. On pictures of the festival of the Lenaea, the image of the god is shown as a dressed pillar with a mask, set behind a sacrificial table where female worshippers ladle wine from stamnoi into skyphoi. These women are dressed but wear garlands like the dwarf in Munich. It is also not uncommon to see women with models of male genitals. On a krater by the Pan Painter, a naked woman carries a phallus as large as herself, adorned with an eye on the glans as on the Munich skyphos.[68] In a cup in the Villa Giulia, two women, one dressed, the other naked, but both like the dwarf with a sakkos binding their hair, dance around a tall phallic pillar.[69] These models were used on many festal occasions like the Country Dionysia, but when associated with naked women they may refer to the festival of the Haloa, a fertility cult

[63] By the Clinic Painter: men courting women: cup, Florence, Museo Etrusco, 81.602; ARV 810.24; CVA Italy 30, Florence 3, pl. 103.1–2. Man courting a youth: cup, Paris, Cabinet des Médailles, 812; ARV 811.37; A. De Ridder, Catalogue des vases peints de la Bibliothèque Nationale (Paris, 1902), 471, fig. 110; cup, Berkeley, University, 8.922; ARV 811.38; CVA USA 5, California 1, pl. 26.

[64] See e.g. the dead hare on the skyphos in Paris, Louvre, A 479; G. Koch-Harnack, Knabenliebe und Tiergeschenke (Berlin, 1983), 90–1, fig. 24. See also the hydria in Salerno, Museo Nazionale, 1371; ARV 188.67, Para 341, Add² 188; A. Greifenhagen, Neue Fragmente des Kleophradesmalers (Heidelberg, 1972), 42–4, pl. 26.2.

[65] Cup, Brunswick, Maine, Bowdoin College, 30.1; ARV 328.114, Add² 216; Boardman, ARFH i, fig. 225.

[66] On offering baskets, see L. Deubner, 'Hochzeit und Opferkorb', JDAI 40 (1925), 210–23; J. Schelp, Das Kanoun. Der griechische Opferkorb (Würzburg, 1975).

[67] On sacrificial tables, see C. Goudineau, ἱεραὶ τράπεζαι, MEFR 79 (1967), 77–134, esp. 97–103 on the Lenaea.

[68] Krater, Berlin, SM (East) 3206; ARV 551.10, Add² 257; Deubner, Att. Feste, pl. 4.2; Boardman, ARFH i, fig. 342.

[69] Cup, Rome, Villa Giulia, 50404; ARV 1565.1, Add² 388; Deubner, Att. Feste, pl. 4.1.

dedicated to Demeter and to Dionysos.[70] Ancient authors tell us that part of this festival was confined to women, who held a secret komos in the telesterion of Eleusis. Models of sexual organs were carried, and perhaps women danced around them on the ἅλως, 'the round threshing floor'.[71] As Robertson has suggested, the painter probably aimed to caricature these secret female performances; he represented a stunted woman with a deformed body to suggest that their nakedness could reveal ugly forms.[72]

The image-play implied by the painter may be more subtle. The painter may use a dwarf figure to parody a literary anecdote. The dwarfs dancing on a cup in Todi (G 20; pl. 53.2) are performing unbridled antics recalling the famous dance which made the Athenian Hippokleides lose his marriage; one walks on all fours, holding a cup in his hand, while his companion stands on his hands and shakes his legs in the air. Herodotus gives a detailed account of the story.[73] Cleisthenes, tyrant of Sicyon (665–565 BC), wanted to give the best possible husband to his daughter Agariste, and invited all deserving suitors to Sicyon for a trial of their qualities as potential son-in-law. The period of probation ended with a sumptuous banquet where Cleisthenes was to declare his choice. An Athenian, Hippokleides, seemed to be specially in favour, but his behaviour at the end of the dinner ruined his chances. He called to the flute-player, began dancing and soon asked for a table:

When it was brought, he mounted upon it and danced first of all some Laconian figures, then some Attic ones; after which he stood on his head upon the table, and began to toss his legs about. Cleisthenes, notwithstanding that he now loathed Hippokleides for a son-in-law, by reason of his dancing and his shamelessness, still, as he wished to avoid an outbreak, had restrained himself during the first and likewise during the second dance; when, however, he saw him tossing his legs in the air, he could no longer contain himself, but cried out: 'Son of Tisandrus, thou hast danced thy wife away!' 'What does Hippokleides care?' was the other's answer.[74]

Herodotus adds that this reply became a famous proverb. It was repeated not only at Athens but remained in favour in Roman times.[75] The Todi Painter may have chosen to depict one of his dwarfs with legs up in the air to suggest that famous retort. It might be noted that the Erlangen dwarf (G 9; pl. 47.1), who dances on a table too, is called Hippokleides, probably as a reference to the anecdote, as Lippold suggests.[76]

[70] On the Haloa, see Deubner, *Att. Feste*, 60–7; H. W. Parke, *Festivals of the Athenians* (London, 1977), 98–100; A. C. Brumfield, *The Attic Festivals of Demeter and their Relation to the Agricultural Year* (New York, 1981), 104–31; E. Simon, *Festivals of Attica, an Archaeological Commentary* (Madison, Wis., 1983), 35–6.

[71] Simon, ibid. 35.

[72] M. Robertson, 'A Muffled Dancer and Others',

in A. Cambitoglou (ed.), *Studies in Honour of Arthur Dale Trendall* (Sydney, 1979), 130.

[73] Hdt. 6. 126–30.

[74] Trans. G. Rawlinson (Chicago, 1952).

[75] *Suda*, s.v. Οὐ φροντὶς Ἱπποκλείδη. See also various authors in the *Corp. Paroem. Graec.* (e.g. Zenob. 5. 31, Diogen. 7. 21).

[76] G. Lippold, 'Zu den Imagines Illustrium', *MDAI(R)* 52 (1937), 44–7.

SOCIAL AND RELIGIOUS STATUS

What can we infer from iconography about the status of short-statured people at Athens? The community may have reacted in two possible ways: either by rejecting them, because their physical malformation was perceived as a disturbing phenomenon threatening the order of the world, or, as in Egypt, by integrating them within the socio-religious system of the state. In other words, did dwarfs have marginal positions, such as that of slaves or professional jesters, or did they have the same status as normal citizens?

Most scholars adopt the first theory. The Greek concept of physical beauty was defined in terms of proportion, of συμμετρία and of εὐρυθμία, of 'commensurability' between the parts of the body.[77] The ideal set of proportions, which probably derived from Egyptian traditions, was fixed in workshop manuals as early as the sixth century BC.[78] Diodorus reports that at that time Theodorus and Telecles of Samos made a statue of a Pythian Apollo, which: 'in accordance with the technique of the Egyptians, [was] divided into two parts by a line which runs from the top of the head, through the middle of the figure to the groin, thus dividing the figure into two equal parts.'[79]

Dwarfs, with their imperfect, ill-proportioned bodies, may have been regarded as inferior beings. They appear in the myth of the pygmies and of the Cercopes as primitive, impious, and asocial human beings, which suggests that Athenians may have assimilated them to the monstrous species who lived at the dawn of the world and were progressively eliminated. Empedocles describes these imperfect primordial beings thus: 'heads . . . without necks, bare arms . . . without shoulders, . . . bull-headed men, . . . with male and female nature combined.'[80]

It is perhaps only by chance that no scornful remark about dwarfs has yet been found. Athenians might have treated them as living curiosities, as the Sybarites did according to Timaeus: 'Another national custom arising from their luxurious habits was to keep tiny manikins, ἀνθρωπάρια μικρά, and owlish jesters, σκωπαῖοι . . . also Maltese lap-dogs, which accompany them even to the gymnasia' and 'took delight in Melitê puppies and human beings who were *less than human*'.[81]

Similar practices may have existed in classical Athens. It is often a very delicate task to define the status of a figure in a vase painting. Himmelmann pointed out the following criteria for the identification of slaves which seem, at first, to apply well to dwarfs:

[77] See J. J. Pollitt, *The Ancient View of Greek Art* (New Haven, Conn. and London, 1974), esp. 14–23, 143–54, 160–2.

[78] See e.g. the renowned Canon by Polyclitus; H. Diels and W. Kranz, *Die Fragmente der Vorsokratiker*[6], i (Berlin, 1951), 40, B 2.

[79] Diod. Sic. 1. 98. 5–9 (trans. Pollitt, n. 77 above, 13). See also Arist. *Part. An.* 4. 10. 686[b], 7–8: 'In man, the size of the trunk is proportionate to the lower portions' (trans. Peck).

[80] Empedocles, fr. 50(57)–52(61) (trans. M. R. Wright, *Empedocles: The Extant Fragments* (New Haven, Conn. and London, 1981)).

[81] *FGrH* 556, F 49 = Ath. 12. 518e–f (trans. C. B. Gulick (Loeb, 1933); italics added).

slaves are usually shown smaller than their master; they are often naked, probably to indicate their physical value; they may be caricatured by comic physical characteristics or poses; their occupation is servile: they care for their master or for his possessions.[82] Short, often naked, malformed, dwarfs seem by nature predisposed to belong to this social group. On several vases, they also seem to have servile positions: as personal attendants, and as dancers in komos contexts, which associates them to the world of private entertainment.

Personal Attendants

In the archaic period, statuettes of kourotrophic dwarf demons had been dedicated for the protection of small children. So far, we have found no expression of this belief in later periods, apart from fourth-century terracottas of old, silenos-like pedagogues with dwarfish proportions.[83] No vase painting shows a real (or mythical) dwarf caring for babies or for the young, except two scenes which depict dwarfs in the company of an adolescent and of young men.

On a cup in Ferrara, a dwarf, characterized by his overlarge head and his short limbs, follows a child as short as himself (G 7; pl. 46.1). The dwarf is probably the servant of the child. The two figures are contrasted by their morphologies and clothing; the dwarf is naked and exhibits his malformed body, the child is well proportioned and wrapped in his cloak. The scheme is frequent in vase paintings: the master walks forward, followed by his servant who carries some personal things. When the slave is male, his status is often stressed by his nakedness. On the Ferrara cup, the dwarf replaces the familiar pedagogue, the old trustworthy slave who accompanies the child to school and carries his material, here perhaps the bundle depicted on his left shoulder (or is it his own cloak?). A similar scene is depicted on a lekythos in New York: the child, also draped in a long cloak, is escorted by an old slave with white hair, dressed in a short garment, who leans on a walking-stick and holds the boy's lyre.[84] The pedagogue also supervised the work of the child at school and at home, but no depiction shows a dwarf in this function too.[85]

The dwarf walking with two young men on a pelike in Boston may have a similar escorting role (G 8; pl. 46.2). He is older than his two companions, as his beard and his half-bald head indicate, and he leads by the collar a large dog with menacing teeth, as

---

[82] N. Himmelmann, 'Archäologisches zum Problem der griechischen Sklaverei', *Abh. Mainz*, 13 (1971), 614–57.

[83] See above 204.

[84] Lekythos, New York, Parke-Bernet Gallery; F. A. G. Beck, *Album of Greek Education* (Sydney, 1975), fig. 68. See also the amphora, Baranello Museum, 85; Rühfel, *Kinderleben*, 45, fig. 23.

[85] For ancient sources on pedagogues (usually foreigners), see E. Schuppe, *RE* xviii. 2 (1942), s.v. Paidagogos, 2375–85. See also Plato, *Alc.* 1. 122 a–b about the old Thracian pedagogue whom Pericles offered to Alcibiades. Plut. *De lib. educ.* 1. 6, *Mor.* 4a, prevents parents from this common practice: 'so that the children may not be contaminated by barbarians and persons of low character', and adds the proverb: 'If you dwell with a lame man, you will learn to limp' (trans. F. C. Babbitt (Loeb, 1927)).

though he ensured their protection. The group is perhaps going to the gymnasium, where the dwarf would care for the two men.[86] Dogs were allowed to enter; on a hydria in Berlin, for example, a Maltese dog is sniffing the ground of a palaestra,[87] and two larger animals, resembling that kept by the dwarf, are seated near their masters on a cup fragment in Rome.[88] The anatomy of the dwarf is very carefully rendered, yet not in a detrimental way, which would imply that he has an inferior status. The three men are individualized by distinct hairstyles, clothing, and attributes. The young man who walks in front has curly hair and is wrapped in a cloak which hides his arms; his companion has a fringe with long locks, he wears a normal long himation which leaves his right arm free, and holds a walking-stick. The dwarf is distinguished by his beard, moustache, and side-whiskers; he carries a chlamys on his shoulder, and also holds a knotty stick. His abnormal morphology appears as a supplementary individualizing element serving the realistic interest of the painter. The dwarf could be not a slave but a citizen. The scene brings to mind many similar pictures of young Athenians talking, in twos or threes, in a non-specific place, sometimes indicated as the palaestra. Most often, the scene is depicted on the obverse of the vase, as a minor subject. On the Boston pelike, the painter seems to have varied the conventional scheme to integrate the unusual figure of the dwarf, and the group constitutes the principal scene of the vase. The dwarf is perhaps going to the gymnasium, like a normal Athenian. He turns his head back and exchanges a look with the man who walks behind him, a look which suggests a reciprocal relationship, without scorn. The company of the dog, which may be his, stresses his integration into a familiar setting.

The scene depicted on a pyxis in the Louvre also associates a dwarf with young people (G *d* 1; pl. 41). The dwarf stands behind a group of young men and girls who seem to take part in a ball-game. The dwarf is naked, like the boys, whereas the girls wear long dresses. Two figures hold sticks evoking the clubs or bats used by boys on an archaic marble statue base at Athens.[89] A tall woman in a short chiton, holding a ball, dominates the scene: she wears a helmet and may be identified with Athena.[90] The scene may refer to the ball-game played by young girls, the Arrephoroi, and possibly also by boys, for Athena on the Acropolis.[91]

[86] For representations of athletes with young servants, see Rühfel, *Kinderleben*, 62–71, figs. 36–41. See e.g. the krater by Euphronios in Berlin, SM (West) 2180; *ARV* 13.1, *Add*² 152; Boardman, *ARFH* i, fig. 24.1–3; Rühfel, *Kinderleben*, fig. 36. See also the 'black' slave on an aryballos fr., Athens, Acropolis Museum, Acr. 874; B. Graef and E. Langlotz, *Die antiken Vasen von der Akropolis zu Athen*. ii (Berlin, 1933), 82, pl. 75; Raeck (n. 55 above), fig. 75.

[87] Hydria, Berlin, SM (West) F 2178; *ARV* 362.24; A. Greifenhagen, *Antike Kunstwerke*² (Berlin, 1966), 48, no. 48.

[88] Cup fr., Rome market; *ARV* 375.66; unpub. (Beazley archive).

[89] Athens, NM 3477 (team of young boys only); Harris (n. 62 above), 101–2, pl. 47; J. Boardman, *Greek Sculpture: The Archaic Period* (London, 1978), fig. 241.

[90] See S. Waiblinger, in *CVA* France 26, Louvre 17, 35–6.

[91] A ball-alley was named after the Arrephoroi on the Acropolis (with a bronze statue of Isocrates as a boy and hockey-player); Parke (n. 70 above), 143, pl. 61; Deubner, *Att. Feste*, 15 n. 3.

In two scenes, dwarfs attend women and substitute for their female servants, as though they were not seen as full men. On a pelike in Agrigento, a naked dwarf follows a woman dressed in a chiton and a long cloak (G 6; pl. 45). As maids do, he is balancing on his head a basket with a flat bottom, surmounted by a heap of food(?) and covered with a cloth.[92] The scene is very uncommon. Women are usually accompanied by female, and not by male slaves. Are the pair coming back from the market, or going to a party? Flat baskets may be full of fruit and loaves,[93] but they are not shown at a symposion. Or are they going to visit a tomb? On funerary lekythoi, the Sabouroff Painter depicted female servants carrying similar flat baskets containing offerings for the dead.[94] Yet these scenes appear on lekythoi only, and the maid brings accessories for the funerary cult, such as a phiale and a jug for libation. A fourth possibility is that the pair are going out to a festival. On a fragment of a chous in Greifswald, a naked boy carries a similar basket, flat at the base and round at the top, in a scene which may be related to an Athenian festival, as is suggested by the decorated jug standing at the right.[95] In the four interpretations, the dwarf has the same ambiguous status; he is naked, like a male servant, but in a female function: he is attached to a woman and behaves like a maid.

The short man on the Agora cup also takes the place of a maid (G 13; pl. 49.1). He stands beside a woman seated on a chair, her feet resting on a foot-stool. They are in her bedroom, as the head of her couch, bearing a large pillow, indicates. The position of the small man is unusual. Servants either stand behind the seat of their mistress, like the black slave holding out a mirror on a lekythos in Bucharest,[96] or they come towards her, bringing some toilet object, or they lace her shoes, like the girl on a pyxis by the Eretria Painter.[97] Visitors face their hostess. On the Agora cup, the servant stands in line with his mistress; the closeness of the two figures and their converging looks suggest that they have a special relationship. The smallness of the dwarf evokes that of the Erotes. In red-figure, these little winged servants are depicted helping their mistress, sometimes Aphrodite herself,[98] at her toilet, bringing a piece of her attire, like real slaves. These scenes became a genre only in the second half of the fifth century, but the

[92] See e.g. the young female servant carrying on her head a wine-skin behind her drunken mistress; skyphos, The J. P. Getty Museum, M. and W. Bareiss coll. 337; Rühfel, *Kinderleben*, 74, fig. 44.

[93] See e.g. the boy with a flat basket with food on an oinochoe, Tampa Flood Museum, Noble coll.; *ARV* 276.78 ter, 1635, 1705; S. P. Murray, *The Joseph Veach Noble Collection, Tampa Flood Museum of Art*[2] (Tampa, 1986), 45, no. 66.

[94] See e.g. the lekythos, Oxford, Ashmolean Museum, 1966.771; *ARV* 850.264, *Para* 424; *Select Exhibition of Sir J. Beazley and Lady Beazley's gifts to the Ashmolean Museum 1912–1966* (London, 1967), 99, no. 366, pl. LII.

[95] Chous fr., Greifswald, University, 361; A.

Hundt, K. Peter, *Greifswalder Antiken* (Berlin, 1961), 68, pl. 40. See also the small Pan carrying a basket full of food behind a woman holding a phiale on a krater, Vienna, Kunsthistorisches Museum, IV 175; K. Schauenburg, 'Pan in Unteritalien', *MDAI(R)* 69 (1962), 31, no. 69, pl. 16.1.

[96] Lekythos, Bucharest, G. and A. Magheru coll.; *CVA* Rumania 2, Bucharest 2, pl. 39.3, 7–9.

[97] Pyxis, London, BM E 774; *ARV* 1250.32, *Add*[2] 354; Rühfel, *Kinderleben*, 74–5, fig. 45.

[98] Cup, Germany, private; H. Hornbostel, *Kunst der Antike* (Mainz am Rhein, 1977), 330–1, no. 283. See also the hydria, Adolphseck, Schloss Fasanerie, 38; *CVA* Germany 11, Schloss Fasanerie 1, pl. 29.3.

theme may have existed earlier. An early fifth-century cup by Makron shows Aphrodite surrounded by four winged Erotes, her head covered with a veil like the woman on our cup.[99] This substitution Eros/dwarf, Aphrodite/ordinary woman suggests that the dwarf was seen as the human equivalent of the small Erotes for Aphrodite—not quite a man, but not as innocent as a child. His physical anomaly guaranteed his devotion and his respect. Such a slave may not have been suitable for a proper Athenian lady. Although she sits gracefully, holding a mirror, in a richly furnished room, the woman may well be a hetaira, as Robertson judiciously notes.[100] The attractive transparence of her chiton and the presence of her bed enhance her ambiguity. Would a painter depict an honest woman in such an enticing dress, in her bedroom, in the company of a dwarf, and on a drinking-vessel? Three scenes show women at home, the head covered with a similar short veil; two are also holding a mirror.[101] These respectable women, however, wear a himation across their chiton, and no bed is indicated; three pictures are found on vases made for women (one pyxis, one red-figure lekythos, and one alabastron) and not on party vessels. For Robertson, this short veil might be compared with that worn by brides, but it is conventionally pulled up with one hand, and brides are never depicted on cups.[102] The Agora veil resembles too that worn by a flute-player on a krater in Copenhagen, which would place the woman in the category of professional entertainers.[103] The painter perhaps honoured the beauty of a well-known hetaira in showing her in an elegant pose, with the traditional attributes of a lady at home (the short veil, the mirror), and with her original attendant. The festal garland on the head of the dwarf suggests that he accompanied her to the parties where the cup which bears their depiction was used. His dress and his walking-stick indicate his mobility, and may imply that the small man was also her discreet messenger.

Dwarf servants may have fulfilled other roles. For example, the dwarf on the Peytel aryballos (G 5; pl. 44b) may be the doctor's attendant. A funerary stela, now in Basle, of about the same period depicts the doctor seated, facing a young attendant holding a cupping-device.[104] The dwarf may similarly be shown performing his daily task, collecting the fees brought to the practitioner, here a dead hare. He perhaps also

[99] Cup, Berlin, SM (West) 2291; *ARV* 459.4, *Add²* 244; E. Simon, *Die Götter der Griechen*³ (Munich, 1985), 244, fig. 232.

[100] M. Robertson, 'An Unrecognized Cup by the Kleophrades Painter?', in L. Robert et al., *Stele. Tomos eis Mnemen Nikolaou Kontoleontos* (Athens, 1980), 129. For the bride, see H. Thompson, 'Activities in the American Zone of the Athenian Agora', *AJA* 37 (1933), 293; for the hetaira: G. Karo, 'Archäologische Funde von Mai 1932 bis Juli 1933', *AA* (1933), 203–4.

[101] Pyxis, Athens, NM TE 1623; *ARV* 572.88 bis, *Para* 391; *AD* 18 (1963), B1, pl. 33 (top left).

Lekythos, Honolulu, 2892; *ARV* 844.153, *Add²* 296; Boardman, *ARFH* ii, fig. 52. Alabastron, Athens, NM 479; *ARV* 727.19; *CVA* Greece 2, Athens 2, pl. 19.1 and 4.

[102] See e.g. the bride on a loutrophoros, Athens, NM, –; *ARV* 1127.14, *Para* 453; C. H. E. Haspels, 'Deux fragments d'une coupe d'Euphronios', *BCH* 54 (1930), 440, pl. 24.

[103] Krater fr., Copenhagen, National Museum, 13365; *ARV* 185.32, *Add²* 187; Boardman, *ARFH* i, fig. 131.2.

[104] Basle, Antikenmuseum, BS 236; E. Berger, *Das Basler Arztrelief* (Basle, 1970), esp. 158 n. 403.

entertained the clients waiting for their turn. He may have fetched the plants used for the ointment kept in small pots like the aryballos which bears his depiction. He was perhaps also regarded as a good-luck figure, most welcome in a clinic, and served as a living advertisement for the young doctor.[105] The use of such servant does not really conform to the ideal chastity and reserve prescribed by Hippocrates, but some eccentricity was permitted: 'for excess of strangeness will win you ill-repute, but a little will be considered in good taste.'[106] The dwarf may also have belonged to a client, as Koch-Harnack suggests,[107] or be a patient himself.[108]

Some dwarfs seem to have attended warriors. The only evidence dates to the fourth century. On two similar loutrophoroi in marble, a dwarf, wearing a helmet and almost hidden behind his shield, stands behind his departing master (G 24–5; pl. 55.2). There is no other depiction of a dwarf relating to war or to hunting. Pugnacious dwarfs are shown only as mythical pygmies, or are today misinterpreted as pygmies.[109]

More generally, there is no representation of dwarfs engaged in arduous activities, apart from a Corinthian pinax showing a small man with short limbs (a dwarf?) digging in a hole with a pickaxe (G d 2; fig, 15.1). A bird is flying above the cave and a man stands in front. Miners might tend to be short-statured, but the early date of the depiction (c.630 BC) and its clumsiness do not permit a certain identification of the figure.

Entertainers

Dwarfs performed other activities. In the symposion Athenians enjoyed professional entertainers who played music, sang, juggled, and danced. Managers like the Syracusan described by Xenophon, trained slaves or freedmen in all these arts and hired his troupe. These performers produced very skilful turns, throwing hoops 'whirling into the air, observing the proper height to throw them so as to catch them in a regular rhythm', and turning somersaults into a hoop 'set all around with upright swords'.[110] Professional musicians and dancers are often depicted in vase paintings, more infrequently acrobats.[111] Managers must have realized that the presence of dwarfs could increase the interest of their shows. There is no literary reference to such performance, but five scenes, depicted on wine vessels (three cups, one skyphos, and one stamnos), may show dwarf entertainers (G 3, 9, 15, 17, 20; pls. 42, 47.1, 50.1–2, 53.2).

---

[105] So for R. Herbig, 'Verkannte Paare', in R. Lullies (ed.), *Festschrift zum 60. Geburtstag von B. Schweitzer* (Stuttgart, 1954), 271 n. 17.

[106] *Precepts*, 10 (trans. W. H. S. Jones (Loeb, 1923)). Cf. *The Oath*: 'I will be chaste and religious in my practice' (trans. J. Chadwick and W. N. Mann (Harmondsworth, 1950)).

[107] Koch-Harnack (n. 64 above), 129.

[108] See below 233–4.

[109] On their identification as mythical figures, see above 184.

[110] Xen. *Symp*. 2. 1, 8, 11, 14 (trans. O. J. Todd (Loeb, 1923)); *Anth.Pal*. 9. 139.1–2. See L. B. Lawler, *The Dance in Ancient Greece* (London, 1964), 127 ff., and P. Maas, *RE* vii (1912), s.v. Γελωτοποιοί, 1019–21 (with further lit. refs).

[111] See e.g. the cup, London, BM E 68; *ARV* 371.24, *Add²* 225; Boardman, *ARFH* i, fig. 253.1–3 (dancing and music).

FIGURE 15.1. (G *d* 2) Pinax fr. (h. 10.5 cm.; w. 7.8 cm.). Berlin, SM (West) F 831

On the first two pictures, dwarfs are dancing, but no precise setting is indicated. On the tondo of a cup in Hamburg, two dwarfs are dancing, facing each other (G 15; pl. 50.1). Their pose conforms to an archaizing scheme, one arm raised, one leg bent behind, which dates back to sixth-century komos scenes. On an amphora in Leiden, for example, the performers dance in pairs too, and their movements are fixed in the same rigid attitudes.[112] The dwarfs on the Louvre skyphos show two moments of the same dance (G 3; pl. 42): on one side, the small man jumps and extends his arms, then crouches on the other side, keeping his arms in the same position.[113]

In three other scenes, a few elements suggest the location of the performance. On a cup in Todi, two dwarfs execute acrobatic turns on a three-legged dining table, which indicates that they are dancing at a symposion (G 20; pl. 53.2). The performance may also take place in a public location, as the scene depicted on a cup from Al-Mina suggests (G 17; pl. 50.2); a naked dwarf dances on a block-like podium which resembles the traditional βῆμα of artistic competitions.[114] The Erlangen dwarf also exhibits himself on a table, as is shown by the position of his body in relation to the height of the musicians' arms surrounding him (G 9; pl. 47.1). His name—Hippokleides—seems to

---

[112] Amphora, Leiden, Rijksmuseum, I 1971/3.1; *CVA* The Netherlands 4, Leiden 2, III H, pl. 54.

[113] Cf. the similar steps of the komasts on the psykter in Paris, Louvre, G 58; *ARV* 21.6; *CVA* France 12, Louvre 8, III, I c, pl. 58.3, 6, 9. See also the similar position of the arms of the young komast on a chous, Heidelberg, University, 66/1; Rühfel, *Kinderleben*, 134, fig. 73.

[114] Cf. the naked komast beside a βῆμα on a skyphos in Bonn, University, 1563; *CVA* Germany 1, Bonn 1, pl. 22.2 and 5.

confirm this interpretation since, as Beazley noted, entertainers always liked pseudonyms refering to well-known persons;[115] thus Apollodorus reports that in the fourth century two courtesans adopted the name of Phryne, Praxiteles' famous model.[116] The name of Hippokleides must have been especially attractive to a dancer since his celebrity was due to an insolent celebration of the pleasure of dancing.[117] The performance of the dwarf does not, however, take place in ordinary circumstances but in a Dionysiac context, and the dwarf is probably not a simple entertainer, but a member of the thiasos.[118]

It may be noted that, apart from the Erlangen dancer, these dwarfs dance quite alone. No picture associates the dwarfs with normal-sized persons. This absence contrasts with the numerous representations of drinking-parties, where young servants dance or play music in the middle of drunken komasts.[119] The dwarfs are also always naked, and appear most often in pairs (G 3, 15, 20; pls. 42, 50.1, 53.2), perhaps in order to enhance the attractiveness of their performance.

In fact, apart from these few pictures of dancing dwarfs, there is no representation of dwarfs as entertainers, or in a theatrical context.[120] I mention only the comic figure on the Vlastos oinochoe.[121] A man dances on a stage, indicated by a ladder and scaffolding, before two spectators. He holds a kind of sickle which identifies him as Perseus. The man is perhaps a dwarf in the guise of Perseus who performs a pantomime, as Beazley and Trendall suggest, or a satyric play, according to Brommer.[122] However, the proportions of the actor do not really denote a pathological condition, but stress the grotesque character of the performance.

Did plays show dwarfs mimicking the legendary battle of the pygmies, as happened at Rome according to Statius?[123] Freyer-Schauenburg notes that the cup fragment in Berlin might refer to an actual performance (G 45; pl. 60.1). The round bellies of the pygmies evoke the costumes of Corinthian padded dancers, and the central figure presents a large frontal head which resembles a mask.[124] No other picture or text, however, alludes to such performance in archaic and classical Athens.

This gap in the material may be due to the fact that dwarfism, and physical defects in

---

[115] J. D. Beazley, 'Excavations at Al-Mina', *JHS* 59 (1939), 11.

[116] *FGrH* 244, F 212 = Ath. 13. 591c. See also A. Raubitschek, *RE* xx. 1 (1941), s.v. Phryne, 893–4.

[117] Cf. n. 73 above.

[118] See below 238.

[119] For small young servants at symposion, see e.g. the cup in Toledo, Museum of Art, 64.126; *ARV* 402.12 bis, *Add²* 231; *CVA* USA 17, Toledo 1, pl. 56. For young dancers, see e.g. the girl in the Brygos cup (n. 111 above); Boardman, *ARFH* i, fig. 253, 1.

[120] Cf. their absence from works of drama (above 214).

[121] Chous, Athens, Vlastos coll.; *ARV* 1215.1, *Add²* 348; F. Brommer, *Satyrspiele* (Berlin, 1944), fig. 19; T. B. L. Webster, *Greek Theater Production* (London, 1970), pl. 14.

[122] Brommer, ibid. 25–8; Beazley in *ARV* 1215.1; A. D. Trendall, 'The Felton Painter and a Newly Acquired Comic Vase by his Hand', in F. Philipp and J. Stewart (eds.), *Essays and Studies in Honour of Sir Daryl Lindsay* (Oxford, 1964), 47.

[123] Stat. *Silv.* 1. 6. 51–64.

[124] B. Freyer-Schauenburg, 'Die Geranomachie in der archaischen Vasenmalerei', in *Wandlungen. Studien zur antiken und neueren Kunst, E. Homann-Wedeking gewidmet* (Waldsassen, 1975), 78–9.

general, were subject to religious associations which limited their representation to specific contexts, in particular Dionysiac. It is only in the context of Mysteries performed in the Kabeirion sanctuary near Thebes that dwarfish figures, very likely inspired by the Egypto-Phoenician Pataikoi, pervade iconography in the fourth century BC. The comic aspect of these figures was intimately linked with ritual efficacy.[125]

Citizens

The fate of the dwarfs born in free Athenian families remains to be defined. No Athenian law stated that a man could be deprived of his citizenship because of a physical deficiency. The creation of a canon of beauty did not imply legal measures against physical minorities, and we know that the eugenics prescribed by Plato and Aristotle were never applied. Many texts show, on the contrary, that Athenians were aware that the ideal man was a forgery; Socrates notes: 'when you copy types of beauty, it is so difficult to find a perfect model that you combine the most beautiful details of several, and thus contrive to make the whole figure look beautiful.'[126]

Several pictures demonstrate that dwarfs were integrated into the polis, yet not quite as average-sized persons. Three groups of pictures are distinguished here: in the first one, dwarfs are shown in familiar public places, in the second they are involved in domestic festivities, and in the third they appear in a religious context.

Public Places

A first survey of the pictures reveals that most dwarfs differ neither in their clothing nor in their behaviour from other citizens. They are not characterized by some special attire denoting, for example, a foreign origin. Like normal Athenians, they carry on their shoulder a chlamys, wear headbands, and use walking-sticks.[127]

In many scenes, it is possible to substitute the figure of a normal-sized adult for the dwarf figure without changing the meaning of the picture. Dwarfs go out with dogs, like the three dwarfs on the rhyton in Ferrara (G 4; pl. 43a–c) and the dwarf on the Boston pelike (G 8; pl. 46.2).[128] They are admitted in the palaestra, as depicted on the pelike in St Petersburg (G 10; pl. 47.2); one kicks a punch-bag, the other holds an unclear elongated object, perhaps a bundle of the thongs wound round the hand by boxers as gloves[129] or a phormiskos, a bag for holding knucklebones. A dwarf also appears in a clinic on the Peytel aryballos (G 5; pl. 44b). As mentioned above, he might

[125] For material, see. K. Braun and T. E. Haevernick, *Bemalte Keramik und Glas aus dem Kabirenheiligtum bei Theben, Das Kabirenheiligtum bei Theben*, iv (Berlin, 1981). See also above ch. 13 *sub* Hephaistos and Small Demons.

[126] Xen. *Mem.* 3. 10. 3 (trans. E. C. Marchant (Loeb 1923)).

[127] Chlamys: G 4, 8; pls. 43, 46.2. Headbands: G

11–12, 16, 22–3; pls. 48.1–2, 51, 54, 55.1. Walking-sticks: G 8, 13, 19; pls. 46.2, 49.1, 53.1.

[128] Cf. e.g. the amphora, Laon Museum, 37.1022; *ARV* 1587.7; *CVA* France 20, Laon 1, III, I, pl. 29.5–6 (draped young man, holding a stick, walking behind a dog with a necklace as on the Boston pelike).

[129] See Gardiner (n. 57 above), 402–4, fig. 133.

be not a servant but an ordinary patient.[130] Achondroplasts can suffer from many secondary complications, such as pains in the legs and in the back; the hare carried on his shoulder could be his own payment for a consultation.[131] His presence does not incite any contemptuous attitude on the part of the other clients. He is talking normally to a man who bends and seems to lend him an attentive ear. A significant detail must be noted: the penis of the dwarf is infibulated, a practice peculiar to free men, so far as I know.[132] If he is not a patient, he need not be the doctor's or a patient's slave. He may be a free man, perhaps the assistant of the doctor.

Domestic Festivities

Four pictures show dwarfs indulging in drinking-parties like any well-built Athenian. The komos of two dwarfs on a skyphos in Yale conforms to a conventional scheme (G 22; pl. 54a–b). On one side, one leg raised and arms outstretched, a naked dwarf dances before a skyphos set on the ground. On the other side, a dwarf, the hair bound by a fillet, stretches his arms out forwards and horizontally, fixing his eyes on a vessel similarly set before him. Normal-sized komasts are shown engaging in similar steps, as depicted on a cup in New York; on one side, a man dances before a skyphos set on the ground, on the other side, a dancer outstretches his arms towards a flautist this time.[133] The dwarf dancers on the Louvre skyphos (G 3; pl. 42a, b) and on the cup in Hamburg (G 15; pl. 50.1) may be taking part in the komos not as entertainers, but as normal guests too. Apart from their physical anomaly, nothing distinguishes them from normal revellers. The komasts on the Louvre skyphos are, moreover, infibulated. On a cup once in Munich, a dwarf is running, holding an unclear object (G 11; pl. 48.1). He is naked, with a headband, and exhibits a long phallus. This scene has no known parallel. Athletes run naked too, but they do not wear headbands. This attribute may refer to a drinking-party. The shape of the object held in his hand is reminiscent of that of a small bowl—or is it a ball? The dwarf is perhaps performing one of the inventive games inflicted in the course of parties by symposiarchs. Could they not order 'stammerers to sing, or bald men to comb their hair, or the lame to dance on a greased wine-skin'?[134]

However, a full equality of status with normal-sized Athenians cannot be positively asserted. Some restrictions must be made. No scene shows dwarf komasts dancing among normal Athenians, nor at a symposion with other table-companions. These scenes would demonstrate clearly the integration of short people into Athenian society

---

[130] Cf. above 229–30.

[131] Cf. the rachitic boy who is examined by the doctor on the Jason stela (2nd cent. BC); London, BM Sc. 629; Berger (n. 104 above), 78–9, fig. 99.

[132] For material, see L. Stieda, 'Die Infibulation bei den Griechen und den Römern', *Anatomische Hefte*, 19 (1902), 232–309; E. J. Dingwall, *Male Infibulation* (London, 1925).

[133] Cup, New York, MMA 16.174.42; *ARV* 338.46; Richter/Hall, pl. 65. See also the komasts dancing before skyphoi on the ground in Vienna, Kunsthistorisches Museum 213; *ARV* 904.65; *CVA* Austria 1, Vienna 1, pl. 20.1–2.

[134] Plut. *Quaest. Conv.* 1. 4, *Mor.* 621e (trans. P. A. Clement and H. B. Hoffleit (Loeb, 1969)).

by placing them at the side of servants or of guests. Only a fragment in Tübingen depicts a dwarf who may be reclining on a kline, as Binsfeld notes (G 21; pl. 53.3).[135] The man stretches his right arm as does a symposiast on a cup in Paris,[136] perhaps holding out a cup to be filled by a servant. The object at the right may be part of a pillow. Unfortunately, the picture is too fragmentary and cannot constitute unambiguous testimony.

A few scenes suggest that dwarfs were assimilated in some ways to children. Thus a chous in Dresden depicts a naked dwarf holding out a skyphos in his left hand, and leaning on a knotty walking-stick; a deep food-basked is hanging on the wall (G 14; pl. 49.2). This pose resembles that of normal komasts, such as the reveller standing on a cup once in the Mitchell collection.[137] Yet the dwarf is painted on a small chous, a vessel reserved for children; he is naked like a youth, while normal komasts in a similar pose are usually dressed in a chlamys. As on most choes, a small jug, crowned with a garland, is set before the short man,[138] while normal komasts often stand beside kraters large in proportion to their own size. The behaviour of the dwarfs depicted on a pelike in Laon also differs from that of normal adults (G 12; pl. 48.2). The dwarfs eat in a very unusual way: they take food from a large vessel put on the ground where it was probably prepared. The scene has no close parallel, except a pelike by the Geras Painter showing two satyrs preparing some food.[139] Like the two dwarfs, the satyrs wear headbands; one satyr is pouring liquid from an oinochoe into a similarly large vessel, but with a flat bottom, where his companion kneads something with his hands, perhaps some kind of unbaked cake. A cup in the Louvre may throw some light on the substance put in these vessels. A man bends to a large bowl resting on legs; he holds in his left hand a pile of limp material which he seems to take out of the receptacle.[140] For Sparkes, he is scraping rolls of dough to bake bread.[141] Thus the two dwarfs may be two greedy cooks tasting the dough just before baking, or two mischievous thieves stealing food in a kitchen. In both interpretations, the dwarfs have a non-canonical activity which could be performed by children.

---

[135] Binsfeld (n. 56 above), 12–13.

[136] Cup, Paris, Cabinet des Médailles, 585; *ARV* 372.28, *Add²* 225; J. D. Beazley, 'Brygan Symposia', in G. E. Mylonas and D. Raymond (eds.), *Studies Presented to D. M. Robinson*, ii (St Louis, 1953), 77, pl. 26h.

[137] Cup, once in London, Mitchell coll.; *ARV* 178.5; Boardman, *ARFH* i, fig. 125. Cf. the revellers in a similar pose and with similar attributes (stick, skyphos, food-basket) on the cup in New York, MMA 21.88.150; *ARV* 379.150; Richter/Hall, pl. 44.50. See also the cup in New York, MMA 16.174.43; *ARV* 376.91; Richter/Hall, pl. 44.44.

[138] A young satyr walks towards a similar small jug bearing an ivy garland on a chous in Würzburg,

M. von Wagner Museum, H 4108/L 601; Rühfel, *Kinderleben*, 161, fig. 94. See also the child dancing before a similar jug on a chous in Heidelberg (n. 113 above).

[139] Pelike, Berkeley, University, 8.4583; *ARV* 286.10, *Add²* 209; J. C. Nickel, in J. K. Anderson and R. V. West (eds.), *Poseidon's Realm: Ancient Greek Art from the Lowie Museum of Anthropology* (Berkeley, 1982), 62–6, no. 51. See also D. A. Amyx, 'A New Pelike by the Geras Painter', *AJA* 49 (1945), 508–16, esp. 516, on unbaked cakes.

[140] Cup, Paris, Louvre, C 10918; *ARV* 467.130, *Add²* 245; B. A. Sparkes, 'The Greek Kitchen: Addenda', *JHS* 85 (1965), 163, pl. 30.5.

[141] Sparkes, ibid. 163.

The scene on the Yale skyphos also bears an element which could be revealing. One of the dwarfs seems to have a fillet across his chest (G 22; pl. 54a);[142] Baur notes that three points appear behind his shoulder near his beard, but this detail is not visible on the photograph.[143] This fillet could be an ivy garland, or some kind of crepundia, which would again assimilate dwarfs to children.

Religious Context

The unusual and disturbing appearance of dwarfs seems to have induced slight social marginality, but it also gave them, as was the case for Egypt, a very special place in the religious system of the community. Archaic and classical vase paintings show that they received a fitting position within Dionysiac cult. They were incorporated into the cult of the most disquieting god in Athens, Dionysos, who led his attendants into the liminal states of ecstasy and enthusiasm, changing women into maenads and men into satyrs. The association of dwarfs with fertility, manifest in the form of kourotrophic demons, found thus a new expression in the worship of a god of vegetation intimately connected with fertility rituals.[144] This incorporation contributed to a positive approach to the unusual aspect of dwarfs which was valued for its ritual efficacy and was made reassuring.

More specifically, dwarfs seem to have been assimilated to satyrs. Their physical resemblance is striking. As noted above, the facial features of achondroplasts and of satyrs are very similar.[145] Archaic and classical artists may have been inspired by the unusual faces of achondroplastic dwarfs, with their funny snub noses and thick lips, to render the comic unreality of Dionysos' companions; the conventional baldness of dwarfs was, in return, probably influenced by that of satyrs.

This similar physical form induced similar behaviour. In several scenes of everyday life, dwarfs interchange with satyrs: they are shown aping normal humans (G 10; pl. 47.2), dancing wildly, throwing their legs up in the air and crawling, as do satyrs (G 20; pl. 53.2).[146] Their sexual behaviour is also very similar. The dwarf on an oinochoe in Oxford dances towards his female partner like a lascivious demon (G 18; pl. 52a, b). He is ithyphallic, and a flying phallus expresses his desire. His hopping step with out-stretched arms evokes the unsuccessful gesticulations of satyrs attempting to catch a maenad like the one on a kantharos once at Castle Goluchow.[147] On the Oxford

[142] See e.g. the garlands across the chests of young komasts on a cup at Basle, Antikenmuseum, BS 463; *ARV* 147.16, *Add²* 179; Boardman, *ARFH* i, fig. 105.

[143] P. V. C. Baur, *Catalogue of the Rebecca Darlington Stoddart Collection of Greek and Italian Vases in Yale University* (New Haven, Conn., 1922), 107.

[144] On this aspect of Dionysos, see principally H. Jeanmaire, *Dionysos. Histoire du culte de Bacchus* (Paris, 1951); Nilsson *GGR³* i. 564–601; Burkert, *GrRel,* 161–7 (with earlier bibliog.).

[145] See above 173.

[146] See e.g. the satyrs on the psykter, London, BM E 768; *ARV* 446.262, *Add²* 241; Boardman, *ARFH* i, fig. 299. 1. See also similar caprioles on a cup, Brussels, Musées Royaux, A 723; *ARV* 317.15, *Add²* 214; *CVA* Belgium 2, Brussels 2, III, I c, pl. 11. 1a–c.

[147] Kantharos, once Goluchow, Czartoryski, 76; *CVA* Poland 1, Goluchow, pl. 35c; *ARV* 764.7. See also the gesture of the satyr on an oinochoe, Oxford, Ashmolean Museum, 1918.63; *ARV* 1215.1; *CVA*

oinochoe, the dwarf is not really assaulting the woman, but mimics that action; the scene happens on the level of play. The depiction of the phallus flying towards the woman symbolizes the erotic nature of their relationship. The attitude of the woman, stooping forward on an outstretched leg, her weight resting on that flexed behind, does not express fright, but evokes the dancing step performed by maenads with satyrs. Pictures of muffled dancers are very rare in red-figure, as Oakley notes.[148] The author mentions two other depictions by the Phiale Painter.[149] A krater from Chiusi is particularly interesting: satyrs advance towards a female veiled dancer as does the dwarf in Oxford. The dress of the woman resembles that of maenads too; the arms of maenads may be completely hidden in sleeves or himatia, and sometimes a portion of their face is covered too.[150]

By placing dwarfs on the side of satyrs, the Oxford scene may imply that dwarfs were credited with the same sexual potency. This comparison is not entirely positive. Despite their spectacular virility, satyrs are unfortunate lovers, always inflamed but seldom embraced by the object of their desire; they must therefore be contented with auto-erotic fulfilment. Dwarfs were perhaps similarly rejected by women, and not regarded as rivals by normal-sized men. This depreciating image is also conveyed by the belief reported by Aristotle that dwarfs have a big penis.[151] Aristotle adds in other passages that animals with a large organ, like donkeys, are less fertile because sperm cools on its way.[152] The large but not erect phallus of the old dwarfish Geras may also refer to this notion.[153] In the Roman period, the figure of the short and ugly god Priapus incarnates this paradox: he suffers from a pathological, painful, and ceaseless erection.[154]

The place reserved to dwarfs in the thiasos may be defined more precisely. Their physical anomaly seems to have been regarded as the sign of an immediate relationship with the supernatural world of Dionysos, and it made them a human counterpart of

Great Britain 3, Oxford 1, III, I, pl. 42.3: he stretches his arms to catch a maenad who escapes with her arms hidden in her cloak; there is a similar scene on a kantharos, Munich, Antikensammlungen, 2560; ARV 832.38; CVA Germany 6, Munich 2, pl. 93.1.

[148] J. H. Oakley, The Phiale Painter (Mainz am Rhein, 1990), 38.

[149] Krater from Chiusi, now lost; ARV 1018.67; Oakley, ibid. no. 67, pl. 52a–b. Krater, Rome, Vatican, Astarita, 42; ARV 1018.68; Oakley, ibid. no. 68, pls. 52c–d, 53–4. Cf. the krater depicting five muffled female dancers with a flute-player in South Hadley, Mass., Mt. Holyoke College; ARV 1074.1; C. M. Galt, 'Veiled Ladies', AJA 35 (1931), 374, fig. 1.

[150] See e.g. the seated maenads, wrapped in their cloaks, who face dancing satyrs on a cup in Oxford,

Ashmolean Museum, 1924.2; ARV 865.1, Add² 299; C. Bérard and C. Bron, 'Le jeu du satyre', in C. Bérard et al., La Cité des images (Lausanne and Paris, 1984), 141, fig. 200a–b.

[151] See above n. 35.

[152] Arist. Gen.An. 1. 7. 718a; 2. 8. 748a–b. On ithyphallic asses, see H. Hoffmann, ὕβριν ὄρθιαν κνωδάλων, in D. Metzler et al. (eds.), Antidoron. Festschrift für J. Thimme zum 65. Geburtstag (Karlsruhe, 1983), 61–73, esp. 61–4 and 66.

[153] On dwarf Geras, see H. A. Shapiro, 'Notes on Greek Dwarfs', AJA 88 (1984), 391–2.

[154] H. Herter, RE xxii. 2 (1954), s.v. Priapos, 1923 ff.; M. Olender, in Y. Bonnefoy (ed.), Dictionnaire des mythologies, ii (Paris, 1981), s.v. Priape, le dernier des dieux, 311–14.

satyrs. While ordinary mortals are transformed by painters into satyrs to show their
passage into the mythical thiasos,[155] dwarfs keep their full human form in the mythical
world. On a krater in private hands, a dwarf dances on a table, holding a tympanon (G
23; pl. 55.1). At his right, a maenad accompanies him with the music of a flute, while at
his left a thoughtful satyr and Dionysos himself, leaning on his thyrsos, contemplate his
performance. All three are crowned with ivy, and a branch hangs above the dwarf to
emphasize the Dionysiac context. The satyr and the maenad refer to real humans
transported into the mythical thiasos by the performance of the rites. Only the dwarf
has no supernatural characteristics, such as a tail or animal ears, as though he were a
human substitute for a satyr. An Italiote krater in Milan shows an almost identical scene
(pl. 39.3).[156] A flute-player and Dionysos, holding a thyrsos, look at a dancer, this time
a satyr, on a trapeza. The satyr has the same pose as the dwarf: he rests the weight of
his body on his right leg, which is slightly flexed, and holds out a jug in his left
hand. Dionysos wears the same heavy taenia as the dwarf on the krater in a private
collection.[157] A krater once in the Paris market depicts an analogous scene, but the satyr
dancing on the trapeza wears a pair of ithyphallic drawers like a satyr-player transformed from a full human into a supernatural being.[158]

   The dwarf dancing on a table on the Erlangen stamnos also takes part, not in a
simple symposion, but in a Dionysiac ritual (G 9; pl. 47.1). The context is marked by
Dionysiac elements; ivy branches are wrapped around his neck and his raised arm, while
his accompanists (of indeterminate sex) are playing the flute and crotala. The table is
probably not a simple piece of furniture, but a ἱερά τράπεζα. Again the painter did not
need to travesty the dwarf as a satyr, as though the malformation of the short man
naturally endowed him with a demonic nature.

   This motif of the dance on a table appears in the second half of the fifth century; it
may indicate a foreign influence upon Dionysiac religion, especially that of the Thracian
and Phrygian god Sabazios. This god, characterized like Dionysos by frenzied and
enthusiastic rituals, seems to have been quickly assimilated to the Greek god.[159] Usually,
a dancer dressed as an Oriental with a long tunic and trousers, sometimes with a
Phrygian hat, performs on a table a specific dance, called 'oklasma', defined by an
alternative squatting/standing step and by hands joined above the head.[160] The dance
takes place in the presence of Dionysos and to the music of flutes and tympanon played
by maenads or satyrs. On a stamnos at Athens, Dionysos himself, indicated by his

[155] Cf. Bérard/Bron (n. 150 above), 127–45.
[156] Krater, Milan, Museo Archeologico, C 408;
CVA Italy 51, Milan, coll. HA 2, IV D, pl. 6.3–4.
[157] See also the headband worn by Dionysos on a
stamnos by the Phiale Painter, Basle market; Oakley
(n. 148 above), no. 84 bis, pl. 67. See also that worn
by a komast on a skyphos at Berlin, SM (West) 3219;

ARV¹ 520; A. Greifenhagen, Antike Kunstwerke²
(Berlin, 1966), no. 50.
[158] Krater, Paris market; ARV 1053.39; A. Cook,
Zeus, i (Cambridge, 1914), pl. 39.1.
[159] Nilsson, GGR³ i. 836–9 (with earlier bibliog.).
[160] Lawler (n. 110 above), 121.

thyrsos and crowned with ivy, is dressed as an Oriental.[161] This cross-influence shows the remarkable capacity of the Dionysiac cult to absorb all that might reinforce its vitality. Greek and Eastern schemes mingle together on a few scenes. Thus the satyr on the Milan krater dances a traditional sikinnis on the table (pl. 39.3). Similarly the Erlangen dwarf dances on a table, but in a Dionysiac context (G 9; pl. 47.1). His gestures are typically those of a man expressing Dionysiac inspiration; he makes the gesture of ἀποσκοπεῖν, gazing into the distance, with one arm raised to the forehead, and his head bent, as do dancing satyrs and komasts.[162] The dwarf on the cup from Al-Mina has also the pose of an Oriental (G 17; pl. 50.2); he dances with hands joined above his head (although not quite completely joined together), one foot raised behind, like an Asiatic.[163] The vessel's find-spot would support this hypothesis, since many representations of oklasma have been found precisely at Al-Mina.

The assimilation of dwarfs with satyrs may be complete. Besides figures of well-built satyrs, we find a few pictures of miniature ones. They are differentiated from youths by their large beards, although these beards may also indicate that youths have from the start the characteristics of adults. The earliest known depiction is a lekane from Boeotia dating to 570 BC (pl. 39.4).[164] The main scene represents a komos where human attendants mix with divine beings indicated by their horse tails. Before a flute-player stands a satyr, abnormally short, with a long beard, holding a type of rhyton.[165] Another miniature demon is depicted in a wine-making scene on a black-figure lekythos in Malibu;[166] the dwarf satyr stands on top of the head of a drunken maenad; he plays the double-flute and defecates at the same time, an uncurbed activity which emphasizes the unbridled Dionysiac context. Most black-figure pictures of dwarf satyrs are found in scenes of mythical grape-harvest. Dionysos is usually sitting or reclining under the vine-arbour, while a crowd of miniature demons infiltrate the vineyard and pick grapes.[167]

[161] Stamnos, Athens, NM 14500; L. B. Lawler, 'The Maenads: A Contribution to the Study of the Dance in Ancient Greece', *MAAR* 6 (1927), 69–112, pl. 19.3.

[162] See the description of this gesture in Ath. 14. 629f, and Pliny, *NH* 35. 138. See e.g. the satyr in the cup in Taranto, Museo Nazionale; *ARV* 1253. 66; unpub. (Beazley archive). See also the komast in the cup in Frankfurt, Liebieghaus, 1522; *ARV* 1280.65, *Add²* 358; *CVA* Germany 30, Frankfurt 2, pl. 66.4.

[163] Krater, Aleppo Museum 60; *ARV* 1333.11, *Add²* 365; H. Metzger, *Les Représentations dans la céramique attique du IVes.* (Paris, 1951) (BEFAR 172), pl. 21.2. For material, see Metzger, ibid. 148–53.

[164] Lekane, Berlin, SM (West) 3366; *CVA* Germany 33, Berlin 4, pl. 202.1.

[165] For a discussion of the scene, see E. Bielefeld, 'Ein boiotischer Tanzchor des 6. Jh. v. Chr.', in W. Müller (ed.), *Festschrift für F. Zucker zum 70.*

Geburtstag (Berlin, 1954), 25–35, pls. v–vi; H. A. Brijder, 'A Pre-Dramatic Performance of a Satyr Chorus by the Heidelberger Painter', in H. A. Brijder et al., *Enthousiasmos: Essays on Greek and Related Pottery Presented to J. M. Hemelrijk* (Amsterdam, 1986), 76 ff.

[166] *The J. P. Getty Museum. Greek vases. Molly and Walter Bareiss collection* (Malibu, Calif., 1983), checklist no. 12.

[167] See e.g. the kyathos once at Castle Ashby 34; *ABV* 709.1; *CVA* Great Britain 15, Castle Ashby, pl. 24.1–3. These short satyrs are especially diminutive on an amphora by Exekias in Boston, MFA 63.952; *Para 62; CVA* USA 14, Boston, III, H, pl. 12.3. I note that the supernatural nature of the small satyr depicted on the kylix in Naples, N. Sant. 172, *CVA* Italy 20, Naples 1, III He, pl. 22, 3, is indicated by his animal ears only, while his companions have well developed tails too.

Representations of dwarf satyrs alone are more frequent in red-figure. On a chous in Athens, a small animated satyr dances on the back of a well-built companion (pl. 40.1);[168] another jumps from a chair into the arms of a normal-sized satyr on a lekythos in Boston (pl. 40.2).[169] These smaller demons do not seem to have a specific role in the thiasos. This notion of dwarf satyrs appears also in South Italian iconography, especially in representations of Pan.[170]

Thus dwarfs seem to have found a privileged place in the Dionysiac world. Liminal people, they were accepted and valued by the god the most generous towards human oddity. Dionysos is associated with other marginal beings, such as the lame Hephaistos. He is the only god who managed to convince Hephaistos to return to Olympos; he made him drunk and brought him back in a very merry and inebriated state.[171] Two dwarfs even take part in that Return in an unusual version depicted on a skyphos in Paris (G 113; pl. 75a–c). On one side, a naked dwarf rides on the neck of a donkey before which a second dwarf stands, holding a club; the animal is loaded with wine vessels and baskets full of grapes which recall the wine just drunk. Before them, two bearded men in long dresses dance with snakes round their shoulders.[172] Perhaps Dionysos himself is one of them. On the reverse, Hephaistos follows, so drunk that he must be carted away. He is sitting, not on a mule, but in a kind of wheel-chair like that ridden by the god, in a more dignified pose, on a red-figure cup in Berlin;[173] instead of having wings, it is pulled by singular tall birds. This scene is best paralleled by a Corinthian amphoriskos, where two padded human figures in the guise of satyrs with erect phalloi welcome Hephaistos riding a mule, Dionysos, and his troop of satyrs (fig. 13.5). These padded beings have a dwarfish appearance, and might also bear witness to a lost version of the Return which associated dwarfs with the Dionysiac thiasos.[174] Yet, at Corinth, the padded actors are separated by a tree from the mythical thiasos, while on the skyphos in Paris the dwarfs are involved in the myth without being transformed into supernatural beings. It is also probably not a coincidence if the only picture of a female dwarf is

[168] Chous, Athens, NM 12139; Deubner, *Att. Feste*, pl. 33.3.

[169] Lekythos, Boston, MFA 00.351; *ARV* 723.1, *Add²* 282. See also the childish bearded satyr carried on the shoulders of a satyr playing the pipe on a krater, Los Angeles, County Museum, 50.8.31; *ARV* 540; *Add²* 256. For other short satyrs, see e.g. the one dancing near Dionysos reclining on a cup, Florence, Museo Etrusco, 73749; *ARV* 355.39, *Add²* 221; *CVA* Italy 30, Florence 3, III, I, pl. 84.1.

[170] For material, see Schauenburg (n. 95 above). See e.g. the Apulian plate in Taranto, Museo Nazionale, 5166; *CVA* Italy 15, Taranto 1, IV, Dr, pl. 7.3. Krater, Bologna, Museo Civico, 813; *CVA* Italy 12, Bologna 3, IV Gs, pl. 1.1–2.

[171] On iconographic and literary sources, see T. Carpenter, *Dionysian Imagery in Archaic Greek Art* (Oxford, 1986), 13 ff.

[172] On snakes in Dionysian rituals, see e.g. C. Bérard and J.-L. Durand, 'Entrer en imagerie', in C. Bérard *et al.*, *La Cité des images* (Lausanne and Paris, 1984), fig. 21c–f, and fig. 27.

[173] Cup, Berlin, SM (West) 2273; *ARV* 174.31, *Add²* 184; Boardman, *ARFH* i, fig. 120.

[174] Cf. the dwarfish demons in a similar pose on a proto-attic krater, once in Berlin, SM A 32; *CVA* Germany 2, Berlin 1, pl. 20. The motif of the stones grasped in their hands could derive from that of the tambourine held by Bes; see P. Blome, 'Phönizische Dämonen auf einem attischen Krater', *AA* (1985), 573–9, and W. Helck, '"Phönizische Dämonen" im frühen Griechenland', *AA* (1987), 445–7.

found in a scene referring to a Dionysiac ritual, and associated with a fertility symbol (G 16; pl. 51).[175]

SOUTH ITALY

Dwarfs are depicted on a few South Italian vases (G 26–39; pls. 56–7). Most of them are Apulian, and are attributed to the Felton Painter, an original artist who had 'a real flair for the grotesque', as Trendall notes;[176] he treated several well-known themes, such as the contest between Marsyas and Apollo (G 27; pl. 56.3a, b), in a humorous way, and depicted a number of pleasing ludicrous beings, such as dwarfs, satyrs, and Papposileni.[177]

The position of dwarfs in these pictures contrasts with their privileged status in fifth-century Athenian society. The first striking difference is that dwarfs are characterized by their close relation to comedy. They are not realistically rendered, but caricatured. They are pot-bellied, with long slender legs ending with embarassingly large feet (G 27, 33–4; pls. 56.3, 57.1–2), and oversized genitals which look like the artificial ones of actors (e.g. G 26, 31; pls. 56.1–2); their large, massive heads have coarse facial features (G 27–8, 31, 34; pls. 56.2–3, 57.2) sometimes accentuated by long noses (G 26; pl. 56.1) and protruding jawbones. These details evoke only remotely achondroplastic features, and no medical diagnosis can be attempted on these pictures. The comic appearance of the dwarfs, often enhanced by ludicrous poses and absurd airs, resembles that of the burlesque actors of the Phlyax farce. Some figures are wrapped in himatia (G 27, 33–4; pls. 56.3, 57.1–2), which resemble the baggy clothing of Phlyax actors, but they exhibit neither the traditional close-fitting trousers nor grinning masks.[178]

The boundary between comedy and pathology is thus more vague than in classical vase-painting, and the identification of 'real' dwarfs must be regarded as being, overall, uncertain. The distinction between actors, caricatures and real dwarfs is often impossible to draw. For Trendall, the presence of a mask indicates an actual stage representation; malformed figures which are not masked are identified as dwarfs or caricatures, but some dwarf figures could be stage-inspired caricatures.[179]

Some scenes evoke parodic Phlyax plays. The best example is the scene on an oinochoe in Melbourne, which depicts an episode of the musical contest between Apollo and the satyr Marsyas (G 27; pl. 56.3a, b).[180] On the left, the flute-player Marsyas, who

---

[175] Cf. the depiction of an ugly woman in a Dionysiac context on a lekythos, Athens, NM 1129; *ABV* (586), *Add²* 139; Haspels, *ABL* 266.1; Boardman, *ABFH*, fig. 277 ('satyrs torture a woman').

[176] Trendall (n. 122 above), 50.

[177] Trendall, ibid. 45–52; Trendall/Cambitoglou, *RVAp* i. 172–80, esp. 172.49, 174–5.61–3.

[178] On the clothing of Phlyax actors, see e.g. M.

Bieber, *The History of the Greek and Roman Theater²* (Princeton, 1961), 129–46; Trendall, *PhV²*.

[179] Cf. Trendall, *PhV²* 86.196; id. (n. 122 above), 49–50.

[180] See C. W. Clairmont, 'Studies in Greek Mythology and Vase-Painting. Apollo and Marsyas', *YCS* 15 (1957), 161–78, esp. 166–8, nos. 23–31 (type C: 'the contest is still undecided'); K. Schauenburg, 'Marsyas', *MDAI(R)* 65 (1958), 42–66; Trendall (n.

has finished his performance, leans on a pillar with a worried look, as if divining his future defeat and his punishment; in the middle, Apollo plays the lyre and looks at a seated young man on the right, who might be the satyr's pupil, Olympos. Two dwarfs stand at each end of the scene, and mimic the attitudes of the protagonists. Unlike Attic vase-painters, the Felton Painter stressed their grotesque aspect: they are draped in very short himatia which hide their arms and their lower faces, but reveals their large genitals and their skinny legs. The dwarf on the left is bald, and both squint.[181] The dwarfs running on an oinochoe in Turin (G 35) could also refer to a Phlyax representation. As Cambitoglou notes, these armed grotesques may caricature Diomedes and Odysseus pursuing Dolon.[182] Paired or isolated figures of running dwarfs may refer to similar comic pursuits (G 28, 30–2, 38; pl. 56.2).

A few vessels depict scenes of everyday life, where dwarfs have the role of servants. On an amphora on the Milan market, a dwarf(?) stands beside a horse, like a groom (G d 29); another dwarf, described as holding a stick and a cithara, follows a child on an oinochoe in private hands (G d 36), perhaps like the dwarf on the Ferrara cup (G 7; pl. 46.1). The dwarf on a pelike in Taranto might be associated with entertainment (G 33; pl. 57.1).[183] He stands before a man reclining, surrounded by two seductive girls, while an Eros flies over the bed. A similar scene, but showing a mythical symposion, is depicted on oinochoe in Toledo (G 34; pl. 57.2); the dwarf stands before the reclining Dionysos, and blinks with a malicious expression at the onlooker. Dwarf servants might also be identified in the figure running with a 'shopping basket' on an askos in Basle (G 26; pl. 56.1)[184] and in the dwarf carrying an undetermined object, perhaps a vessel, on an oinochoe in Laon (G 39).

A privileged link between dwarfs and Dionysos is also suggested by South Italian representations. Short people may stand in a symposion scene (G 33; pl. 57.1) or may be depicted with the god, as on the Toledo oinochoe (G 34; pl. 57.2).[185] This Dionysiac affinity is also indicated by their integration in the world of the Phlyakes who, as actors, were closely associated with the Dionysiac cult, and probably belonged to the thiasos.[186]

122 above), 45–7, n. 4 (with further refs.); K. Schauenburg, 'Der besorgte Marsyas', *MDAI(R)* 79 (1972), 317–22, nn. 1–2 (with add. refs.).
   [181] Trendall (n. 122 above), 46.
   [182] A. Cambitoglou, 'The Felton Painter in Sydney', in E. Böhr and W. Martini (eds.), *Studien zur Mythologie und Vasenmalerei: K. Schauenburg zum 65. Geburtstag* (Mainz and Rhein, 1986), 146.

[183] See Schauenburg (n. 180 above), 318 (for a real, not mythical, symposion).
   [184] Trendall, *PhV*² 88.202.
   [185] See R. Hurschmann, *Symposienszenen auf unteritalischen Vasen* (Würzburg, 1985), 108–9.
   [186] On the Phlyakes and Dionysos, see Bieber (n. 178 above), 143–4, figs. 531–4.

16

Conclusion

The condition of dwarfs in archaic and classical Greece is very ambivalent. A few Greek myths, which have left only scanty traces in literature and iconography, involve dwarf creatures. These stories express forms of rejection: they transmit the image of liminal beings with comical, wily, and disquieting aspects. As robbers, the Cercopes brothers stand outside the ordered world; they are associated with both wit and fraud, as stressed by their names, Sillos ('wag') and Tryballos ('sycophant'). The ludicrous pygmies, who are relegated to the end of the known world, appear as primitive and impious beings. Deities associated to smithcraft may have a grotesque, risible appearance, like the Kabeiroi; they may also be slanderous and malevolent, like the Telchines, who can cast the evil eye.

In all these stories dwarfs relate in an ambiguous way to the animal world. Some dwarfs are equated with animals, like the puny pygmies who hardly resist the attacks of birds, or they transform into animals, like the Cercopes, who are changed into monkeys, or they have a mixed appearance, like the Telchines. They are also closely associated with childhood in the myths of pygmies and of Cercopes. These pejorative notions passed into the common language. The names Cercops and Pataikos became nicknames for cheats and hypocrites, while Telchines became synonymous for defamers.

Dwarfs were also credited with positive qualities. As discoverers of metallurgy, the Telchines, Kabeiroi, and Daktyloi are regarded as helpers of mankind. They are associated with mystery ceremonies, like those of the Kabeiroi at Samothrace which offered protection at sea. In the form of kourotrophic demons, dwarfs also ensured the protection of children, and, more generally, of fertility. These qualities stress their links with the lame smith Hephaistos, who is also the divine obstetrician who delivered Athena from the skull of Zeus.

We have only glimpses of the position of dwarfs in daily life. Our knowledge is limited by the paucity of evidence, which is essentially pictorial and covers mainly the classical period. The absence of legal restrictions suggests that their status as full humans was in principle recognized. Iconographic sources show that they had their place within Athenian society; we find them in the streets of Athens, walking with young men, in the palaestra, in a clinic, taking part in the komos and in Dionysiac rituals; some are infibulated like normal-sized Athenian citizens. Their physical abnormality does not appear as a handicap. Dwarfs are almost all muscular, supple and brisk, like the two dwarfs hopping on a skyphos in the Louvre (G 3; pl. 42), or gambolling on a table at Todi (G 20; pl. 53.2); pictures of fighting pygmies reflect the same idea, as do the pictures of dwarf Cercopes and small animated satyrs (G 104–12; pls. 40.1–2, 71–4, figs. 13.2–4).

Yet we cannot assert that dwarfs were regarded as entirely normal citizens. Aristotle compares them to children, and declares that they suffer from various mental deficiencies, such as poor reasoning and memory. Iconography confirms that dwarfs may not have received a full adult status. Some are assimilated to children, as on the Dresden chous (G 14; pl. 49.2) and on the pelike in Laon (G 12; pl. 48.2), some replace female figures, as on the Agrigento pelike and on the Agora cup (G 6, 13; pls. 45, 49.1). Their liminal status is marked by their assimilation to satyrs. In Dionysiac scenes, dwarfs appear as living witnesses to the mythical thiasos; they substitute for satyrs while keeping their human form. They stand between the real and the transcendental world, as privileged intermediaries to the deity. Their integration within the Dionysiac cult made their physical abnormality symbolically acceptable; in return, Dionysiac worship must have benefited from their assimilation which reinforced the belief in a transcendental world.

Iconographic gaps suggest rejection. No mythical model stands for female dwarfs who seem to have suffered from social exclusion. They do not appear in vase-paintings, apart from the bold scene on the Munich skyphos (G 16; pl. 51), as though female malformation was felt unpleasant or disquieting. This absence may also bear witness to some sympathetic feelings, since their representation would have implied making them objects of popular derision, as happened with the woman on the Munich skyphos. The absence of depictions of the more severe conditions, such as pseudo-achondroplasia or hypothyroidism, may indicate a similar unease towards crippling conditions, perhaps mingled with compassion.

A number of dwarfs also appear in servile positions. Their rarity made them precious, and they seem to have entered the same category of luxury servants as did Blacks, for example.[1] Their unusual anatomies attracted great attention, and are often very carefully rendered. As costly possessions, these dwarfs were not engaged in crafty or low-ranking activities; they cared for members of the family, especially for women, children, and adolescents (G 6, 7, 8, 13; pls. 45–6, 49.1). This domestic role may derive from their association with kourotrophic functions in the archaic period. They could also entertain their master at the symposion (G 3, 15, 17, 20; pls. 42, 50.1–2, 53.2). Some Athenians seem to have been conspicuous for having such original attendants, like the doctor on the Peytel aryballos (G 5; pl. 44c) and the woman, possibly a hetaira, on the Agora cup (G 13; pl. 49.1). These dwarf slaves were perhaps Greeks, but captured in war or kidnapped and sold at Athens, or they may have been born in slavery. Some may have come from foreign countries, as suggested by the negroid features of the woman in

---

[1] On Blacks, see W. Raeck, *Zum Barbarenbild in der Kunst Athens* (Bonn, 1981), 181–2, 216–17.

Munich (G 16; pl. 51). These slaves could be seen as exotic possessions: as Blacks evoked Egypt and its divine Ethiopians, real dwarfs may have been presented by skilful traders as members of a mysterious foreign population, perhaps pygmies.

Many questions on concrete aspects of dwarfs' position in society remain unsolved. Could they exert the same rights as normal-sized citizens? No dwarf figure of primary importance has been found on funerary objects, such as stelae or lekythoi. Dwarfs are also excluded from the heroic repertory. They do not appear in hunting, nor in war scenes, partly because they could not enter the battle field with normal-sized companions, partly also because war was not easily subject to derision. The only pictures of fighting dwarfs are those of mythical pygmies. How did they earn their living? They are not shown engaged in a specific form of work, apart from that of attendant (e.g. G 5; pl. 44). Could they marry? They are never shown as true couples with normal-sized (or short-statured) people. When depicted with a normal-sized woman, dwarfs either have a subordinate position (G 6, 13; pls. 45, 49.1), or a satyr-like relationship (G 18; pl. 52). The only pairs are of male dwarfs (G 3, 10, 12, 15, 20; pls. 42, 47.2, 48.2, 50.1, 53.2).

The respectful attitude towards dwarfs, due to religious associations, seems to have begun to vanish by the end of the fifth century. In South Italian vase-painting they became purely grotesque and ludicrous creatures. They slowly left the world of ritual performance for that of professional entertainment, to which they have been assigned down to the present day.

General Conclusion

This study has shown that in ancient Egypt and Greece, dwarfs had a specific place in the socio-religious system of the community. They were regarded as full humans, and perhaps also as lifelong children. Their physical abnormality was accepted and valued because of religious associations specific to each culture. In Egypt, dwarfs were regarded as manifestations of the sun-god Re and of Horus; in Greece, they were equated with satyrs and included in the world of Dionysos. These associations are reflected in the different models used to depict them in each culture: in Egypt their connections with youthful solar deities, such as Horus, are suggested by their ideal youthful faces and flat-topped heads, while in Greece snub noses, half-bald heads, and large genitals stress their resemblance to satyrs.

A number of similar functions which emerged and developed independently may be identified in both cultures. Both in Egypt and Greece, dwarfs seem to have acted as intermediary beings between men and gods. This role is best illustrated by their function as dancers in ritual contexts, such as the burials of the Apis and Mnevis bulls in Egypt (E 84; pl. 26.2), or in the mythical thiasos in Greece (G 9, 23; pls. 47.1, 55.1). Their comical appearance also seems to have had a pacifying effect. In both cultures, dwarfs contribute to the pacification of an angry deity. Egyptian Bes-gods accompany the return of Hathor-Tefenet to Egypt (pl. 9.1a, b), while Greek dwarfs escort Hephaistos on his way back to Olympos (G 113; pl. 75a–c). In both myths, the appeasing and rejoicing role of dwarfs complements that of wine. Ptah, patron of craftsmen, could be depicted as a dwarf in the form of Ptah-Pataikos. This motif is paralleled by Hellenic beliefs in lame or dwarf smiths, such as Hephaistos or the Telchines, but these are unrelated; Ptah is shown as a dwarf not because dwarfs are associated with metallurgy, but because of their solar and regeneration symbolism. Clear traces of Egyptian influence on Greek religion may be seen in the archaic period only in the emergence of kourotrophic demons deriving from Bes and Ptah-Pataikoi.

In both cultures the acceptance of dwarfs is expressed by the fact that their disorder is never shown as a physical handicap. Even dwarfs with club-feet are not presented as handicapped people. Very crippling conditions, or conditions involving severe mental disorders, such as hypothyroidism, are not shown. To represent such people might have created uneasy or unpleasant feelings, or it would have implied making fun of them, which was not compatible with the respect due to human beings who were accepted within the culture. These gaps are also partly related to the traditionally positive character of Egyptian and Greek iconography.

In both cultures short people seem thus to have achieved a degree of integration deemed impossible for medieval and post-medieval Europe.[1] Their acceptance seems to have been more successful in Egypt than in Greece. This is suggested by the contrasts between the roles of dwarfs in myth in the two cultures. In Egypt dwarfs became very popular, mass-produced, apotropaic figures, especially in amuletic forms, while in Greece they are absent or hardly identifiable. Only pygmies are clearly shown and described as dwarfs, but they are located in far-away countries, not in Greece itself. This lack of interest may reveal some rejection of physical abnormality. It is probably related to the greater importance of physical perfection in Greece than in Egypt, which is also suggested by the tendency to gloss over other unusual physical marks, such as the lameness of Hephaistos.

The acceptance of dwarfs in Egypt and classical Greece contrasts with the centuries of exclusion which followed in the Hellenistic and Roman periods, when attitudes to physical malformation changed drastically. In Regal and Republican Rome, human physical anomalies were first seen as portentous, and eliminated;[2] under the Empire these negative religious associations vanished, and 'monsters' became part of a profit-earning business. The uneasiness created by their appearance still provoked rejection, but it was counterbalanced by laughter. All kinds of deformities, including cretinism and severe disorders, appear in a wide range of works, from small terracotta and bronze statuettes to wall-paintings and mosaics. In this mass production, dwarfism is no longer distinguished by special religious associations and appears as a malformation among many others. Dwarfs are shown with an exaggerated realism, often with additional grotesque features, such as overlarge genitals, and grimacing faces.[3] Many dwarfs earned a living by exhibiting themselves as entertainers in shows which staged, for example, fights between pygmies and cranes.[4] Wealthy Romans and emperors delighted in them, as later did European courts.[5] In the Empire, dwarfs were even produced

---

[1] Cf. P. Darmon, 'Autrefois les nains', *L'Histoire*, 19 (1980), 9: 'Riche ou pauvre, noble ou roturier, nain de foire ou nain de cour, toute l'histoire témoigne, depuis l'Antiquité, de l'impossible insertion sociale des nains.'

[2] See e.g. Cic. *Leg.* 3. 19; Dion. Hal. *Ant. Rom.* 2. 15; Seneca, *De Ira*, 1. 15. 1. For further sources, see esp. M. Schmidt, 'Hephaistos lebt. Untersuchungen zur Frage der Behandlung behinderter Kinder in der Antike', *Hephaistos*, 5–6 (1983–4), 145–50.

[3] On this change, see J. P. Cèbe, *La Caricature*

et la parodie dans le monde romain antique des origines à Juvénal (Paris, 1966) (BEFAR 206), 345–71, esp. 355. For iconographic material, see W. E. Stevenson, *The Pathological Grotesque Representation in Greek and Roman Art*, Ph.D. thesis (University of Pennsylvania, 1975); Dasen, 'Dwarfism', 273 ff.

[4] Stat. *Silv.* 1. 6. 51–64. Cf. the professional 'clown' in Lucian, *The carousal or the Lapiths*, 18–19.

[5] See e.g. Mart. 1. 43, 6. 39, 14. 212; Luxorius, 10, 24. On emperors: Suet. *Aug.* 83; *Tib.* 61; *Dom.* 4; S. H. A., *Heliogab.* 34. 2–5.

artificially; young children were crippled by twisting and cutting their limbs, or by keeping them in boxes which stunted their growth.[6] This treatment of people as commodities is associated with the presence of large-scale slavery, absent in Egypt and classical Greece, and with the transformation of attitudes to human dignity.

[6] Seneca, *Controv.* 10. 4; Longinus, *Subl.* 44. 5.

IV

CATALOGUES

The letter *d* beside a catalogue number indicates that the identification of the dwarf is doubtful. The bibliography given in the catalogue entry is selective; it includes works of reference and earlier publications, with preference given to the most accessible illustrations and to a commentary. In the description, it is assumed that Egyptian figures have shaven heads, and Greek ones short hair; only special hairstyles (wigs, beards etc.) and baldness are mentioned. Measurements are given in centimetres throughout.

E: *Egypt*

The monuments are grouped chronologically, according to location, moving south of Alexandria, and within each site in PM order: *Giza*: WF (West Field), EF (East Field), GIS (cemetery GIS), CF (Central Field). *Saqqara*: NSP (North of Step Pyramid), ESP (East of Step Pyramid), WSP (West of Step Pyramid), NTP (North of Teti Pyramid), UPC (Uni Pyramid Cemetery).

RELIEFS AND DRAWINGS

EARLY DYNASTIC PERIOD (2920–2770 BC)

Abydos

Tomb complex of King Djer

1 Limestone stela (h. 36; w. 23). Berlin, SM 18136.
   Kaplony, *Inschriften*, i. 213, 215, 298, st. 220; Rupp, 'Zwerg', 299, 2C.
   E. Amélineau, *Les Nouvelles Fouilles d'Abydos, 1897–1898*, i (Paris, 1904), tomb 96, 229–30; A. Scharff, *Die Altertümer der Vor- und Frühzeit Ägyptens*, ii (Berlin, 1929), 22, no. 43, pl. 9.
   Dwarf standing. Found with skeletal remains (see S 5).
   Inscr.: *šdj* (name). Cf. Ranke, *PN* i. 331. 15.

2 Frr. of a quartzite bowl with incised figure (h. 9; w. 6). Boston, MFA 01.7292.
   Rupp, *Zwerg*, 299, 1C; Weeks, *Anatomical Knowledge*, 192–3, 12, fig. 25.
   W. M. F. Petrie, *Abydos*, i (London, 1901) (MEEF 22), 5, no. 11, pl. IV; Ruffer, *Palaeopathology*, 40, pl. VIII, fig. 2.
   Dwarf in a kilt, standing, holding an uncertain object (a stick or a strip of cloth?) in his right hand (not visible on the photograph).
   Pl. 17.1

3 Limestone stela (h. 66; w. 30). Philadelphia, University Museum, 9499.
   Kaplony, *Inschriften*, i. 184, 215, st. 58; Rupp, 'Zwerg', 299, 1L(?); Weeks, *Anatomical Knowledge*, 194, 14.
   W. M. F. Petrie, *The Royal Tombs of the Earliest Dynasties*, ii (London, 1901) (MEEF 21), pls. XXVI, XXVIII, no. 58; Ruffer, *Palaeopathology*, 44, pl. IX, fig. 16; Dasen, *Dwarfism*, 261, fig. 2d.
   Dwarf standing.
   Inscr.: *dd* (name). See Ranke, *PN* i. 402. 3.

Tomb complex of King Den

d   4   Limestone stela (h. 24). Cairo Museum, CG 14604.

           Kaplony, *Inschriften*, i. 197, st. 210; Rupp, 'Zwerg', 300, 3C; Weeks, *Anatomical Knowledge*, 194, 15.

           W. M. F. Petrie, *The Royal Tombs of the First Dynasty*, i (London, 1900) (MEEF 18), pl. XXXII, no. 7; J. E. Quibell, *Archaic Objects*, i (Cairo, 1905) (CGC), 291, no. 14604.

           Dwarf(?) in a kilt, standing.

           Inscr.: *ḥp* (name). See Ranke, *PN* i. 236. 24.

d   5   Limestone stela (h. 32.5; w. 18). Hanover, Kestner Museum, 1935.200.33.

           Kaplony, *Inschriften*, i. 197, st. 201; Rupp, 'Zwerg', 300, 4C.

           Petrie, op. cit. pl. XXXII, no. 17 (incomplete); I. Woldering, *Bildkatalog des Kestner-Museums Hannover*, i, *Ausgewählte Werke der ägyptischen Sammlung* (Hanover, 1958), 61, no. 8 (with fig.).

           Dwarf(?) in a kilt, standing.

           Inscr.: *stj-jnpw* (or *sr-jnpw*) (name).

      6   Limestone stela (h. 25.5; w. 16.5). Leiden, Rijksmuseum, F 1927/1.1.

           Kaplony, *Inschriften*, i. 195, st. 186; Rupp, 'Zwerg', 299, 2C.

           Petrie, op. cit. pl. XXXII, no. 9; A. Klasens, *OMRO*, 37 (1956), 11–34, fig. 5; Dasen, 'Dwarfism', 261, fig. 2c.

           Dwarf in a kilt(?), standing.

           Inscr.: *(w)dj-wsḫ* (name).

           Pl. 17.2

d   7   Limestone stela (h. 45; w. 19). Paris, Louvre, E 21700.

           Kaplony, *Inschriften*, i. 197, st. 205; Rupp, 'Zwerg', 299, 1C.

           Petrie, op. cit., pl. XXXII, no. 8.

           Dwarf(?) in a kilt, standing.

           Inscr.: *sjm3-nṯr* (name).

Tomb complex of King Semerkhet

      8   Limestone stela (h. 45; w. 24). London, BM 35018.

           Kaplony, *Inschriften*, i. 182, st. 37; Rupp, 'Zwerg', 300, 2C; Weeks, *Anatomical Knowledge*, 193, 3.

           W. M. F. Petrie, *The Royal Tombs of the First Dynasty*, i (London, 1900) (MEEF 18), 13, 27, pls. XXXI, XXXV, no. 37; A. J. Spencer, *Catalogue of Egyptian Antiquities in the British Museum*, v, *Early Dynastic Objects* (London, 1980), 16, no. 12, pls. 6–7.

           Dwarf in a kilt, standing. Found with skeletal remains (see S 7).

           Inscr.: *nfrt* (or *nfr. tj*?) (name). See Ranke, *PN* i. 194.1.

           Pl. 17.3

9   Limestone stela (h. 25; w. 21.5). Philadelphia, University Museum, 9186.
    Kaplony, *Inschriften*, i. 182, st. 36; Weeks, *Anatomical Knowledge*, 193, 13.
    Petrie, op. cit. 13, 27, pls. XXXI, XXXV, no. 36; D. O'Connor, *Expedition*, 29/1
    (1987), 34–5, fig. 13.
    Dwarf in a kilt, standing.
    Inscr.: Same as above no. 8.

Unknown Provenance

10 Limestone cylinder (w. 5.5). Cairo Museum, CG 14518.
    J. E. Quibell, *Archaic Objects*, i (Cairo, 1905) (CGC), 279, pl. 59; Sourdive, *La
    Main*, 472, fig. a.
    Four dwarfs, seen in front, standing, tête-bêche, above two full-sized men with
    crocodiles, lizards, a scorpion, and a bee(?).
    Fig. 9.1

OLD KINGDOM (2575–2134 BC)

Dynasty 4

Giza

11 Tomb of Queen Meresᶜankh III. (EF).
    a. PM iii². 1. 197, room I (3). G 7530.
    D. Dunham and W. K. Simpson, *The Mastaba of Queen Mersyankh III* (Boston,
    1974), 16, fig. 8.
    Female dwarf, naked, carrying an obliterated object on her head or shoulder, in a
    line of female bearers.
    Fig. 9.2
    b. PM iii². 1. 199, chapel, north wall. G 7540.
    Dunham/Simpson, op. cit., 3, pl. 13a.
    Female dwarf, in a dress(?), carrying an obliterated object (a chest?) on her left
    shoulder, in a line of female bearers.

12 Tomb of Nebemakhet. LG 86 (CF).
    PM iii². 1. 231, room II (6) I (no mention of dwarfs); Weeks, *Anatomical
    Knowledge*, 184–5, FI.
    S. Hassan, *Giza*, iv (Cairo, 1943), 140–1, fig. 81.
    Three dwarfs in kilts, seated, making jewellery.

13 Tomb of Debheni. LG 90 (CF).
    PM iii². 1. 236, room I (3) V; Rupp, 'Zwerg', 300, 1C; Weeks, *Anatomical
    Knowledge*, 176, D1.
    Lepsius, *Denkmaeler*, 2, pl. 36c; Hassan, op. cit., 170, 172, fig. 119, pl. XLIX;
    Brunner-Traut, *Tanz*, 19–20, fig. 4; Dasen, 'Dwarfism', 261, fig. 3a.

Female dwarf, naked, standing before a tall box at the end of a line of girls dancing and clapping.
Pl. 18.1, Fig. 9.16

Maidum

d   14   Tomb of Nefermaᶜet (h. 100, w. 117). Copenhagen, Ny Carlsberg Glypt., AEIN 1133.
PM iv. 93, façade (no mention of dwarf); Rupp, 'Zwerg', 300, 1H.
W. M. F. Petrie, *Medum* (London, 1892), 26, pl. XXIV; M. Mogensen, *La Glyptothèque Ny Carlsberg, La collection égyptienne* (Copenhagen, 1930), 87–8, pl. XCI; O. Koefoed-Petersen, *Catalogue des bas-reliefs et peintures égyptiens* (Copenhagen, 1956), 14–15, no. 2, pl. X; Smith, *Art and Architecture*, 81, fig. 73.
Dwarf(?) or young, naked(?), with two monkeys and an ibis.

Dynasty 5

Giza

15   Tomb of Kanufer. G 2150 (WF).
PM iii². 1. 77, interior offering room, I (c); Rupp, 'Zwerg', 300, 8C; Weeks, *Anatomical Knowledge*, 179, E1.
G. A. Reisner, *A History of the Giza Necropolis*, i (Cambridge, Mass., 1942), 444, fig. 263; Vandier d'Abbadie, 'Singes', 165–6, fig. 23.
Dwarf, naked, carrying a monkey on his head, at the end of a line of bearers.
Pl. 18.2

16   Tomb of Senedjemib: Mehi. G 2378 (WF).
PM iii². 1. 88, II, hall (6) III; Weeks, *Anatomical Knowledge*, 185, F2.
Lepsius, *Denkmaeler*, 2, pl. 74a.
Four dwarfs in kilts, seated in pairs, making jewellery.

17   Tomb of Wehemka. D 117 (WF). Hildesheim, Pelizaeus-Museum, 2970.
PM iii². 1. 115, chapel (3).
H. Kayser, *Die Mastaba des Uhemka* (Hanover, 1964), 37 and 70.
Dwarf in a kilt, carrying a basket on his head and a pair of sandals, in a line of female bearers.
Inscr.: nfr-ḥww (name). See Ranke, *PN*, i. 199. 7.
Fig. 9.3

18   Tomb of Seshemnufer I. G 4940 (WF).
PM iii². 1. 142, chapel (4)–(6) (no mention of dwarf); Weeks, *Anatomical Knowledge*, 170, B2.
Lepsius, *Denkmaeler*, 2, pl. 27.
Female dwarf, naked(?), carrying a box on her head. Above, hunchbacked female attendant in a long dress, carrying a box in her hands. The ornaments of both

figures (pectoral with pendant, bracelets and anklets) are painted in a green now partly lost.

Pl. 19.1a, b

19   Tomb of Nesutnufer. G 4970 (WF).

PM iii². 1. 144, chapel (5)–(7); Rupp, 'Zwerg', 300, 9C–10C; Weeks, *Anatomical Knowledge*, 164–5, A2, fig. 9.

H. Junker, *Giza*, iii (Vienna, 1938), 166, fig. 27, 179–80, pl. v.

Two dwarfs servants standing below two Nubian attendants:

a. In a kilt, holding a head-rest and a flywhisk (or a backrest?). Remains of a later copying grid, possibly from the Late Period.

Inscr.: rᶜ-dd.f-ᶜnḫ (name); *jsww* (title).

b. Naked, holding sandals and a tall staff.

Inscr.: ᶜnḫ-jwd-s(w) (name); *jsww* (title).

Fig. 9.4

20   Tomb of ᶜAnkhmaᶜreᶜ (EF).

a. PM iii². 1. 206, room I (6). G 7837.

Weeks, *Anatomical Knowledge*, 182, E13; G. A. Reisner, *BMFA* 34/206 (1936), 98 (with fig.); D. K. Simpson, in *Festschrift E. Edel* (Bamberg, 1979), 499, fig. 3.

Dwarf, naked, leading a dog under the litter carrying the deceased.

b. PM iii². 1. 206, room II (7). G 7843.

G. A. Reisner, *A History of the Giza Necropolis*, i (Cambridge, Mass., 1942), 351–2.

Unpublished. 'Men and dwarf with various articles, dog and monkey' (PM ibid.).

d   21   Tomb of Tjenti (EF).

PM iii². 1. 210 (dwarf not mentioned); Weeks, *Anatomical Knowledge*, 183, E17 ('Saqqara').

W. S. Smith, *A History of Egyptian Sculpture and Painting in the Old Kingdom* (London, 1946), 189.

Unpublished. 'Part of a presentation of animals . . . which includes a dwarf leading a bull' (Smith, op. cit.)

22   Tomb of Itisen (CF).

PM iii². 1. 252, pillared offering-room (2) IV (no mention of dwarf); Weeks, *Anatomical Knowledge*, 181, E9.

S. Hassan, *Giza*, v (Cairo, 1944), 266–7, figs. 122–3, 269; Vandier d'Abbadie, 'Singes', 158, fig. 10; Sourdive, *La Main*, 32, 34, pl. xii, fig. 3.

Dwarf in a kilt, holding a staff, at the end of a line of attendants leadings pet animals (a pair of dogs, a monkey, and a baboon).

23   Tomb of Ireru (CF).

PM iii². 1. 280, west chapel, between false-doors; Weeks, *Anatomical Knowledge*, 183 E20.

S. Hassan, *Giza*, iii (Cairo, 1941), 65–7, fig. 57, pl. XXII.
Dwarf (proportionate) in a kilt, leading an ox before a scribe.
Fig. 9.12

24   Tomb of Wepemnefert (CF).
PM iii². 1. 282, IV, offering room (6) IV; Weeks, *Anatomical Knowledge*, 185, F3.
S. Hassan, *Giza*, ii (Cairo, 1936), 198–9, fig. 219, pls. LXXIV–LXXVI; P. Montet,
Revue Archéologique, 40 (1952), fig. 5.
Four dwarfs in kilts, one pair seated, the other squatting, making jewellery.
Inscr.: 'Pull this strongly which is in your hand. Will you delay the work at the
beginning of a nice day?' and 'What is the matter? Look! The metal is beside you'
(above the left pair). 'Make haste with this necklace, in order that it be finished',
and 'As surely as Ptah loves you, I should like to finish it today' (above the right
pair).
Pl. 19.2a, b, Fig. 9.14

25   Tomb of Queen Khentkaus I. LG 100 (CF).
PM iii². 1. 289, chapel (no mention of dwarf); Weeks, *Anatomical Knowledge*, 179,
E2.
S. Hassan, *Giza*, iv (Cairo, 1943), 22–3, fig. 29.
Dwarf (left arm and torso preserved) in a kilt, holding a staff (or sandals).

Abusir

d 26   Pyramid complex of Sahureᶜ, frr. of relief. Braniewo (Poland), Lyceum
Hosianum.
PM iii². 1. 332.
L. Borchardt, *Das Grabdenkmal des Königs Sa3hu-reᶜ*, ii (Leipzig, 1913), 59, pl. 51;
L. Klebs, *Die Reliefs des Alten Reiches* (Heidelberg, 1915), 33.
Man called Ty with his dog and his dwarf (elbow partly preserved before the head
of the dog).

d 27   Pyramid complex of Neuserreᶜ.
PM iii². 1. 33.
L. Borchardt, *Das Grabdenkmal des Königs Ne-user-Reᶜ* (Leipzig, 1907), 81;
Sourdive, *La Main*, 95.
Unpublished. 'Von den Szenen, die im Schlafzimmer spielen, sind nur die
krummen Beine eines Zwerges neben einem Tisch erhalten' (Borchardt, op. cit.).

Saqqara

28   Tomb of Kaemnefert (NSP). Boston, MFA, 04.1761.
PM iii². 2. 467–8 (no mention of dwarf); Weeks, *Anatomical Knowledge*, 180, E4.
W. S. Smith, *Ancient Egypt as Represented in the Museum of Fine Arts, Boston*
(Boston, 1960), 61. Unpublished.

Dwarf in a pointed kilt, holding a staff and leading a monkey behind the litter carrying the deceased.

29 Tomb of Ty. D 22 (NSP).
a. PM iii². 2. 470, II, pillared hall (6)–(7); Rupp, 'Zwerg', 300, 6.
G. Steindorff, *Das Grab des Ti*, ii (Leipzig, 1913), pl. 15; L. Epron and F. Daumas, *Le tombeau de Ti*, i (Cairo, 1939) (MIFAO 65, i), pls. XVI, XVIII; Vandier d'Abbadie, 'Singes', 159–60, fig. 13; Sourdive, *La Main*, 25–6, pl. IX; Dasen, 'Dwarfism', 265–6, fig. 3d.
Dwarf (proportionate) in a pointed kilt, leading a monkey and a dog, below the litter carrying the deceased.
Inscr.: *Ppj* (name). See Ranke, *PN* i. 131. 12; *jwḥw*, an unclear title ('animal's leader'?) See *Wb* i. 57.15; Sourdive, *La Main*, 94–5.
Pl. 20.1
b. PM iii². 2. 475, VI, inner hall (44) VII; Rupp, 'Zwerg', 300, 5C; Weeks, *Anatomical Knowledge*, 180, E3.
Steindorff, op. cit., pl. 115; P. Montet, *Les Scènes de la vie privée dans les tombeaux égyptiens de l'Ancien Empire* (Strasburg, 1925), pl. XIII, 2; M. A. Murray and K. Sethe, *Saqqara Mastabas*, ii (London, 1937) (ERA 11), 26, pl. IV, 11; H. Wild, *Le Tombeau de Ti*, ii (Cairo, 1953) (MIFAO 65, II, 1), pls. XCIV, CXXVI; Vandier d'Abbadie, 'Singes', 159–60, fig. 12; Sourdive, *La Main*, 27–9, pl. X; Dasen, 'Dwarfism', 266, fig. 3f.
Dwarf in a pointed kilt, holding a baton ending in a hand, leading a monkey. In the sub-register below a naked hunchback leads two dogs.
Pl. 20.2

30 Tomb of Kaemrehu, relief (h. 97; w. 235) (NSP). Cairo Museum, CG 1534. PM iii². 2. 486, chapel (2) IV; Rupp, 'Zwerg', 300, 3C–4C; Weeks, *Anatomical Knowledge*, 185–6, F4.
Montet, op. cit. 283, fig. 39; L. Borchardt, *Denkmäler des alten Reiches*, i (Cairo, 1937) (CGC), 235, pl. 48; *Musée égyptien du Caire*, no. 59.
Two dwarfs in kilts, making jewellery, the left one kneeling, the right one seated.
Inscr.: 'Smelting (or: casting) gold'. See *Wb* iv. 330.2 ('knotting'?).

31 Tomb of Nikauhor. S 915 (NSP).
PM iii². 2. 498, offering room, East wall; Weeks; *Anatomical Knowledge*, 165, A4.
J. E. Quibell, *Excavations at Saqqara 1907–1908* (Cairo, 1909), 115, pl. LXII, 2.
Dwarf in a kilt, holding sandals at the end of a line of attendants.
Inscr.: Partly obliterated, perhaps *šms(w)* 'retainer' (title).

32 Tomb of Kaᶜaper (NSF). Present location of the relief unknown.
PM iii². 2. 501, chapel (4); Rupp, 'Zwerg', 300, 1C; Weeks; *Anatomical Knowledge*, 176, D2.
H. G. Fischer, *JNES* 18 (1959), 250–1, figs. 2 and 8, pl. V; Vandier d'Abbadie,

'Singes', 170, fig. 33; J. Baines and J. Málek, *Atlas of Ancient Egypt* (Oxford, 1980), fig. on 204.

Dwarf in a kilt, raising his arms as if singing, facing left towards a flautist. Below him, a monkey in the same pose.

Fig. 9.18

33   Tomb of Neferirtenef. D 55 (ESP). Brussels, Musées Royaux, E 2465.
PM iii². 2. 584, chapel (8); Rupp, 'Zwerg', 301 (1).
B. van De Walle, *La Chapelle funéraire de Neferirtenef* (Brussels, 1978), 53, pl. 12.
Dwarf in a pointed kilt, leading a monkey and a dog, behind larger figures.
Inscr.: *šms(w)* 'retainer' (title).
Fig. 9.10

34   Tomb of Ptahhotpe II. D 64 (WSP).
PM iii². 2. 600, offering-room (16) I; Rupp, 'Zwerg', 301, 15c–18c; Weeks, *Anatomical Knowledge*, 186, f6.
P. F. E. Paget and A. A. Pirie, *The Tomb of Ptah-hetep* (London, 1898) (BSAE 2), 27, pl. xxxv; *Atlas*, iii. 4, pl. 2; Vandier, *Manuel*, iv, pl. iii, fig. 61.
Four dwarfs in kilts, making jewellery, the left pair seated, the right pair standing.
Pl. 21

35   Tomb of Queen Nebet (UPC).
PM iii². 2. 624, room II (8); Weeks, *Anatomical Knowledge*, 167, a10.
Z. Y. Saad, *ASAE* 40 (1940), 683, pl. lxxix; id., ibid. 42 (1943), 152, pls. xiv–xv.
Female dwarf in a long dress, wearing collar, necklace with an amulet and wristlets, carrying a box on her left shoulder and holding sandals, behind two female attendants.

36   Tomb of Nufer (UPC).
a. PM iii². 2. 640, offering-room (1)–(2) IV.
A. M. Moussa and H. Altenmüller, *The Tomb of Nefer and Ka-hay* (Mainz am Rhein, 1971) (AV 5), 25, pl. 9.
Four dwarfs in two sub-registers, wearing pointed kilts and making jewellery: above, the left figure stands and the right figure squats; below, both are squatting.
b. PM iii². 640, offering-room (4) I.
Moussa/Altenmüller, op. cit., 29, pl. 24a.
Dwarf, naked, holding a scribe's palette. Before him, a table with pen case or rolls of papyri and two scribes writing.
Fig. 9.5

37   Tomb of Niᶜankhkhnum and Khnumhotpe (UPC).
PM iii². 2. 643, outer hall (18) (no mention of dwarf).
A. M. Moussa and H. Altenmüller, *Das Grab des Nianchchnum und Chnumhotep* (Mainz am Rhein, 1977) (AV 21), 129, pl. 60.

Dwarf, naked, carrying a box, below the litter of Ni^cankhkhnum.

Inscr.: name: *nj-sw-qd* (J. Baines) or *qd(w)n.s* (Moussa/Altenmüller, ibid.).

See Ranke, *PN* i. 179. 9.

Pl. 22.2

Hammamiya

d 38   Tomb of Kakhent.

a. PM v. 8, hall (15)–(16) (no mention of dwarf).

E. Mackay *et al.*, *Bahrein and Hemamieh* (London, 1929) (BSAE 47), 34, pl. XXII.

Dwarf in a kilt leading an ox.

b. PM v. 8, hall (19)–(20) (no mention of dwarf).

Mackay *et al.*, op. cit., 35, pl. XXIV.

Dwarf(?), proportionate, in a kilt, standing at the prow of the boat, holding a sceptre.

Inscr.: *Krsj* (name), 'scribe' (title).

Hagarsa

39   Tomb of Kaemnefert.

PM v. 35, passage (1)–(2); Weeks, *Anatomical Knowledge*, 164, A1, fig. 8.

W. M. F. Petrie, *Athribis* (London, 1908) (BSAE 14), 2, pl. I; Ruffer, *Palaeopathology*, 41, pl. IX, fig. 9.

Female dwarf naked, carrying a tall object (box or chest?) on her head, at the end of a line of female servants.

Fig. 9.6

Dynasty 6

Giza

d 40   Tomb of Hetepniptah. G 2430 (WF).

PM iii². 1. 94, chapel (2); Weeks, *Anatomical Knowledge*, 181, E10.

Lepsius, *Denkmaeler: Ergänzungsband*, x (a).

Dwarf(?) proportionate, naked, leading a dog and a monkey under the litter carrying the deceased.

41   Tomb of the dwarf Seneb (WF). Cairo Museum, JdE 51297.

PM iii². 1. 101–3, I, offering-room (3) false-door; Rupp, 'Zwerg', 301, 14C; 302, 19C; Weeks, *Anatomical Knowledge*, 174C.

H. Junker, *Giza*, v (Vienna, 1941), 3–124, figs. 5b, 7, 14a, b, 15, 18, 20, 22; N. Cherpion, *BIFAO* 84 (1984), 34–54, pls. I–X.

For the statues of Seneb and his titles, see E 113.

Eight scenes show the high official Seneb in various occupations:

a. Seated, in a long dress made of an animal skin, with a short curly wig, holding a fly-whisk(?) and a long stick, being censed by his son.

b. Standing, in a kilt, with a long wig, leaning on a long stick and facing two servants bringing linen.

Fig. 9.19a

c. Squatting, in a kilt, holding a long stick, in a boat being paddled.

d. Seated in a sailing-boat, in a kilt, with a broad collar, holding the rigging.

e. Standing in a boat, in a kilt, with a broad collar, pulling papyrus.

Fig. 9.19b

f. Seated in a pavilion, in a kilt, holding a long stick and a piece of cloth(?), facing a scribe and two dogs.

g. Carried in a litter, wearing a broad collar.

Fig. 9.19c

h. Seated, in a kilt, with a curled wig and a broad collar, holding a long stick and a sceptre, facing three standing scribes.

Fig. 9.19d

d 42   Tomb of Inpuhotpe (WF).

PM iii². 1. 107, chapel.

H. Junker, *Giza*, ix (Vienna, 1950), 167–8, fig. 75, pl. xv.

Two dwarfs(?), proportionate, leading two oxen.

43   Tomb of Khentkaus (WF).

PM iii². 1. 149, room II (2) I; Weeks, *Anatomical Knowledge*, 165, A3.

H. Junker, *Giza*, vii (Vienna, 1944), 76, fig. 31.

Dwarf, in a pointed kilt, holding a fly-whisk(?), facing seated couple.

Fig. 9.7

44   Tomb of Nufer: Idu. G 5550 (WF). Hildesheim, Pelizaeus-Museum, 2390.

PM iii². 1. 165, chapel, south wall; Rupp, 'Zwerg', 302, 15C; Weeks, *Anatomical Knowledge*, 180, E6.

H. Junker, *Giza*, viii (Vienna, 1947), 82–3, fig. 35; Vandier d'Abbadie, 'Singes', 166, fig. 24; K. Martin, *Reliefs des alten Reiches*, i (Mainz am Rhein, 1978) (Corpus Ant. Aegypt., Pelizaeus-Museum Hildesheim 3), 86–8.

Dwarf in a pointed kilt, standing below the deceased and his wife seated on a chair, carrying a monkey on his head.

Inscr.: *mrrj* (name). See Ranke, *PN* i. 162. 22, and ii. 399.

45   Tomb of Nunetjer (Gis).

PM iii². 1. 217, chapel II; Rupp, 'Zwerg', 302, 16C; Weeks, *Anatomical Knowledge*, 176–7, D3.

Brunner-Traut, *Tanz*, 28, fig. 10; H. Junker, *Giza*, x (Vienna, 1951), 136, figs. 44, 46; H. Hickmann, *45 Siècles de musique dans l'Égypte ancienne* (Paris, 1956), pl. xva.

Female (or male?) dwarf in a kilt (or belt with strips), with a headband of flowers, holding a sistrum, dancing between two different groups of female dancers.
Fig. 9.17

d 46 Tomb of Seshemnufer: Tjeti (Gis).
PM iii². 1. 227, room I; Weeks, *Anatomical Knowledge*, 183, E19.
H. Junker, *Giza*, xi (Vienna, 1953), 136, no. 9, 250–1, 254, fig. 100, pl. xxva; Vandier d'Abbadie, 'Singes', 162, fig. 17.
Dwarf (head and part of torso preserved), leading a dog and a monkey below the litter carrying the deceased.
Inscr.: *nfr-wdn.t* (name). see Ranke, *PN* i. 195. 27.
jmj-r3 sšr 'overseer of linen' (title).

d 47 Tomb of Niᶜankhkhnum (CF).
PM iii². 1. 248, between false-doors; Weeks, *Anatomical Knowledge*, 183, E21.
S. Hassan, *Giza*, vi (Cairo, 1950), 139, fig. 131.
Dwarf(?) (proportionate) in a kilt, leading an ox.

Saqqara

48 Tomb of Khentika: Ikhekhi (NTP).
PM iii². 2. 510, room IX (38); Weeks, *Anatomical Knowledge*, 167, A11.
T. G. H. James and M. R. Apted, *The Mastaba of Khentika called Ikheki* (London, 1953) (ASE 30), 28–9, pl. xxxi.
Dwarf in a pointed kilt, handing a box to the deceased and holding a strip of cloth (?) in his right hand.
Inscr.: *ᶜnḫ.f* (name). See Ranke, *PN* i. 67. 2.
jmj-r3-pr 'the steward' (title). James/Apted. op. cit., 59, no. 186.
Fig. 9.8

49 Tomb of ᶜAnkhmaᶜhor (NTP).
a. PM iii². 2. 513, room II (7) III; Weeks, *Anatomical Knowledge*, 187, F10.
J. Capart, *Une Rue des tombeaux à Saqqarah*, ii (Brussels, 1907), pl. xxxiii; *Atlas*, iii. 59, pl. 34; A. Badawi, *GBA* 86 (1975), 134, fig. 8; id. *The Tomb of Nyhetep-Ptah at Giza and the Tomb of ᶜAnkmaᶜhor at Saqqara* (Berkeley, Calif., 1978), 23, fig. 32, pls. 35 and 39.
Three pairs of dwarfs in kilts making jewellery, seated and squatting; the right pair have pointed kilts; in the left one, the last figure is damaged (hands only preserved).
b. PM iii². 2. 513, room II (8); Rupp, 'Zwerg', 302, 17C; Weeks, *Anatomical Knowledge*, 181, E8.
Capart, op. cit. pl. xli; Vandier d'Abbadie, 'Singes,' 165, fig. 22; Badawi, op. cit., 24, fig. 33.

Dwarf in a kilt, holding a basket of fruit and the leash of a monkey which squats on his shoulders.

50 Tomb of Mereri (NTP).
PM iii². 2. 518, outer hall (3); Rupp, 'Zwerg', 302, 15C; Weeks, *Anatomical Knowledge*, 165–6, A6.
Z. Y. Saad, *ASAE* 43 (1943), 454, pl. 39b; W. V. Davies *et al.*, *Saqqara Tombs*, i. *The Mastabas of Mereri and Wernu* (London, 1984) (ASE 36), 7 (2), 11, pls. 5 and 8b.
Dwarf in a pointed kilt holding a box. Before him, the deceased in a punt.
Inscr.: *rdj*. See Ranke, *PN* i. 227. 30.
jmj-r3-sšr 'overseer of linen', *ḥm-k3* 'Ka-servant' (titles).

51 Tomb of Kagemni Memi. LS 10 (NTP).
PM iii². 2. 523, room IV (22); Rupp, 'Zwerg', 302, 17C; Weeks, *Anatomical Knowledge*, 183, E18.
F. W. v. Bissing, *Die Mastaba des Gem-ni-kai*, i (Berlin, 1905), 16, pls. XXII–XXIII; *Atlas*, iii. 14, pl. 9; Vandier, *Manuel*, iv, pl. XV, fig. 167; Vandier d'Abbadie, 'Singes', 161, fig. 16.
Dwarf in a pointed kilt leading two dogs and a monkey, below the litter carrying the deceased.

52 Tomb of Mereruka (NTP).
a. Chapel of Mereruka: PM iii². 2. 528, room III (20) VI; Rupp, 'Zwerg', 301, 2C–4C; Weeks, *Anatomical Knowledge*, 187, F9.
P. Duell (ed.), *The Mastaba of Mereruka*, i (Chicago, 1938) (OIP 31), pls. 29, 30, 32–3; *Atlas*, iii. 55, pl. 33; Dasen, 'Dwarfism', 261, pl. 2c.
Two pairs of dwarfs, in kilts, making jewellery.
Dialogue: Above the pair on the left: 'It is very good, my companion.'
Above the pair on the right: 'Hurry up, get it done' (trans. J. Baines).
Pl. 22.1a, b
b. Chapel of Mereruka: PM iii². 2. 532, room XII (74); Rupp, 'Zwerg', 301, 1C; Weeks, *Anatomical Knowledge*, 180–2, E7, E12 (two dogs not mentioned).
P. Duell (ed.), *The Mastaba of Mereruka*, ii (Chicago, 1938) (OIP 39), pls. 157–8; *Atlas*, iii, pl. 12; Vandier, *Manuel*, iv, pl. XIV, fig. 165; Vandier d'Abbadie, 'Singes', 160–1, fig. 14.
Two naked dwarfs leading one monkey and three dogs below the litter carrying the deceased. On the left, one holds a baton ending in a hand.
Fig. 9.11
c. Chapel of Waꜥtetkhethor: PM iii². 2. 535, room V (108) (no mention of dwarfs).
Atlas, iii. 16, pl. 11; Vandier, *Manuel*, iv. pl. XIII, fig. 160.
Register with twelve female dwarfs in long skirts bringing offerings.
Inscr.: title: *šḫdt-sḏ3wt* 'supervisor of the treasury' (beside the fifth woman from the right).

Above the last chest from the right: 'the box of bearing myrrh'; above the third from the right: 'the box of bearing oil'.
Pl. 23a, b
d. Chapel of Meryteti: PM iii². 2. 536, room III (no mention of dwarf); Weeks, *Anatomical Knowledge*, 169, B1, fig. 10.
Dwarf in a pointed kilt in a row of offering bearers.
Fig. 9.9

53 Tomb of Hesy, Hapi (NTP).
Unpublished. Forthcoming study by M. Abd el-Razeq, Cairo.
a. Above the false-door, west wall of portico.
Dwarf standing, in a pointed kilt, behind a man catching birds.
b. Western façade of the tomb.
Dwarf in a pointed kilt, leading a dog and a monkey in a row of servants. In the register below, deceased in a papyrus boat.

54 Tomb of Nikauisesi (NTP).
Unpublished. Forthcoming study by M. Abd el-Razeq, Cairo.
a. Dwarf in a pointed kilt standing behind his lord.
Inscr.: jrj.n-ptḥ(?) (name), *jmj-r3(?)* 'overseer' (title).
b. Dwarf in a pointed kilt with three dogs and a monkey, standing below the litter carrying the deceased.
Inscr.: (damaged) *jrj(?)* (name).
c. Dwarf(?) hunchback in a kilt, holding a box and the leash of a dog, standing below the chair where the deceased is seated.
Inscr.: jtj (name), *sḏ3 wtj* 'the treasurer' (title).

55 Tomb of Kairer (UPC).
PM iii². 2. 631. room II (9) I (no mention of dwarfs).
J. P. Lauer, *Saqqara. The Royal Cemetery of Memphis* (London, 1976), 77, pl. 68.
Two pairs of dwarfs in pointed kilts making jewellery, seated on two tables.

56 Tomb of Akhtihotpe (UPC). Paris, Louvre, E 10958.
PM iii². 2. 635–6, chapel (3) VIII (no mention of dwarf).
Vandier, *Manuel*, v. 315; id., ibid. v. 2, pl. XVIII, fig. 142.
Dwarf standing, in a pointed kilt, in a scene of bird-catching.

57 Tomb of Ni^cankhnesut, relief (h. 57.5; w. 210). Precise position unknown.
Formerly Lucerne, Kofler–Truniger coll., A 90.
PM iii². 2. 696; Weeks, *Anatomical Knowledge*, 182, E14 and 16.
H. W. Müller, *Aegyptische Kunstwerke, Kleinfunde und Glas in der Sammlung E. und M. Kofler–Truniger, Luzern* (Berlin, 1964) (MÄS 5), 56–7 (with fig.); Vandier d'Abbadie, 'Singes', 162–3, fig. 18; A. Brodbeck, in H. Schlögl (ed.), *Geschenk des Nils. Aegyptische Kunstwerke aus Schweizer Besitz* (Basle, 1978), no. 122 (with fig.); J. Málek, *SAK* 8 (1980), 205.

Two dwarfs in kilts. On the left, the dwarf holds a bag of linen and leads a monkey and a cheetah (or a panther); the right dwarf holds sandals and leads two dogs.

Inscr.: (above the dwarf on the left) *ḥtp* (name), *sḏȝwtj*, 'treasurer' (title). See Ranke, *PN* i. 257. 22.

Pl. 24.1

Saqqara(?)

58   Fr. of relief, perhaps base of door jamb from a mastaba (h. 26; w. 25).
Formerly in the G. Michailidis collection. England, private.
Kaplony, *Inschriften, Suppl.*, 33, no. 1067, pl. 5; Rupp, 'Zwerg', 299, 4C; Weeks, *Anatomical Knowledge*, 166, A9, pl. XXVIII.
Female dwarf(?), naked with a necklace (or neckline to her dress), standing, carrying a box, followed by a smaller woman.
Inscr.: *sꜥnḫw(t)-ḥwt-ḥrw(?)* (name).

Dishasha

59   Tomb of Inti
a. PM iv. 121, hall (1) (no mention of dwarf); Rupp, 'Zwerg', 301, 5C; Weeks, *Anatomical Knowledge*, 186, F7.
W. M. F. Petrie, *Deshasheh* (London, 1898) (MEEF 15), 8, pl. XIII; Ruffer, *Palaeopathology*, 41, pl. VIII, fig. 5 (incomplete scene).
Three dwarfs in kilts at the end of a line of offering-bearers. On the left, one carries a box on his head and a collar in his hand, on the right, two others (only the back of the one on the extreme right is preserved) are engaged in a very damaged scene of metalworking.
Fig. 9.13a
b. PM iv. 122, hall (6); Rupp, 'Zwerg', 301, 5C; Weeks, *Anatomical Knowledge*, 174–5, C1.
Petrie, op. cit., 7, pl. V; Ruffer, *Palaeopathology*, 41, pl. VIII, fig. 6; Vandier, *Manuel*, v. 67, fig. 44.
Dwarf in a kilt, standing on the poop of a boat, holding an object of uncertain use in his right hand, with his left hand on the head of an oarsman.
Fig. 9.13b

Beni Hasan

60   Tomb of Ipi (no. 481).
PM iv. 161.
J. Garstang, *The Burial Customs of Ancient Egypt* (London, 1907), 38–9; Sourdive, *La Main*, 34–6, pl. XIII and cover.
Dwarf in a pointed kilt, holding a baton ending in a hand, standing with a dog

below the chair where the deceased is seated. Cf. the dwarf's coffin found in a pit in the tomb (S 14).

El-Sheikh Sa'id

61   Tomb of Serfka.

a. PM iv. 188, outer hall (3); Rupp, 'Zwerg', 301, 19C; Weeks, *Anatomical Knowledge*, 180, E5, fig. 23.

N. de G. Davies, *The Rock Tombs of Sheikh Saïd* (London, 1901) (ASE 10), 12–13, pl. IV; Ruffer, *Palaeopathology*, 41, pl. VIII, fig. 7; Vandier d'Abbadie, 'Singes', 167, fig. 25.

Dwarf in a kilt, holding the leash of a monkey, standing under the chair of the deceased.

Fig. 9.15a

b. PM iv. 188, outer hall (3) (no mention of dwarfs); Rupp, 'Zwerg', 301, 20C–23C; Weeks, *Anatomical Knowledge*, 186, F8.

Davies, op. cit., 13, pl. IV.

Four dwarfs in kilts(?) making jewellery. On the left, a standing dwarf hands a necklace to a squatting man; on the right, two dwarfs are sitting opposite a third one.

Fig. 9.15a

c. PM v. 188, outer hall (5); Rupp, 'Zwerg', 301, 25C; Weeks, *Anatomical Knowledge*, 181, E11, fig. 23c.

Davies, op. cit., 12, pl. VI; Ruffer, *Palaeopathology*, 41, pl. VIII, fig. 8.

Dwarf in a kilt, holding a baton ending in a hand (or is it the dwarf's hand ?) and leading a dog.

Fig. 9.15b

Deir el-Gabrawi

62   Tomb of Ibi.

a. PM iv. 244, hall (12–13); Rupp, 'Zwerg', 301, 9C–12C; Weeks, *Anatomical Knowledge*, 187, F11 ('Aba').

N. de G. Davies, *The Rocks Tombs of Deir el-Gebrawi*, i (London, 1902) (ASE 11) 19, pl. XIII; Ruffer, *Palaeopathology*, 40, pl. VIII, fig. 3.

Two pairs of dwarfs in pointed kilts making jewellery.

b. PM iv. 244, entrance to shrine (19)–(21) (no mention of dwarf); Rupp, 'Zwerg', 302, 18C; Weeks, *Anatomical Knowledge*, 167, A12 ('Ibi').

Davies, op. cit., 23, pl. XVII; Ruffer, *Palaeopathology*, 40–1, pl. VIII, fig. 4; C. Lilyquist, *Ancient Egyptian Mirrors from the Earliest Times through the Middle Kingdom* (Munich and Berlin, 1979) (MÄS 27), fig. 110.

Dwarf in a pointed kilt, holding a mirror, standing behind a chest below the chair of the deceased.

Hierakonpolis (Kôm el-Ahmar)

63   Tomb of Nen^cankhpepy.

PM v. 197, hall (2) 'men with carrying-chair' (dwarf not mentioned); Weeks,
Anatomical Knowledge, 170, B3, 227, fig. 11.
'South wall, left end, top. As yet unpublished. A clearly rachitic achondroplast
stands atop the sedan-chair of Men-Pepi' (Weeks, op. cit.).

Unknown provenance

64   Fr. of relief (h. 35.5; w. 31.5). Hanover, Kestner-Museum, 1935.200.201.
Bulletin van de Vereeniging tot Bevordering der Kennis van de Antieke Beschaving, 13/2
(1938), 33–5, fig. 12.
Dwarf on the top of the cabin at the back of a boat, holding an indistinct object in
the left hand.
Pl. 24.2

FIRST INTERMEDIATE PERIOD (2134–2081? BC)

Dynasty 9/10

Nag' el-Deir

65   Tomb of Meru.
Not in PM; Weeks, *Anatomical Knowledge*, 184, E22.
W. E. Smith, *A History of Egyptian Sculpture and Painting in the Old Kingdom*
(London, 1946), 225, fig. 91; C. N. Peck, *Some Decorated Tombs of the First
Intermediate Period at Naga ed-Dêr*, Ph.D. diss. (Brown University, 1959), 113, pl.
13.
Dwarf in a kilt, leading a gazelle.

MIDDLE KINGDOM (2081–1640? BC)

Dynasties 11/12

Beni Hasan

66   Tomb of Amenemhet (no. 2).
PM iv. 144, hall (20) (no mention of dwarf).
P. E. Newberry, *Beni Hasan*, i (London, 1893) (ASE 1), 31 and 37, row 7 (no
mention of dwarf in the text), pl. XII; P. Richer, *Le Nu dans l'art*, i (Paris, 1925),
191, fig. 218.
Female dwarf (proportionate) in a long dress with a large necklace, two bracelets
and long hair, holding a strip of cloth(?) in her right hand and a fan in her left
hand, standing before two harpists.
Fig. 9.20

67 Tomb of Khety (no. 17).
PM iv. 158, hall (18); Rupp, 'Zwerg', 303, 1C; Weeks, *Anatomical Knowledge*, 171, B10, fig. 12.
P. E. Newberry, *Beni Hasan*, ii (London, 1893) (ASE 2), 61, pl. XVI; Ruffer, *Palaeopathology*, 42–3, pl. IX, figs. 11–12; Vandier, *Manuel*, iv, 327–8, fig. 156.
Dwarf in a pointed kilt, followed by a club-footed man of the same size and in the same dress.
Inscr.: *nmw* 'dwarf' (above the dwarf). See *Wb* ii. 267.4.
dnb 'the crippled one' (above the club-footed man). See *Wb* v. 576.2–6.

68 Tomb of Baqt I (no. 29).
PM iv. 160, hall (15)–(16); Rupp, 'Zwerg', 303, 2C, 1 Or, 2s, 3 Or; Weeks, *Anatomical Knowledge*, 172, B11, fig. 13.
Newberry, op. cit., 36, pl. XXXII; Ruffer, *Palaeopathology*, 42, pl. IX, figs. 10 and 15; Vandier, *Manuel*, v. 207, fig. 98.2.
Dwarf in a pointed kilt, followed by a club-footed man and by a hunchback of similar sizes and in similar dresses.
Inscr.: *nmw*, as on no. 67 (above the dwarf). *Dnb*, as on no. 67 (above the club-footed man). *Jw*, a term which may perhaps refer to a hump (above the hunchback). See *Wb* i. 43.11.
Fig. 9.21

d 69 Tomb of Baqt II (no. 33).
PM iv. 160–1, hall (7)–(8); Weeks, *Anatomical Knowledge*, 173, B14.
Newberry, op. cit., 40. Badly defaced, unpublished.
'Seated figure of Baqt. Below his chair are his two dogs. Four large figures. A dwarf. Standing figure of Baqt. Hunting and agricultural scenes' (Newberry, op. cit., 40).

Abydos

70 Limestone stela (h. 59; w. 49). Cairo Museum, CG 20459.
Rupp, 'Zwerg', 302, 1C.
H. O. Lange and H. Schäfer, *Grab- und Denksteine des Mittleren Reiches*, ii (Berlin, 1908) (CGC), 58–9; id., ibid. iv (Berlin, 1902), pl. 32.
Dwarf in a kilt, holding in his right hand a strip of cloth(?), his left hand to his chest, standing below the chair of the deceased.
Inscr.: *jmnmḥ3t-snb* (name).
Pl. 24.3

d 71 Limestone stela (h. 65; w. 52). Cairo Museum, CG 20725.
Rupp, 'Zwerg', 303 (1).
Lange/Schäfer, op. cit., ii. 357; iv, pl. 54.
Dwarf(?), leading an ox.

72 Limestone stela (h. 160; w. 105). Leiden, Rijksmuseum, AP 63.
P. A. A. Boeser, *Beschreibung der aegyptischen Sammlung des niederländischen Reichsmuseums der Altertümer in Leiden*, i, *Stelen* (The Hague, 1909), 3, no. 5, pl. IV.
Dwarf in a kilt, standing in a line of offering-bearers.

NEW KINGDOM (1539–1075? BC)

Dynasty 18: Amenophis III

Soleb

73 Temple of Amenophis III. Pylon, register R 28.
J. Leclant, *Annuaire du Collège de France, 1979–1980* (Paris, 1980), 533; Sourdive, *La Main*, 124, 127 n. 7. Unpublished.
Dwarf dancing with four female(?) dancers from Punt in a *sed*-festival scene.
Inscr.: *jhb* dancers (top register); from *Pwnt* (side, bottom).

Amenophis IV/Akhenaten

Hermopolis

74 Fr. of relief. Germany, private.
R. Hanke, *Amarna-Reliefs aus Hermopolis. Neue Veröffentlichungen und Studien* (Hildesheim, 1978), 21–2, no. 14, 218, fig. 6; 219, fig. 7.
Female dwarf with four court women.

El-ʿAmarna

75 Tomb of Panehesi (northern group).
a. PM iv. 218, façade, door lintel (1)–(2); Rupp, 'Zwerg', 303, 2C–5C; Weeks, *Anatomical Knowledge*, 173–4, B18.
N. de G. Davies, *The Rock Tombs of el Amarna*, ii (London, 1905) (ASE 14), 13–14, pl. V.
Scene with all figures repeated in mirror image: two female dwarfs, in long dresses, standing behind the royal family worshipping the Aten.
Inscr.: Badly damaged. See no. 77a.
Fig. 9.27
b. PM iv. 218, entrance to outer hall (4); Rupp, 'Zwerg', 303, 7C–8C; Weeks, *Anatomical Knowledge*, 173–4, B18.
Davies, op. cit., 13–14, pl. VIII.
Two female dwarfs, in long dresses, standing behind the queen's sister.
Pl. 25

76 Tomb of May (southern group).
PM iv. 225, entrance (3) (dwarfs not mentioned); Rupp, 'Zwerg', 303, 9C, 10C; Weeks, *Anatomical Knowledge*, 173–4, B18.

N. de G. Davies, *The Rock Tombs of el Amarna*, v (London, 1908) (ASE 17), 2, pls. III, XXXVI.

Two female dwarfs in long dresses, standing behind the queen's sister.

Inscr.: Badly damaged. See no. 77a.

77 Tomb of Ay (southern group).
a. PM iv. 228. entrance (3)–(4); Rupp, 'Zwerg', 303, 11C, 12C; Weeks, *Anatomical Knowledge*, 173–4, no. B18.
N. de G. Davies, *The Rock Tombs of el Amarna*, vi (London, 1908) (ASE 18), 18, pls. XXVI, XXXI (detail).
Two female dwarfs in long dresses standing behind the queen's sister.
Inscr.: 'The vizier of the queen, *r nḥḥ* (for ever)'(above the dwarf on the left). 'The vizier of his mother, *p3 rˁ* (the Sun)' (above the dwarf on the right).
(Davies, op. cit., 18).
b. PM iv. 229, hall (6)–(8); Rupp, 'Zwerg', 303, 13C, 14C; Weeks, *Anatomical Knowledge*, 173–4, B18.
Davies, op. cit., 21, esp. n. 2, pl. XXVIII.
Two female dwarfs in long dresses. In the register below, the queen's sister.

78 Tomb of Tutu (southern group).
PM iv. 222, pillared hall (5)–(7) (no mention of dwarf); Rupp, 'Zwerg', 303, 15C; Weeks, *Anatomical Knowledge*, 173–4, B18.
Davies, op. cit., 10, pl. xvii.
Female dwarf in a long dress(?), standing behind the queen's sister.

79 Fr. of relief. Cairo Museum, -.
Weeks, *Anatomical Knowledge*, 173, B17, fig. 15.
U. Fiedler, R. Watermann, *Zeitschrift für menschl. Vererbungs- und Konstitutionslehre*, 33 (1956), 506, fig. 1.
Dwarf in a kilt (or a belt with strips?) with locks of hair (or a cap?), raising his right arm.

Thebes

d 80 Sandstone block (talatat) (approx. l. 52; h. 22; w. 26). Luxor Museum, 0119.17306.
J. Gohary, in R. W. Smith and D. B. Redford (eds.), *The Akhenaten Temple Project*, i, *Initial Discoveries* (Warminster, 1976), 67, pl. 85. 4; Sourdive, *La Main*, 125 (with fig.).
Two bald men kneeling behind two short men (dwarfs?) in long kilts with feline masks(?).

Dynasty 20

81 Papyrus. Turin Museum, 55001.

J. A. Omlin, *Der Papyrus 55001 und seine satyrisch-erotischen Zeichnungen und Inschriften* (Turin, 1973), 50, pl. XIII, no. 1.

Ithyphallic dwarf in a kilt, carrying a bag, in an erotic scene.

Fig. 9.25

THIRD INTERMEDIATE PERIOD (1075?–750? BC)

Dynasty 22

Tell Basta (Bubastis)

82 Relief from the temple of Bastet, festival hall of Osorkon III. Philadelphia, University Museum, E 226.

PM iv. 29, first hall, north half (no mention of dwarf); Rupp, 'Zwerg', 304, 1P–3P.

E. Naville, *The Festival Hall of Osorkon II in the Great Temple of Busbastis* (London, 1892) (EES 20), 30–1, pl. XX, 5; A. Nibbi, *Ancient Egypt and Some Eastern Neighbours* (New Jersey, 1981), 210, pl. 18.

Four dwarfs (proportionate) in kilts holding a long cane in the right hand, in a line of priests.

Inscr.: possibly $s^c\check{s}3$, 'guard' (*Wb* iv. 55.14–18) or 'chiefs ($h3t$) of the numerous ($^c\check{s}3$) s dwarfs' (H. G. Fischer, *ZÄS* 105 (1978), 49).

Fig. 9.26

LATE PERIOD (750?–332BC)

Dynasty 26

Unknown Provenance

d 83 Obelisk (h. 23; w. 5.3). Brooklyn Museum, 50.169.

H. De Meulenaere, *CdE*, 40/80, 1965, 254–5; K. Martin, *Ein Garantsymbol des Lebens* (Hildesheim, 1977), 114, 241, fig. 10; H. De Meulenaere, J. Yoyotte, *BIFAO*, 83, 1983, 113, no. 7; Dasen, 'Dwarfism', 266, fig. 3e.

Female dwarf(?) with a short trunk.

Inscr.: Speech by Osiris, foremost of the Westerners, to *dg*-Neith'.

Pl. 26.1

Dynasty 30

Saqqara (NTP)

84 Granite sarcophagus of the dwarf Djeho (h. 110; w. 110; l. 180). Cairo Museum, CG 29307.

Rupp, 'Zwerg', 304, 2C.

G. Maspero, *Sarcophages des époques persane et ptolémaique*, i (Cairo, 1914)
(CGC), pl. 22, no. 29307; id. and H. Gauthier, ii (Cairo, 1939) (CGC), 1–17; W.
Spiegelberg, *ZÄS* 64 (1929), 76–83; Brunner-Traut, *Tanz*, 36; H. de Meulenaere,
Le Surnom égyptien à la Basse Epoque (Istanbul, 1966) 8, no. 20; Dasen, 'Dwarfism',
258, pl. 1b, fig. 2a; J. Baines, *DEA* 78 (1992) forthcoming.

Dwarf, naked, standing.

Inscr.: a. *Around the upper border of the lid*: see Spiegelberg, op. cit., 80; Baines, op.
cit.

b. *On the lid, above the dwarf figure*: see Spiegelberg, op. cit., 82; Baines, op. cit.

Pl. 26.2

Thebes(?)

d 85   Frr. of two leather membranes of a round tambourine (diam. *c.*40). Oxford,
Ashmolean Museum, 1890.543.

L. Manniche, *AcOr* 35 (1973), 29–34, pls. I–II.

a. Four registers with gods and dancing figures.

b. Osiris, Isis, and a small female figure (dwarf?) turning her back to the goddess.

Akhmim

86   Two leather membranes of a round tambourine (Diam. 25). Cairo Museum, CG
69351 and 69352.

L. Borchardt, 'Die Rahmentrommel im Museum zu Kairo', in *Mélanges Maspero*, i
(Cairo, 1935–8) (MIFAO 66), 1–6; H. Hickmann, *Instruments de musique* (Cairo,
1949) (CGC), 111, pl. LXXX; id. *45 Siècles de musique dans l'Égypte ancienne* (Paris,
1956), 21, pl. XCVIIC (membrane a); H. Wild, *Kush*, 7 (1959), 81–3, fig. 4.

a. Black female dwarf, in a dress, dancing on a kind of pedestal before seated Isis;
behind the dwarf, a woman beats a tambourine.

b. Bes (or dwarf in the guise of Bes) dancing on a kind of pedestal before seated
Isis; behind Bes, a woman beats a tambourine.

Fig. 9.28a, b

GRAECO-ROMAN PERIOD

Ptolemy IX Soter II (88–81 BC)

Edfu

87   Birth-house. Court, intercolumnar wall, interior.

PM vi. 171, II, court (35).

E. Chassinat, *Le Mammisi d'Edfou* (Cairo, 1939) (MIFAO 16), 189–90, 192–3, pls.
X. 2, XLIX (1).

Dwarf in a dress (or cloak), standing behind Ptolemy IX.

STATUARY

LATE PREDYNASTIC PERIOD (Gerzean) (*c.*3200 BC)

El-Ballas

88   Ivory (h. 5.6). London, UC 15135.
W. M. F. Petrie, *Prehistoric Egypt* (London, 1920) (BSAE 31), 6, 9, pl. II, no. 18.
Female dwarf, naked, standing, hands on hips, with protruding ears. Eyes
hollowed, pubic hair indicated by incised points. Right arm, left forearm, and
feet missing.

89   Ivory (h. 5.7). London, UC 15136.
Petrie, op. cit., 6, 9, pl. II, no. 19.
Female dwarf as above no. 1. Erased face, forearms missing.

90   Calcite (h. 3.6; w. 1.8). London, UC 15142.
Petrie, op. cit., pl. II, no. 25.
Dwarf, naked, standing. Loop for suspension in the neck.

Naqada?

91   Ivory (h. 5.2). Baltimore, WAG 71.531.
Weeks, *Anatomical Knowledge*, 191, 6; Hornemann, *Types*, 830.
G. Steindorff, *Catalogue of the Egyptian Sculpture in the Walters Art Gallery*
(Baltimore, 1946), 19, pl. 1.4; Smith, *Art and Architecture*, 29–30, fig. 7 (bottom,
right); R. H. Randall, *Masterpieces of Ivory from the Walters Art Gallery* (New York,
1985), 42–3, no. 14, pl. 6.
Female dwarf naked, standing, hands on hips, with protruding ears. Eyes
hollowed. Right arm and feet missing.

92   Ivory (h. 4.2). Baltimore, WAG 71.532.
Weeks, *Anatomical Knowledge*, 190, 3; Hornemann, *Types*, 831. Steindorff, op. cit.,
19, pl. 1.3; Smith, *Art and Architecture*, 29–30, fig. 7 (top left); Randall, op. cit.,
42–3, no. 12, pl. 5; Dasen, 'Dwarfism', 260, pl. 1c.
Female dwarf as above no. 91. Pubic hair indicated by incised points.
Pl. 27.1

93   Ivory (h. 5.7). Baltimore, WAG 71.533.
Rupp, 'Zwerg', 298, 2c; Weeks, *Anatomical Knowledge*, 190, 4.
Steindorff, op. cit., 19, pl. 1.5; Smith, *Art and Architecture*, 29–30, fig. 7 (top
middle); Randall, op. cit., 42–3, no. 15, pl. 8.
Dwarf, naked, with a beard(?), holding a child whose arms and head are missing.
Eyes hollowed.
Hole through the back to the groin.

94   Ivory (h. 4.6). Baltimore, WAG 71.534.
Weeks, *Anatomical Knowledge*, 190, 5; Hornemann, *Types*, 346.

Steindorff, op. cit., 19, pl. 1.2; Smith, *Art and Architecture*, 29–30, fig. 7 (top right); Randall, op. cit., 42–3, no. 13, pl. 5.

Dwarf, naked, standing, hands on hips, with protruding ears. Eyes hollowed. Peg at the base.

Pl. 27.2

Unknown provenance

95 Ivory (h. 5.5). London, BM 32144.

Rupp, 'Zwerg', 298, 4C.

E. A. W. Budge, *A Guide to the Egyptian Collections in the British Museum* (London, 1909), 24, 26, fig. 8.

Female dwarf as above no. 91. Right forearm and foot missing.

EARLY DYNASTIC PERIOD (2920–2770 BC)

Kafr Tarkhan

d 96 Limestone (h. 5.8). London, UC 15183.

W. M. F. Petrie, *Prehistoric Egypt* (London, 1920) (BSAE 31), 10, pl. VIII, no. 36; A. Scharff, *Die Altertümer der Vor- und Frühzeit Ägyptens*, ii (Berlin, 1929), 48, fig. 31.

Dwarf(?), naked, with protruding ears, squatting with left arm between legs.

Abusir el-Meleq

97 Black steatite (h. 5.7). Berlin, SM 19080.

Weeks, *Anatomical Knowledge*, 192, 11; Honemann, *Types*, 969.

A. Scharff, *Die archaeologischen Ergebnisse des vorgeschichtlichen Gräberfeldes von Abusir el-Meleq* (Leipzig, 1926), 62, no. 433, pl. 39; id. op. cit., 47–8, no. 73, pl. 15.

Female(?) dwarf, naked with a belt, standing, hands crossed on her chest.

Abydos

d 98 Limestone. Cairo Museum, - (room 42, 'M 65').

Rupp, 'Zwerg', 299, 1.

W. M. F. Petrie, *Abydos*, ii (London, 1903) (MEEF 24), 27, pl. X, no. 213; Ballod, *Prolegomena*, 37, fig. 11.

Dwarf(?), naked, standing, arms crossed in front. Head and lower part of legs missing.

d 99 Gneiss (h. 4.2; w. 1.7). Lucerne, Kofler–Truniger coll., A 35.

H. W. Müller, *Ägyptische Kunstwerke, Kleinfunde und Glas in der Sammlung E. und M. Kofler–Truniger* (Berlin, 1964) (MÄS 5), 31 (with fig.); H. Schlögl, in *Geschenk des Nils. Aegyptische Kunstwerke aus Schweizer Besitz* (Basle, 1978), 27, no. 82 (with fig.).

Dwarf(?), naked(?), standing, with arms crossed on his chest.

d 100   Limestone (h. 8.1; w. 4.2). Lucerne, Kofler–Truniger coll., A 38.
Müller, op. cit., 31 (with fig.); Schlögl, op. cit., 27–8, no. 84 (with fig.).
Dwarf(?), naked, with a belt, standing, with arms crossed on his chest. Legs missing.

d 101   Green glazed ware (h. 6; w. 3.5). Lucerne, Kofler–Truniger coll., A 39.
Müller, op. cit., 32 (with fig.); Schlögl, op. cit., 28, no. 85 (with fig.).
Female dwarf(?), naked, standing, with arms crossed on her chest.

d 102   Green glazed ware (h. 8.1). Lucerne, Kofler–Truniger coll., A 40.
Müller, op. cit., 32 (with fig.); Schlögl, op. cit., 28, no. 86 (with fig.).
Dwarf(?), naked, standing, hands on hips.
Pl. 27.3

d 103   Fired clay, fr. (h. 15.2). Oxford, Ashmolean Museum, E 275.
W. M. F. Petrie, *Abydos*, ii (London, 1903) (MEEF 24), 26–7, pl. XLII (general context).
Trunk and part of right leg of a dwarf(?).

d 104   Limestone. Present location unknown.
Rupp, 'Zwerg', 298, 1.
Petrie, op. cit., 25, pl. v, no. 48.
Female dwarf(?), naked, standing, with arms crossed on her chest.

Hierakonpolis, Temple Enclosure, Main Deposit

d 105   Blue glazed ware (h. 7.8). Oxford, Ashmolean Museum, E 10.
Rupp, 'Zwerg', 299 (2); Weeks, *Anatomical Knowledge*, 192, 10.
J. E. Quibell, *Hierakonpolis*, i (London, 1900) (ERA 4), 7, pl. XVIII, no. 7; J. E. Quibell and F. W. Green, *Hierakonpolis*, ii (London, 1902) (ERA 5), 38; B. Adams, *Ancient Hierakonpolis*, ii (Warminster, 1974), 132.
Female dwarf(?), standing, with long hair hanging in a lock over each shoulder and down back, wearing a dress and a necklace, with arms crossed on her chest.

106   Ivory (h. 13.3). Oxford, Ashmolean Museum, E 298.
Hornemann, *Types*, 881.
Quibell, op. cit., 6–7; Quibell/Green, op. cit., 33–4 (general); J. Capart, *Primitive Art in Egypt* (London, 1905), 170, fig. 132, no. 15; Adams, op. cit., ii. 136; J. Málek, *In the Shadow of the Pyramids* (London, 1986), fig. on 37.
Female dwarf, standing, in a dress, with long puffed wig hanging in a lock over each shoulder and down back. Left arm and right forearm missing. Peg at the base.
Pl. 27.4

107 Ivory (h. 11.9). Oxford, Ashmolean Museum, E 299.

Quibell, op. cit., 6–7; Quibell/Green, op. cit., 33–4 (general); Capart, op. cit., 170, fig. 132, no. 16; Adams, op. cit., ii. 136.

Female dwarf, standing, in a dress(?), with long puffed wig hanging in a lock over each shoulder and down back. Left arm and foot missing. Peg at the base.

108 Ivory (h. 15.3). Oxford, Ashmolean Museum, E 333/346.

Rupp, 'Zwerg', 299, 1C; Weeks, *Anatomical Knowledge*, 194, 16; Hornemann, *Types*, 882.

Quibell, op. cit., 7, pl. XI (top); Quibell/Green, op. cit., 34, 37; Ruffer, *Palaeopathology*, 43–4, pl. X, fig. 1; Capart, op. cit., 170, fig. 132, no. 22; Adams, op. cit., ii. 138.

Female dwarf, naked, standing on a small circular pedestal, with long hair hanging in a lock over each shoulder and down back, right arm on the chest. Left arm and side missing.

Pl. 28.1a, b

d 109 Glazed ware. Present location unknown.

Rupp, 'Zwerg', 298, 5C; Weeks, *Anatomical Knowledge*, 192, 9.

Quibell, op. cit., 7, pl. XVIII, no. 19; Quibell/Green, op. cit., 38.

Dwarf(?) naked, standing, hands on hips, with shaven head and protruding ears.

Elephantine, Temple of Satet

d 110 Glazed ware (h. 7.9). Aswan, Elephantine Museum, K 1024.

G. Dreyer, *Elephantine*, viii, *Der Tempel der Satet. Die Funde der Frühzeit und des alten Reiches* (Mainz am Rhein, 1986) (AV 39), 61, 99, no. 1, pl. 11.

Dwarf(?), naked, with a wig(?), standing, hands on hips.

Unknown Provenance

d 111 Ivory (h. 7.5). Berlin, SM (East), 14.441.

Hornemann, *Types*, 1246.

J. Capart, *Primitive Art in Egypt* (London, 1905), 168–9, fig. 131; A. Scharff, *Die Altertümer der Vor- und Frühzeit Ägyptens*, ii (Berlin, 1929), 50–1, no. 79, pl. 16; S. Wenig, *Die Frau im alten Ägypten* (Leipzig, 1967), 45–6, no. 4.

Female dwarf(?) naked, with wig, holding a child.

OLD KINGDOM (2575–2134 BC)

Dynasties 5/6

Giza

d 112 Limestone statue (h. 48). Cairo Museum, JdE 37.719. From the brick mastaba of the dwarf Petpennesut (WF).

PM iii^2. 1. 55; Hornemann, *Types*, 134.

R. Engelbach, *ASAE* 38 (1938), 285, 699, pl. XXXVIII; W. S. Smith, *A History of Egyptian Sculpture and Painting in the Old Kingdom* (London, 1946), 58, 64, pl. 25d.
Dwarf(?) in a kilt, standing on an inscribed base.
Inscr.: *Pt-pn-nswt* (name).

113 Tomb of the dwarf Seneb (WF) (for the false-door reliefs see E 41, fig. 9.19a–d).
a. Wooden statue of Seneb (h. *c*.30), now disintegrated. The stone chest which contained the statue is in Hildesheim, Pelizaeus-Museum, 3115.
PM iii². 1. 102, chapel, II, small serdab.
H. Junker, *Giza*, v (Vienna, 1941), 104–5, 121, pl. 8a–c.
Seneb standing, holding a staff in his raised left arm, a sceptre in his right.
b. Fragment of granite base of statue of Seneb. Present location unknown.
PM iii². 1. 102, chapel, III, large serdab south of offering-room.
Junker, op. cit., 121–2, fig. 29B, pl. 20.
Seneb, probably seated.
Inscr.: *Snb* 'Seneb' (name); titles (from right to left):
1. *wr-ʿj* 'great one of the litter'
2. *ḥrp* 𓀀 *sšrw* 'overseer of the dwarfs in charge of linen'
3. *ḥm-nṯr w3djt* 'priest of the (goddess) Wadjet'
4. *ḥrp ḥwwt nt nt* 'overseer of the administration of the crown of Lower Egypt'
5. *mrjj nb.f* 'beloved of his lord'
c. Limestone statue group found in a stone chest (h. 34; w. 22.5; th. 25).
Cairo Museum, JdE 51.280.
PM iii². 1. 102–3, east façade, IV, serdab north of false-door of wife.
Rupp, 'Zwerg', 302, 19C; Hornemann, *Types*, 1463–4.
Junker, op. cit., 107–14, frontispiece, pl. IX, fig. 29A; Vandier, *Manuel*, iii. 80, 137–8, pl. XLVIII. 5; E. L. B. Terrace and H. G. Fischer, *Treasures of the Cairo Museum* (London, 1970), 65–8, pl. III; C. Aldred, *Egyptian Art* (London, 1980), 76–7, fig. 37; N. Cherpion, *BIFAO* 84 (1984), 35–54, pl. XI; *Musée égyptien du Caire*, no. 39; Dasen, 'Dwarfism', 263, pl. 2a, b.
Seneb in a kilt, seated on a chair with his wife; beneath him, where his legs would be, his son and daughter.
Inscr.: *Snb* (name); titles (after Junker, op. cit., 12–18):
1. *ḥrp* 𓀀 *sšrw* 'overseer of the dwarfs in charge of linen'
2. *jmj-r3 jwḥw* 'overseer of the *jwḥw* (animal tenders?)'
3. *jmj-r3 mr pr ʿ3* 'overseer of weaving in the palace'
4. *wr-ʿj* 'great one of the litter'
5. *ḥrp ḥwwt nt nt* 'overseer of the administration of the crown of Lower Egypt'
6. *ḥrp ḥwwt nt mw* 'overseer of the administration of the *mw*'
7. *ḥrp ʿprw kzw* 'overseer of the crew of the *kz* ships'
8. *sd3wtj-nṯr Wn-ḥr-b3w* 'keeper of the God's seal of the *Wn-ḥr-b3w* boat'
9. *ḥm-nṯr ḥwfw*, 'priest of Khufu'

10. *ḥm-nṯr Rᶜḏd.f* 'priest of Raᶜdjedef'

11. *ḥm-nṯr w3ḏjt* 'priest of the (goddess) Wadjet'

12. *ḥm-nṯr k3 wr ḥntj Sṯpt* 'priest of the great bull which is at the head of *Sṯpt*'

13. *ḥm-nṯr k3-Mrḥw* 'priest of the bull *Mrḥw*'

14. *smr* 'companion'

15. *smr pr* 'companion of the house'

16. *ḥrp ᶜḥ* 'controller of the palace'

17. *nb jm3ḥ ḥr nb.f* 'possessor of the venerated state by his lord'

18. *mrjj nb.f* 'beloved of his lord'

19. *mrrw nb.f nb rᶜ* 'beloved of his lord every day'

20. *jrj mrrt nb.f* 'who does what his lord likes'

Frontispiece; pl. 28.2

See J. Pirenne, *Histoire des institutions et du droit privé de l'ancienne Egypte*, ii (Brussels, 1934), 423, no. 98; K. Baer, *Rank and Title in the Old Kingdom* (Chicago, 1960), 123–4, no. 441.

114 Limestone (h. 4.7). Cairo Museum, JdE 72.144. From tomb G 7715 (EF).
PM iii². 1. 203. Weeks, *Anatomical Knowledge*, 166, A7.
W. S. Smith, *A History of Egyptian Sculpture and Painting in the Old Kingdom* (London, 1946), 100 (k 3), pl. 28b; J. H. Breasted, *Egyptian Servant Statues* (New York, 1948), 58, no. 3, pl. 50b; Vandier, *Manuel*, iii. 97, pl. XXXIX. 5.
Dwarf, naked, standing on a base, carrying four pots.

115 Limestone (h. 21.4). Chicago, Oriental Institute, 10627. From the tomb of Nikauinpu (WF?).
PM iii². 1. 300; Weeks, *Anatomical Knowledge*, 165, A5; Hornemann, *Types*, 19.
Smith, op. cit., 100 (k 2); Breasted, op. cit., 58, no. 2, pl. 50a; Vandier, *Manuel*, iii. 97, pl. XXXIX. 1.
Dwarf, naked, standing on a base, carrying a bag on his left shoulder.
Pl. 29.1

116 Limestone (h. 12.5). Chicago, Oriental Institute, 10641. From the tomb of Nikauinpu (WF?).
PM iii². 1. 301; Weeks, *Anatomical Knowledge*, 177, D5; Hornemann, *Types*, 381.
Smith, op. cit., 101 (l 3), pl. 27e; Breasted, op cit., 87, no. 2, pl. 81b; Vandier, *Manuel*, iii. 97, pl. XL. 8–9; Dasen, 'Dwarfism', 265, pl. 1d.
Dwarf, naked, sitting on a base, playing a harp.
Pl. 29.3a, b

Saqqara

117 Limestone (h. 46). Cairo Museum, CG 144.
PM iii². 2. 722–3; Rupp, 'Zwerg', 301, 24C; Hornemann, *Types*, 10.
L. Borchardt, *Statuen und Statuetten von Königen und Privatleuten*, i (Berlin, 1911)

(CGC), 105–6, no. 144, pl. 32; Ruffer, *Palaeopathology*, 38–9, pl. VII; Smith, op. cit., 57; Vandier, *Manuel*, iii. 61, pl. XVIII. 1.

Dwarf standing on a pedestal, in a pointed kilt.

Inscr.: ḫnmḥtp (name); *jmj-r3-sšr* 'overseer of linen', ḥm-k3 'Ka-servant' (titles).

Pl. 29.2

Unknown Provenance

118　Ivory (h. 8.9). Baltimore, WAG 71.504.

R. H. Randall, *Masterpieces of Ivory from the Walters Art Gallery* (New York, 1985), 42–3, no. 16, pl. 9.

Dwarf, naked, standing, with hands clasped before his chest. Hole in the head (for a dowel?) and through the hands to the back. Lower legs missing.

Pl. 30.1

MIDDLE KINGDOM (2081–1640? BC)

Dynasties 11/12

Beni Hasan

119　Seal, bronze (h. 3). Cairo Museum, –.

Weeks, *Anatomical Knowledge*, 172, B12.

J. Garstang, *The Burial Customs of Ancient Egypt* (London, 1907), 235, no. 691, fig. 230; B. Schrumpf-Pierron, *Aesculape*, 24/9 (1934), 237, fig. 26.

Dwarf, naked, standing.

Deir el-Bersha

120　Wood (h. 11). Cairo Museum, JdE 34.298. From the tomb of Sithedjhotep.

PM iv. 184; Weeks, *Anatomical Knowledge*, 172–3, B13, B15, fig. 14.

A. Kamal, *ASAE* 2 (1901), 34; F. Regnault, *Bull. Soc. Fr. Hist. Méd.* 13 (1914), 141, fig. 2; B. Schrumpf-Pierron, *Aesculape*, 24/9 (1934), 230, fig. 8.

Female dwarf, naked, standing, with a plait hanging on her left shoulder. Left forearm, right plait, and legs below knees missing.

121　Wood (h. 9.2). Cairo Museum, JdE 34.299. From the tomb of Sithedjhotep.

PM iv. 184; Weeks, *Anatomical Knowledge*, 173, B15.

Kamal, op. cit., 34; Regnault, op. cit., 141–2, fig. 3.

Female dwarf, naked, standing, with a long wig hanging in a lock over each shoulder.

Fig. 9.22

Dynasties 12/13

El-Lisht, South Pyramid Cemetery

122　Ivory (h. 6.5). Cairo Museum, JdE 63.858; New York, MMA, 34.1.130. From the tomb of Hepy.

Rupp, 'Zwerg', 303 1P–4P; Weeks, *Anatomical Knowledge*, 177, D6, 236, fig. 20; Hornemann, *Types*, 347.

A. Lansing, *BMMA* 29 (1934), part ii. 30–6, figs. 31–3; Hayes, *Scepter*, i. 222–3, fig. 139; Brunner-Traut, *Tanz*, 35, fig. 12; R. H. Randall, *Masterpieces of Ivory from the Walters Art Gallery* (New York, 1985), 35, figs. 11–12; *Musée égyptien du Caire*, no. 90; Dasen, 'Dwarfism', 266–7, pl. 3c.

Four dwarfs, naked, with garlands and collars around the neck and the torso, with uplifted arms or clasped hands. Three are set on a common base (now in Cairo), the fourth one (now in New York), stands on a square tenon.

Pls. 30.2, 31a, b

123   Glazed ware (h. 9.1). New York, MMA 24.1.47. From a tomb on the west side of the pyramid of Senwosret I.

Hornemann, *Types*, 1248.

A. Lansing, *BMMA* 19 (1924), part ii. 37–8, and 35, fig. 2 (middle row, second to left); B. J. Kemp and R. S. Merillees, *Minoan pottery in Second Millennium Egypt* (Mainz am Rhein, 1980), 167, pl. 22 (upper row, first to left).

Female dwarf, in a dress, with a puffed wig, standing, holding a child in her left arm.

Pl. 30.3

124   Glazed ware (h. 7.4). New York, MMA 24.1.48. From the same tomb group as above.

Hornemann, *Types*, 1260.

Lansing, op. cit., fig. 2 (middle row, second to right); Kemp/Merillees, op. cit., pl. 22 (upper row, first to right).

Female dwarf, in a dress, with a puffed wig, holding a child in her left arm.

Pl. 30.4

d 125   Glazed ware (h. 5.5). New York, MMA 24.1.49. From the same tomb group as above.

Hornemann, *Types*, 1264.

Lansing, op. cit., fig. 2 (middle row, middle figure); Kemp/Merillees, op. cit., pl. 22 (upper row, middle figure).

Female dwarf, in a dress, with a puffed wig, squatting, holding a child in her left arm.

El-Lisht, North Pyramid Cemetery

126   Glazed ware (h. 3.5). New York, MMA 15.3.60. From tomb 333.
Weeks, *Anatomical Knowledge*, 170, B4. Unpublished.
Dwarf, naked, standing on a base. Upper body missing.

127   Wood (h. 5.4). New York, MMA 15.3.414. From tomb 809.
J. Bourriau, *Pharaohs and Mortals: Egyptian Art in the Middle Kingdom* (Cambridge, 1988), 123. Unpublished.

Female dwarf, naked, standing. Upper body missing, hole drilled feet.

128   Glazed ware (h. 5.2). New York, MMA 22.1.125. From tomb 883(?).
Weeks, *Anatomical Knowledge*, 170, B5. Unpublished.
Dwarf, naked, standing on base, his left hand on his belly and his right hand in his mouth.
Pl. 32.1

129   Glazed ware (h. 4.5). New York, MMA 22.1.177. From tomb 885.
Rupp, 'Zwerg', 303 (1)–(15); Weeks, *Anatomical Knowledge*, 167, A13.
Hayes, *Scepter*, i. 222. Unpublished.
Dwarf, naked, standing on a base, carrying a pot on each shoulder.
Pl. 32.2

d 130   Glazed ware (h. 5.1). New York, MMA 22.1.179. From tomb 885.
Rupp, 'Zwerg', 303 (1)–(15); Weeks, *Anatomical Knowledge*, 168, A17.
Hayes, *Scepter*, i. 222. Unpublished.
Dwarf(?) naked, standing, holding an object in front of him. Legs missing.

131   Glazed ware (h. 8.6). New York, MMA 22.1.286. From tomb 964.
Rupp, 'Zwerg', 303 (1)–(15); Weeks, *Anatomical Knowledge*, 167, A14;
Hornemann, *Types*, 919.
Hayes, *Scepter*, i. 222; B. J. Kemp and R. S. Merillees, *Minoan Pottery in Second Millennium Egypt* (Mainz am Rhein, 1980), 138. Unpublished.
Dwarf, naked, standing on a base, holding an object in front of him.
Pl. 32.3

Surface Finds

d 132   Glazed ware (h. 8.6). New York, MMA 22.1.1140.
Rupp, 'Zwerg', 303 (1)–(15); Weeks, *Anatomical Knowledge*, 167, A15.
Hayes, *Scepter*, i. 222. Unpublished.
Dwarf(?), naked, squatting on a base, holding a tambourine(?) on his knees. Head missing.
Pl. 33.1

133   Glazed ware (h. 4.8). New York, MMA 22.1.1163.
Weeks, *Anatomical Knowledge*, 170, B6. Unpublished.
Female dwarf, naked, standing on a base, with a long wig hanging over each shoulder and down back.
Pl. 32.4

134   Glazed ware (h. 6.2). New York, MMA 1972.48.
Kemp/Merillees, op. cit., 139. Unpublished.
Female dwarf, naked, standing on a base. Shaven hair, necklace, bracelets, nails and pubic hair painted in black.
Pl. 33.2

El-Riqqa

d 135   Limestone (h. 17.4). New York, MMA 14.4.47. From cemetery A.
Weeks, *Anatomical Knowledge*, 171, B8.
R. Engelbach, *Riqqeh and Memphis*, vi (London, 1915) (BSAE 26), 16, pl. XII, nos.
8–9; Hayes, *Scepter*, i. 210.
Dwarf(?), in a long kilt, with a wig.
Inscr.: (back pillar) *nj-ḥrw* (name).

El-Lahun (Kahun)

d 136   Limestone (h. 30). London, UC 16520.
W. M. F. Petrie, *Kahun, Gurob and Hawara* (London, 1890), 26; Raven, *Pataekos*,
14, pl. 2.
Two dwarfs(?), naked, standing back to back, supporting a cup with raised arms.

d 137   Limestone (h. 15). London, UC 16522.
W. M. F. Petrie, *Illahun, Kahun and Gurob* (London, 1891), 11. Unpublished.
Head of dwarf(?) supporting a lamp on his head with his hands.

d 138   Limestone (h. 42.3). London, UC 16523/6.
Petrie, op. cit. Unpublished.
Dwarf(?), naked, standing, bearing a lamp(?) on his head. Feet missing.
Pl. 34.1

d 139   Limestone (h. 22). London, UC 16526.
Petrie, op. cit. Unpublished.
Torso of dwarf(?), part of a lampstand(?).

d 140   Limestone (h. 18.5). London, UC 16527.
Petrie, op. cit. Unpublished.
Torso of dwarf(?), part of a lampstand(?), holding his belly with both hands.

d 141   Glazed ware (h. 3.3). London, UC 16684.
W. M. F. Petrie, *Kahun, Gurob and Hawara* (London, 1890), 31, pl. VIII, no. 9; B.
J. Kemp and R. S. Merillees, *Minoan Pottery in Second Millennium Egypt* (Mainz am
Rhein, 1980), 139.
Head of dwarf(?) playing double pipe.

d 142   Clay block (14.5 × 14.5 × 22.5). Manchester Museum, 280.
W. M. F. Petrie, *Illahun, Kahun and Gurob* (London, 1891), 11, pl. VI, no. 10; A.
R. David, *The Pyramid Builders of Ancient Egypt* (London etc., 1986), 134, pl. 10;
Raven, *Pataekos*, 14.
Two dwarfs(?), naked, back to back, supporting a cup with raised arms.

143   Limestone. Present location unknown.
Petrie, op. cit., 11, pl. VI, no. 9; Raven, *Pataekos*, 14.
Female dwarf, naked, standing, supporting with both hands a dish on her head.
Fig. 9.24

El-Haraga

144   Glazed ware (h. 5.1). London, UC 18745. From tomb 55.
R. Engelbach, *Harageh* (London, 1923) (BSAE 28), 12, pl. xiv, no. 9.
Dwarf, naked, standing, hands raised. Legs missing below knees.
Pl. 33.3

El-Badari

d 145   Glazed ware (h. 3.7). Present location unknown. From tomb 4909 (of a very young child).
G. Brunton, *Qau and Badari*, i (London, 1927) (BSAE 44), 41, 64, pl. xxix. 15.
Dwarf(?), naked. Upper and lower limbs missing.

Abydos

146   Wood, painted (h. 18.4). Liverpool University, E 7081. From locus 352.
J. Bourriau, *Pharaohs and Mortals: Egyptian Art in the Middle Kingdom* (Cambridge, 1988), 122–3, no. 115.
Female dwarf, naked, standing, with hair divided into three plaits. She holds a child in her left arm. Back of head, left foot, and object held in the right hand missing.
Pl. 34.2

147   Glazed ware, frr. (h. 3.5). Oxford, Ashmolean Museum, E 3287. From tomb 416.
Kemp/Merillees, op. cit., 138–9, fig. 44, pls. 13–14.
Fragments of a dwarf, naked, standing on a base, his right arm raised and his left arm on his belly; necklace and pendant painted in black.

Thebes

d 148   Ivory (h. *c.*5). Present location unknown. From a tomb in the Ramesseum.
J. E. Quibell, *The Ramesseum* (London, 1898) (BSAE 2), 3, pl. ii, nos. 1–2.
Dwarf(?), naked, standing, carrying a calf on his shoulders.

Luxor?

149   Lead. Present location unknown. Formerly Cairo, I. Bey coll.
B. Schrumpf-Pierron, *Aesculape*, 24/9 (1934), 234–5, fig. 16.
Female dwarf, naked, standing.

Byblos, Deposit in the Temple of the Obelisks

The figurines are all in Beirut, National Museum; they are cited here by Dunand number: M. Dunand, *Fouilles de Byblos*, i, *1926–1932*, Atlas (Paris, 1937); id., *Fouilles de Byblos*, ii, *1933–1938* (Paris, 1958).

150   Glazed ware (h. 6.8). Dunand 15309. Standing with an oversized phallus.

151 Glazed ware (h. 6.6). Dunand 15310. Standing, with an oversized phallus.

152 Glazed ware (h. 6.4). Dunand 15311, pl. xcvii. Kneeling.

153 Glazed ware (h. 6.6). Dunand 15312, pl. xcvii; Hornemann, *Types*, 334. Standing, right hand to mouth.

154 Glazed ware (h. 5.9). Dunand 15313. Standing.

155 Glazed ware (h. 6.6). Dunand 15314. Squatting.

156 Glazed ware (h. 9.1). Dunand 15315. Standing with an oversized phallus.

d 157 Glazed ware (h. 5.4). Dunand 15316, pl. xcvi. Kneeling, holding a vessel(?).

d 158 Glazed ware (h. 6.2). Dunand no. 15317. Kneeling, holding a vessel(?).

159 Glazed ware (h. 6.2). Dunand 15318. A monkey(?) sits on his shoulders. Legs missing.

160 Glazed ware (h. 7.5). Dunand 15319, pl. xcvii. Squatting.

d 161 Glazed ware (h. 6). Dunand 15320, pl. xcvi. Squatting, holding an elongated object. Head missing.

162 Glazed ware (h. 5.1). Dunand 15321. Kneeling. Upper body missing.

d 163 Glazed ware (h. 3.8). Dunand 15322. Squatting(?), holding a vessel(?). Legs missing.

164 Glazed ware (h. 6.9). Dunand 15323. Standing(?). Lower legs missing.

d 165 Glazed ware (h. 6.5). Dunand 15324. Holding an elongated object.

d 166 Glazed ware (h. 6.1). Dunand 15325. Holding a vessel(?).

d 167 Glazed ware (h. 4.8). Dunand 15326. Squatting, holding an object. Head missing.

168 Glazed ware (h. 5.2). Dunand 15327, pl. xcviii. Kneeling, right hand to head. Oversized phallus. Lower legs missing.

d 169 Glazed ware (h. 5.7). Dunand 15329. Squatting, holding an object.

170 Glazed ware (h. 8.7). Dunand 15330. Female(?), standing, with puffed wig, the arms raised to her head.

171 Glazed ware (h. 5.9). Dunand 15331. Female(?), squatting, with puffed wig, left hand to her head.

172 Glazed ware (h. 7.9). Dunand 15332, pl. xcvii. Female(?), standing, with puffed wig, the arms raised to her head.

d 173 Glazed ware (h. 3.8). Dunand 15333, pl. xcvii. Kneeling.

d 174 Glazed ware (h. 6.7). Dunand 15336. Holding a quadruped. Lower legs missing.

d 175 Glazed ware (h. 3.8). Dunand 15337, pl. xcvii. Squatting(?), arms raised to head. Legs missing.

d 176 Glazed ware (h. 5.6). Dunand 15338, pl. xcvii. Half-kneeling, carrying a quadruped on his shoulders.

d 177   Glazed ware (h. 4.9). Dunand 15339. Carrying a quadruped on his shoulders. Lower legs missing.

d 178   Glazed ware (h. 5). Dunand 15340, pl. xcviii. Squatting, holding a vessel.

d 179   Glazed ware (h. 5.2). Dunand 15342. An animal(?) sits on his shoulders. Lower legs missing.

   180   Glazed ware (h. 6.2). Dunand 15343. Standing, holding in his right hand a dish or a tambourine(?), in his left a vessel.

d 181   Glazed ware (h. 3.5). Dunand 15344, pl. xcviii. Holding a vessel. Lower body missing.

d 182   Glazed ware (h. 6.1). Dunand 15345, pl. xcviii. Standing, the hair divided into three strands. Lower legs missing.

   183   Glazed ware (h. 10.5). Dunand 15347. Standing, carrying a quadrupéd on his shoulders.

d 184   Glazed ware (h. 8.4). Dunand 15348. Standing, with hair divided into three strands.

d 185   Glazed ware (h. 6). Dunand 15349, pl. xcviii. Squatting, with puffed wig(?). Lower legs missing.

d 186   Glazed ware (h. 7.1). Dunand 15350. Standing.

d 187   Glazed ware (l. 7.1). Dunand 15354, pl. xcvi. Lying on the back.

d 188   Glazed ware (l. 5.4). Dunand 15355. Lying on the back.

d 189   Glazed plaque (h. 6.5; l. 3.6). Dunand 15356. Standing, seen in profile.

d 190   Glazed ware (l. 5.3). Dunand 15359. Lying on the back. Oversized phallus.

Unknown Provenance

d 191   Glazed ware (h. 5.2). Boston, MFA 11.1524.
       Hornemann, *Types*, 1339.
       B. J. Kemp and R. S. Merillees, *Minoan Pottery in Second Millennium Egypt* (Mainz am Rhein, 1980), 139.
       Dwarf(?), naked, kneeling, carrying a calf(?) on his shoulders.
       Fig. 9.23

   192   Glazed ware (h. 7.2). Cambridge, Fitzwilliam Museum, E 60.1984.
       J. Bourriau, *Pharaohs and Mortals: Egyptian Art in the Middle Kingdom* (Cambridge, 1988), 121–2, no. 114.
       Dwarf naked, standing. Legs missing.

   193   Glazed ware (h. 6.5). Cambridge, Fitzwilliam Museum, E FG 37.
       Unpublished.
       Dwarf naked, squatting, with hands over ears; necklace and belt painted in black.
       Pl. 33.4

194 Glazed ware (h. 6.5). London, BM 22882.
E. Denison Ross (ed.), *The Art of Egypt through the Ages* (London, 1931), 150, fig. 1 (third from left).
Dwarf, naked, standing.

195 Ivory (h. 6.9). London, BM 58409.
Weeks, *Anatomical Knowledge*, 171, B9. Unpublished.
Dwarf naked, standing on a base, his right arm on his chest.
Pl. 35.1a, b

d 196 Glazed ware (h. 5). London, BM 65679.
Kemp/Merillees, op. cit., 139; Denison Ross (ed.), op. cit., 150, fig. 1.
Dwarf(?), naked, squatting, drinking from a vessel.

197 Wood (h. 10.9; w. 4.3). Paris, Louvre, E 14696.
Vandier, *Manuel*, iii. 239, pl. LXXXII. 1; A.-P. Leca, *La Médecine égyptienne au temps des pharaons* (Paris, 1971), pl. 7; E. Delange, *Catalogue des statues égyptiennes du Moyen Empire, 2060–1560 av. J. C.* (Paris, 1987), 170–2.
Dwarf, in a kilt, standing, the right fist clenched, the left palm turned outwards.
Inscr.: 'An offering which the King gives to Ptah and Sokar (that they may give) an offering of bread and beer, foods and nourishments, ointment and clothing for the Lady of the house, Itasenbet(?) Mereryt, the justified' (After Delange, op. cit., 171; the plinth probably does not belong to the statue).
Pl. 34.3a, b

NEW KINGDOM (1539–1075? BC)

Dynasty 18

El-ᶜAmarna

198 Boxwood (h. 5.9). Boston, MFA 48.296.
Weeks, *Anatomical Knowledge*, 168, A18.
B. v. Bothmer, *BMFA* 47/267 (1949), 9–11, figs. 1–2, 4.
Dwarf in a kilt, standing, carrying a jar on his left shoulder.
Inscr.: on the body of the vase, cartouches of King Akhenaten and of Queen Nefertiti.
Pl. 35.2

d 199 Black stone (h. 11). London, BM 37201.
Weeks, *Anatomical Knowledge*, 168, A19; Hornemann, *Types*, 338.
v. Bothmer, op. cit., 10. Unpublished.
Dwarf(?) in a kilt, standing, carrying a jar on his left shoulder.

d 200 Calcite (h. 19.5). New York, MMA 17.190.1963.
Weeks, *Anatomical Knowledge*, 169, A20; Hornemann, *Types*, 339.
Hayes, *Scepter*, ii. 314, 316, fig. 198.

Dwarf(?) in a pleated kilt, standing, carrying a jar on his left shoulder.
Pl. 35.3

Balabish

d 201   Figure vase, calcite. Ithaca, Cornell University Museum, -.
Weeks, *Anatomical Knowledge*, 178, D7.
G. A. Wainwright, *Balabish* (London, 1920) (MEEF 37), 56, pl. xx; W. R.
Dawson, *Ann. Med. Hist.* 9/4 (1927), 320, fig. 24; E. Brunner-Traut, in A.
Kuschke and E. Kutsch (eds.), *Festschrift K. Galling* (Tübingen, 1970), 39, no. 10.
Female dwarf(?), naked, with a belt, holding a lute.

Thebes, Tomb of Tutankhamun

202   Calcite boat in pylon-shaped tank (h. 67.5; w. 70). Cairo Museum, JdE 535.
Weeks, *Anatomical Knowledge*, 175, C2.
H. Carter, *The Tomb of Tut-ankh-Amen*, iii (London, 1933), 127–30, pls. XLI,
LXXIVb; B. Schrumpf-Pierron, *Aesculape*, 24/9 (1934), 227, fig. 5, 229, fig. 7; H.
Murray and M. Nuttal, *A Handlist to H. Carter's Catalogue of Objects in
Tutⁱankhamun's Tomb* (Oxford, 1963), 17, no. 578 (Tutⁱankhamun's tomb series I);
H. Carter, *The Tomb of Tutankhamen* (London, 1972), 207–8 (with fig.).
Female dwarf, naked, with a curled wig and bracelets, standing on the prow of the
boat, holding a pole.
Pl. 36a–c

Serabit el-Khadim (Sinaï), Temple of Hathor

203   Figure vase, calcite (h. 18). Oxford, Ashmolean Museum, 1911.407.
W. M. F. Petrie, *Researches in Sinaï* (London, 1906), 137–8; E. T. Leeds, *JEA* 8
(1922), 2, pl. II. 3.
Dwarf, naked, standing, carrying a large vase on his back.
Pl. 37.1

Unknown Provenance

204   Figure vase, terracotta (h. 9). London, BM 29935.
E. A. W. Budge, *The Mummy: A Handbook of Egyptian Funerary Archaeology*[2]
(Cambridge, 1925), 392, pl. XXVIII.
Dwarf, naked, standing, carrying a pot on his left shoulder.
Pl. 37.3

205   Figure vase, calcite (h. 20). London, BM 30459.
Hornemann, *Types*, 888.
Budge, op. cit., pl. XXVIII; E. Brunner-Traut, in A. Kuschke and E. Kutsch (eds.),
Festschrift K. Galling (Tübingen, 1970), 38–9, no. 9, pl. 6.

Female dwarf, naked, with a belt, holding a lute.
Pl. 37.2

206   Figure vase, calcite (h. 10). London, UC 15758.
Unpublished.
Dwarf, naked, standing, carrying a large amphora. Feet missing.
Pl. 37.4

LATE PERIOD (750?–332 BC)

Unknown Provenance

d 207   Bronze (h. 10). Cairo Museum, JdE 27.708.
Rupp, 'Zwerg', 304, 4P, figs. 1–2.
G. Daressy, *ASAE* 4 (1903), 124–5; W. R. Dawson, *Ann. Med. Hist.* 9/4 (1927),
320, fig. 25; B. Schrumpf-Pierron, *Aesculape*, 24/9 (1934), 236, fig. 22.
Dwarf(?), standing, naked, holding a lost object in the right hand. Lower legs
missing.
Inscr.: Ḏd-ḥr (name).

G: Greek World

DRAWINGS AND RELIEFS

Within each section the material is arranged chronologically and according to location. Athenian vases attributed by Sir John Beazley (*ABV*, *ARV*) are listed after his catalogue numbers, non-attributed vases are arranged in museums' alphabetic order.

A. HUMAN DWARFS

ARCHAIC PERIOD (7th–6th century BC)

Attico-Boeotian

d 1 Pyxis (h. 3.9). Paris, Louvre, CA 1707. From Thebes.
CVA France 26, Louvre 17, 35–6, pl. 32.5.
Dwarf(?) naked, with a beard and a wreath around his neck, standing below a floral crown, between men and women taking part in a ball game.
Pl. 41a–c

Corinthian

d 2 Pinax fr. (h. 10.5; w. 7.8). Berlin, SM (West) F 831.
A. Furtwängler, *Antike Denkmaeler*, i (Berlin, 1887), pl. 8.3b; id., *Beschreibung der Vasensammlung im Antiquarium*, i (Berlin, 1885), 90, no. 831.
A. Dwarf (or miniature malformed slave), naked(?), digging in a hole with a pickaxe. Before the cave stands a draped bearded man with boots, hair bound by a fillet; above the cave flies a human-headed bird.
B. Boat.
Fig. 15.1

CLASSICAL PERIOD

Athenian (attributed)

3 Skyphos (h. 7.5). Paris, Louvre, G 617. From Capua.
ARV 768.33, *Add²* 287 (Manner of Sotades P.).
E. Pottier, *Vases antiques du Louvre*, iii (Paris, 1922), 294, pl. 157; P. Ghiron-Bistagne, *Recherches sur les acteurs dans la Grèce antique* (Paris, 1976), 150–2, figs. 59–60; Dasen, 'Dwarfism', 271–2, pl. 5c.

A. Dwarf, naked, infibulated, balding, with a beard, dancing.
B. Idem.
Cover; Pl. 42a, b

4   Rhyton, donkey-head (l. 20.5). Ferrara, Museo Nazionale, 2561. From Spina,
    Valle Trebba, T 392.
    ARV[1] 730.5 (Manner of the Eretria P.), *ARV*[2]-.
    S. Aurigemma and N. Alfieri, *Il Museo Nazionale di Spina in Ferrara* (Ferrara,
    1935), 130–1, pl. 71; S. Aurigemma, *Scavi di Spina*, i (Rome, 1960), 223–4, pl.
    224; H. Hoffmann, *Attic Red-Figured Rhyta* (Mainz am Rhein, 1962), 36, no. 97; A.
    Lezzi-Hafter, *Der Eretria-Maler. Werke und Weggefährten* (Mainz am Rhein, 1988),
    268, 348, no. 259, pl. 171a–d.
    Three dwarfs, naked, balding, with beards, the left shoulder covered by a
    chlamys, run with a dog.
    Pl. 43a–c

5   Aryballos (h. 9). Paris, Louvre, CA 2183.
    ARV 813.96, *Para* 420, *Add*[2] 291–2 (Clinic P.).
    E. Pottier, *MonPiot*, 13 (1906), 149–66, pl. 13; Boardman, *ARFH* i, fig. 377; E.
    Berger, *Das Basler Arztrelief* (Basle, 1970), 75, 77, figs. 91–5; *Médecine antique. IVe
    colloque international hippocratique* (Lausanne, 1981), 57–9, no. 23 (with fig.); Dasen,
    'Dwarfism,' 270–1, pl. 4c.
    Dwarf, naked, infibulated, balding, with a beard, carrying a hare on his left
    shoulder between two draped men in a surgery.
    Pl. 44a–d

6   Pelike (h. 15). Agrigento, Museo Civico, ex Giudice 638.
    ARV 854.i4 (reverse by the Sabouroff P., painter of the obverse unknown).
    Unpublished.
    A. Dwarf, naked, with a beard, carrying a basket on his head behind a woman.
    B. woman.
    Pl. 45

7   Cup fr. (Dm. 19.2). Ferrara, Museo Nazionale, 20363. From Spina, Valle Pega, T
    19 C.
    ARV 934.67 *bis* (Curtius P., Penthesilea workshop). Unpublished.
    I: Dwarf, naked, balding, with a beard, the left shoulder covered by a chlamys,
    walking behind a draped youth.
    Pl. 46.1

8   Pelike (h. 24.1). Boston, MFA 76.45. From Capua.
    ARV 1011.13, *Para* 440, *Add*[2] 314 (Dwarf P.).
    Caskey/Beazley, i. 52, no. 59, pl. 27; E. Paribeni, *EAA* v (1963), 329, fig. 443;
    Boardman, *ARFH* ii, fig. 120.

Dwarf naked, balding, with a beard, the left shoulder covered by a chlamys, walking with a dog between two draped youths.
Pl. 46.2a, b

9   Stamnos fr. (h. 8.5; w. 12.5). Erlangen, University, I 707.
    ARV 1039.6 (Peleus P.).
    G. Lippold, *MDAI(R)* 52 (1937), 44–7, pl. 14; Dasen, 'Dwarfism', 271, pl. 4d.
    Dwarf naked, balding, with a beard, with ivy tendrils around his chest and left arm, dancing on a table before two musicians with pipes and krotala (hands only preserved).
    Inscr.: [ΗΙΠΠΟ] ΚΛΕΙΔΕΣ
    Pl. 47.1

10 Pelike (h. 14). St Petersburg, Hermitage, b 1621 (740; St. 1611).
    ARV 1134.11, *Add*² 333 (Manner of the Washing P.).
    E. N. Gardiner, *Athletics of the Ancient World* (Oxford, 1930), 84–5, fig. 50; F. A. G. Beck, *Album of Greek Education* (Sydney, 1975), 36, no. 206, pl. 39.
    Two dwarfs, naked, balding, with beards: one is kicking a punching-bag before his companion standing, holding an undetermined object in his right hand (a bag for knucklebones?).
    Pl. 47.2

11 Cup. Formerly Munich, Preyss coll. From Campania (Curti or Capua).
    ARV 1135.21 (Manner of the Washing P.). Unpublished.
    I: Dwarf, naked, with a short beard, hair bound by a fillet, running with an undetermined object in his left hand.
    Pl. 48.1

12 Pelike (h. 12). Laon Museum, 37.1031.
    ARV 1142.2 (P. of Munich 2358, related to the Washing P.).
    CVA France 20, Laon 1, III, I, 25, pl. 33.6.
    A. Two dwarfs, naked, balding, hair bound by fillets, eating at a large vessel.
    B. Draped youth.
    Pl. 48.2

(unattributed)

13 Cup fr. (h. 7.5; Dm. of medallion 10.7). Athens, Agora Museum, P 2574. From the agora.
    E. Vanderpool, *Hesperia*, 15 (1946), 282, no. 40, pl. 31; M. Robertson, in *Stele: Tomos eis Mnemen N. Kontoleontos* (Athens, 1980), 125–9, pl. 43a (Kleophrades P.); Dasen, 'Dwarfism', 271, pl. 5a.
    I: Draped dwarf, with a moustache, crowned with a wreath, holding a stick beside a seated woman, head covered by a veil, holding a mirror.
    Pl. 49.1

14 Chous (h. 9.5). Dresden, Albertinum, ZV 1827. From Attica.
Brommer, *VL* 548 Ba 1.
G. Van Hoorn, *Choes and Anthesteria* (Leiden, 1951), 125, no. 505, fig. 240 (akin to
the Sotades P.).
Dwarf naked, with a beard, holding a stick and a skyphos.
Pl. 49.2

15 Cup fr. (h. 4.5; l. 13.7; w. 8.5). Hamburg, Museum für Kunst und Gewerbe,
1984.456.
W. Hornbostel *et al.*, *Aus Gräbern und Heiligtümern* (Mainz am Rhein, 1980),
142–3, no. 83 (with fig.).
I: Two dwarfs naked, balding, with beards, dancing.
Pl. 50.1

16 Skyphos (h. 6). Munich, Antikensammlungen, 8934.
M. Robertson, in A. Cambitoglou (ed.), *Studies in Honour of A. D. Trendall*
(Sydney, 1979), 129–34, pl. 34.3–4 (Kleophon P.); Dasen, 'Dwarfism', 271, pl.
5b.
A. Female dwarf naked, with a headband adorned with branches, holding out a
skyphos.
B. Table with skyphos before a winged phallic pillar crowned with a basket.
Pl. 51a, b

17 Cup fr. (Dm. 9.5). Oxford, Ashmolean Museum, 1938.312. From Al-Mina.
J. D. Beazley, *JHS* 59 (1939), 10–11, fig. 30.
I: Dwarf naked, with a beard, dancing on a platform.
Pl. 50.2

18 Oinochoe (h. 18.3). Oxford, Ashmolean Museum, 1971.866.
Robertson, op. cit., 129–34, pl. 34.1–2 (Phiale P.); J. H. Oakley, *The Phiale
Painter* (Mainz am Rhein, 1990), 38–9, 94, R3 quater, pl. 143.
Dwarf naked, dancing towards a woman muffled in her himation.
Pl. 52a, b

19 Cup fr. (h. 4.6; Dm. 14.6). Paris, Louvre, CA 5909.
Unpublished.
I. Dwarf naked, with a beard, standing with a walking-stick.
Pl. 53.1

20 Cup fr. (Dm. 23.5; Dm. of medallion 13.5). Todi, Museo Civico, 471.
CVA Italy 16, Musei Comunali Umbri 1, III, I, d, 4–5, pl. 4, fig. 4; G. Fabrini, in
Todi Preromana (Perugia, 1977), 63, pl. 41a (P. of Todi 474).
I: Two dwarfs naked, with beards, dancing on a table.
Pl. 53.2

21 Oinochoe(?) fr. (h. 3.7). Tübingen, University, 1616.
C. Watzinger, *Griechische Vasen in Tübingen* (Reutlingen, 1924), 39, no. 39, pl. 19.

Upper part of a draped dwarf with a beard, hair bound by a fillet, perhaps lying on a kline.
Pl. 53.3

22   Skyphos (h. 7.6). Yale University, R. Darlington Stoddart Coll., 160. From Greece.
P. V. C. Baur, *Catalogue of the R. Darlington Stoddart Collection of Greek and Italian Vases in Yale University* (New Haven, Conn., 1922), 107, no. 160, pl. 13.
A. Dwarf naked, balding, with a beard, dancing before a skyphos set on the ground.
B. Id., but hair bound by a fillet.
Pl. 54a, b

23   Bell krater (h. 29.1). Zurich market, Arete Gallery.
Arete, Galerie für antike Kunst, *Griechische Schalen und Vasen, Liste 20*, Zurich (n.d.), no. 37 (Circle of the Dinos and Meidias P.).
Dwarf naked, with a large headband, holding a tympanon, dancing on a table before Dionysos, a satyr, and a maenad with pipes.
Pl. 55.1

FOURTH CENTURY

Athenian

24   Loutrophoros (h. 89). Athens, NM 2563.
C. Blümel, *MDAI(A)* 51 (1926), 57–8, fig. 1.
Dwarf with a beard, a helmet and a large shield, standing behind a departing warrior.
Pl. 55.2

25   Loutrophoros (h. 164). Athens, NM -.
Blümel, op. cit., 58–9, figs. 2 and 3.
As above no. 24.

South Italy

Apulian

26   Askos (h. 12.5). Basle, Antikenmuseum, Z 303.
Trendall, *PhV²* 88.202 (Felton P.); id., in *Essays and Studies in Honour of Sir Daryl Lindsay* (Oxford, 1964), 49, fig. 32.
Dwarf, naked, running with a basket.
Pl. 56.1

27   Oinochoe (h. 17.3). Melbourne, National Gallery, 90/5.
Trendall/Cambitoglou, *RVAp* i. 172.49; Trendall, *PhV²* 86.195, pl. XIIIa (Felton P.); id., *Red Figure Vases of South Italy and Sicily* (London, 1989), 79, fig. 136.

Two dwarfs, wrapped in himatia, standing around Marsyas, Apollo, and Olympos.
Pl. 56.3a, b

28 Oinochoe (h. 20). Milan market (formerly Oria, Pasanisi coll.).
Trendall/Cambitoglou, *RVAp* i. 177.90; Trendall, *PhV*² 86.196, pl. xiid (Felton P.).
Two naked dwarfs running.

d 29 Amphora. Milan market (formerly Oria, Pasanisi coll.).
Trendall/Cambitoglou, *RVAp* ii. 1032.107, pl. 399, 5–6 (P. of BM F 339).
Dwarf(?), naked, standing beside a horse.

30 Oinochoe (h. 18). Taranto, Museo Nazionale, I.G. 52571. From Taranto.
Trendall/Cambitoglou, *RVAp* i. 177.89; Trendall, *PhV*² 87.198 (Felton P.). A. D. Trendall, in *Essays and Studies in Honour of Sir Daryl Lindsay* (Oxford, 1964), 49, fig. 31; C. Belli, *Il tesoro di Taras* (Milan and Rome, 1970), 56–7, fig. 174.
Dwarf, naked, running.

31 Oinochoe (h. 20). Taranto, Ragusa coll., 7.
Trendall/Cambitoglou, *RVAp* i. 177.93; Trendall, *PhV*² 86.197, pl. xiic ('Ragusa coll. 10')(Felton P.?).
Dwarf, naked, running.
Pl. 56.2.

32 Oinochoe. Taranto, Ragusa coll., 120.
Trendall/Cambitoglou, *RVAp* i. 177.92 (Felton P.?).
Dwarf, naked, running. Unpublished.

33 Pelike. Taranto, Ragusa coll., 127.
Trendall/Cambitoglou, *RVAp* i. 173.54 (Felton P.).
K. Schauenburg, *MDAI(R)* 79 (1972), 318, pl. 133.2.
Dwarf, wrapped in himation, standing before a young man reclining and two women.
Pl. 57.1

34 Oinochoe (h. 21). Toledo, Museum of Art, 67.136.
Trendall/Cambitoglou, *RVAp* i. 172.50; Trendall, *PhV*² 85.193, pl. xiiib (Felton P.).
Schauenburg, op. cit., 318, pls. 131.2; 132; R. Hurschmann, *Symposienszenen auf unteritalischen Vasen* (Würzburg, 1985), 108–9, 174, pl. 13 (top left).
Dwarf, wrapped in himation, standing before Dionysos reclining and a maenad; a maenad and a satyr dance behind Dionysos.
Pl. 57.2

35 Oinochoe. Turin, private.
Trendall/Cambitoglou, *RVAp* i. 177.91 (Felton P.?).

Zurich, *Koller, sale catalogue*, 32 (Oct.–Nov. 1974), no. 3771, pl. 62.1; A. Cambitoglou, in E. Böhr and W. Martini (eds.), *Studien zur Mythologie und Vasenmalerei. K. Schauenburg zum 65. Geburtstag* (Mainz am Rhein, 1986), 146, pl. 26.5.

Dwarf, naked, with boots, running with sword and shield, followed by a dwarf with chiton, pilos, boots, and holding a shield (Diomedes and Odysseus pursuing Dolon?).

d 36 Oinochoe. Zurich market, Arete Gallery.
Trendall/Cambitoglou, *RVAp* i. 209.140 (Workshop of the Iliupersis P.): 'Youth with bird followed by dwarf attendant, with stick in right hand and cithara in left.' Unpublished.

37 Oinochoe (h. 20). New York market, Atlantis Gallery.
Atlantis Antiquities, Greek and Roman Art (New York, 1990), no. 8 (Felton P.).
Dwarf, seen in front, wrapped in himation, crouching, raising his dress.

38 Oinochoe. Zurich market, Nefer Gallery.
Cambitoglou, op. cit., 146, pl. 26.6 (Felton P.?).
Dwarf, naked, with boots, running, holding a shield or a tambourine.

Campanian

39 Skyphos (h. 13.5). Laon Museum, 37.1075.
Trendall, *PhV*^{2} 88.201 (Capua Silen P.); id., *LCS*, 286.437, pl. 116.6.
Dwarf, naked, standing, holding an undetermined object.

B. PYGMIES

ARCHAIC PERIOD (7th–6th century BC)

Athenian (attributed)

40 Volute krater (h. 66). Florence, Museo Etrusco, 4209. From Chiusi.
ABV 76.1 (682), *Para* 29, *Add*^{2} 21 (Kleitias); Brommer, *VL* 546 A1.
A. Furtwängler and K. Reichhold, *Griechische Vasenmalerei*, i (Munich, 1900), pl. 3.9; O. Waser, Roscher, iii. 2 (1902–9), 3291, fig. 1; G. Becatti, *EAA* vi (1965), 166–8, figs. 181–2; Boardman, *ABFH* 34, fig. 46.8; *Bollettino d'Arte*, 1 serie speciale (Rome, 1981), 146–7, figs. 102–5.
Foot: Nineteen pygmies, with slings, curved sticks, and clubs, fighting against fourteen cranes. The pygmies are naked, apart from two in short chitons(?); three have beards and one wears a pilos. Five pygmies are riding goats.
Pl. 58a–d

41 Aryballos (h. 7.8). New York, MMA 26.49. From Attica.
ABV 83.4 (682), *Para* 30, *Add*^{2} 23 (Nearchos); Brommer, *VL* 546 A2.

G. M. A. Richter, *AJA* 36 (1932), 272–5, pls. 10–11; E. Homann-Wedeking, *La Grèce archaïque* (Paris, 1968), 115 (with fig.), 116–17; Boardman, *ABFH* 35, fig. 50.

Rim: Ten naked pygmies, with clubs, fighting against eight cranes.

Pl. 59.1a–c

42 Cup (h. 16.8; Dm. 28.4). Taranto, Museo Nazionale, I.G. 4435. From Taranto.

ABV 159.1, *Para* 67 (Antidoros); Brommer, *VL* 547 A3.

J. C. Hoppin, *A Handbook of Greek Black-Figured Vases* (Paris, 1924), 53 (with fig.); *CVA* Italy 35, Taranto 3, III, H, 6–7, pl. 30.2.

A. Pygmies, naked, fighting against cranes (the scene is incomplete because of missing frr.).

B. Boar hunt.

Pl. 59.2

43 Cup (h. 16; Dm. 27.4). Würzburg, Martin von Wagner Museum, L 414.

ABV 160 (Antidoros); Brommer, *VL* 547 A4.

E. Langlotz, *Griechische Vasen in Würzburg* (Rome, 1968), 77, no. 414, pl. 114; E. Simon (ed.), *Führer durch die Antikenabteilung des Martin Von Wagner Museums der Universität Würzburg* (Mainz am Rhein, 1975), 111; Dasen, 'Dwarfism', 270, pl. 4a.

A. Seven naked pygmies, with clubs(?), fighting against seven cranes. One pygmy has a beard.

B. Six naked pygmies, with clubs(?), fighting against seven cranes.

(unattributed)

d 44 Lekythos fr. (h. 4.1). Athens, Agora Museum, P 1719. From the N. slope of Acropolis.

C. Roebuck, *Hesperia*, 9 (1940), 221–2, no. 203, fig. 42.

Head and outstretched arm of a pygmy, crane, right hand of another pygmy.

45 Cup fr. (Dm. *c.*18). Berlin, SM (West) F 1785.

Brommer, *VL* 547 A5.

A. Furtwängler, *Beschreibung der Vasensammlung im Antiquarium*, i (Berlin, 1885), 298, no. 1785; B. Freyer-Schauenburg, in *Wandlungen. Studien zur antiken und neueren Kunst, E. Homann-Wedeking gewidmet* (Waldsassen, 1975), 78–9, pl. 15c.

Three naked pygmies, with beards, holding clubs, fighting against three cranes.

Pl. 60.1

46 Cup fr. (h. 2). Berlin, SM (West), Brommer coll., 249.

M. Krumme, *Kunst und Archäologie. Die Sammlung Brommer* (Berlin, 1989), 54, no. 249.

Pygmy in a chiton, with a club, holding the neck of a crane.

Pl. 60.2

47 Hydria (h. 34). Paris, Louvre, F 44. From Etruria.
Brommer, *VL* 547 A6.
CVA France 9, Louvre 6, III, He, pl. 65.5–6; G. Koch-Harnack, *Knabenliebe und Tiergeschenke* (Berlin, 1983), fig. 58.
Shoulder: A. Seven naked pygmies, with short curved knives and clubs, fighting against five cranes. Three pygmies have beards.
B. Wedding cart.
Pl. 60.3

48 Hydria (h. 31.1). Rome, Villa Giulia, 50425.
Brommer, *VL* 547 A7.
P. Mingazzini, *Catalogo dei vasi della collezione Castellani*, i (Rome, 1930), 201–2, no. 438, pl. 49.4.
Shoulder: Three naked pygmies, with beards, holding swords, fighting against four cranes.
Body: Komos.
Pl. 61.1a, b

Boeotian

d 49 Skyphos (h. 4.7). Athens, Kanellopoulos Museum, 548.
J.-J. Maffre, *BCH* 99 (1975), 430, 437–8, no. 7, fig. 7a–d.
A. Pygmy in a chiton, running with a club between two cranes.
B. Id.

Corinthian

d 50 Skyphos, fr. (h. 3.8; l. 4.1). Athens, Agora Museum, P 13853. From the agora.
Unpublished.
Pygmy in a chiton, holding the neck of a crane.

d 51 Fr. (h. 33). Athens, NM –. From Perachora.
T. J. Dunbabin, in *Perachora*, ii (Oxford, 1962), 92, no. 788, pl. 35, fig. 6.
Part of body and lower arms of a man holding a spear which pierces a large bird.

d 52 Fr. (h. 40). Athens, NM –. From Perachora.
Dunbabin, op. cit., 96–7, no. 847, pls. 23, 57.
A large bird facing three men, one with a shield, two with spears; tail and hindleg of a dog.

53 Clay altar (h. 13.2). Corinth Museum, M.F. 8953. From Corinth.
M. Hamilton Swindler, *AJA* 36 (1932), 512–20, pl. F, figs. 1–2; O. Broneer, *Hesperia*, 16 (1947), 214–23, pl. 21; C. M. Robertson, *Greek Painting* (Geneva, 1959), 81, fig. on 79.
A. Pygmy, naked, with a club, grasping the neck of a crane.
B. Lion.
Pl. 61.2

d 54 Fr. (h. 8). Leiden, Rijksmuseum, I.1905/1.31. From Rhodes.

J. P. J. Brants, *Description of the Ancient Pottery Preserved in the Department of Greek and Roman Antiquities*, ii (The Hague, 1930), 11, no. 9, pl. xii; T. J. Dunbabin, in *Perachora*, ii (Oxford, 1962), 97.

Large bird and two kneeling archers facing a third archer (bow only preserved).

Laconian

d 55 Cup fr. (h. *c*.3.5; w. *c*.5). Samos, Vathy Museum, K 176. From Samos. W. Technau, *MDAI(A)* 54 (1929), Beil. xvi, 1; B. Shefton, *BSA* 49 (1954), 307, no. 11 (Hunt P.); C. M. Stibbe, *Lakonische Vasenmaler des 6. Jhs. v. Chr.* (Amsterdam and London, 1972), 247, 280, no. 206a, pl. 68.3; M. Pipili, *Laconian Iconography of the Sixth Century BC* (Oxford, 1987), 3, fig. 5, 39–40, 114, no. 100.

I: Pygmy(?), naked, lying on the ground, attacked by a bird; Heracles and the hydra.

East Greek or with East Greek Connections

d 56 Amphora (h. 32.4). Formerly Castle Ashby. From Etruria.

Brommer, *VL* 548 c2.

J. D. Beazley, *PBSR* 11 (1929), 1–2, pl. 1.3; E. Walter-Karidy, *Samos*, vi. 1 (Bonn, 1973), 144, no. 932, pl. 129; E. Simon and M. and A. Hirmer, *Die griechischen Vasen* (Munich, 1976), 61–2, pl. 16; *CVA* Great Britain 15, Castle Ashby, 1–2, pls. A, 1–3.

A. Two small, naked men riding cranes and holding clubs.

B. Dionysos and satyrs.

Pl. 62.1

57 Amphora frr. Munster, University, 292/293 (formerly Berlin, Rubensohn coll.). From Cyprus(?).

Brommer, *VL* 548 c1.

R. M. Cook, *BSA* 34 (1933–4), 18, k2, pl. ix; E. Walter-Karydi, *Samos*, vi. 1 (Bonn, 1973), 63, pl. 84, no. 613; B. Freyer-Schauenburg, in *Wandlungen. Studien zur antiken und neueren Kunst, E. Homann-Wedeking gewidmet* (Waldsassen, 1975), 77, pl. 15a–b.

Two naked pygmies with clubs and three cranes; komos.

58 Diadem, gold (l. 22). Once in Rhodes Museum. From Rhodes, Ialysos necropolis, tomb 10.

L. Laurenzi, *Clara Rhodos*, 8 (1936), 112–13, no. 8 (A 1934), figs. 100–1.

Eight naked pygmies, with clubs, fighting against five cranes.

CLASSICAL PERIOD

Athenian (attributed)

d 59 Cup (Dm. 41.9). London market, Sotheby.

ARV 66, 127 quater, *Para* 328 (Oltos P.).

Sotheby, sale catalogue (26 Nov. 1968), 47, no. 122 (with fig.).

A. Two naked men, with clubs, fighting against a crane.

B. Warrior with horse and groom.

60   Rhyton, hound-head (h. 17). St Petersburg, Hermitage, b 1818 (679; St. 360).
ARV 382.188 (1649), *Add²* 228 (Brygos P.); Brommer, *VL* 547 B1.
H. Hoffmann, *Attic Red-Figured Rhyta* (Mainz am Rhein, 1962), 10, no. 9, pl. II.4;
Boardman, *ARFH* i, fig. 258; I. V. Stal, *Klio*, 68 (1986), 353, figs. 2–3; Dasen,
'Dwarfism', 270, pl. 4b.
A. Two naked pygmies, with beards and soft pointed caps, attacking a crane; they
hold a club and a sword.
B. Pygmy, naked, with a beard, a Scythian cap and boots, attacked by two cranes;
he holds a club and a bow, the left arm covered with an animal skin.
Pl. 62.2a, b

61   Hydria (h. 22.5). Bologna, Museo Civico, 169. From Bologna.
ARV 571.80 (Earlier Mannerists); Brommer, *VL* 547 B2.
G. Pellegrini, *Catalogo dei vasi greci delle necropoli felsinee* (Bologna, 1912), 57–8,
fig. 35, no. 169.
Three naked pygmies, balding, with clubs, fighting against three cranes.
Pl. 63.1

62   Hydria, fr. (h. 8.8). Athens, Agora Museum, P 8892. From the Agora.
ARV 587.63, *Add²* 263 (Earlier Mannerists); Brommer, *VL* 547 B3.
R. D. Lamberton and S. I. Rotroff, *Birds of the Athenian agora* (Princeton, 1985)
(Picture book 22), 20, fig. 38.
Two naked pygmies fighting against two cranes; on the left the pygmy has a
round shield.
Pl. 63.2

63   Rhyton, figure vase (h. 30.3). Boston, MFA 03.799. From Nola.
ARV 737.125 (Karlsruhe P.)
F. G. Lo Porto, in *Locri Epizefiri*, xvi (Naples, 1977), 741, pl. CXII.
Pygmy, naked, balding, with a beard, carrying a dead crane.

64   Rhyton, figure vase (h. 29.8). Basle, Antikenmuseum (private coll.).
ARV 766.2 bis (1669), *Para* 415 (Manner of Sotades); Brommer, *VL* 547 B13.
Auktion MM 34 (May 1967), no. 117, pl. 29; K. Schefold and F. Jung, *Die
Urkönige, Perseus, Bellerophon, Herakles und Theseus* (Munich, 1988), 169, fig.
205a–c.
Pygmy, naked, with a beard and curly hair, carrying a dead crane.

65   Rhyton, figure vase (h. 30). Erlangen, University, P I 864. From Italy.
ARV 766.3, *Para* 415, *Add²* 286 (Manner of Sotades); Brommer, *VL* 547 B4.
E. Buschor, 'Das Krokodil des Sotades', *MJBK* 11, 1/2 (1919), 23–5, figs. 36–7.

H. Sichtermann, *Die griechische Vase* (Berlin, 1963), 66, fig. 11.
Pygmy, naked, carrying a dead crane.

66   Rhyton, figure vase (h. 29.7). Bonn, University, 545.
      ARV 766.4 (Manner of Sotades); Brommer, *VL* 547 B5.
      Buschor, op. cit., 18, fig. 28; *CVA* Germany 1, Bonn 1, 25, pl. 24.2, 5.
      Pygmy, naked, with a beard, carrying a dead crane.
      Pl. 63.3a, b

67   Rhyton, boar-head (h. 24). Compiègne, Musée Vivenel, 898. From Nola.
      ARV 767.16 (Manner of Sotades); Brommer, *VL* 547 B7.
      O. Waser, Roscher, iii. 2 (1902–9), 3295, fig. 5; *CVA* France 3, Compiègne, pl.
      18.16, 23 and pl. 20.10; H. Hoffmann, *Attic Red-Figured Rhyta* (Mainz am Rhein,
      1962), 21, no. 46; Boardman, *ARFH* ii, fig. 107.
      A. Pygmy, naked, infibulated, balding, with a beard and curly hair, holding a
      club, fighting against a crane.
      B. Id.
      Pl. 64a, b

68   Rhyton, ram-head and boar-head. Formerly Hamilton coll.
      ARV 767.19 (Manner of Sotades); Brommer, *VL* 547 B8.
      T. Panofka, *Die griechischen Trinkhoerner und ihre Verzierungen ans Licht gestellt*
      (Berlin, 1851), pl. 1.12; Waser, op. cit., 3295, fig. 6; Hoffmann, op. cit., 20, no.
      38.
      Four naked pygmies, balding, with beards and curly hair, fighting against two
      cranes; they hold spears, the left arms covered with animal skins.

69   Rhyton, figure vase, fr. Reggio Museum, -. From Locri.
      ARV 773.2 (Sotades Potter): 'Of the plastic figure, the pygmy's right foot
      remains.' Unpublished.

70   Amphora (h. 30.2). Brussels, Musées Royaux, R 302.
      ARV 1044.7 (Epimedes P.); Brommer, *VL* 547 B9.
      CVA Belgium 2, Brussels 2, III, I, d, pls. 7.3, 8.5; Boardman, *ARFH* ii, fig. 148.
      Pygmy, naked, balding, with a beard, fighting against a crane; he holds a club, the
      left arm covered with an animal skin.
      Pl. 65.1

71   Rhyton, figure vase (negro eaten by a crocodile) (h. 25.5). Ruvo, Museo Jatta,
      1408. From Ruvo.
      ARV 1551.19, *Para* 505, *Add*[2] 388 (Group of Class W); Brommer, *VL* 547 B12.
      H. Sichtermann, *Griechische Vasen in Unteritalien aus der Sammlung Jatta in Ruvo*
      (Tübingen, 1966), pls. 44, 45.1; L. Bugner (ed.), *The Image of the Black in Western
      Art* (New York, 1976), 273, fig. 371.

Pygmy, naked, with a beard and a pilos, fighting against a crane; he holds a club, the left arm covered with an animal skin.
Pl. 65.3a, b

(unattributed)

72   Bell krater frr. ((a) h. 10; (b) h. 11). Athens, Agora Museum, P 244.
L. Talcott *et al., Hesperia*, suppl. 10 (1956), 65, no. 318, pl. 9.
A. Right arm of pygmy holding a club, crane attacking a pygmy whose right hand only is preserved.
B. Crane grasping the arm of a pygmy covered with his chlamys.

73   Mug (h. 8). Athens, NM 1355. From Tanagra.
Brommer, *VL* 547.1. Unpublished.
Pygmy, naked, infibulated, balding, with a beard, fighting against a crane with a club.

74   Pyxis (lid) (h. 5.5). Athens, NM 17714.
S. Papaspyridis and N. Kyparissis, *AD* 11 (1927–8), 92, fig. 2.
I. Pygmy, naked, with a beard, running with a club, the left arm covered with an animal skin.
Pl. 65.2

75   Rhyton, boar-head (h. 24). Berlin, SM (East) F 2758. From Ruvo.
Brommer, *VL* 547.4.
T. Panofka, *Die griechischen Trinkhoerner und ihre Verzierungen ans Licht gestellt* (Berlin, 1851), pl. 2.11–12; O. Waser, Roscher, iii. 2 (1902–9), 3295, fig. 7; H. Hoffmann, *Attic Red-Figured Rhyta* (Mainz am Rhein, 1962), 21, no. 49, pl. x.1 (Manner of Sotades).
A. Pygmy, naked, infibulated, balding, with a beard, fighting against a crane with a club.
B. Id.
Pl. 66.1a, b

76   Askos (h. 9). Hamburg, Museum für Kunst und Gewerbe, 1984.457.
H. Hoffmann, *Sexual and Asexual Pursuit* (London, 1977), 13, no. 135, pl. 11.3; W. Hornbostel *et al., Aus Gräbern und Heiligtümern* (Mainz am Rhein, 1980), 150–1, no. 87 (with fig.).
A. Pygmy, naked, balding, with a beard, holding a bow.
B. Pegasus.
Pl. 66.2

77   Cup fr. (Dm. 17.9; h. 5.5). Leipzig, Karl-Marx University, T 548.
Brommer, *VL* 548 Ba 2.
F. Hauser, *JDAI* 11 (1896), 196, fig. 50; E. Paul, *Antike Keramik* (Leipzig, 1982), no. 52 (with fig.).

I. Pygmy, naked, balding, with a beard and a headband, running with a club.
Pl. 67.1

78 Lekythos (h. 19). Paris, Louvre, TH 16.
Brommer, *VL* 548.7. Unpublished.
Pygmy, naked, infibulated, balding, fighting against a crane; he holds a club, the left arm covered with an animal skin.
Pl. 67.2

79 Askos (h. 6.4). Prague, Museum of Applied Arts, Z 260/1 K 18.
J. Frel, *Choix de vases attiques en Tchécoslovaquie, notamment au Musée National de Prague* (Prague, 1959) (*Sbornik Narodniho Musea v Praze* 13), 253, no. 62, pl. 11; H. Hoffmann, *Sexual and Asexual Pursuit* (London, 1977), 13, no. 137.
A. Pygmy, naked, balding, with a beard, holding a club or a sling, the left arm covered with an animal skin.
B. Crane.
Pl. 68.1a, b

80 Cup (h. 5.4; Dm. 19.3). Rome, Vatican, 35072 (formerly Astarita coll. 117).
Brommer, *VL* 548.8. Unpublished.
Pygmy, naked, balding, with a beard(?) and boots, fighting against a crane; he holds a club, the left arm covered with an animal skin.
Pl. 68.2

81 Askos fr. Rome, Vatican, Astarita coll.
H. Hoffmann, *Sexual and Asexual Pursuit* (London, 1977), 13, no. 138.
Unpublished (Beazley archive).
Pygmy, naked, with a beard, fighting against a crane; he holds a sword, the left arm covered with an animal skin.

82 Oinochoe (h. 10.6). Zurich market, Koller Gallery.
The Ernest Brummer Collection, Ancient art, II, Auction sale, October 1979 at Zurich (Zurich, 1979), 334–5, no. 698 (with fig.).
Pygmy, naked, with a crested helmet, a round shield and a spear, fighting against a crane.
Pl. 69.1

Boeotian

83 Cup (h. 11). Athens, NM 12419.
P. Pelagatti, *EAA* suppl. (1973), 146, fig. 151.
Pygmies, naked(?), with clubs, the arms covered with animal skins, fighting against cranes.

FOURTH CENTURY

Athenian (attributed)

84   Pelike (h. 29.3). St Petersburg, Hermitage, P 1877.64 (KAB 51e). From Panticapaeum (Kerch).
ARV 1466.106 (Group G); Brommer, *VL* 547 B10.
O. Waser, Roscher iii (2), 1902–9, 3295, fig. 12; K. Schefold, *Untersuchungen zu den Kertscher Vasen* (Berlin and Leipzig, 1934), 49, no. 451, pl. 24.1; I. V. Štal, *Klio*, 68 (1986), 359, figs. 11–12; Boardman, *ARFH* ii, fig. 411.
Normal-sized pygmy, naked, with a beard and a pilos, attacked by two cranes; he holds a sword, the left arm covered with an animal skin.

85   Pelike (h. 24.5). Brussels, Musées Royaux, A 726.
ARV 1474.3 (the Helbig reverse-group); Brommer, *VL* 547 B11.
CVA Belgium 3, Brussels 3, III, I, e, 4, pl. 4.12b.
A. Normal-sized pygmy, naked, with a beard, hair bound by a fillet, fighting against two cranes; he holds a thyrsos, a pelta, the left arm covered with an animal skin.
B. Youths
Pl. 69.2

(unattributed)

86   Bell krater (h. 22). Athens, NM 12599.
Brommer, *VL* 547.6. Unpublished.
Pygmy, naked, with a beard and a wreath, the left arm covered with an animal skin, holding a club, attacked by two cranes.

87   Pelike. Jalta Museum, JaKM 505.
I. V. Štal, *Klio*, 68 (1986), 363, figs. 16–17.
Normal-sized pygmy, naked, with a beard and a pilos, the left arm covered with a chlamys, attacked by two cranes.

88   Pelike. Kerch Museum, KMAK 24.
Štal, op. cit., 362, figs. 14–15.
Two normal-sized pygmies, with beards and pilos, fight a crane; they hold clubs, the arms covered by animal skins.

89   Pelike (h. 22.5). St Petersburg, Hermitage, b 3323.
Štal, op. cit., 361, fig. 13.
Normal-sized pygmy, naked, with a beard and a pilos, the left arm covered with a chlamys, attacked by two cranes.

90   Pelike (h. 25.5). St Petersburg, Hermitage, b 5261 (St. 1814b).
Brommer, *VL* 547.5.

O. Waser, Roscher, iii. 2 (1902–9), 3295, fig. 11; Schefold, op. cit., 49, no. 450; Štal, op. cit., 356–7, figs. 9–10.

Pygmy, naked, with a petasos, fighting against two cranes; he holds a club and a round shield, the left arm covered with an animal skin.

91 Pelike (h. 27.7). St Petersburg, Hermitage, b 9046.
    Unpublished.
    Normal-sized pygmy, naked, with a beard and a pilos, the left arm covered with a chlamys, attacked by two cranes.

92 Pelike (h. 28). St Petersburg, Hermitage, P 18361 (St. 1927, KAB 37).
    Brommer, *VL* 547.3.
    Schefold, op. cit., 44, no. 383, pl. 5.2; Štal, op. cit., 355–6, figs. 4–7.
    Four naked pygmies, with beards and headbands, fighting against three cranes; they hold swords, peltae, the left arms covered with animal skins.
    Pl. 70.1

93 Pelike (h. 25.1). Vienna, Kunsthistorisches Museum, IV 3221.
    Brommer, *VL* 547.2.
    CVA Austria 2, Vienna 2, III, I, pl. 85.1.
    Pygmy, naked, fighting against two cranes; he holds a sword and a round shield, the left arm covered with an animal skin.

Boeotian, Kabeirion style

94 Kantharos frr. ((a) h. 8.5; (b) h. 8.8). Athens, NM 10530.
    P. Wolters and G. Bruns, *Das Kabirenheiligtum bei Theben*, i (Berlin, 1940), 111–12, M29, pl. 12; K. Braun and T. E. Haevernick, *Das Kabirenheiligtum bei Theben*, iv, *Bemalte Keramik und Glas* (Berlin, 1981), 63, no. 324.
    A. Two naked pygmies fighting against two cranes; one pygmy holds a spear.
    B. Bellerophon, Pegasus, and Chimaera.

95 Kantharos frr. ((a) h. 5.3; (b) h. 6.8). Athens, NM 10530.
    Wolters/Bruns, op. cit., 115, S12, pl. 45.11; Braun/Haevernick, op. cit., 63, no. 325.
    A. Hunter with a stick.
    B. Pygmies and cranes.

96 Kantharos (h. 9.8). Athens, NM 12880.
    Wolters/Bruns, op. cit., 108, M8, pl. 54.2; Braun/Haevernick, op. cit., 62, no. 299.
    A. Pygmy, naked, with a club, the left arm covered with an animal skin, fighting against two cranes.
    B. Ivy.

97   Kantharos (h. 20.9). Berlin, SM 3159.
      Wolters/Bruns, op. cit., 108, M7, pl. 29.3–4; Braun/Haevernick, op. cit., 64, no.
      354.
      Four pygmies with chlamys and petasos, holding clubs and spears, fighting five
      cranes. One pygmy is riding a donkey.
      Fig. 13.1

98   Kantharos (h. 20.5). Boston, MFA 99.534.
      A. Fairbanks, *Catalogue of Greek and Etruscan vases*, i (Cambridge, Mass., 1928),
      197, no. 564, pl. 70; Wolters/Bruns, op. cit., 99, K15; Braun/Haevernick, op. cit.,
      65, no. 367.
      A. Naked pygmy, holding a spear, running towards a crane.
      B. Naked pygmy, holding a rhyton, facing a goat.

d 99   Skyphos (h. 17.7). London, BM B 77.
      Wolters/Bruns, op. cit., 109, M17, pl. 54.3; Braun/Haevernick, op. cit., 67, no.
      399.
      A. Naked pygmy(?) running with outstretched arms to catch a crane.
      B. Peleus bringing Achilles to Chiron.

d 100   Kantharos fr. (h. 4.3; w. 3.7). Kabeirion, K 2018.
      Braun/Haevernick, op. cit., 40, no. 48, pl. 2.21.
      Head of pygmy with pilos, beak of crane.

d 101   Kantharos fr. (h. 4.5; w. 5.4). Kabeirion, K 3096.
      Braun/Haevernick, op. cit., 40, no. 49, pl. 2.20.
      Right outstretched arm of pygmy(?) holding a spear or a stick.

d 102   Kantharos fr. (h. 9.5; w. 10.4). Kabeirion, K 352, K 407.
      Braun/Haevernick, op. cit., 47, no. 136, pl. 8.7.
      Pygmy(?), naked, holding a branch. Behind him a bird.

South Italian

103   Rhyton, figure vase (h. 20.8). Brussels, Musées Royaux, A 745.
      E. Buschor, 'Das Krokodil des Sotades', *MJBK* 11, 1/2 (1919), 18; *CVA* Belgium
      3, Brussels 3, IV D, 1, pl. 1.3a–b.
      Pygmy, naked, with curly hair, carrying a dead crane.
      Pl. 70.2

C. DWARF CERCOPES

ARCHAIC PERIOD

Athenian (attributed)

104   Amphora (h. 44). Boulogne, Musée des Beaux-Arts, 413. From Vulci.
      ABV 370.137 (696), *Add*[2] 99 (Leagros Group); Brommer, *VL* 98 A8.

E. Gerhard, *Auserlesene griechische Vasenbilder*, ii (Berlin, 1843), pl. 110; *Heraion*, ii. 188, fig. 39; *Hommes, dieux et héros de la Grèce* (Rouen, 1982), 230–1, no. 94, fig. 94b; F. Brommer, *Herakles*, ii (Darmstadt, 1984), 29, fig. 10.

A. Heracles carrying two small, well-proportioned Cercopes, with beards(?), dressed in chitons(?).

B. Apollo and goddess.

Fig. 13.2

105 Amphora. Florence, Museo Etrusco, 3871.

ABV 383.2, *Para* 168, *Add*² 101 (Acheloos P.); Brommer, *VL* 98 A2.

Heraion, ii. 189, fig. 40; Brommer, op. cit., 30, pl. 7a.

A. Heracles carrying two small malformed Cercopes, naked, with long hair hanging down.

B. Heracles and Apollo struggling for the stag.

Pl. 71.1

(unattributed)

106 Amphora fr. (h. 3.6; w. 5.4). Göttingen, University, H 48.

Brommer, *VL* 98 A7.

Leg of Heracles and upper body of a well-proportioned Cercops, naked, with long hair hanging down.

107 Lekythos (h. 31.2). Palermo, Museo Nazionale, N.I. 1865 (2635).

Brommer, *VL* 98 A6.

Haspels, *ABL*, 50; *Heraion*, ii. 191, fig. 43; Brommer, op. cit., 30, fig. 11.

Heracles carrying two small, well-proportioned Cercopes, naked, with beards.

Pl. 71.2

Corinthian

108 Pinax fr. (h. 6; w. 4.6). Berlin, SM (West) F 767.

A. Furtwängler, *Beschreibung der Vasensammlung im Antiquarium*, i (Berlin, 1885), 79–80, no. 767; *Heraion*, ii. 187, fig. 36.

Heracles carrying a small, well-proportioned Cercops, naked, with a beard.

Fig. 13.3

Laconian

109 Cup fr. (h. 11.8; Dm. 18.1). Tocra Museum, 934. From Tocra.

Brommer, *VL* 99 C2.

J. Boardman and J. Hayes, *Excavations at Tocra, 1963–65,* i (Oxford, 1966), 84, no. 934, pl. 57 (Manner of Arkesilas P.); C. M. Stibbe, *Lakonische Vasenmaler des sechsten Jahrhunderts v. Chr.* (Amsterdam and London, 1972), 163, 285, no. 286 (Rider P.); M. Pipili, *Laconian Iconography of the Sixth Century BC* (Oxford, 1987), 10, no. 22, fig. 14.

I: Heracles carrying two small, well-proportioned Cercopes, with beards and long hair, dressed in chitons.
Pl. 72.1

CLASSICAL PERIOD

Athenian

110　Volute krater (h. 44). Munich, Antikensammlungen, 2382. From Sicily.
ARV 287.27, *Add*² 209 (Geras P.); Brommer, *VL* 99 B1.
Heraion, ii. 195, fig. 46; Boardman, *ARFH* i, fig. 181; K. Schefold and F. Jung, *Die Urkönige, Perseus, Bellerophon, Herakles und Theseus* (Munich, 1988), 174, fig. 213.
Heracles carrying two small, well-proportioned Cercopes, naked, with beards.
Pl. 72.2.

FOURTH CENTURY

South Italy

Apulian

111　Krater. Catania, Museo Civico, Biscari coll. 735. From Camarina.
Brommer, *VL* 99 D1.
Trendall, *PhV*² 31, no. 25; O. Navarre, *DA* iv (1907), 437, fig. 5634; M. Bieber, *The History of the Greek and Roman Theater*² (Princeton, 1961), 133, fig. 486; F. Brommer, *Herakles*, ii (Darmstadt, 1984), 30–1, fig. 12.
Heracles bringing two very small, naked Cercopes in baskets to Eurystheus.
Fig. 13.4

Lucanian

112　Pelike (h. 28.5). Malibu, The J. Paul Getty Museum, 81.AE.189.
Christie, sale catalogue (5 May 1979), 34, no. 118, pl. 50 (Creusa or Dolon P.); J. Frel and M. Jentoft-Nilsen, in M. E. Mayo (ed.), *The Art of South Italy: Vases from Magna Graecia* (Richmond, 1982), 67, no. 7; Brommer, op. cit., 31, pl. 8; id., in J. Frel and S. Knudsen (eds.), *Greek Vases in the J. Paul Getty Museum*, ii (Malibu, Calif., 1985), 203–4, fig. 25; K. Schefold and F. Jung, *Die Urkönige, Perseus, Bellerophon, Herakles und Theseus* (Munich, 1988), 176.
Heracles carrying two small, malformed Cercopes, naked, with overlarge genitals.
Pls. 73, 74a, b

D. HEPHAISTOS

CLASSICAL PERIOD

Athenian (unattributed)

113 Skyphos (h. 15). Paris, Louvre, F 410.
E. Pottier, *Catalogue des vases antiques de terre cuite*, iii (Paris, 1906), 814; L. Kahil, *RA* (1972), 279 n. 3.
A. Two dwarfs naked, with beards; one rides a donkey(?), the other holds a club before two men in long chitons, hair bound by mitra-turbans, who dance with snakes.
B. Return of Hephaistos(?) on a winged chariot.
Pl. 75a–c

FOURTH CENTURY

Apulian

114 Amphora (h. 88). Foggia Museum, 132723. From Arpi.
Trendall/Cambitoglou *RVAp* ii. 924–5.90, pl. 360.2 (Arpi P.); A. D. Trendall, *Red Figure Vases of South Italy and Sicily* (London, 1989), 101, fig. 264.
Dwarf Hephaistos coming to set free Hera from the magic throne in the presence of Athena, Zeus, Eros, Aphrodite, Ares, and a woman.
Pl. 76

STATUARY

The figurines are all of East Greek type; they are listed according to their finding place, moving west of Greece, anticlockwise around the Mediterranean. Within each area (Greek mainland, Aegean islands and Cyprus, South Sicily, North Africa, etc.), places are grouped in alphabetic order.
Type I: dwarf standing, naked, the legs slightly flexed, the hands laid on his belly, often marked with creases. Variant: he wears a hat or a helmet on his head, and carries a shield or a basket.
Type II: dwarf carrying on his left shoulder a small figure, a child most likely.

ARCHAIC PERIOD

Greek mainland

Argos

115 Terracotta (h. 6). Athens, NM 14243. From Heraion.
C. Waldstein, *The Argive Heraeum*, ii (Boston and New York, 1905), 28, no. III, fig. 45.
Type I.

Delphi

116 Terracotta (h. 7). Delphi Museum, 2997. From Cassotis.
P. Perdrizet, *Fouilles de Delphes*, v (Paris, 1908), 202, no. 652, pl. XXIII, 17.
Type I. Top of head damaged.

117 Terracotta (h. 6.5). Delphi Museum, -. From Marmaria.
Perdrizet, op. cit., 202–3, no. 653, fig. 891.
Type II. Legs below knees missing, head and trunk of child broken.

118 Terracotta lamp, fr. Delphi Museum, -.
Perdrizet, op. cit., 186, no. 490, fig. 786.
Feet of dwarf (?) standing on a lamp (cf. below G 201).

Perachora

119 Terracotta fr. (h. 4.5). Athens, NM 17118. From the sanctuary of Hera Limenia.
J. H. Jenkins, in H. Payne *et al.*, *Perachora*, i (Oxford, 1940), 254, no. 293, pl. 114
('three other frr. from the bodies of grotesques such as 293 were found').
Head only.

Tanagra

120 Terracotta (h. 8). Paris, Louvre, CA 946.
S. Mollard-Besques, *Catalogue raisonné des figurines et reliefs en terre cuite grecs,
étrusques et romains*, i (Paris, 1954), 20, B113, pl. 15.
Type II.

Aegean islands and Cyprus

Aegina

121 Terracotta (h. 8.9). Aphaia temple reserves, T 8. From the sanctuary of Aphaia.
U. Sinn, in R. Hägg *et al.*, *Early Greek Cult Practice. Proceedings of the Fifth
International Symposium at the Swedish Institute at Athens, 26–29 June, 1986*
(Stockholm, 1988), 152–3, fig. 5.
Type I, with pilos or helmet. Sinn mentions ten other figurines, seven of type I (T
9, T 13–18), one with pilos or helmet (T 10), and two of type II (T 11, 12).

122 Terracotta (h. 6.5). Aphaia temple reserves, T 314. From the sanctuary of Aphaia.
A. Furtwängler *et al.*, *Aegina. Das Heiligtum der Aphaia* (Munich, 1906), 380, no.
66, pl. 110.14.
Type I. The author mentions frr. of three other figurines (h. 4.5–6.5).

123 Terracotta (h. 8.5). Aegina Museum(?). From tomb 5.
V. Kallipolitis, *AD* 19 (1964) (1966), 78, Σαρκοφ, 5; 1964 β₁, pl. 77, top right.
Type II, with pilos or helmet.

124 Terracotta (h. 7.4). Copenhagen, National Museum, 467.

N. Breitenstein, *Catalogue of terracottas, Cypriote, Greek, Etrusco-Italian and Roman (Danish National Museum)* (Copenhagen, 1941), 14–15, no. 132, pl. 13.
Type II, child missing.

125   Terracotta. Once in Kassel, Habich coll. Present location unknown.
Winter, *TK* i. 213. 7 (fig.); J. Boehlau, *Aus ionischen und italischen Nekropolen* (Leipzig, 1898), 155–6, fig. 72.
Type I, standing on a vase in the shape of an ox.
Fig. 13.6

Amorgos

126   Terracotta (h. 7.3). Syra Museum, –.
A. Laumonier, *Exploration archéologique de Délos*, xxiii (Paris, 1956), 92, no. 212.
Unpublished.
Type I.

Cyprus

127   Terracotta (h. 12.6). Copenhagen, National Museum, 3744.
C. Blinkenberg, *Lindos*, i (Berlin, 1931), 561.
Type I.

Cyprus: Amathus

128   Terracotta (h. 14.3). London, BM A 152.
A. S. Murray *et al.*, *Excavations in Cyprus* (London, 1900), 114, 120, fig. 165.4; Blinkenberg, op. cit., 561; Wilson, *Levant*, 94, pl. xva.
Type I.

129   Terracotta (h. 14.3). Nicosia Museum, –. From tomb 9.
Blinkenberg, op. cit., 561; E. Gjerstad *et al.*, *The Swedish Cyprus Expedition*, ii (Stockholm, 1935), 57, pl. XVII. 10.
Type I.

Cyprus: Larnaca

130   Terracotta (h. 12.5). Paris, Louvre, AM 46.
A. Caubet, *RLouvre*, 19 (1969), 11–12, fig. 7.
Type I.

131   Terracotta (h. 8.2). Paris, Louvre, AM 739.
Caubet, op. cit., 12 n. 25. Unpublished.
Type I. Legs missing.

132   Terracotta (h. 8). Paris, Louvre, MNB 130.
Caubet, op. cit., 12 n. 25. Unpublished.
Type I. Legs missing.

133   Terracotta (h. 12.5). Paris, Louvre, N 3310.
      G. Perrot and C. Chipiez, *Histoire de l'art dans l'antiquité*, iii (Paris, 1885), 76, 78,
      fig. 27; Caubet, op. cit., 12 n. 25; Wilson, *Levant*, 94–5, pl. XVIA (left).
      Type I.

Cyprus: Salamis

134   Terracotta, figure vase. Salamis Museum(?).
      A. P. Di Cesnola, *Salaminia (Cyprus)*[2] (London, 1884), 243, fig. 289; Winter, *TK*
      i. 213. 1e.
      Type I.

Delos

135   Terracotta (h. 7.3). Mykonos Museum, 337.
      A. Laumonier, *Exploration archéologique de Delos*, xxiii (Paris, 1956), 92, no. 212,
      pl. 25.
      Type I.

136   Terracotta (h. 7.5). Mykonos Museum, 338.
      Laumonier, op. cit., 92, no. 213, pl. 25.
      Type I.

Melos

137   Terracotta (h. 7.5). London, BM 89.
      Winter, *TK* i. 213. 3B; R. A. Higgins, *Catalogue of the Terracottas in the British
      Museum*, i (London, 1954), 57, no. 89, pl. 18.
      Type I.
      Pl. 77.1a, b

138   Terracotta (h. 9). London, BM 93.
      Winter, *TK* i. 213. 6 (with fig.); Higgins, op. cit., 57 no. 93, pl. 18.
      Type II.
      Pl. 77.2.

139   Terracotta (h. 8). London, BM 94.
      Winter, *TK* i. 213. 3C; Higgins, op. cit., 58, no. 94, pl. 18.
      Type I, with pilos or helmet.
      Pl. 77.3a, b

Paros (Sanctuary of Artemis)

140   Terracotta (h. 9.5). Paros Museum, Δ 37a.
      A. Laumonier, *Exploration archéologique de Delos*, xxiii (Paris, 1956), 92, no. 212;
      O. Rubensohn, *Das Delion von Paros* (Wiesbaden, 1962), 141, no. 34, pl. 25, T34.
      Type I.

141 Terracotta (h. 6). Paros Museum, *Δ* 37b.
Laumonier, op. cit., 92, no. 212; Rubensohn, op. cit., 141, no. 34. Unpublished.
Type I.

Rhodes

142 Terracotta (h. 7.5). Paris, Louvre, CA 12.
S. Mollard-Besques, *Catalogue raisonné des figurines et reliefs en terre cuite grecs, étrusques et romains*, i (Paris, 1954), 38, B221, pl. XXVIII.
Type I.

143 Terracotta, figure vase (h. 14). Paris, Louvre, MNB 1766.
L. Heuzey, *Les Figurines antiques de terre cuite du Musée du Louvre* (Paris, 1883), 231–2, no. 45; Winter, *TK* i. 213. 1 (with fig.).
Type I.

Rhodes: Camirus

144 Terracotta (h. 7.5). Lausanne, Musée Cantonal, 3858.
S. G. Zervos, *Rhodes, capitale du Dodécanèse* (Paris, 1920), 134, fig. 307.
Type I.

145 Terracotta, figure vase (h. 15). London, BM 86.
Winter, *TK* 213. 1b; R. A. Higgins, *Catalogue of the Terracottas in the British Museum*, i (London, 1954), 56, no. 86, pl. 18; id. *Greek Terracottas* (London, 1967), xxii. 36.9, pl. 14B.
Type I, with sandals.
Pl. 78.1

146 Terracotta (h. 17). London, BM 88.
Winter, *TK* i. 213.2 (with fig.); Higgins, op. cit., 56–7, no. 88, pl. 18.
Type II, with petasos, and a square purse hanging from his right wrist.
Pl. 78.2

Rhodes: Lindos (Acropolis, Sanctuary of Athena?)

147 Terracotta (h. 7.7). Copenhagen, National Museum, 10610.
C. Blinkenberg, *Lindos*, i (Berlin, 1931), 561–2, no. 2314, pl. 108.
Type I.

148 Terracotta (h. 6.9). Copenhagen, National Museum, 10611.
Blinkenberg, op. cit., 561–2, no. 2316. Unpublished.
Type I. Upper part of body missing.

149 Terracotta (h. 7.4). Istanbul, Archaeological Museum, –.
Blinkenberg, op. cit., 561–2, no. 2315, pl. 108.
Type I. Legs below knees missing.

150   Terracotta. Istanbul, Archaeological Museum, -.
       Blinkenberg, op. cit., 561–2, no. 2317, pl. 108.
       Type I, with pilos or helmet. Lower part of body missing.

151   Terracotta (h. 10). Istanbul, Archaeological Museum, -.
       Blinkenberg, op. cit., 561–2, no. 2318, pl. 108.
       Type II, with petasos; a cloak is painted in yellow on the back of the dwarf. Feet
       missing.

Rhodes: Ialysos Necropolis

152   Terracotta, figure vase (h. 17.5). Rhodes Museum, 6488. From tomb 93.
       CVA Italy 10, Rodi 2, II, pl. 485.3 ('donna steatopigica, forse gravida').
       Type I.

153   Terracotta. Rhodes Museum, 14792.
       G. Jacopi, *Clara Rhodos*, 6–7 (1933), 297, no. 25, fig. 26.
       Type II. Child and feet missing.

Samos

154   Terracotta (h. 19). Kassel, Staatl. Kunstsammlungen, S 55. From W. necr., tomb
       21 (girl).
       J. Boehlau, *Aus ionischen und italischen Nekropolen* (Leipzig, 1898), 39, 155 ff., pl.
       XIII, 4; U. Sinn, *Antike Terrakotten* (Kassel, 1977), 25, no. 15, pl. 5.
       Type I.
       Pl. 78.3a, b

155   Wood (h. 21.9). Samos, Vathy Museum, H 43. From Heraion.
       G. Köpcke, *MDAI(A)* 82 (1967), 109–12, Beil. 52–4, 81.2; R. Hampe and E.
       Simon, *The Birth of Greek Art: From the Mycenaean to the Archaic Period* (London,
       1981), 230, fig. 346; H. Kyrieleis, *Führer durch das Heraion von Samos* (Athens,
       1981), 42, fig. 30; Sinn, *Antidoron*, 90–1, fig. 8.
       Type II.
       Pl. 79.1a–c

156   Terracotta (h. 8.1). Samos, Vathy Museum, RB 77–643, 1. From Heraion.
       Sinn, *Antidoron*, 87 ff., figs. 3, 6a; id., *MDAI(A)* 100 (1985), 151–3, no. 33, pl.
       38.1–6.
       Type I, carrying a rectangular basket on top of head.
       Pl. 79.2

Italy

Cumae

157   Terracotta (h. 6.5). Formerly E. Stevens coll. Naples, Museo Nazionale(?). From a
       tomb.

E. Gàbrici, *Monumenti Antichi*, 22 (1913), 546–7, pl. LXXII, 6 (mentions two similar statuettes in that collection).
Type I. Feet missing.

Locri

158 Terracotta (h. 12.5). Reggio Museum, –.
Winter, *TK* i. 213. 5c. Unpublished.
Type I.

Reggio

159 Terracotta (h. 8.5). Reggio Museum, 131.
Winter, *TK* i. 213. 5b. Unpublished.
Type I.

160 Terracotta (h. 7). Reggio Museum, 132.
Winter, *TK* i. 213. 5b. Unpublished.
Type I.

Taranto

161 Terracotta (h. 14). Taranto, Museo Nazionale, 209.
Winter, *TK* i. 213. 5d. Unpublished.
Type I.

162 Terracotta (h. 14.7). Taranto, Museo Nazionale, 4960. Montedoro, via S. Lucia, tomb 10.
F. G. Lo Porto, *Bollettino d'Arte*, 47 (1962), 166–7, figs. 21g, 24; G. Penso, *La Médecine romaine* (Paris, 1984), fig. 128 (wrongly described as 'femme obèse').
Type I.
Pl. 79.3

Sicily

Catania (Sanctuary of Demeter)

163 Terracotta (h. 7.6). Catania, Museo Comunale, K 445.
G. Rizza, *Bollettino d'Arte*, 45 (1960), 258, fig. 22.3.
Type I.

164 Terracotta (h. 9.2). Catania, Museo Comunale, K 455.
Rizza, op. cit., 258, fig. 22.4.
Type I, with pilos or helmet.

165 Terracotta, figure vase (h. 12). Catania, Museo Comunale, K 473.
Rizza, op. cit., 258, fig. 22.13.
Type I.

166   Terracotta, figure vase (h. 10). Catania, Museo Comunale, –.
       Winter, *TK* i. 213. 1h. Unpublished.
       Type I.

Gela

167   Terracotta (h. 7.5). London, BM 90. From the sanctuary of Demeter.
       Winter, *TK* i. 213. 3D; R. A. Higgins, *Catalogue of the Terracottas in the British
       Museum*, i (London, 1954), 57, no. 90, pl. 18.
       Type I.

Megara Hyblaea

168   Terracotta (h. 6). Syracuse, Museo Archeologico Regionale, 7946. From tomb 86
       (five children).
       Winter, *TK* i. 213. 3f; S. Cavallari and P. Orsi, *Megara Hyblaea* (Rome, 1892), col.
       226, pl. vi.4.
       Type I.

169   Terracotta, figure vase (h. 13). Syracuse, Museo Archeologico Regionale, 7947.
       From the same tomb as above.
       Winter, *TK* i. 213. 1f; Cavallari/Orsi, op. cit., pl. vi.6; B. Pace, *Arte e civiltà della
       Sicilia antica*, iii (Genoa etc., 1945), 670, fig. 181.
       Type I.

170   Terracotta (h. 9). Syracuse, Museo Archeologico Regionale, –. From the same
       tomb as above.
       Cavallari/Orsi, op. cit., pl. vi.3; Winter, *TK* i. 213. 5.
       Type I, with pilos (or helmet?).

171   Terracotta (h. 7). Syracuse, Museo Archeologico Regionale, –. From tomb 767.
       E. Caruso, NSA (1892), 251. Unpublished.
       Type I.

Morgantina

172   Terracotta (h. 9). Morgantina Museum, 69.624. From necr. II, tomb 17.
       M. Bell, *Morgantina Studies*, i (Princeton, 1981), 15–16, 129, no. 48b (wrongly
       labelled as 48a), pl. 11.
       Type II. Arms and legs of child missing.

173   Terracotta (h. 7.3). Morgantina Museum, 69.852. From necr. VI, tomb 2.
       Bell, op. cit., 15–16, 129, no. 48a (wrongly labelled as 48b), pl. 11.
       Type II. Child missing.

174   Terracotta (h. 7.3). Morgantina Museum, 69.853. From necr. VI, tomb 2.
       Bell, op. cit., 15–16, 129, no. 49, pl. 11.
       Type I.

Selinus

175 Terracotta (h. 4.5). Palermo, Museo Nazionale, 183. From the sanctuary of Demeter Malophoros.
E. Gàbrici, *Monumenti Antichi*, 32 (1927), 220. Unpublished.
Type I.

176 Terracotta (h. 7.5). Palermo, Museo Nazionale, 4131. From the tomb of a youth.
V. Tusa, *Sicilia archeologica*, 11 (1970), 16, fig. 9a–c.
Type I.

177 Terracotta (h. 7.8). Palermo, Museo Nazionale, 4132. From the sanctuary of Demeter Malophoros.
Gàbrici, op. cit. Unpublished.
Type I, with pilos or helmet.

178 Terracotta (h. 7.4). Palermo, Museo Nazionale, -. From the sanctuary of Demeter Malophoros.
Gàbrici, op. cit., 220, pl. XLI.2.
Type I.

Syracuse

179 Terracotta, figure vase (h. 16.5). Syracuse, Museo Archeologico Regionale, 12551. Del Fusco necropolis, tomb 118.
Winter, *TK* i. 213. 1g; P. Orsi, *NSA* (1893), 480 (with fig.).
Type I.

180 Two terracottas (h. 6; 6.8). Syracuse, Museo Archeologico Regionale, 51565. From tomb 68 (child).
S. L. Agnello, *NSA*, ser. 8/3 (1949), 202–3, tomb 68, no. 7 ('Due figurine apotropaiche di Phtà'). Unpublished.
Type I.

181 Terracotta, figure vase (h. 13). Syracuse, Museo Archeologico Regionale, 52182. From tomb 46.
G. V. Gentili, *NSA*, ser. 8/5 (1951), 310–11, tomb 46, no. 3, fig. 44.1.
Type I.

182 Four terracottas (h. *c.*8.5). Syracuse, Museo Archeologico Regionale, -. Del Fusco necropolis, tomb 126 (child).
P. Orsi, *NSA* (1893), 481–2, tomb 126 ('Quattro figurine di Bes fittili . . . , di cui una con beretto cristato'). Unpublished.
Type I; one figure wears a pilos or helmet.

North Africa

Carthage

183   Terracotta (h. 10). Carthage Museum, 895.12.
      Wilson, *Levant*, 94, pl. xvb.
      Type II.
      Pl. 80.1

184   Terracotta (h. 7.7). Carthage Museum, 898.87. From Douimès necropolis.
      R. P. Delattre, *Musée Lavigerie de St-Louis de Carthage* (Paris, 1900), 110–11, no. 8,
      pl. xvi.8.
      Type I.

Tocra (Sanctuary of Demeter and Kore?)

185   Terracotta (h. 7.4). Tocra Museum, 48.
      J. Boardman and J. Hayes, *Excavations at Tocra 1963–5*, i (Oxford, 1966), 154, no.
      48, pl. 100; J. Boardman, *The Greeks Overseas*[2] (London, 1980), 147, fig. 180.
      Type I. Feet partly missing.

186   Terracotta (h. 7.4). Tocra Museum, 49.
      Boardman/Hayes, op. cit., 154, no. 49, pl. 100.
      Type I.

Egypt

Naucratis

187   Terracotta (h. 6.5). London, BM 92.
      Winter, *TK* i. 213. 1c ('C 592'); R. A. Higgins, *Catalogue of the Terracottas in the
      British Museum*, i (London, 1954), 57, no. 92, pl. 18.
      Type I. Feet missing.

Asia Minor and Bosporus

Apollonia

188   Terracotta (h. 5). Paris, Louvre, CA 1761.
      A. Caubet, *RLouvre*, 19 (1969), 10, fig. 6.
      Type I.

Berezan

189   Terracotta (h. 7.2). St Petersburg, Hermitage, B 82.273.
      Drevnie pamjatniki kultury na territori SSSR (St Petersburg, 1986), 131, pl. 3, fig. 2.
      Type I.

190   Terracotta (h. 7.5). St Petersburg, Hermitage, B 251.
      Anticnaja Koroplastika (St Petersburg, 1976), 20, no. 31. Unpublished.
      Type I.

Dadja

191 Terracotta. Istanbul, Archaeological Museum, -.
G. Mendel, *Catalogue des figurines grecques de terre cuite. Musées impériaux ottomans* (Constantinople, 1908), 585, no. 3500. Unpublished.
Type I.

Elaious (Thrace)

192 Terracotta (h. 9.5). Paris, Louvre, ELE 604. From the necropolis.
S. Mollard-Besques, *Catalogue raisonné des figurines et reliefs en terre cuite grecs, étrusques et romains*, i (Paris, 1954), 48, B305, pl. 34.
Type I.

Ephesus (Artemision)

193 Terracotta, figure vase (h. 3.5). London, BM 87.
R. A. Higgins, *Catalogue of the Terracottas in the British Museum*, i (London, 1954), 56, no. 87, pl. 18.
Type I(?). Head only preserved.

Mylasa

194 Terracotta (h. 7). Smyrna Museum, 1363.
Winter, *TK* i. 213. 3a ('zwei Exemplare'). Unpublished.
Type I.

Neandria

195 Terracotta, figure vase. Formerly Calvert coll.
Winter, *TK* i. 213. 1d. Unpublished.
Type I.

Nymphaeum

196 Terracotta mould (h. 9.5). St Petersburg, Hermitage, Nf 41.959.
Archeologia SSSR, G 1–11 (1970), pl. 34, no. 4.
Type I.

Unknown Provenance

197 Terracotta (h. 8). Berlin, once Antiquarium, 7797.
Winter, *TK* i. 213. 3h. Unpublished.
Type I, with pilos or helmet.

198 Terracotta (h. 19). Berlin, once Antiquarium, 8402. From Thebes?
Winter, *TK* i. 213. 4 (with fig.).
Type I.

199   Terracotta (h. 9). Bologna, Museo Civico, G 842.
       Unpublished.
       Type I.

200   Terracotta (h. 7.5). Bonn, University, D 838. From Selinus?
       N. Himmelmann, *Antiken aus dem Akademischen Kunstmuseum* (Bonn, 1971), 48,
       no. 53; Sinn, *Antidoron*, 90, fig. 2.
       Type I.

201   Terracotta, frr. (of a lamp?) (h. 8.6). Brussels, Musées Royaux, 105 A/B.
       P. Perdrizet, *Fouilles de Delphes*, v (Paris, 1908), 186, fig. 786a; Sinn, *Antidoron*, 94
       n. 54.
       Type I.

202   Terracotta (h. 9.2). Frankfurt, Liebieghaus, 454.
       A. Furtwängler, *ARW* 10 (1907), 324, pl. 1; F. Eckstein and A. Legner, *Antike
       Kleinkunst im Liebieghaus* (Frankfurt am Main, 1969), no. 35, pl. 35; P. C. Bol,
       Liebieghaus: Ancient Art Guide to the Collection (Frankfurt am Main, 1981), 26–7,
       fig. 30; Sinn, *Antidoron*, 91, fig. 5.
       Type II.
       Pl. 80.2

203   Terracotta (h. 8.9). Hanover, Kestner Museum, 1899. 67e.
       U. Liepmann, *Griechische Terrakotten, Bronzen, Skulpturen* (Hanover, 1975), 44,
       T15 (with fig.).
       Type II.

204   Terracotta (h. 7.5). London, BM 91.
       R. A. Higgins, *Catalogue of the Terracottas in the British Museum*, i (London, 1954),
       57, no. 91, pl. 18.
       Type I.

205   Terracotta (h. 11). Marseilles, Musée Borely, 2714.
       W. Froehner, *Musée de Marseille, Catalogue des antiquités grecques et romaines* (Paris,
       1897), no. 1214. Unpublished.
       Type I, with a broken pilos or helmet.

206   Terracotta (h. 7.5). Munich, Antikensammlungen, 7563.
       J. Sieveking, *MJBK* 2 (1925), 288, fig. 34; Sinn, *Antidoron*, 88, 91, fig. 4.
       Type I, with helmet and shield.
       Pl. 80.3

207   Terracotta (h. 12.5). Oxford, Ashmolean Museum, 1971.1016.
       *Ancient Life in Miniature: An Exhibition of Classical Terracottas from Private Collections
       in England* (Birmingham, 1968), no. 235, pl. 18.
       Type I.

208   Terracotta (h. 9). Paris, Louvre, AM 185. From Rhodes?
      S. Mollard-Besques, *Catalogue raisonné des figurines et reliefs en terre cuite grecs,
      étrusques et romains*, i (Paris, 1954), 38, B222, pl. XXVIII.
      Type I. Legs missing.

209   Terracotta (h. 7.4). Paris, Louvre, E 20913. From Egypt?
      A. Caubet, *RLouvre*, 19 (1969), fig. 4.
      Type I.

210   Wood (h. 24.5). Paris, Louvre, N 846. From Naucratis?
      Caubet, op. cit., 7–12, figs. 1–3; U. Sinn, *MDAI(A)* 100 (1985), 153, pl. 39. 1–3.
      Type I.
      Pl. 80.4

211   Terracotta, figure vase (h. 11.5). Paris, Louvre, S 4388.
      Caubet, op. cit., 10, fig. 5.
      Type I.

212   Terracotta, figure vase (h. 14). England, private.
      *Ancient Life in Miniature: An Exhibition of Classical Terracottas from Private Collections
      in England* (Birmingham, 1968), 16, pl. 3, fig. 22; *Ancient Glass Jewellery and
      Terracottas from the Collection of Mr and Mrs J. Bomford, Ashmolean Museum* (Oxford,
      1971), 42, no. 107, pl. 41.
      Type I.

213   Terracotta (h. 9.5). Netherlands, private. From Rhodes?
      Klassieke Kunst uit particulier Bezit (Leiden, 1975), no. 257. figs. 111–12.
      Type I, with pilos or helmet.

S: *Skeletal Remains*

EGYPT

PREDYNASTIC PERIOD

El-Mustagidda (Badarian culture, *c*.4500 BC)

Cemetery 2200/3500, tomb 3510

1   Skull, clavicles, vertebrae, upper limbs (ulnae, radius). Formerly London, Museum of the Royal College of Surgeons. Present location unknown. E. W. A. H. Jones, *J. of Anat.* 66 (1932), 569–73 (achondroplasia); G. Brunton, *Mostagedda and the Tasian Culture* (London, 1937), 42 ('Upper part of a male dwarf'); A. Bleyer, *Ann. Med. Hist.* 2 (1940), 306 (achondroplasia); Ortner/Putschar, *Pathological Conditions*, 331.
Short-limbed dwarfism (pseudoachondroplasia, multiple epiphyseal dysplasia?).
Fig. 2.1a–f

Unknown provenance

2   One femur (max. l. 22.8), two tibiae (max. l. 19.3). University of Cambridge, Dept. of Physical Anthropology.
D. R. Brothwell, in *Diseases in Antiquity*, 432 (achondroplasia). Unpublished.
Short-limbed dwarfism (achondroplasia?).

EARLY DYNASTIC PERIOD (2920–2770 BC)

Saqqara

Tomb complex of King Wadj(?) 3504

3   Subsidiary grave 58. Skeleton of male adult. Cairo University, Dept. of Anatomy. W. B. Emery, *Great Tombs of the First Dynasty*, ii (Cairo and London, 1954), 36, no. 58, fig. 14, pl. xxv ('rickets'); id., *Archaic Egypt* (Harmondsworth, 1961), pl. 23.
Short-limbed dwarfism (achondroplasia?).
Pl. 2.2

Abydos

Tomb complex of King Djer

d   4   Chamber 61. Complete skeleton. L. of wooden coffin: 82; w. 55. Present location unknown.

E. Amélineau, *Les Nouvelles Fouilles d'Abydos, 1897–1898*, i (Paris, 1904), 103–4; Kaplony, *Inschriften*, i. 216. Unpublished.

Dwarf or child.

5 Chamber 96. Complete skeleton. L. of wooden coffin: 114; w. 37. Present location unknown.

E. Amélineau, *Le Tombeau d'Osiris* (Paris, 1899), 64; id. *Les Nouvelles Fouilles d'Abydos, 1897–1898,* i (Paris, 1904), 229–30; ii (Paris, 1905), 730; Kaplony, *Inschriften*, i. 216–17. Unpublished.

Achondroplasia? The stela of a dwarf was found in the same chamber (See E 1).

6 One humerus. Present location unknown.

W. M. F. Petrie, *The Royal Tombs of the Earliest Dynasties*, ii (London, 1901) (MEEF 21), 24, pl. VIA, no. 14; D. R. Brothwell, *Digging up Bones*[3] (London, 1981), 167, fig. 6.18/A3 (achondroplasia); id., in *Diseases in Antiquity*, 432; Ortner/Putschar, *Pathological Conditions*, 331–2.

Short-limbed dwarfism (achondroplasia?).

Tomb complex of King Semerkhet

7 Chambers M and L. Bones of two dwarfs: (1) long bones; (2) skull, left humerus, right femur, tibiae (young adult). London, BM, Natural History, AF 11.4/427; 11.4.462.

W. M. F. Petrie, *The Royal Tombs of the First Dynasty*, i (London, 1900) (MEEF 18), 13, pl. LX; D. Randall-McIver, *The Earliest Inhabitants of Abydos* (Oxford, 1901), pl. VII; D. R. Brothwell, in *Diseases in Antiquity*, 433, fig. 8a; Ortner/Putschar, *Pathological Conditions*, 331–2, figs. 518–21 (as dating wrongly to the 4th dyn.).

Achondroplasia. The stela of a dwarf was found in chamber M (See E 8, pl. 17.3).

Tomb complex of King Qaʿa

d 8 Chamber 5. Skeleton. Present location unknown.

W. M. F. Petrie, *The Royal Tombs of the First Dynasty*, i (London, 1900) (MEEF 18), pl. LX: 'dwarf(?)'; Kaplony, *Inschriften*, i. 374. Unpublished.

d 9 Chamber 17. Skeleton. Present location unknown.

W. M. F. Petrie, *The Royal Tombs of the First Dynasty*, i (London, 1900) (MEEF 18), pl. LX: 'dwarf(?)'; Kaplony, *Inschriften*, i. 374. Unpublished.

Hierakonpolis (Kom el-Ahmar)

10 Skeleton. Present location unknown.

J. E. Quibell and F. W. Green, *Hierakonpolis*, ii (London, 1902) (BSA 5), 26: 'In one tomb the skeleton of a dwarf was found.' Unpublished.

Unknown Provenance

11 Two humeri. Minimum age of about 14, possibly young adult. London, BM, Natural History, AF 11.3/75.

D. R. Brothwell, *Digging up Bones*[3] (London, 1981), 167, fig. 6.18/A1 (achondroplasia); Ortner/Putschar, *Pathological Conditions*, 336–7, figs. 526–7.

Chondrodysplasia, possibly mucopolysaccharidoses (Hunter's syndrome or Morquio's syndrome).

12 Nine skulls, adults, all but one female. Present location unknown.

H. D. Smith, *Biometrika*, 8 (1921), 262–6; A. T. Sandison and C. Wells, in *Diseases in Antiquity*, 527.

Infantile characteristics, without negroid features, presumably hypopituitarism.

OLD KINGDOM (2575–2134 BC)

Dynasty 5

Abydos

13 Skeleton. Formerly London, Museum of the Royal College of Surgeons. Present location unknown.

W. R. Dawson, *Ann. Med. Hist.* 9 (1927), 319. Unpublished.

Dynasty 6

Beni Hasan

14 Tomb of Ipi. Male skeleton. Present location unknown.

PM iv. 61, no. 481; J. Garstang, *The Burial Customs of Ancient Egypt* (London, 1907), 40–1, fig. 28 (funerary equipment); Dawson, *Pygmies*, 186. Unpublished.

Short-limbed dwarfism. Cf. the depiction of a dwarf in the same tomb (E 60).

MIDDLE KINGDOM (2081–1640? BC)

Beni Hasan

d 15 Tomb 487. Sarcophagus of Seneb. Female skeleton. Estimated stature: 140 (4 ft. 9 in.). Present location unknown.

Garstang, op. cit., 41, 113–14, 226, pl. v. Unpublished.

Hypopituitarism?

Asyut

16 Tomb 6. Sarcophagus of Ankhef. Male skeleton. Estimated stature: 92. Present location unknown.

E. Chassinat and C. Palanque, *Une Campagne de fouilles dans la nécropole d'Assiout* (Cairo, 1911) (MIFAO 24), 12–15, esp. 14; Sourdive, *La Main*, 91 n. 10. Unpublished.

Short-limbed dwarfism (achondroplasia?).

NEW KINGDOM (1539–1075? BC)

Dynasty 18

Thebes, Tomb Chamber in the Temple of Tuthmosis IV

17   Female skull (20–25 years old). Formerly London, Royal College of Surgeons.
     Present location unknown.
     W. M. F. Petrie, *Six Temples at Thebes* (London, 1897), 7–8; C. G. Seligmann,
     Man, 12 (1912), 17–18, pl. B (cretinism); A. Keith, *J. Anat. Physiol.* 47 (1913),
     195–200, figs. 7–11; D. R. Brothwell, *Digging up Bones*[3] (London, 1981), 168, fig.
     6.16/B; Ortner/Putschar, *Pathological Conditions*, 331.
     Achondroplasia.

Dynasty 21

Speos Artemidos, near Beni Hasan

18   Skeleton of an infant. L. of cartonnage case: 73. London, BM 41603.
     J. Garstang, *The Burial Customs of Ancient Egypt* (London, 1907), 244 ('bones of a
     monkey'), fig. 219 (cartonnage case); W. R. Dawson and P. H. K. Gray, *Catalogue
     of Egyptian Antiquities in the British Museum*, i, *Mummies and Human Remains*
     (London, 1968), 13–14, no. 24, pl. VIIb (cartonnage case); P. H. K. Gray, *Clin.
     Radiol.* 20 (1969), 106–8, figs. 1–6; Ortner/Putschar, *Pathological Conditions*, 338.
     Osteogenesis imperfecta.

GREECE

Gouvalari (Pylos area) (c. 1500 BC)

19   Tomb 7. Female skeleton (35–40 years old). Estimated stature: 138.
     C. S. Bartsocas, *Hippocrates Magazine*, 2 (1977), no. 2, 157–60, fig. 1; id. in
     Skeletal Dysplasias (New York, 1982), 11–12, fig. 10; M. D. Grmek, *Les Maladies
     à l'aube de la civilisation occidentale* (Paris, 1983), 111.
     Congenital absence of clavicles. Cleidocranial dysplasia?

GLOSSARY OF MEDICAL TERMS

Definitions are taken from various sources (see medical bibliography).

Acquired: refers to a condition or disorder that is not attributable to hereditary genetic causes.

Club-foot: see **Talipes**.

Congenital: refers to a condition or disorder that is present at birth, and that may or may not be genetic.

Coxa: the hip joint; **coxa valga**: the angle between the neck and the shaft of the femur is abnormally increased; **coxa vara**: the angle between the neck and the shaft of the femur is abnormally decreased.

Dominant: in genetics, refers to a condition which is shown in the individual by the action of only one of a pair of genes.

Dysplasia: abnormal development of skin, bone, or other tissues.

Endocrine gland: gland responsible for the production of hormones.

Equinus: see **Talipes**.

Equinovarus: see **Talipes**.

Foramen magnum: hole in the base of the skull through which the spinal cord passes.

Genu: the knee; **genu valgum**: abnormal in-curving of the knees ('knock-knee'); **genu varum**: abnormal out-curving of the knees ('bow-leg').

Goitre: swelling of the neck due to enlargement of the thyroid gland.

Hydrocephalus: abnormal increase in the amount of cerebrospinal fluid within the cavities of the brain, causing enlargement of the skull.

Hypotonia: state of reduced tension in muscle.

Infantilism: persistence of childlike physical or psychological characteristics into adult life.

Kyphosis: backward curvature of the upper spine ('round back').

Lordosis: inward curvature of the lower spine ('hollow back').

Lumbar: relating to the loin.

Metabolic: relating to the biochemical changes that take place within the body and enable its continued growth and functioning.

Myxoedema: dry firm waxy swelling of the skin and subcutaneous tissues, including weight gain and mental dullness; clinical syndrome due to hypothyroidism.

Paraplegia: paralysis of both legs.

Pituitary gland: the master **endocrine gland** situated at the base of the skull.

Recessive: refers to a condition shown in the individual only when both of a pair of genes are effective.

Scoliosis: lateral deviation of the backbone ('hunchback').

Talipes ('club-foot'): a congenital deformity of the foot, which is twisted out of shape or position; **talipes equinovarus**: the sole of the foot is twisted downwards and inwards; **talipes equinus**:

the sole of the foot is plantar flexed, so that the person walks on the toes; **talipes valgus**: the sole of the foot is twisted outwards; **talipes varus**: the sole of the foot is turned inwards.

Thoracolumbar: relating to the chest and the loin.

Thyroid gland: **endocrine gland** situated in the base of the neck.

BIBLIOGRAPHY

MEDICAL TYPOLOGY AND PALAEOPATHOLOGY

BERGSMA, D. (ed.), *Birth Defects Compendium*[2] (Basingstoke and London, 1979).

BLEYER, A., 'The Antiquity of Achondroplasia', *Ann. Med. Hist.* 2 (1940), 306.

BROTHWELL, D. R., *Digging up Bones*[3] (London, 1981).

——, and SANDISON, A. T. (eds.), *Diseases in Antiquity* (Springfield, Ill. 1967).

CAVALLI-SFORZA, L. L. (ed.), *African Pygmies* (London etc., 1986).

FRASIER, S. D., *Pediatric Endocrinology* (New York, 1980).

GARDNER, R. J. M., 'A New Estimate of the Achondroplasia Mutation Rate', *Clin. Genet.* 11 (1977), 31–8.

GRAY, P. H. K., 'A Case of Osteogenesis Imperfecta, Associated with Dentinogenesis Imperfecta, Dating from Antiquity', *Clin. Radiol.* 20 (1969), 106–8.

GRMEK, M. D., *Les Maladies à l'aube de la civilisation occidentale* (Paris, 1983).

HORTON, W. A. *et al.*, 'Standard growth curves for achondroplasia', *J. Pediatr.* 93 (1978), 435–8.

JOB, J. C., PIERSON, M., *Pediatric Endocrinology* (New York, 1981).

JONES, E. W. A. H., 'Studies in Achondroplasia, *J. of Anat.* 66 (1932), 565–77.

KEITH, A., 'Abnormal Crania-Achondroplastic and Acrocephalic', *J. of Anat. and Physiol.* 47 (1913), 189–206.

KUNZE, J., and NIPPERT, I., *Genetics and Malformations in Art* (Berlin, 1986).

MCKUSICK, V. A., *Heritable Disorders of Connective Tissue*[4] (St Louis, 1972).

MANCHESTER, K., *The Archaeology of Disease* (Bradford, 1983).

MAROTEAUX, P., 'The Chondrodystrophies Detectable at Birth', in F. C. Frasier and V. A. McKusick (eds.), *Congenital Malformations. Proceedings of the Third International Conference, The Hague, 7–13 Sept. 1969* (Amsterdam and New York, 1970), 222–6.

—— *Diseases of Children* (Philadelphia, 1979).

MERIMEE, T. J. *et al.*, 'Dwarfism in the Pygmy', *N. Engl. J. Med.* 305/17 (1981), 965–8.

MOLL, H., *Atlas of Pediatric Diseases* (Philadelphia, 1976).

MORSE, D. *et al.*, 'Tuberculosis in Ancient Egypt', *Amer. Review Resp. Diseases*, 90 (1964), 524–41.

MURDOCH, J. L. *et al.*, 'Achondroplasia: A Genetic and Statistical Survey', *Ann. Hum. Genet.* 33 (1970) 227–44.

NEHME, A.-M. E. *et al.*, 'Skeletal Growth and Development of the Achondroplastic Dwarf', *Clin. Orthop.* 116 (1976), 8–23.

ORTNER, D. J., and PUTSCHAR, W. G. J., *Identification of Pathological Conditions in Human Skeletal Remains* (Washington, 1981).

RIMOIN, D. L., and LACHMAN, R. S., 'The Chondrodysplasias', in A. E. H. Emery and D. L. Rimoin (eds.), *Principles and Practice of Medical Genetics*, i (Edinburgh, 1983), 703–35.

SELIGMANN, C. G., 'A Cretinous Skull of the Eighteenth Dynasty', *Man*, 12 (1912), 17–18.

SILVERMAN, F. N., 'De l'Art du diagnostic des nanismes et du diagnostic des nanismes dans l'art', *J. Radiol.* 63/2 (1982), 133–40.

SMITH, D. W., *Recognizable Patterns of Human Malformations*[2] (Philadelphia, 1976).

—— *Growth and its Disorders* (Philadelphia, 1977).

SMITH, H. D., 'A Study of Pygmy Crania, Based on Skulls Found in Egypt', *Biometrika*, 8 (1912), 262–6.

SMITH, R., *Biochemical Disorders of the Skeleton* (Boston and London, 1979).

—— 'Disorders of the Skeleton', in *Oxford Textbook of Medicine*, ii (Oxford, 1983), 17.30–36.

SMITH, R. *et al.*, *The Brittle Bone Syndrome* (London, 1983).

SPRANGER, J. W. *et al.*, *Bone Dysplasias. An Atlas of Constitutional Disorders of Skeletal Development* (Philadelphia and Toronto, 1974).

WYNNE-DAVIES, R. *et al.*, *Atlas of Skeletal Dysplasias* (Edinburgh, 1985).

EGYPT

EL-AGUIZY, O., 'Dwarfs and Pygmies in Ancient Egypt', *ASAE* 71 (1987), 53–60.

ALBRIGHT, W. F., 'Dwarf Craftsmen in the Keret Epic and Elsewhere in North-West Semitic Mythology', *IEJ* 4 (1954), 1–4.

ALTENMÜLLER, H., *Die Apotropaia und die Götter Mittelägyptens*, Diss. (Munich, 1965).

—— *LÄ* i (1975), s.v. Aha, 96–8.

—— *LÄ* i (1975), s.v. Bes, 720–4.

—— *LÄ* i (1975), s.v. Beset, 731.

—— *LÄ* ii (1977), s.v. Hit, 1226–7.

BADAWI, A. M., 'Le Grotesque: Invention égyptienne', *Gazette des Beaux-Arts*, 66 (1965), 189–98.

BAINES, J., *Fecundity Figures. Egyptian Personification and the Iconology of a Genre* (Warminster, 1985).

—— 'Egyptian Twins', *Orientalia*, 54/4 (1985), 461–82.

—— *LÄ* vi (1986), s.v. Zwilling, 1436–7.

BALLOD, F., *Prolegomena zur Geschichte der zwerghaften Götter in Ägypten* (Moscow, 1913).

BECATTI, G., *EAA* v (1963), s.v. Pataikoi, 986.

BER, A., 'Déité de l'Egypte ancienne, Bès eut-il pour modèle un nain hypothyroïdien?', *Organorama*, 10/4 (1973–4), 24–7.

BISI, A. M., 'Da Bes a Herakles (a proposito di tre scarabei del Metropolitan Museum)', *RStudFen* 8 (1980), 19–42.

BISSING, F. W. v., 'Miszellen. Zur Deutung der "pantheistischen Besfiguren"', *ZÄS* 75 (1939), 130–2.

BONNER, C., *Studies in Magical Amulets, Chiefly Graeco-Egyptian* (London and Ann Arbor, 1950).

BONNET, H., *Reallexikon der ägyptischen Religionsgeschichte* (Berlin, 1952).

BORGHOUTS, J. F., *The Magical Texts of Papyrus Leiden I 348* (Leiden, 1971) (OMRO 51).

BOSSE-GRIFFITHS, K., 'A Beset Amulet from the Amarna Period', *JEA* 63 (1977), 98–106.

BOTHMER, B. v., 'The Dwarf as Bearer', *BMFA* 47/267 (1949), 9–11.

BOURRIAU, J., *Pharaohs and Mortals: Egyptian Art in the Middle Kingdom* (Cambridge, 1988).

BRUNNER-TRAUT, E., 'Gravidenflasche. Das Salben des Mutterleibes', in A. Kuschke and E. Kutsch (eds.), *Festschrift für K. Galling* (Tübingen, 1970), 35–48.

BRUNNER-TRAUT, E., 'Nachlese zu zwei Arzneigefässen', *WdO* 6 (1970–1), 4–6.

—— *LÄ* iii (1980), s.v. Karikatur, 337–9.

BRUYÈRE, B., *Rapport sur les fouilles de Deir el Médineh (1934–1935)*, iii (Cairo, 1939) (FIFAO 16).

CHERPION, N., 'De quand date la tombe du nain Seneb?', *BIFAO* 84 (1984), 35–54.

CRAZZOLARA, P., 'Pygmies on the Bahr el Ghazal', *Sudan Notes and Records*, 16 (1933), 85–8.

DARESSY, G., 'Statuette grotesque égyptienne', *ASAE* 4 (1903), 124–5.

—— *Statues de divinités* (CG 38001–39348) (Cairo, 1905–6).

DAUMAS, F., *Les Mammisis des temples égyptiens* (Paris, 1958).

—— *Les Mammisis de Dendara* (Cairo, 1959).

DAWSON, W. R., 'Pygmies and Dwarfs in Ancient Egypt', *JEA* 24 (1938), 185–9.

—— 'Pygmies, Dwarfs and Hunchbacks in Ancient Egypt', *Ann. Med. Hist.* 9/4 (1927), 315–26.

DELATTE, A., and DERCHAIN, P., *Les Intailles magiques gréco-égyptiennes* (Paris, 1964).

DELPECH-LABORIE, J., 'Enquêtes. Le dieu Bès, nain, pygmée ou danseur?' *CdE* 16/32 (1941), 252–4.

DERCHAIN, P., 'Appendix K. Observations sur les erotica', in G. T. Martin, *The Sacred Animal Necropolis at North Saqqâra* (London, 1981), 166–70.

DRENKHAHN, R., *Die Handwerker und ihre Tätigkeiten im alten Ägypten* (Wiesbaden, 1976).

DZIERŻYKRAY-ROGALSKI, T., and POMIŃSKA, E., 'La Statuette de Ptah-Pathèque des collections du Musée Egyptien du Caire', *Africana Bulletin*, 13 (1970), 109–11.

FIEDLER U., and WATERMANN R., 'Über die Zwerge im alten Ägypten', *Ztschr. menschl. Vererb.- u. Konstitutionslehre*, 33 (1956), 505–13.

FISCHER, H. G., 'Chroniques. Monuments of the Old Kingdom in the Cairo Museum', *CdE* 43/86 (1968), 305–12 ('4. Cairo Cat. 1652', 310–12).

—— 'Five Inscriptions of the Old Kingdom', *ZÄS* 105 (1978), 42–59 ('3. An Overseer of Dwarfs', 47–56).

—— 'The Ancient Egyptian Attitude towards the Monstrous', in A. E. Farkas *et al.* (eds.), *Monsters and Demons in the Ancient and Medieval Worlds, Papers Presented in Honor of Edith Porada* (Mainz am Rhein, 1987), 13–26.

GARSTANG, J., *The Burial Customs of Ancient Egypt* (London, 1907).

GHALIOUNGUI, P., *The House of Life: Magic and Medical Science in Ancient Egypt*[2] (Amsterdam, 1973).

GRAPOW, H. *et al.*, *Grundriss der Medizin der alten Ägypter* (Berlin, 1954–73).

GRENFELL, A., 'The Iconography of Bes, and of Phoenician Bes-Hand Scarabs', *PSBA* 24 (1902), 21–40.

GRIFFITHS, J. G., *LÄ* iv (1982), s.v. Patäke, 914–15.

GUGLIELMI, W., 'Humor in Wort und Bild auf altägyptischen Grabdarstellungen', in H. Brunner *et al.* (eds.), *Wort und Bild* (Munich, 1979), 181–200.

GUNN, B., 'The Egyptian for Short', *RecTrav* 39 (1921), 101–4.

HALL, H. R., 'An Egyptian St Christopher', *JEA* 15 (1929), 1.

HAYES, W. C., *The Scepter of Egypt: A Background for the Study of Egyptian Antiquities in the Metropolitan Museum of Art* (New York, 1953–9).

HELWIN, H., 'Gehörte die Königin von Punt zu den chondrodystrophen Zwergen?', *Gegenbaurs Morph. Jahrb.* 120/2 (1974), 280–9.

HORNUNG, E., *Conceptions of God in Ancient Egypt*[2] (trans. and rev. by J. Baines) (London, 1983; Darmstadt, 1971).

HORNUNG, E., and STAEHELIN, E., *Skarabäen und andere Siegelamulette aus Basler Sammlungen* (Mainz am Rhein, 1967).

HÜCKEL, R., 'Über Wesen und Eigenart der Pataiken', *ZÄS* 70 (1934), 103–7.

JÉQUIER, G., 'Notes et remarques. Nature et origine du dieu Bès', *RecTrav* 37 (1915), 114–20.

JUNKER, H., *Giza*, v, *Die Mastabas des Snb und die umliegenden Gräber* (Vienna, 1941) (DAAW).

KAPLONY, P., *Die Inschriften der ägyptischen Frühzeit* (Wiesbaden, 1963–4).

KEIMER, L., 'Un Bès tatoué?', *ASAE* 42 (1943), 159–61.

KEMP, B. J., and MERILLEES, R. S., *Minoan Pottery in Second Millennium Egypt* (Mainz am Rhein, 1980).

KLEBS, L., *Die Reliefs des alten Reiches* (Heidelberg, 1915).

KRALL, J., 'Ueber den ägyptischen Gott Bes', in O. Benndorf and G. Niemann, *Das Heroon von Gjölbaschi-Trysa* (Vienna, 1889), 72–96.

LANSING, A., 'The Egyptian Expedition 1933–1934', *BMMA* 29, part ii (1934), 1–41.

LECA, A.-P., *La Médecine égyptienne au temps des pharaons* (Paris, 1971).

LEFEBVRE, G., *Essai sur la médecine égyptienne de l'époque pharaonique* (Paris, 1956).

LEGGE, F., 'The Magic Ivories of the Middle Empire', i, *PSBA* 27 (1905), 130–52; ii, ibid. 297–303; iii, *PSBA* 28 (1906), 159–70.

MARTIN, G. T., '"Erotic" Figurines: The Cairo Museum Material', *GM* 96 (1987), 71–84.

MEEKS, D., 'Génies, anges, démons en Égypte', in *Génies, anges et démons* (Paris, 1971) (SourcesOr 8), 17–84.

MICHAILIDIS, G., 'Le Dieu Bès sur une stèle magique', *BIE* 42–3 (1960–2), 65–85.

—— 'Bès aux divers aspects', *BIE* 45 (1963–4), 53–93.

MONTET, P., 'Ptah patèque et les orfèvres', *Revue Archéologique*, 40 (1952), 1–11.

—— 'Ptah patèque et les orfèvres nains', *BSFE* 11 (1952), 73–4.

MORENZ, S., 'Ptah-Hephaistos der Zwerg', in *Festschrift für F. Zucker* (Berlin, 1954), 275–90.

NASTER, P., 'Die Zwerge als Arbeiterklasse in bestimmten Berufen im alten Ägypten', in D. O. Edzard (ed.), *Gesellschaftsklassen im alten Zweistromland und in den angrenzenden Gebieten. XVIII. Rencontre assyriologique internationale, 1970* (Munich, 1972) (ABAW 75), 139–43.

NAVILLE, E., 'Figurines égyptiennes de l'époque archaïque', *RecTrav* 22 (1900), 65–71.

NIBBI, A., 'Punt and Pygmies in the Northern Red Sea', *DE* 2 (1985), 27–36.

PARLASCA, K., 'Zwei ägyptische Bronzen aus dem Heraion von Samos', *MDAI(A)* 68 (1953), 127–36.

PARROT, J., 'Sur l'origine d'une des formes du dieu Phtah', *RecTrav* 2 (1880), 129–33.

PIANKOFF, A., 'Sur une statuette de Bès', *BIFAO* 37 (1937–8), 29–33.

PINCH, G., 'Childbirth and Female Figurines at Deir el-Medina and el-ᶜAmarna', *Orientalia*, 52 (1983), 405–14.

QUAEGEBEUR, J. *et al.*, 'The Memphite Triad in Greek Papyri', *GM* 88 (1985), 25–37.

QUIBELL, J. E., *Excavations at Saqqara (1905–1906)* (Cairo, 1907).

RAVEN, M. J., 'A Puzzling Pataekos', *OMRO* 67 (1987), 7–17.

REGNAULT, F., 'Le Dieu égyptien Bès était-il myxoedémateux?', *Bull. Soc. Anthr. Paris*, 4th. ser. 8 (1897), 434–9.

—— 'Les Nains dans l'art égyptien', *Bull. Soc. Fr. Hist. Méd.* 13 (1914), 137–46.

ROEDER, G., *Ägyptische Bronzefiguren* (Berlin, 1956) (Staatl. Museen zu Berlin, Mitteilungen aus der ägyptischen Sammlung 6).

ROMANO, J. F., 'The Origin of the Bes-Image', *Bull. of the Egyptological Seminar*, 2 (1980), 39–56.

ROMANO, J. F., *The Bes-Image in Pharaonic Egypt*, Ph.D. thesis (New York Univ., 1989).

RUFFER, M. A., 'On Dwarfs and Other Deformed Persons in Ancient Egypt', in R. L. Moodie (ed.), *Studies in the Palaeopathology of Egypt* (Chicago, 1921), 35–48.

RUPP, A., 'Der Zwerg in der ägyptischen Gemeinschaft. Studien zur ägyptischen Anthropologie', *CdE* 40/80 (1965), 260–309.

SAAD, Z., 'Statuette of the God Bes', *ASAE* 42 (1943), 147–52.

SANDMAN HOLMBERG, M., *The God Ptah* (Lund, 1946).

SAUNERON, S., 'Les Songes et leurs interprétations dans l'Égypte ancienne', in *Les Songes et leurs interprétations* (Paris, 1959) (SourcesOr 2), 17–61.

—— 'Le Monde du magicien égyptien', in *Le Monde du sorcier* (Paris, 1966) (SourcesOr 7), 27–65.

—— *Le Papyrus magique illustré de Brooklyn, Brooklyn Museum 47.218.156* (Brooklyn, 1970).

SAUNERON, S., and YOYOTTE, J., 'La Naissance du monde selon l'Égypte ancienne', in *La Naissance du monde* (Paris, 1959) (SourcesOr 1), 17–91.

EL-SAYED, R., 'Deux aspects nouveaux du culte à Saïs', *BIFAO* 76 (1976), 91–100.

—— *La Déesse Neith de Saïs. Importance et rayonnement de son culte* (Cairo, 1982) (BdE 86).

SCHOTT, S., 'Eine Kopfstütze des Neuen Reiches', *ZÄS* 83 (1958), 141–4.

SCHRUMPF-PIERRON, B., 'Les Nains achondroplasiques dans l'ancienne Égypte', *Aesculape*, 24, NS 9 (1934), 223–38.

SEYFRIED, K.-J., *LÄ* vi (1986), s.v. Zwerg, 1432–5.

SILVERMAN, D., 'Pygmies and Dwarves in the Old Kingdom', *Serapis*, 1 (1969), 53–61.

SPIEGELBERG, W., 'Ägyptologische Mitteilungen III. Zu dem Typus und der Bedeutung der als Patäken bezeichneten ägyptischen Figuren', *SBAW* 2 (1925), 8–11.

—— 'Die Weihestatuette einer Wöchnerin', *ASAE* 29 (1929), 162–5.

—— 'Das Grab eines Grossen und eines Zwerges aus der Zeit des Nektanebês', *ZÄS* 64 (1929), 76–83.

STRACMANS, M., 'Les Pygmées dans l'ancienne Égypte', in *Mélanges Georges Smets* (Brussels, 1952), 621–31.

STRICKER, B. H., 'Bes de danser', *OMRO* 37 (1956), 35–48.

TOLSTOÏ, S., *Étude des représentations pathologiques dans l'art égyptien* (Paris, 1939).

TRAN TAM TINH, *LIMC* iii (1986), s.v. Bes, 98–108.

—— *LIMC* iii (1986), s.v. Besit, 112–14.

VANDIER D'ABBADIE, J., 'Les Singes familiers dans l'ancienne Égypte', *RdE* 16 (1964), 147–77; ibid. 17 (1965), 177–88; ibid. 18 (1966), 143–201.

VARTAVAN, C. T. DE, *Bes, The Bow-Legged Dwarf or the Ladies' Companion* (University of London, 1986; unpublished).

VASSAL, P. A., 'La Physio Pathologie dans le panthéon égyptien: Les Dieux Bès et Phtah, le nain et l'embryon', *Bull. Mém. Soc. Anthr. Paris*, 7 (1956), 168–81.

VYCICHL, W., 'Amharique *denk* "nain", Egyptien *d-n-g*', *Annales d'Ethiopie*, 2 (1957), 248–9.

WARD, W. A., 'A Unique Beset Figurine', *Orientalia*, 41 (1972), 149–59.

WATERMANN, R., *Bilder aus dem Lande Ptah und Imhotep. Naturbeobachtung, Realismus und Humanität der alten Ägypter* (Cologne, 1958).

WEBER, W., *Die ägyptisch-griechischen Terrakotten* (Berlin, 1914) (Königl. Museen zu Berlin, Äg. Slg. II).

WEEKS, K. R., *The Anatomical Knowledge of the Ancient Egyptians and the Representation of the Human*

Figure in Egyptian Art, Ph.D. thesis (Yale University, 1970).

—— 'Art, Word, and the Egyptian World View', in K. R. Weeks (ed.), *Egyptology and the Social Sciences* (Cairo, 1979), 59–81.

WERBROUCK, M., 'A propos du dieu Bès', *Egyptian Religion*, 1 (1933), 28–32.

—— 'Les Multiples Formes du dieu Bès, *BMRAH*, 3rd. ser. 11 (1939), 78–82.

WESTENDORF, W., *LÄ* iv (1982), s.v. Missbildung, 148–9.

WILD, H., 'Une Danse nubienne d'époque pharaonique', *Kush*, 7 (1959), 76–90.

—— 'Les Danses sacrées de l'Égypte ancienne', in *Les Danses sacrées* (Paris, 1963) (SourcesOr 6), 33–117.

WILSON, V., 'The Iconography of Bes with Particular Reference to the Cypriot Evidence', *Levant*, 7 (1975), 77–103.

WOLFF, H. F., 'Die kultische Rolle des Zwerges im alten Ägypten', *Anthropos*, 33 (1938), 445–514.

GREECE

ADLER, A., *RE* xi. 1 (1921), s.v. Kobaloi, 931.

BALLABRIGA, A., 'Le Malheur des nains. Quelques aspects du combat des grues contre les pygmées dans la littérature grecque', *REA* 83 (1981), 57–74.

BARTSOCAS, C. S., 'An Introduction to Ancient Greek Genetics and Skeletal Dysplasias', in *Skeletal Dysplasias. Third International Clinical Genetics Seminar, Athens 1982* (New York, 1982), 3–13.

—— 'Goiters, Dwarfs, Giants and Hermaphrodites', in *Endocrine Genetics and Genetics of Growth. Fourth International Clinical Genetics Seminar, Athens 1985* (New York, 1985), 1–18.

BECATTI, G., *EAA* ii (1959), s.v. Caricatura, 342–8.

—— *EAA* v (1963), s.v. Pataikoi, 986.

—— *EAA* vi (1965), s.v. Pigmei, 167–9.

BINSFELD, W., *Grylloi. Ein Beitrag zur Geschichte der antiken Karikatur*, Diss. (Cologne, 1956).

BLINKENBERG, C., 'Rhodische Urvölker', *Hermes*, 50 (1915), 271–303.

—— *Lindos*, i (Berlin, 1931).

BOLKESTEIN, H., *Wohltätigkeit und Armenpflege im vorchristlichen Altertum* (Utrecht, 1939).

BÖHLAU, J., *Aus ionischen und italischen Nekropolen* (Leipzig, 1898).

BRECHT, F. J., *Motiv- und Typengeschichte des griechischen Spottepigramms* (Leipzig, 1930) (Philologus suppl. 22.2).

BRELICH, A., *Gli eroi greci. Un problema storico-religioso* (Rome, 1958).

BROMMER, F., *Hephaistos. Der Schmiedegott in der antiken Kunst* (Mainz am Rhein, 1978).

BUSCHOR, E., 'Das Krokodil des Sotades', *MJBK*, 11 1/2 (1919).

BUXTON, R. G. A., 'Blindness and Limits: Sophokles and the Logic of Myth', *JHS* 100 (1981), 22–37.

CAMBITOGLOU, A., 'The Felton Painter in Sydney', in E. Böhr and W. Martini (eds.), *Studien zur Mythologie und Vasenmalerei. K. Schauenburg zum 65. Geburtstag* (Mainz am Rhein, 1986), 143–7.

CAUBET, A., 'Statuette en bois de démon égypto-grec d'époque archaïque', *RLouvre*, 19 (1969), 7–12.

CÈBE, J. P., *La Caricature et la parodie dans le monde romain antique des origines à Juvénal* (Paris, 1966) (BEFAR 206).

DELCOURT, M., *Stérilités mystérieuses et naissances maléfiques dans l'antiquité classique* (Liège, 1938).

DELCOURT, M., *Hephaistos ou la légende du magicien* (Paris, 1957).

DETIENNE, M., 'Le Phoque, le crabe et le forgeron', in R. Crahay, M. Derwa, and R. Joly (eds.), *Hommages à Marie Delcourt* (1970) (Latomus coll. 114), 219–33.

DEUBNER, L., *Attische Feste* (Berlin, 1932).

ESSER, A., *Das Antlitz der Blindheit in der Antike*[2] (Leiden, 1961) (Janus suppl. 4).

FREYER-SCHAUENBURG, B., 'Die Geranomachie in der archaischen Vasenmalerei. Zu einem pontischen Kelch in Kiel', in *Wandlungen. Studien zur antiken und neueren Kunst, Ernst Homann-Wedeking gewidmet* (Waldsassen, 1975), 76–83.

FRIEDLÄNDER, P., Roscher, v (1916–24), s.v. Telchinen, 234–43.

FURTWÄNGLER, A., 'Zwei griechische Terrakotten', *ARW* 10 (1907), 321–32 (repr. in *Kleine Schriften*, ii (Munich, 1913), 417–26).

GERMAIN, L., 'Aspects du droit d'exposition en Grèce', *Revue Historique de Droit Français et Etranger*, 47 (1969), 177–97.

GIGLIOLI, G. Q., 'Una pelike attica da Cerveteri nel Museo di Villa Giulia a Roma con Herakles e Geras', in G. E. Mylonas and D. Raymond (eds.), *Studies presented to D. M. Robinson*, ii (St Louis, 1953), 111–13.

GRMEK, M. D., 'Les Affections de la colonne vertébrale dans l'iconographie médicale et les arts antiques', *Dossiers histoire et archéologie*, 123 (1988), 52–61.

GUSINDE, M., *Kenntnisse und Urteile über Pygmäen in Antike und Mittelalter*, (Leipzig, 1962) (Nova Acta Leopoldina 162, Bd. 25).

HADZISTELIOU PRICE, TH., *Kourotrophos. Cults and Representations of the Greek Nursing Deities* (Leiden, 1978).

HANDS, A. R., *Charities and Social Aid in Greece and Rome* (London, 1968).

HEMBERG, B., *Die Kabiren* (Uppsala, 1950).

HENNIG, R., 'Der kulturhistorische Hintergrund der Geschichte vom Kampf zwischen Pygmäen und Kranichen', *RhM* 81 (1932), 20–4.

HIMMELMANN, N., 'Archäologisches zum Problem der griechischen Sklaverei', *Abh.Mainz*, 13 (1971), 614–57.

HOFFMANN, H., *Sexual and Asexual Pursuit: A Structuralist Approach to Greek Vase Painting* (Royal Anthropological Institute, Occ. Paper 34; London, 1977).

—— ὕβριν ὀρθιάν κνωδάλων, in D. Metzler et al. (eds.), *Antidoron. Festschrift für J. Thimme zum 65. Geburtstag* (Karlsruhe, 1983), 61–73.

JANNI, P., *Etnografia e mito. La storia dei Pigmei* (Rome, 1978).

LIPPOLD, G., 'Zu den Imagines Illustrium, *MDAI(R)* 52 (1937), 44–7.

MALTEN, L., *RE* viii. 1 (1913), s.v. Hephaistos, 311–66.

—— 'Hephaistos', *JDAI* 27 (1912), 232–64.

METZLER, D., *Porträt und Gesellschaft* (Berlin, 1971).

MINTO, A., *Il vaso François* (Florence, 1960) (AAT).

MONCEAUX, P., 'La Légende des pygmées et les nains de l'Afrique équatoriale', *RH* 47 (1891), 1–64.

OLENDER, M., 'Priape, le dernier des dieux', in Y. Bonnefoy (ed.), *Dictionnaire des mythologies*, ii (Paris, 1981), 311–14.

POTTIER, E., *Une clinique grecque au Ve siècle* (Paris, 1906) (MonPiot 13), 149–66.

PRÉAUX, C., 'Les Grecs à la découverte de l'Afrique par l'Égypte', *CdE* 32/64 (1957), 284–312.

RAECK, W., *Zum Barbarenbild in der Kunst Athens im 6. und 5. Jahrh. v. Chr.*, Diss. (Bonn, 1981).

RICHTER, G. M. A., 'An Aryballos by Nearchos', *AJA* 36 (1932) 272–5.

RIZZO, G. E., 'Caricature antiche', *Dedalo*, 7 (1926–7), 402–18.

ROBERTSON, M., 'A Muffled Dancer and Others', in A. Cambitoglou (ed.), *Studies in Honour of Arthur Dale Trendall* (Sydney, 1979), 129–34.

—— 'An Unrecognized Cup by the Kleophrades Painter?', in *Stele: Tomos eis Mnemen Nikolauou Kontoleontos* (Athens, 1980), 125–9.

SCHAUENBURG, K., 'Der besorgte Marsyas', *MDAI(R)* 79 (1972), 317–22.

SCHMIDT, M., 'Hephaistos lebt. Untersuchungen zur Frage der Behandlung behinderter Kinder in der Antike', *Hephaistos*, 5–6 (1983–4), 133–61.

SEELIGER, K., Roscher, ii. 1 (1890–97), s.v. Kerkopen, 1166–73.

SHAPIRO, H. A., 'Notes on Greek Dwarfs', *AJA* 88 (1984), 391–2.

SINN, U. 'Zur Wirkung des ägyptischen "Bes" auf die griechische Volksreligion', in D. Metzler *et al.* (eds.), *Antidoron. Festschrift für J. Thimme zum 65. Geburtstag* (Karlsruhe, 1983), 87–94.

—— 'Der Kult der Aphaia auf Aegina', in R. Hägg *et al.* (eds.), *Early Greek Cult Practice. Proceedings of the Fifth International Symposium at the Swedish Institute at Athens, 26–29 June, 1986* (Stockholm, 1988), 149–59.

SNOWDEN, F. M., *Blacks in Antiquity* (London, 1970).

ŠTAL, I. V., 'The Myth of the Pygmies on the Black Sea Littoral', *Klio*, 68 (1986), 351–66.

TRENDALL, A. D., The Felton Painter and a Newly Acquired Comic Vase by his Hand', in F. Philipp and J. Stewart (eds.), *Essays and Studies in Honour of Sir Daryl Lindsay* (Oxford, 1964), 45–52.

—— *Phlyax Vases*[2] (London, 1967) (BICS suppl. 19).

WASER, O., Roscher, iii. 2 (1902–9), s.v. Pygmaien, 3283–317.

WILLAMOWITZ-MOELLENDORFF, U. v., Hephaistos, *NGG* (1895), 217–45.

WÖLKE, H., 'Pietro Janni: Etnografia e mito', *Gnomon*, 55 (1983), 97–9.

WÜST, E., *RE* xxiii. 2 (1959), s.v. Pygmaioi, 2064–74.

ZANCANI MONTUORO P., and ZANOTTI-BIANCO, U., *Heraion. Alla foce del Sele*, ii (Rome, 1954).

ZINSERLING, V., 'Physiognomische Studien in der spätarchaischen und klassischen Vasenmalerei', *WZRostock*, 16 (1967), 571–5.

—— 'Die Anfänge griechischer Porträtkunst als gesellschaftliches Problem', *AAntHung* 15 (1967), 283–95.

INDEX OF OBJECTS IN MUSEUMS
AND *IN SITU*

This index includes objects which are illustrated, and objects which are listed in the catalogues (E: Egypt, G: Greek World). It includes also Egyptian monuments with inscriptions discussed in the text.

EGYPT

Museums

GREEK WORLD

Minor Arts

INDEX OF SOURCES

GENERAL INDEX

Plate 1

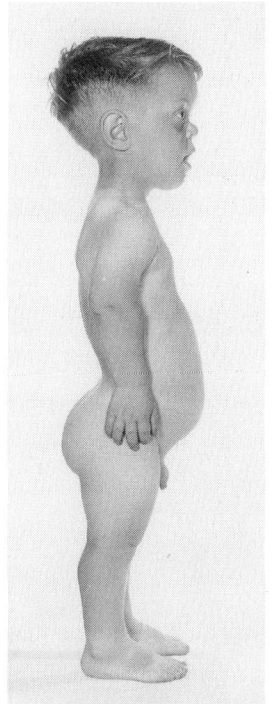

1 Achondroplastic boy

2 Achondroplastic girl

3 Hypochondroplasia 'Das Kölner
Heinzelmännchen'

4 Hypopituitarism 'Adrien'

PLATE 2

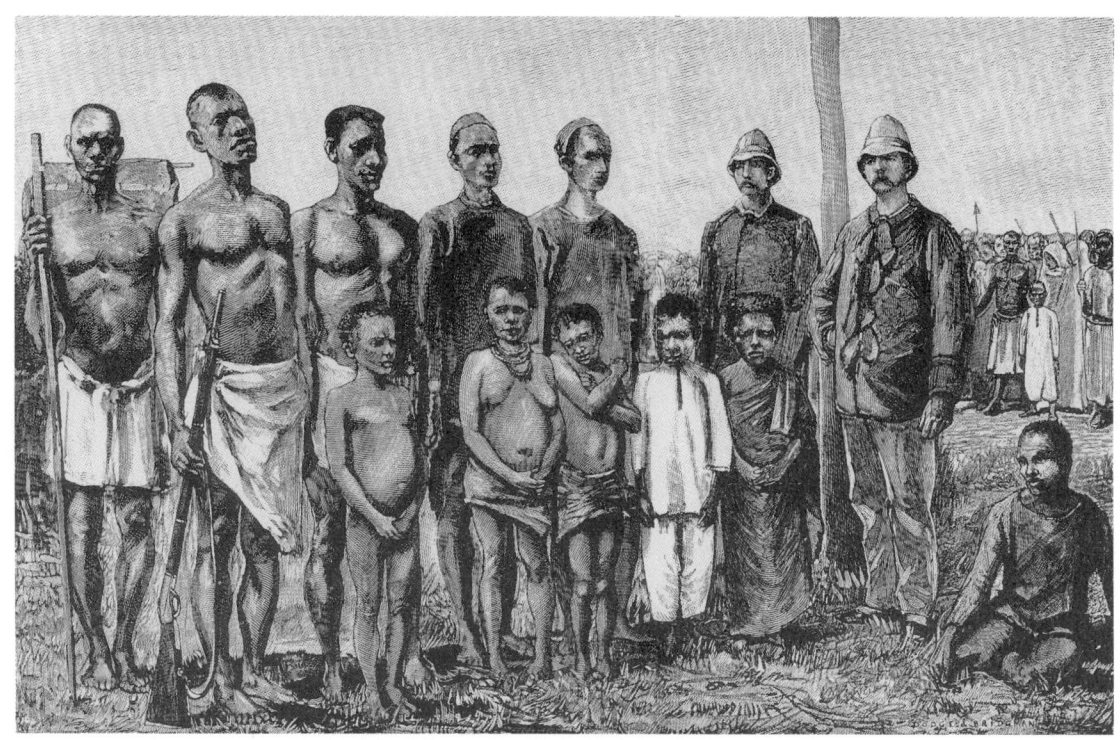

1 Aka pygmies

2 (S 3) Saqqara, tomb complex of King Wadj(?) 3504

PLATE 3

1 Stela. Cairo Museum, JdE 28731

2 Stela. London, BM, 36250

3 Faience. Baltimore, WAG, 48.420

PLATE 4

1 Faience. Brooklyn Museum, 16.426

2 Faience. Brooklyn Museum, 37.912 E

3 Faience. London, BM, 22610

4 Faience. London, BM, 65242

PLATE 5

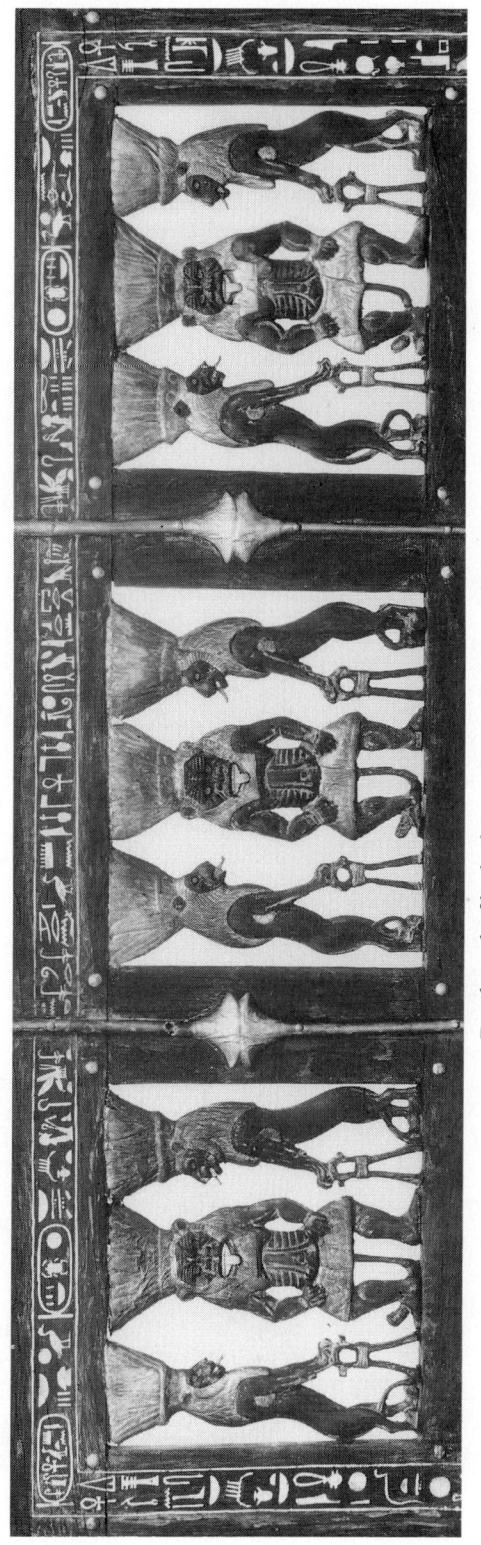

Footboard of bed, ebony. Cairo Museum, JdE 62016

PLATE 6

1 Faience flask. Oxford, Ashmolean Museum, 1890.897

2 Faience. Berlin, SM (West), 7759

3 Faience. Oxford, Ashmolean Museum, 1890.357

PLATE 7

1 Bronze. Athens, NM, 614

2 Bronze. Samos, Vathy
Museum, B 353

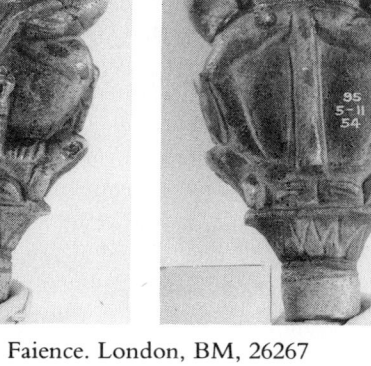

a   3 Faience. London, BM, 26267   *b*

PLATE 8

1 Papyrus. Berlin, SM (East), P 3128

a　　　　*b*　　　　*c*

2 Faience. London, BM, 26316

PLATE 9

a   1   Philae, temple of Hathor   b

2   Dendara, mammisi of Nectanebo   3   Saqqara, 'Bes chambers', room 14

PLATE 10

1 Stela. London, BM, 1178

2 Stela. Brooklyn Museum, 58.98

PLATE 11

a   1 Papyrus. Brooklyn Museum, 47.218.156   *b*

2 Terracotta. London,   3 Terracotta. Leiden, Rijksmuseum,
BM, 61296   F 1975/11.2

PLATE 12

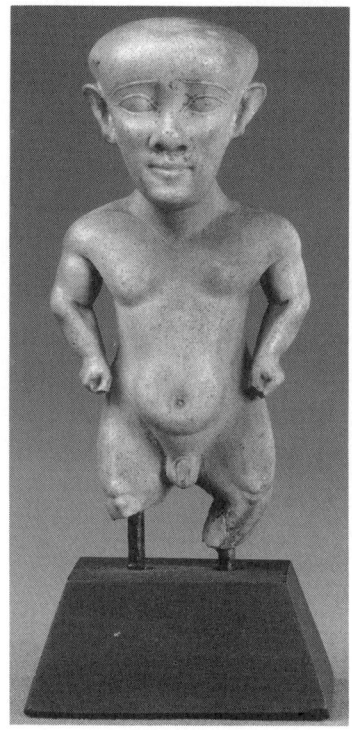 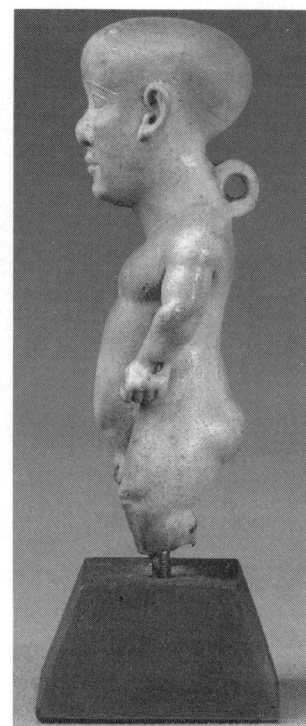 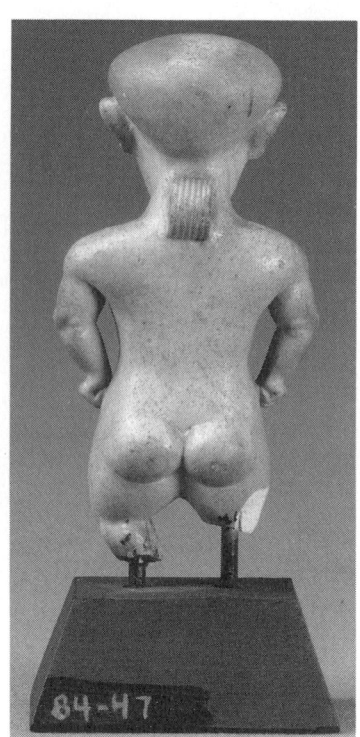

a   *b*   *c*

1 Faience. Virginia Museum, 84.47

2 Faience. London, BM, 11211   3 Faience. London, BM, 63475   4 Faience. London, BM, 11210

PLATE 13

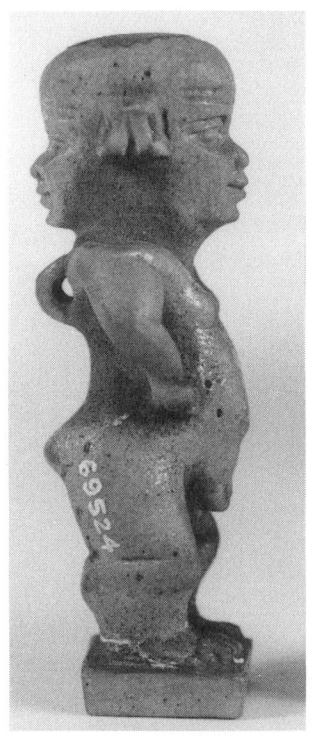

1 Faience. London, BM, 69524

2 Faience. London, Bm, 54000

a

3 Faience. Tokyo, private

b

PLATE 14

a   1 Faience. London, BM, 60205   *b*

2 Faience. Fribourg, Institut
   Biblique, 1609

3 Faience. Fribourg, Institut
   Biblique, 1608

PLATE 15

a 1 Faience. London, BM, 60109 *b*

a 2 Bronze. Oxford, Ashmolean Museum, 1965.172 *b*

PLATE 16

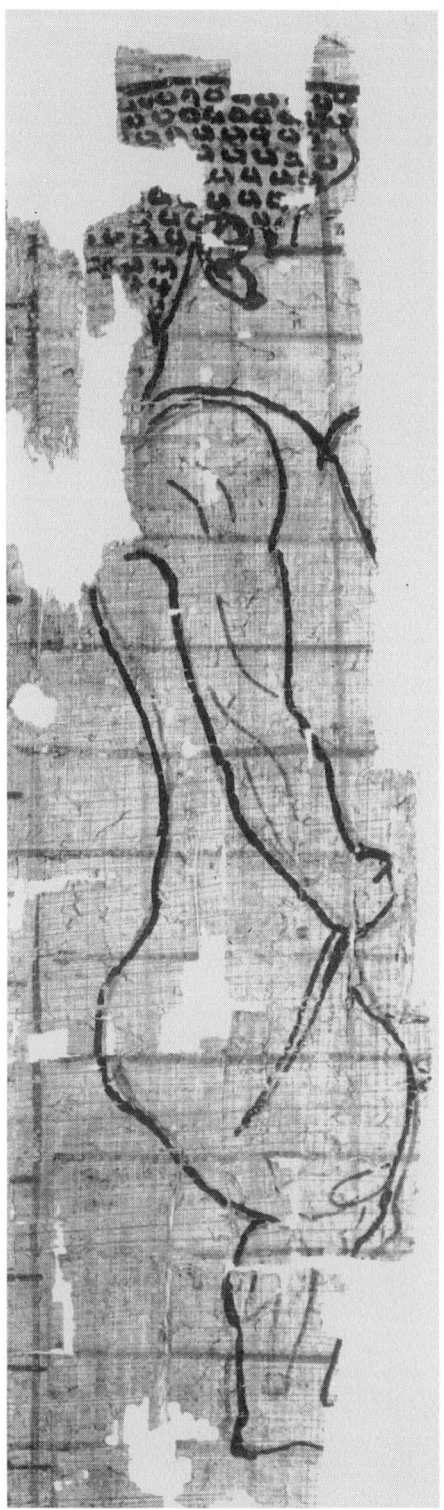

2a

2b

1 Papyrus. Berlin, SM (East), P 13558   2 Faience. London, BM, 11249

PLATE 17

1 (E 2) Bowl fr. Boston, MFA, 01.7292

2 (E 6) Stela. Leiden,
Rijksmuseum, F 1927/1.1

3 (E 8) Stela. London, BM, 35018

PLATE 18

1 (E 13) Giza, tomb of Debheni

2 (E 15) Giza, tomb of Kanufer

PLATE 19

a   1 (E 18) Giza, tomb of Seshemnufer I   *b*

a   2 (E 24) Giza, tomb of Wepemnefert   *b*

PLATE 20

1 (E 29 a) Saqqara, tomb of Ty

2 (E 29 b) Ibid.

PLATE 21

(E 34) Saqqara, tomb of Ptahhotpe II

PLATE 22

1a

1 (E 52 a) Saqqara, tomb of Mereruka

1b

Ibid., detail

2 (E 37) Saqqara, tomb of Ni'ankhkhnum and Khnumhotpe

PLATE 23

a　　　　　(E 52 c) Saqqara, tomb of Mereruka, chapel of Wa'tetkhethor

b　　　　　　　　　　Ibid., detail

PLATE 24

1 (E 57) Relief fr., tomb of Niʿankhnesut. Switzerland, private

2 (E 64) Relief fr. Hanover, Kestner
Museum, 1935.200.201

3 (E 70) Stela. Cairo Museum, CG 20459

PLATE 25

(E 75 b) El'Amarna, tomb of Panehesi

PLATE 26

2 (E 84) Sarcophagus. Cairo Museum, CG 29307

1 (E d 83) Obelisk. Brooklyn Museum, 50.169

PLATE 27

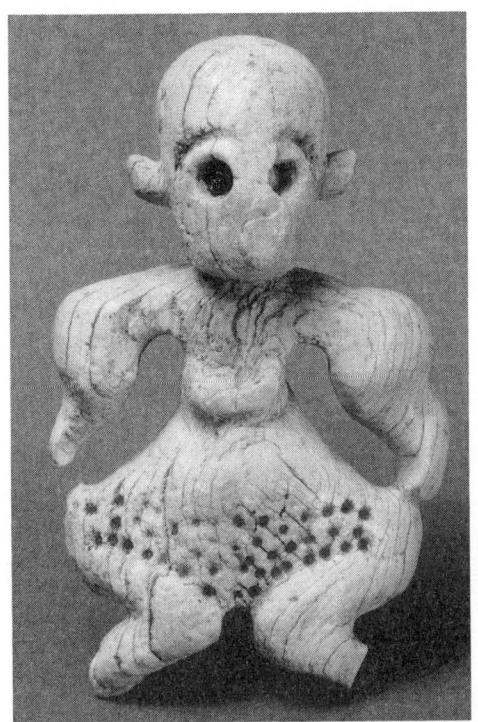

1 (E 92) Ivory. Baltimore, WAG, 71.532

2 (E 94) Ivory. Baltimore, WAG, 71.534

3 (E d 102) Glazed ware. Lucerne, Kofler-
Truniger coll., A 40

4 (E 106) Ivory. Oxford, Ashmolean
Museum, E 298

PLATE 28

a 1 (E 108) Ivory. Oxford, Ashmolean Museum, *b*
E 333/346

2 (E 113) Statue of Seneb, detail. Cairo Museum, JdE 51280

PLATE 29

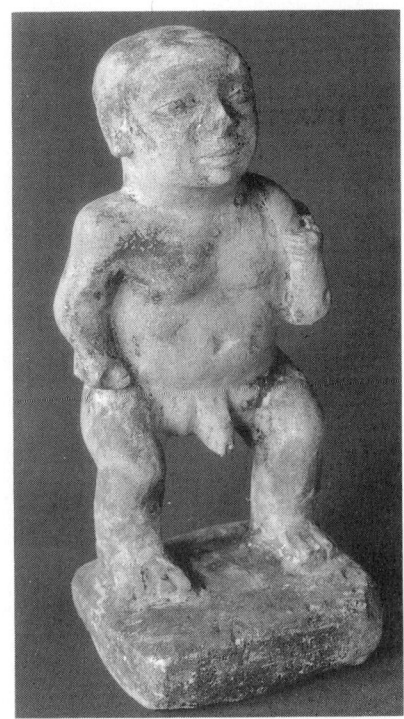

a

b

1 (E 115) Limestone. Chicago,
Oriental Institute, 10627

2 (E 117) Limestone. Cairo
Museum, CG 144

3 (E 116) Limestone. Chicago, Oriental Institute, 10641

PLATE 30

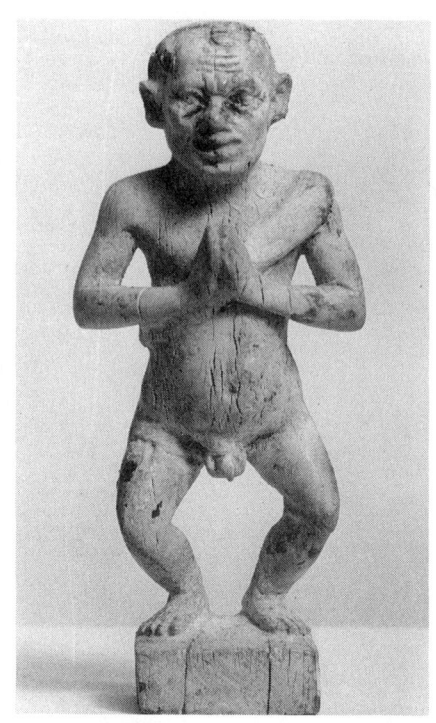

1 (E 118) Ivory. Baltimore, WAG, 71.504

2 (E 122) Ivory. New York, MMA, 34.1.130

3 (E 123) Glazed ware. New York, MMA, 24.1.47

4 (E 124) Glazed ware. New York, MMA, 24.1.48

PLATE 31

a

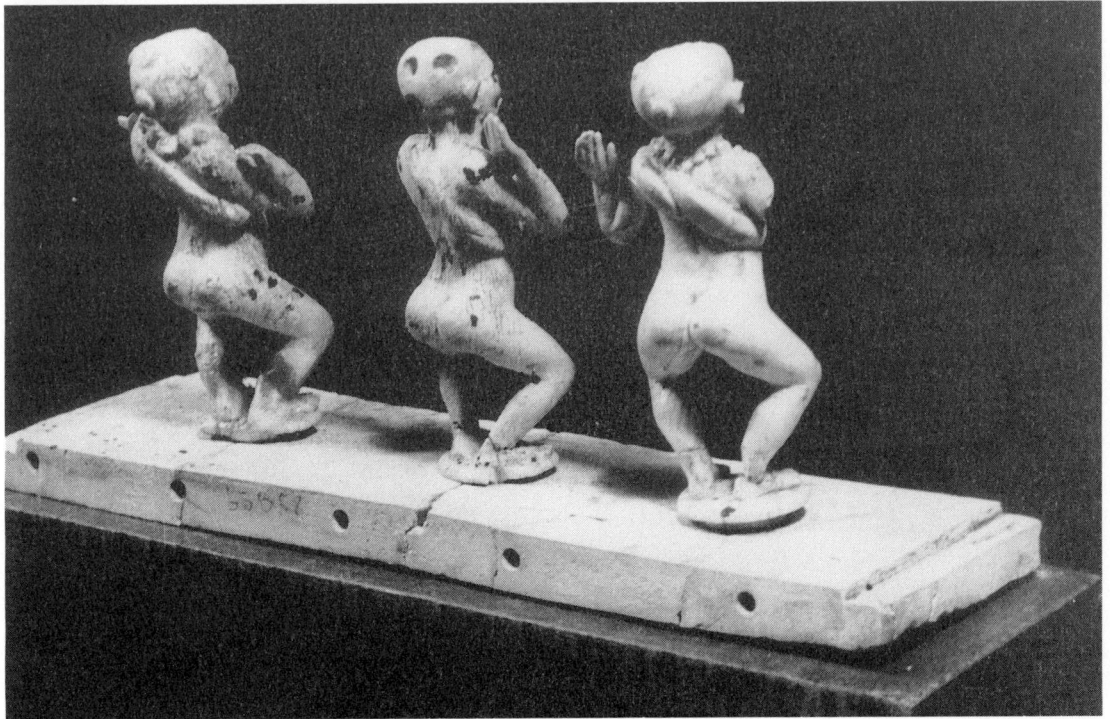

b   (E 122) Ivory. Cairo Museum, JdE 63858

PLATE 32

1 (E 128) Glazed ware. New York,
MMA, 22.1.125

2 (E 129) Glazed ware. New York, MMA,
22.1.177

3 (E 131) Glazed ware. New York,
MMA, 22.1.286

4 (E 133) Glazed ware. New York, MMA,
22.1.1163

PLATE 33

1 (E d 132) Glazed ware. New York,
MMA, 22.1.1140

2 (E 134) Glazed ware. New York, MMA,
1972.48

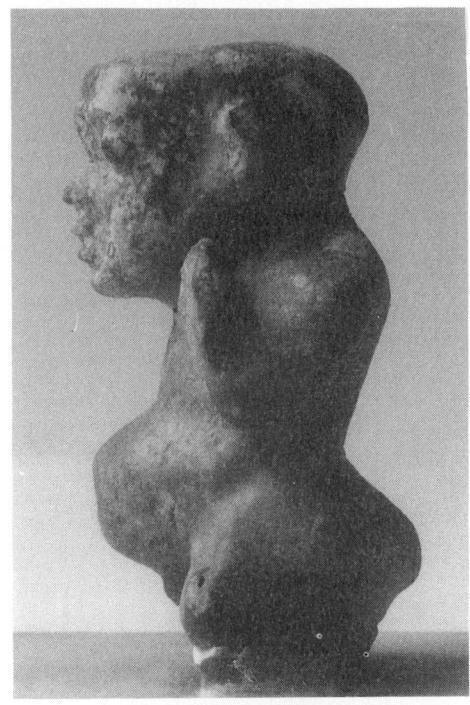

3 (E 144) Glazed ware. London, UC,
18745

4 (E 193) Glazed ware. Cambridge,
Fitzwilliam Museum, E FG 37

PLATE 34

1 (E d 138) Limestone. London, UC, 16523/6

2 (E 146) Wood. Liverpool University, E 7081

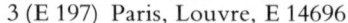

a   3 (E 197) Paris, Louvre, E 14696   *b*

PLATE 35

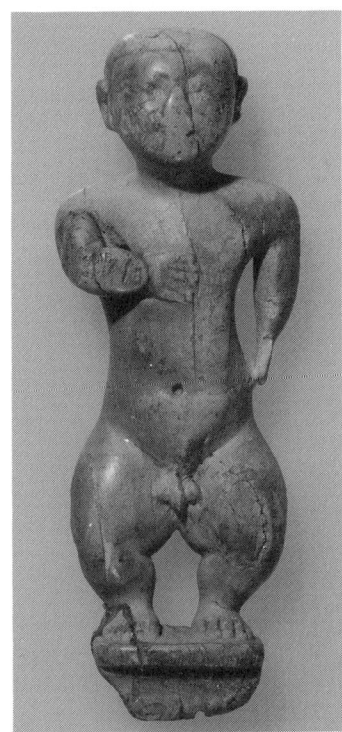
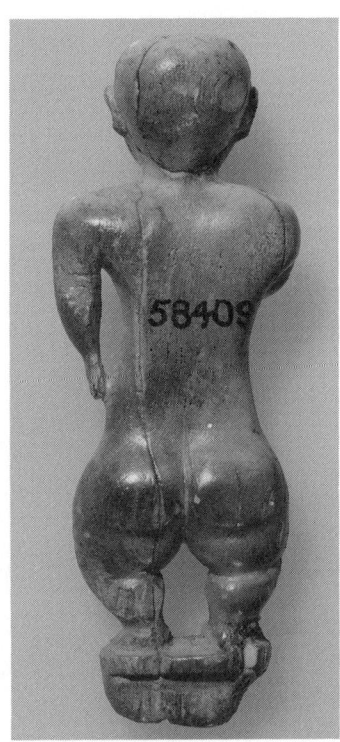

1 (E 195) Ivory. London, BM, 58409

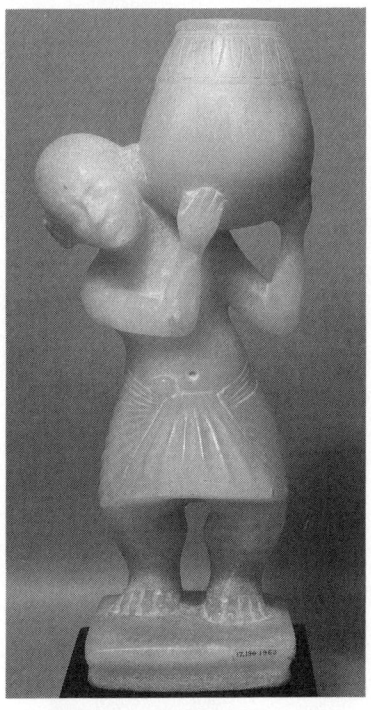

2 (E 198) Boxwood. Boston,
MFA, 48.296

3 (E d 200) Calcite. New York,
MMA, 17.190.1963

PLATE 36

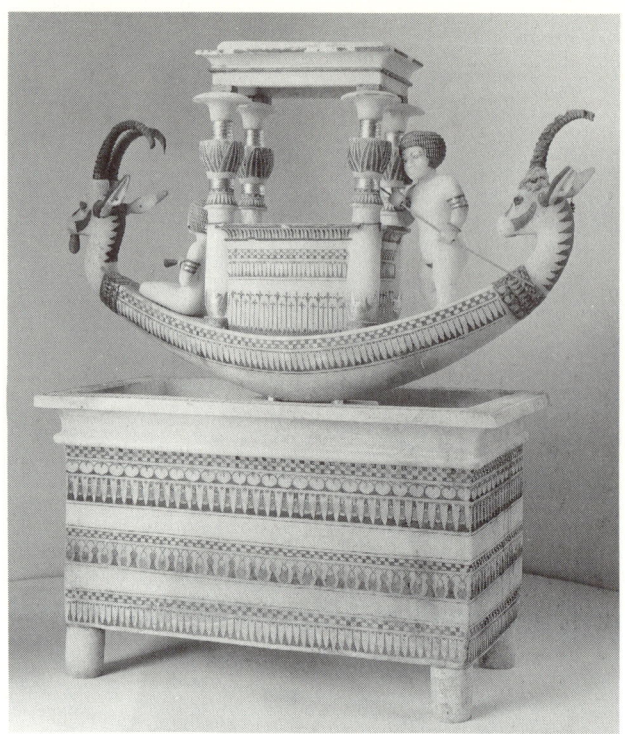

a

b (E 202) Boat, from the tomb of Tutankhamun. Cairo Museum, JdE 535 *c*

PLATE 37

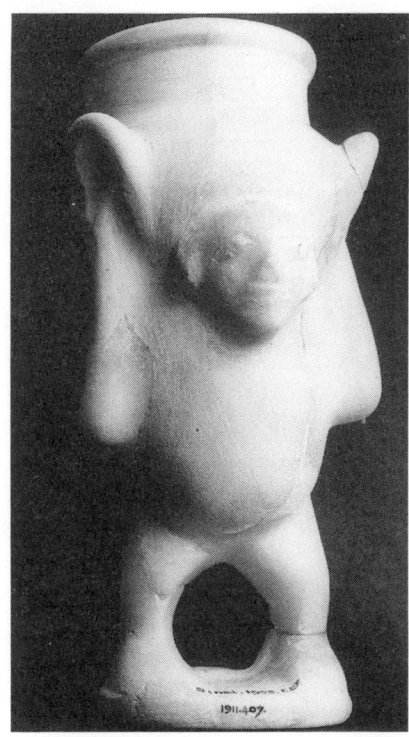

1 (E 203) Figure vase. Oxford,
Ashmolean Museum, 1911.407

2 (E 205) Figure vase. London,
BM, 30459

3 (E 204) Figure vase. London,
BM, 29935

4 (E 206) Figure vase. London,
UC, 15758

PLATE 38

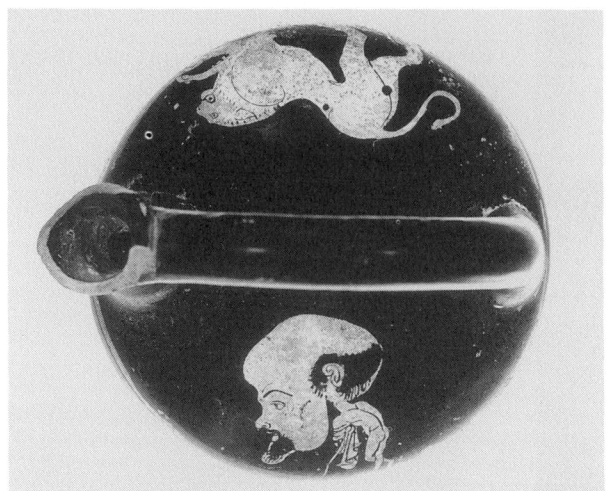

1 Askos. Paris, Louvre, G 610

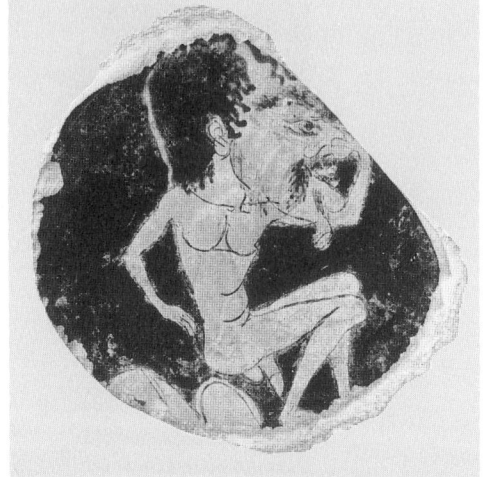

2 Plate fr. Athens, NM, ACR 1073

3 Pyxis fr. Boston, MFA, 10.216

4 Cup. Rome, Vatican, 16552

PLATE 39

1 Skyphos. Boston, MFA, 20.18

2 Chous. Taranto, Ragusa coll., 126

3 Krater. Milan, Museo Archeologico, C 408

4 Lekane. Berlin, SM (West), 3366

PLATE 40

1 Chous. Athens,
NM, 12139

2 Lekythos. Boston,
MFA, 00351

a　　　　　3 Terracotta, figure vase. Paris market　　　　　*b*

PLATE 41

a

b

c   (G 1) Pyxis. Paris, Louvre, CA 1707

PLATE 42

a

b (G 3) Skyphos. Paris, Louvre, G 617

PLATE 43

a

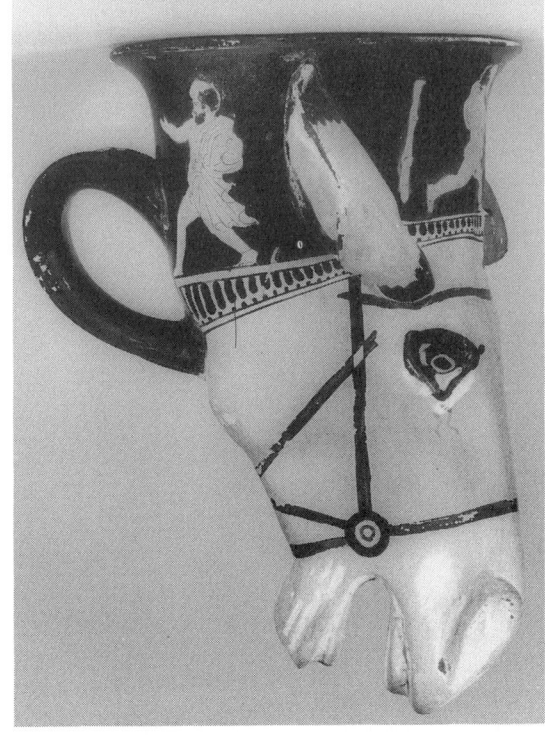

b

(G 4) Rhyton. Ferrara, Museo Nazionale, 2561

c

PLATE 44

a

b

c

(G 5) Aryballos. Paris, Louvre, CA 2183

d

PLATE 45

(G 6) Pelike. Agrigento, Museo Civico, ex Giudice 638

PLATE 46

1 (G 7) Cup fr. Ferrara, Museo Nazionale, 20363

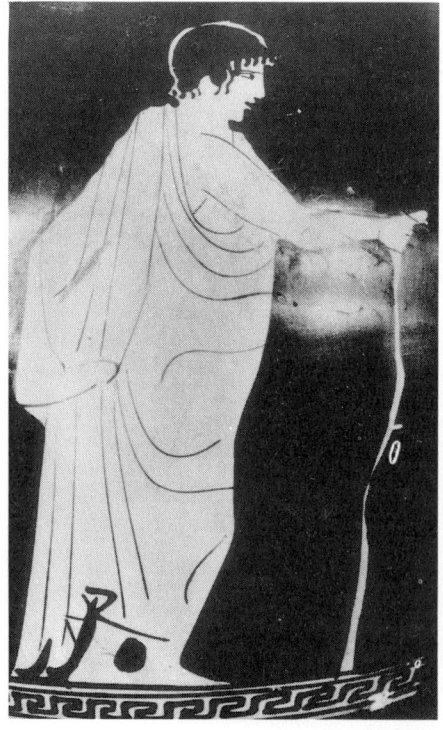

2 (G 8) Pelike. Boston, MFA, 76.45

PLATE 47

1 (G 9) Stamnos fr. Erlangen, University, I 707

2 (G 10) Pelike, St. Petersbourg, Hermitage, b 1621

PLATE 48

1 (G 11) Cup. Formerly Munich, Preyss coll.

2 (G 12) Pelike. Laon Museum, 37.1031

PLATE 49

1 (G 13) Cup fr. Athens, Agora Museum, P 2574

2 (G 14) Chous. Dresden, Albertinum, ZV 1827

PLATE 50

1 (G 15) Cup fr. Hamburg, Museum für

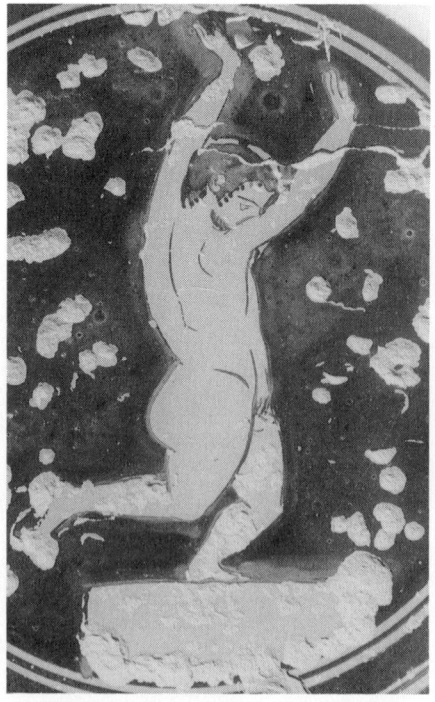

2 (G 17) Cup fr. Oxford, Ashmolean
Museum, 1938.312

PLATE 51

a

b   (G 16) Skyphos. Munich, Antikensammlungen, 8934

PLATE 52

a

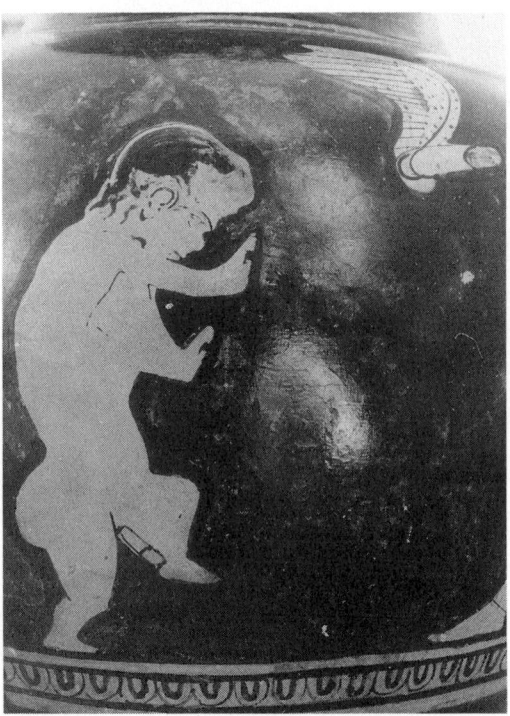

b

(G 18) Oinochoe. Oxford, Ashmolean Museum,
1971.866

PLATE 53

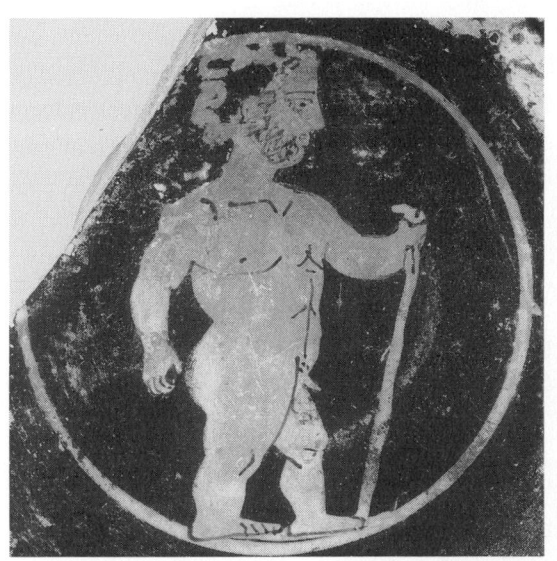

1 (G 19) Cup fr. Paris, Louvre, CA 5909

2 (G 20) Cup fr. Todi, Museo Civico, 471

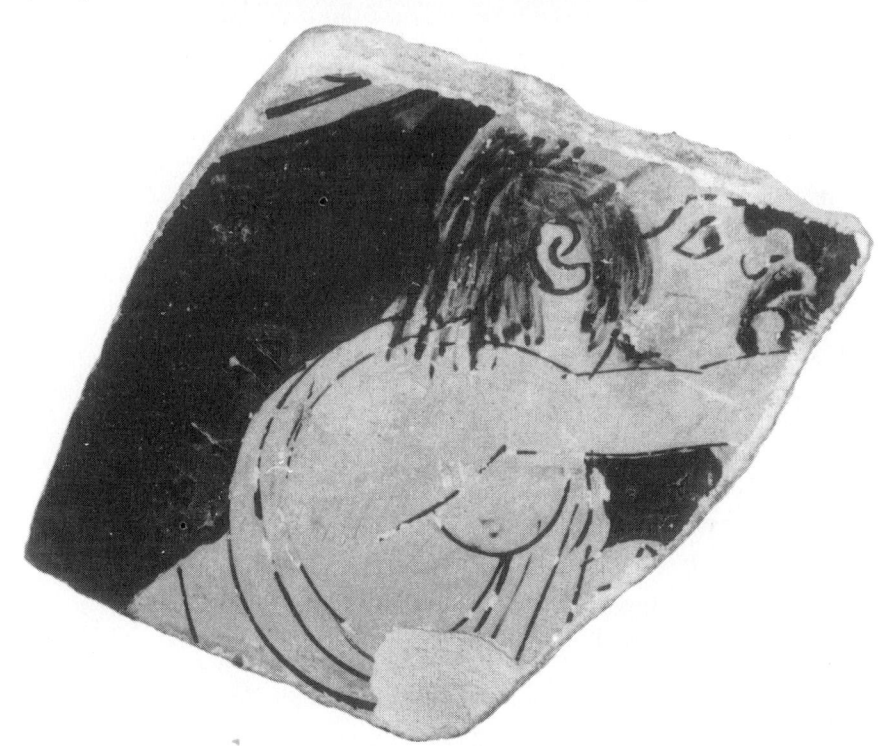

3 (G 21) Oinochoe. Tübingen, University, 1616

PLATE 54

a

b   (G 22) Skyphos. Yale, University, R. Darlington Stoddart coll., 160

PLATE 55

1 (G 23) Bell krater. Zurich market, Arete Gallery

2 (G 24) Loutrophoros. Athens, NM, 2563

PLATE 56

1 (G 26) Askos. Basle, Antikenmuseum, Z 303

2 (G 31) Oinochoe. Taranto,
Ragusa coll., 7

a

3 (G 27) Oinochoe. Melbourne, National Gallery, 90/5

b

PLATE 57

1 (G 33) Pelike. Taranto, Ragusa coll., 127

2 (G 34) Oinochoe. Toledo, Museum of Art, 67.136

PLATE 58

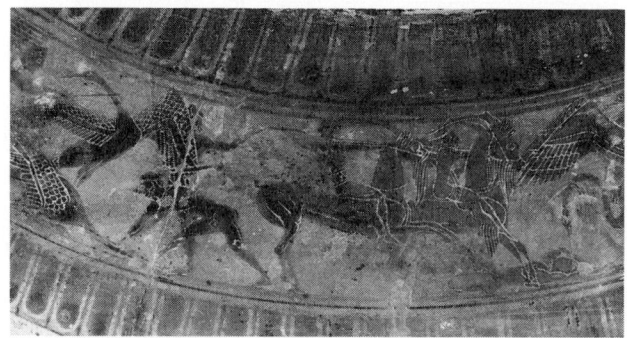

a

b

c

d   (G 40) Volute krater. Florence, Museo Etrusco, 4209

PLATE 59

a

b

c   1 (G 41) Aryballos. New York, MMA, 26.49

2 (G 42) Cup (detail). Taranto, Museo Nazionale, I.G. 4435

PLATE 60

1 (G 45) Cup fr. Berlin, SM (West), F 1785

2 (G 46) Cup fr. Berlin, SM
(West), Brommer coll., 249

3 (G 47) Hydria. Paris, Louvre, F 44

PLATE 61

a   1 (G 48) Hydria. Rome, Villa Giulia, 50425

b

2 (G 53) Clay altar. Corinth Museum, M.F. 8953

PLATE 62

1 (G d 56) Amphora. Formerly Castle Ashby

a

b 2 (G 60) Rhyton. St. Petersbourg, Hermitage, b 1818

PLATE 63

1 (G 61) Hydria. Bologna, Museo Civico, 169

2 (G 62) Hydria fr. Athens, Agora Museum, P 8892

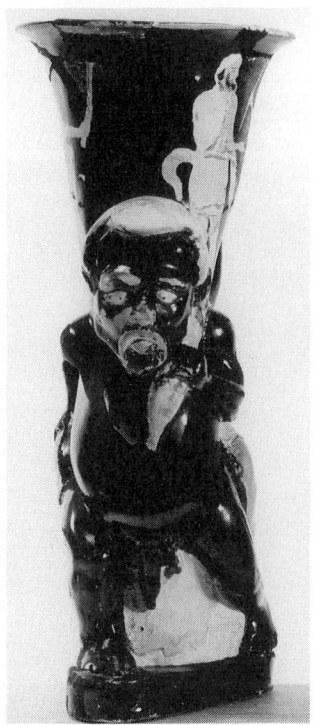

a 3 (G 66) Rhyton. Bonn, University, 545 *b*

PLATE 64

b

(G 67) Rhyton. Compiègne Museum, 898

a

PLATE 65

1 (G 70) Amphora. Brussels, Musées Royaux, R 302

2 (G 74) Pyxis (lid). Athens, NM, 17714

a

3 (G 71) Rhyton. Ruvo, Museo Jatta, 1408

b

PLATE 66

a　　　　1 (G 75)　Rhyton. Berlin, SM (East), F 2758　　　　*b*

2 (G 76)　Askos. Hamburg, Museum für Kunst und Gewerbe, 1984.457

PLATE 67

1 (G 77) Cup fr. Leipzig, Karl–Marx University, T 548

2 (G 78) Lekythos. Paris, Louvre, TH 16

PLATE 68

a   1 (G 79) Askos. Prague, Museum of Applied Arts, Z 260/K 18   *b*

2 (G 80) Cup. Rome, Vatican, 35072

PLATE 69

1 (G 82) Oinochoe. Zürich market

2 (G 85) Pelike. Brussels, Musées Royaux, A 726

PLATE 70

1 (G 92) Pelike, St. Petersbourg, Hermitage, P 18361

2 (G 103) Rhyton, figure vase.
Brussels, Musées Royaux, A 745

PLATE 71

1 (G 105) Amphora. Florence, Museo Etrusco, 3871

2 (G 107) Lekythos. Palermo, Museo Nazionale, N.I. 1865

PLATE 72

1 (G 109) Cup fr. Tocra Museum, 934

2 (G 110) Volute krater. Munich, Antikensammlungen, 2382

PLATE 73

(G 112) Pelike. Malibu, The J. Paul Getty Museum, 81. AE.189

PLATE 74

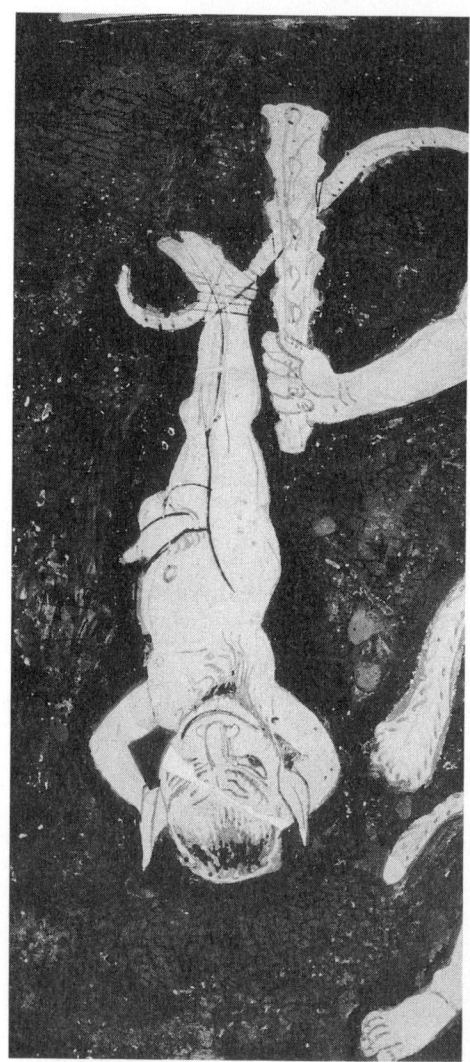

a   (G 112) Details   *b*

PLATE 75

a

b

c　　　　(G 113) Skyphos. Paris, Louvre, F 410

PLATE 76

(G 114) Amphora. Foggia Museum, 132723

PLATE 77

 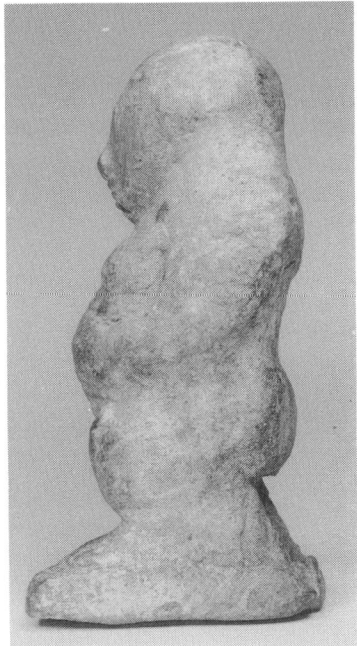 

a   1 (G 137) Terracotta. London, BM, 89   *b*   2 (G 138) Terracotta. London, BM, 93

 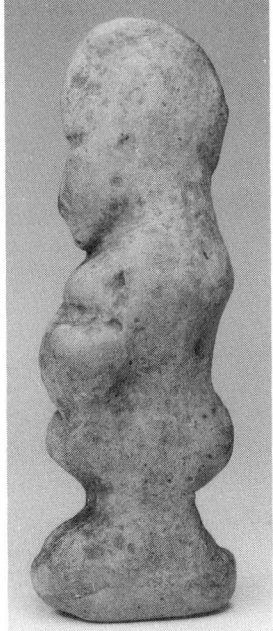

a   3 (G 139) Terracotta. London, BM, 94   *b*

PLATE 78

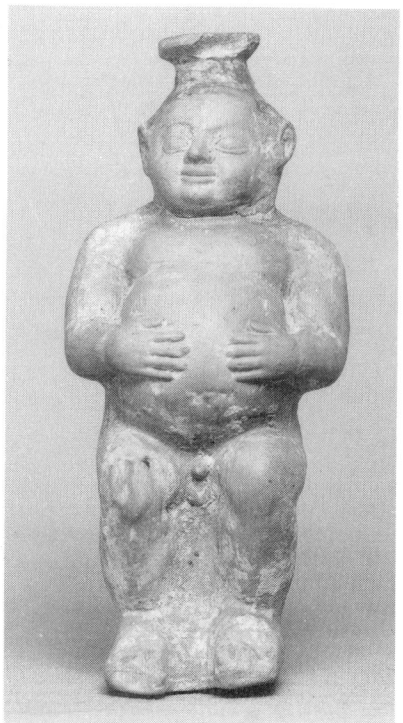

1 (G 145) Terracotta. London, BM, 86   2 (G 146) Terracotta. London, BM, 88

a   3 (G 154) Terracotta. Kassel, Staatliche Kunstsammlungen, S 55   *b*

PLATE 79

a *b* *c*

1 (G 155) Terracotta. Samos, Vathy Museum, H 43

2 (G 156) Terracotta. Samos, Vathy
Museum, RB 77–643, 1

3 (G 162) Terraotta. Taranto,
Museo Nazionale, 4960

PLATE 80

1 (G 183) Terracotta. Carthage
Museum, 895.12

2 (G 202) Terracotta. Frankfurt,
Liebieghaus, 454

3 (G 206) Terracotta. Munich,
Antikensammlungen, 7563

4 (G 210) Wood. Paris,
Louvre, N 846